Digital Filters
Theory and Applications

Digital Filters

Theory and Applications

N. K. Bose

Professor of
Electrical Engineering
and Mathematics
University of Pittsburgh
Pittsburgh, Pennsylvania

North-Holland

Elsevier Science Publishing Co., Inc.
52 Vanderbilt Avenue, New York, New York 10017

Sole distributors outside the United States and Canada:

Elsevier Science Publishers B.V.
P.O. Box 211, 1000 AE Amsterdam, The Netherlands

Library of Congress Cataloging in Publication Data

Bose, N. K. (Nirmal K.), 1940–
  Digital filters .

  Includes bibliographies and index.
  1. Electric filters, Digital.   I. Title.

TK7872.F5B67   1985   621.3815′324   85-20460
ISBN 0-444-00980-9

Current printing (Last digit):

10   9   8   7   6   5   4   3   2   1

Manufactured in the United States of America

To Chandra
Meena
Ena

Contents

Preface

The impact of digital technology on society has been so profound that at undergraduate and graduate levels of instruction in engineering and science, the exposure of students, both to digital techniques, methods of analysis and design of digital processors is expected to increase. This book is intended for upper division undergraduates and first year graduate students, primarily in but not restricted to electrical engineering. The balance between mathematical rigor and engineering design of digital filters should also motivate course offerings in applied mathematics departments based on the contents of this book. The book is also highly suitable for self-study by practicing engineers and scientists who want to acquire the important fundamentals in this rapidly developing area.

The user of the book is expected to have an exposure to a first level course on continuous and discrete-time systems based on a book like Oppenheim, Willsky and Young's, *Signals and Systems*. For use in an one semester undergraduate course, the chapters recommended for emphasis are 1, 2 (Section 2.2 only), 3, 4 and 5. For use in an one semester first year graduate course, the recommended chapters are 2, 3, 4 (Chapters 3 and 4 can, then, be covered at a faster pace) 5, 6, and selected portions of Chapter 7. The whole book can be covered at a satisfactory pace in a two trimester or a two semester sequence. The Introduction should provide sufficient motivation for perusal of the text and a sense of purpose, direction, and expectation.

To make the book self-contained, Chapter 1 presents a concise and compact review of discrete-time systems. In this chapter, the reader should find the presentation of the materials concerning implementation of bilinear transformation along with stability criteria and tests to be informative. Most of Chapter 2 is devoted to a simple exposition of relatively complex research results of recent origin. Excluding the material on Discrete Fourier Transform and its computation via the Fast Fourier Transform (FFT) technique, the remaining topics in Chapter 2 are unique but important in a book of this kind. Since, in its most usual form, the digital filter is a digital machine that performs the filtering process by the numerical evaluation of a linear difference equation in real time under program control, the topic of algebraic computational complexity is of crucial importance in the implementation of the digital filter. In Chapter 2, the basic principles that have led to the development of "faster than FFT" algorithms are described with adequate clarity. The reader, with a background in elementary abstract algebra and number theory, would find the portions of this chapter, which follow Section 2.3, easier to understand.

In Chapter 3, the various types of design methods for infinite impulse response (IIR) digital filters are presented. The design of IIR filters is somewhat facilitated by knowledge of analog filter design techniques. Therefore, wherever necessary, the links between analog and digital filter design techniques are underscored, and the reader is led to the digital design from the analog prototype via transformations. The design of finite impulse response (FIR) digital filters, on the other hand, has no analog counterparts. Chapter 4 is devoted to the exposition of and comparison between the different available schemes for FIR filter design.

In Chapter 5, important problems, inherent in all digital computations, concerned with the effects of quantization, coefficient inaccuracy due to representation by a finite number of bits, multiplication round-off error, together with the problems of adder overflow and limit cycle oscillations are discussed. The objective is to alert the reader to the need for distinguishing between expected filter performance under the practical restrictions of finite word lengths in comparison with the unrealistic hypothesis of infinite precision when representing digital data before and after performing numerical computations. Only certain special cases are singled out for detailed analysis because the task of delineating the performance of all possible structures, under different modes of quantization and different types of arithmetics used for implementation, is somewhat routine, if not conceptually unenriching.

The contents of Chapter 6, like Chapter 2, are concerned with the improvements in computational complexity, achievable by exploiting certain discernible structures that exist in the mathematical characterization of typical problems occurring in digital filtering. The Toeplitz matrix structure in a system of linear equations, for example, is encountered in diverse applications including the modeling of speech parameters and prediction, filtering of seismic data. It is important that the student recognizes the power and usefulness of the Levinson algorithm in those situations. The order of magnitude savings in computational complexity for solving a system of linear equations, characterized by a Toeplitz matrix, becomes especially significant when the Toeplitz matrix is of large order and the generic problem is required to be solved many many times. Similar statements are valid when the characterizing matrix belongs to one of several other distinguishing classes of matrices. In digital filter theory, Hankel matrices and circulants also occur, quite frequently. The student is not only made aware of the availability of fast algorithms for solving Toeplitz and Hankel systems of linear equations, but is also informed about their deeper mathematical properties involving sets of orthogonal polynomials and continued fraction expansions.

The final chapter is concerned with three different classes of digital filters. The first class is obtained by assigning the filter coefficients to belong to the binary field instead of the field of real numbers as done in the earlier chapters; the input and output sequences also belong to the chosen field implying that the digital computations are also implemented in this field. The structure and properties of the impulse response sequence of such filters are of use in certain applications. The second class of filters discussed in Chapter 7 are two-dimensional digital filters, which find a very wide range of applications in spatial and temporal signal processing problems. The third class of digital filters introduced in Chapter 7 is the class of linear but shift-variant filters. The theory of linear shift-variant filters is required in several important topics including multirate sampling theory.

The range of topics covered in this text is very broad. To keep the size of the book within reasonable limits and to eliminate the possibility of "spreading too thin," certain choices had to be made. To wit, the state-space formulation was not included because in digital filtering applications the input-output descriptions and transform domain analysis are more prevalent in contrast to modern control and estimation theory, where state-space concepts occur more naturally. This of course, does not totally overrule the use of state-space techniques in the analysis of digital filter structures. However, the primary objective in digital filtering and signal processing problems is the design of an implementable system satisfying certain design specifications like the filter frequency response. At stake are the vital issues of spatial and temporal computational complexity in implementation and the choice of filter structure for satisfactory performance evaluation under finite arithmetic conditions.

A distinguishing feature of the book is the detailed treatment at an elementary level of various key concepts that are encountered in the development of fast algorithms for implementing designed filters, which are expected to perform complex processing tasks at very high data rates. Fundamentals are emphasized, proper comprehension of which will provide the reader with needed ingredients to tackle practical problems irrespective of the domain of application. From the author's experience, it is very common to witness the use of a fundamental theorem or algorithm in different contexts. One might, for example, learn the use Levinson's algorithm in a course on speech processing without realizing its occurrence and use in other areas. In a worse situation, the same fundamental result may be repeated in different courses developed around different applications. Such inadvertant duplication of time and effort cannot be permitted. Unprecedented technological advances make it necessary to screen for presentation, in a lecture or seminar, only the basics, then expose the student to some of their uses and, importantly, leave the student to explore other avenues where his basic knowledge can be applied. This book strives towards that goal. Every attempt has also been made to expose peripheral topics of interest, which could not be included in the main body of the text, through carefully framed problems and a somewhat expanded list of references.

Portions of this book were used in an undergraduate course and the complete text was very successfully adopted in a first year graduate course offered at the University of Pittsburgh. The advanced materials have been either class tested or used in projects assigned to graduate students. The author is very grateful to all the feedback and constructive criticisms he received from students, colleagues, and several reviewers of the manuscript. I would like to thank my present and former graduate research assistants, particularly K. A. Prabhu (for his contributions to Section 7.2), H. M. Valenzuela (for his overall contributions and for taking up a graduate project on algebraic computational complexity under my supervision), J. P. Guiver (for his helpful comments on Chapter 1), and H. M. Kim (for writing some of the computer programs and working out the solutions to most of the problems given in the book). I am grateful to my colleagues who showed interest in this project; I would like especially to acknowledge Dr. L. F. Chaparro of the University of Pittsburgh for his comments on Chapter 5, Professor S. C. Dutta Roy of I.I.T., New Delhi, for his useful remarks on discrete Hilbert transform and its uses, and Professor E. I. Jury of the University of Miami, Coral Gables. Several reviewers read the complete manuscript and provided helpful comments, and it is a pleasure for me to thank Dr. D. M. Goodman (formerly with Purdue University and currently at Lawrence Livermore National Laboratory), Professor W. K. Jenkins of the University of Illinois and Professor S. R. Parker of Stanford University and the Naval Postgraduate School at Monterey, for their careful perusal of the entire manuscript. The research support over the years extended by the Air Force Office of Scientific Research and the National Science Foundation influenced the presentation of complex results in a form suitable for pedagogical purposes. This is especially underscored in the presentation of many of the results in Chapters 2, 6, and 7, which have appeared in a textbook intended for advanced undergraduates and beginning graduate students, for the first time. It is expected that in the future most of these results appearing as offshoots of recent research will be transferred to the classroom to equip students with up-to-date technological and scientific knowledge. Finally, I would like to express my sincere thanks to Ms. Marie Pelino who typed the entire manuscript with remarkable efficiency and accuracy, and the editorial staff of Elsevier Science Publishing Company, Inc., for their prompt cooperation in the various phases of production of this book.

Digital Filters

Theory and Applications

Introduction and Motivation

Like analog filtering, digital filtering is the process of spectrum shaping of signal waveforms. While the basic analog filter components are resistors, inductors, capacitors, and operational amplifiers, a digital filter is built with adders, multipliers and delays. The input-output behavior of a digital filter is simulated by a difference equation. The origin of digital filtering can be traced to the earliest efforts to simulate analog signal processing schemes on general purpose digital computers. At present, instead of being viewed as extensions of analog filtering principles, digital filtering is considered to be a new way of looking at how linear data processing may be performed on a wide variety of signals, which are discrete both in amplitude and time, in such diverse areas as time series analysis, spatial data transformations, and numerical analysis.

The continued improvement in the speed of computation and the storage capacity of digital computers has motivated an increasing number of applications to which digital filtering techniques may be applied. During the mid 1940s, computer hardware was dependent on the use of vacuum tubes and magnetic drums, and computer performance was limited to about 2-kilobytes of memory and an execution speed of 10 kiloinstructions per second. By the end of the present decade, with the availability of very large-scale-integrated (VLSI) circuits, bubble memories, optical disks, and distributed computing systems, it is expected that "conventional systems should be able to execute on the order of 100 million sequential instructions per second and to access gigabytes (10^9 bytes) of memory." It is projected that by the end of this century, the fifth generation computers will be able to execute between 10^9 instructions per second to 10^{12} instructions per second [1]. With the vast computing power, either already available or forthcoming, the range of applications of digital filter theory will continue to increase and we provide below brief descriptions of some proven applications.

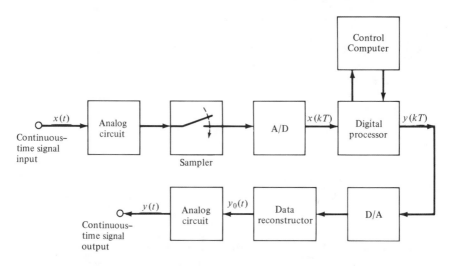

Figure 1. Functional units in a typical system whose input and output are continuous-time (analog) signals, and that uses a digital signal processor to
implement a digital filter.

Many industries already have invested in the production of digital television receivers, where the received analog signal is first digitized, then
processed by VLSI chips and finally converted into an analog signal for
transfer to the picture tube [2]. The block diagram in Figure 1 illustrates the
typical functional units that are encountered in a system, whose input is an
analog signal, and whose output is another analog signal obtained after
digital processing. This digital processing encompasses, in general, the
problems of representation of signals by sequences of numbers or symbols
and the filtering or transformation of these sequences in order to obtain new
sequences, which, according to certain chosen measures, are more desirable.
In Figure 1, the analog input signal $x(t)$ passes through a lowpass analog
filter (or in the case of a digital television receiver, a demodulator, which is
also built from conventional analog circuit elements), before being sampled,
say, every T seconds and, subsequently, coded or quantized in the analog to
digital (A/D) converter. The resulting digital sequence, $\{x(kT)\}$, is filtered
(the filtering unit is controlled by a central control computer which receives
inputs according to an user's settings) to yield an output sequence, $\{y(kT)\}$.
The output sequence $\{y(kT)\}$ is related to the input sequence $\{x(kT)\}$ by
a linear difference equation,

$$y(kT) = \sum_{j=0}^{m} a_j x(kT - jT) + \sum_{j=1}^{n} b_j y(kT - jT),$$

where the coefficients a_j's and b_j's are real constants for the filters to be studied (except in Section 7.4) in this book. The difference equation is obtained from filter design specifications. Usually these specifications are given in the frequency domain. After the filter transfer function satisfying the specifications is obtained, the linear difference equation relating the input-output properties of the filter is written down. The digital signal processor, which is used for implementing the digital filter designed, is then programmed to repetitively evaluate the difference equation by performing the arithmetic operations at very high speeds. Most of the algorithms for digital filtering repeatedly use multiplications and additions, and devices, based on a parallel, pipelined architecture, which can perform over a million high-precision arithmetic computations per second, have been designed. Hardware realization of the digital filter is also possible; the advent of the VLSI era and the expected technological improvements and innovations in the future will underscore the advantages of digital over analog filtering from the standpoints of reduced size, high reliability, low cost and negligible power consumption.

The output sequence, $\{y(kT)\}$, from the digital signal processor, which implemented the filtering process, is passed through a digital to analog (D/A) converter and a data reconstructor, which could be a zero-order hold. This zero-order hold provides an analog output, by holding, between two sampling instants, the incoming digital signal to the value of the immediately preceding sample. Therefore, the output, $y_0(t)$ from the zero-order hold is described by

$$y_0(kT + t) = y_0(kT), \qquad 0 \leq t \leq T.$$

In order to eliminate possible presence of high frequency ripples, $y_0(t)$, could, in turn, be lowpass filtered by simple analog filter circuitry.

In Figure 2, the various operations involved in the digital processing of TV signals are described. The received signal is demodulated, prior to digitization, so that the high bandwidth of the incoming carrier frequency and sidebands may be reduced to a value compatible with digitization rates, which are economically feasible to implement with available technology. The digital processing of the signal is performed by five basic VLSI chips, assisted by a clock generator, a digital amplifier and a central control unit. The five main chips are a video codec (a codec consists mainly of A/D and D/A converters), an audio codec, video and audio processor units, which perform the crucial tasks of filtering and decoding of the video and audio signals, and a deflection control unit, which deals with the sweep synchronization signals. The need for digital filtering occurs both in the video and audio processing units. The video processing unit takes the digital signal from the video codec and then separates this signal into two channels. In the luminance channel, which encodes the brightness on the screen, digital filters are designed to emphasize the high-frequency components of the signal for improving picture sharpness and contrast. The chrominance

4

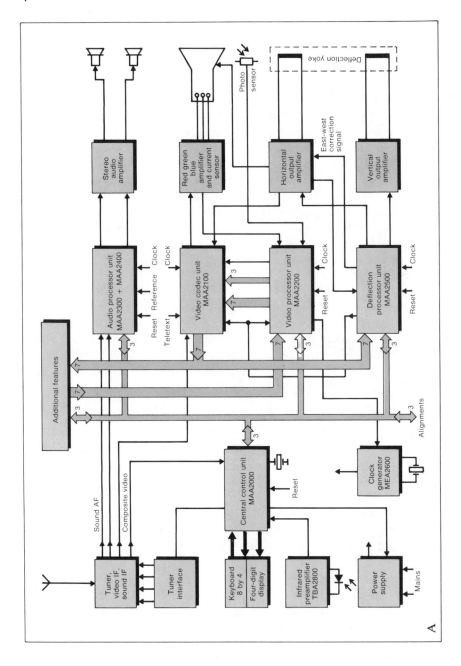

◄Figure 2. Illustration of various digital filtering functions (A–E) in a digital TV receiver. A. The new ITT digital television VLSI chip set processes television signals after they have been demodulated. In this schematic of the structure of a digital TV set demodulated signals from the IF amplifier (upper left) are fed to the audio and video codec chips, which convert the analog signals into digital and pass them along to the audio and video processor units. These units are controlled by the central control computer (center left), which receives inputs from the user via a keyboard or an infrared detector and preamplifier, driven by signals from a remote-control infrared emitter. Processed signals from the video and audio processor units are fed back through the converter, converted into analog, and then fed to the audio amplifier and speakers or to the beam amplifier and electron guns. An additional chip, the deflection processor unit, controls the deflection yoke, which points the electron beams and maintains synchronization between the signals and the two sweeps. (Reprinted with permission from Lerner: Digital TV: Makers bet on VLSI. *IEEE Spectrum,* Feb. 1983:39–43. © 1983, *IEEE.*)

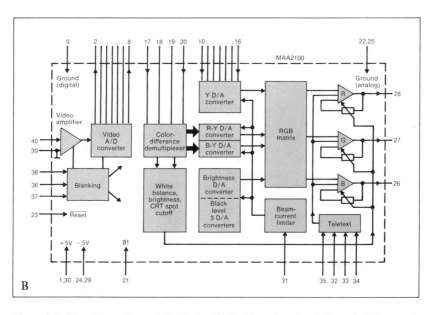

Figure 2 B. The video codec, a 3-D bipolar VLSI chip, takes signals from the IF stage, then converts them into digital in the analog-to-digital converter. It then feeds the signal to the video processor. The processed signals are then converted back into analog by three D/A converters (one for luminance and two for the color-difference signals). These analog signals are then fed into the RGB matrix, an analog circuit that combines them to produce the signals for the red, green, and blue electron guns. The process is controlled by brightness settings passed by the control computer from the user. The electron gun circuits on the chip can also be actuated by teletext input. (Reprinted with permission from Lerner: Digital TV. © 1983 *IEEE.*)

channel, which encodes the relative weightings of the red, green, and blue electron guns, consists of, among other functional units, a digital bandpass filter and a comb filter. Following filtration, the signals from the two channels are sent to the codec for D/A conversion. Similarly, the audio

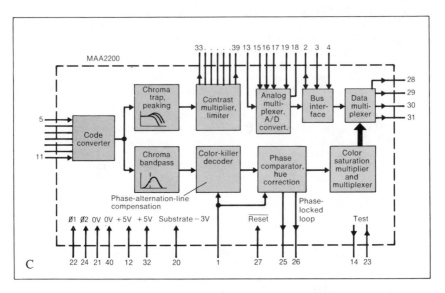

Figure 2 C. The video processor, an NMOS chip, takes the digitized output for the video codec and separates it into two channels. The luminance channel is filtered by a peaking filter, which emphasizes the high-frequency components, improving definition and sharpness. The signal is then passed to a multiplier, which sets the contrast, and from there back to the codec chip. The chrominance channel signal goes through a bandpass filter, an automatic color control circuit, a comb filter, and the color decoder and is then sent on to the codec. (Reprinted with permission from Lerner: Digital TV. © 1983 *IEEE*)

process splits the digital signal from the audio codec, and separates it into two channels, before transmission through a cascade of digital filters designed to control factors like tone, loudness, etc.

The digital processing of TV signals within a set not only improves picture quality via partial filtering out of ghosts (ghosts are mainly linear distortions that occur in transmission path) using digital networks comprised of delay blocks, multipliers, and adders, but also provides high resolution and image storage. The high resolution is achieved via interpolation between the scan lines of a picture while the capability to store entire pictures can allow faster rescanning of each picture on the screen to suppress the power frequency flicker.

Digital electronics has influenced the telephone industry to a very great extent. All-digital telephone network designs including digital switching, signaling and digital transmission offer economic as well as reliability advantages [3]. The need for digital filter design is very great in a digital communications network. The specifications for these filters are very tight since those are required to have sharp cut-off in order to be able to separate voice and message information from undesirable signals. The filters are also

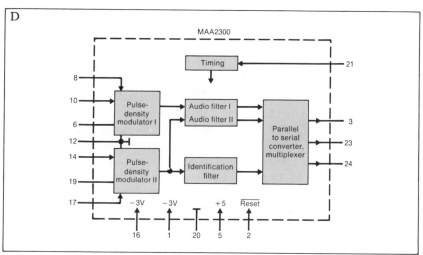

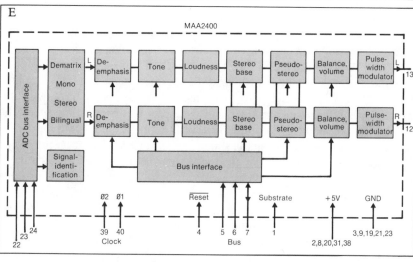

Figure 2 D. The audio A/D converter uses a pulse-density modulator and digital converter. The circuit samples a 4-MHz input signal to produce a 1-bit data stream and then converts this into a 16-bit resolution stream at 35 kHZ. The digital identification filter takes out the identification signal that tells whether the broadcast is mono, stereo, or bilingual. A parallel-to-serial converter multiplies the output to the audio processor, reducing the number of pins required. E. The audio processor takes the signal from the converter and splits it into two channels. It then sends each signal through a series of filters that control stereo balance, tone, loudness, and so on. The coefficients of the filters are controlled by signals from the control computer and are based on user settings. A fast multiplier shared by all the filters is used for the multiplications. (Reprinted with permission from Lerner: Digital TV. © 1983 *IEEE.*)

required to have small passband ripples. Typically, the filter specifications for handling the decoder output of commercial telephone digital equipment consist of a passband ripple not exceeding 0.15 db, the minimum attenuation at 4 KHz (in the transition band) is 15 db and the stop-band attenuation is at least 35 db. Though these tight specifications can be met by hybrid (involving both integrated and thick/thin film technologies on a single substrate and one or more chips) active RC filters, the cost, even in mass-scale manufacturing, is quite high and the advantages of infinite impulse response (you will study the recursive implementation and design of this type of filters in Chapter 3) monolithic (integrated circuits fabricated on a single chip) digital filters are obvious.

Digital filters find applications in the design of echo cancellars for use in long distance satellite or terrestrial voice communication. In local telephony, limited to a range of about 60 Km, a subscriber is linked to a central office by a two-wire line, which serves for communication in both directions. For long distance telephone, it becomes necessary to use four-wire lines, since a separate path becomes necessary for each direction, so that the needs for amplification by booster amplifiers and multiplexing (where a number of calls use portions of one wide-band transmission channel for reasons of economy) can be attended to. A typical long distance telephone circuit is shown in Figure 3A, where the hybrid transformer connects a four-wire long distance circuit to a two-wire local circuit. When the hybrid, which is essentially a Wheatstone bridge, is not balanced, a portion of the speech by a talker at one end, is transmitted back as echo through the hybrid. The principle of echo cancellation [4], shown in Figure 3B is based on the design of an adaptive digital filter designed to match the transfer characteristics of the echo path (the echo path is assumed to be linear but could be quite variable depending upon factors like distance to the hybrid, characteristics of the two-wire end circuit, etc.), and then the synthesized echo is subtracted from the actual echo, which otherwise, would have travelled to the speaker. The impulse response of the echo path is measurable as the response at point b to an unit impulse at point a, as indicated in Figure 3B. Then, a finite impulse response filter, which you will study in Chapter 4 of this text, can be designed to simulate the echo path impulse response. The coefficients of the digital filter may be conveniently varied to provide adaptivity, since the echo path is variable. An entirely digital echo cancellar that serves to eliminate annoying echoes in two-way conversations via satellite, has been built on an inexpensive, silicon integrated circuit chip, whose power consumption is also very low [5].

Over and above echo distortion, voice-frequency transmission for telephone circuits could be adversely affected by transmission loss and attenuation distortion. These degradations are controlled by digital filters functioning as equalizers. Typically, fourth order infinite impulse response filters, realized as two cascades of second order blocks have been used to

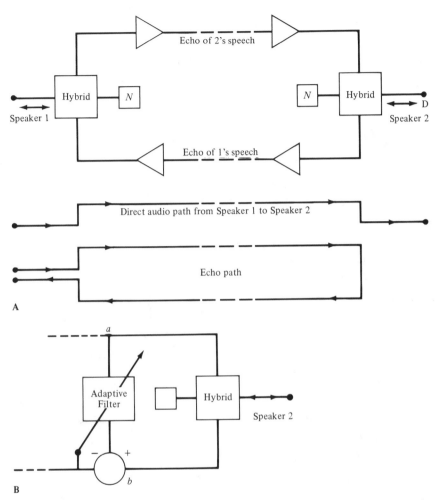

Figure 3 A. A typical long distance telephone circuit. Each line represents a two-wire circuit. The boxes marked N are balancing impedances. In a local call, the two-wire circuit of one customer is connected directly to the two-wire circuit of the other. B. Principle of echo cancellation.

provide a high-quality channel for voice frequency signal transmission [6]. An extremely versatile large scale-integration digital signal processor (dubbed DSP) has been built at Bell Laboratories [7] for providing variable line equalization, gain and echo cancellation in addition to numerous other applications. One of those applications is of special relevance here, because of the different types of digital filters that are required and we describe briefly that application, next.

An all-digital touch-tone receiver (TTR), whose architecture is shown in Figure 4, has been designed and constructed at Bell Laboratories [8]. The TTR is designed to receive any one of 16 digits that may be transmitted by

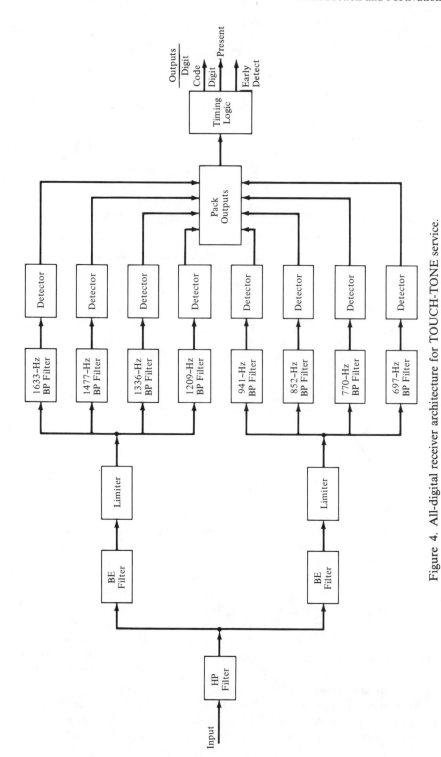

Figure 4. All-digital receiver architecture for TOUCH-TONE service.

simultaneously sending two tones in a voice frequency signaling system. Only a few years back, analog TTRs were standard. However, the digital TTR has multiplexing capabilities and is particularly suited for operation by signals that have been encoded in digital format. The input filter in the TTR is a fourth order high-pass (HP) filter, whose function is to attenuate the 60 Hz power frequency hum and the low-frequency dial tone present, when the initial digit is received. The two band-elimination (BE) filters are each sixth order, designed to provide attenuation in the frequency bands from 600 to 1050 Hz and from about 1200 to 1650 Hz, respectively. The tone-detection band-pass (BP) filters are each second order. The other signal processing units required are those which implement, in digital form, the nonlinear operations of signal limiting and detection.

In addition to the specific problems discussed so far, digital filters occur in numerous other commercial, military and medical applications, when the separation of a signal from noise or other undesirable signals is necessary. In moving target indicator (MTI) radars, the Doppler shift in frequency is used to identify moving targets even when the echo signal from fixed targets is orders of magnitude greater. Since, signals from fixed targets are not shifted in frequency, a filter is used to extract the Doppler frequency shift and reject the clutter frequency, so that a moving target is detected. The advantages of digital filters to filter out fixed targets has been pointed out in [9].

Digital filtering is useful for separating signal from noise in many medical applications [10], [11]. The stroboradiocardiogram permits the generation of a plot characterizing the quantity of blood in the heart during ventricular systole and diastole. The subject is given an intravenous injection of a radioactive tracer, and a collimated scintillation counter placed over the subject's heart provides the measurements prior to processing. The low dosage of the radioactive tracer is responsible for a high amount of noise, which tends to make the recording meaningless. However, lowpass digital filtration of the data has led to significant improvement in the processed signal [10]. The technique, of separation, of course, is successful, when the frequency content of the signal is well separated from the noise frequency spectrum (see Figure 5). There are numerous other application areas of digital filtering. These include speech recognition and processing, and seismic signal processing [12], [13]. The development of a semiconductor chip costing a few hundred dollars and capable of processing a billion instructions per second (about 1000 times faster than the fastest single processor chip, like the DSP developed at Bell Laboratories, available today) could lead to a continuous speech processing system at a reasonable cost. In a typical speech recognition system, spoken words enter the system via a noise-cancelling microphone. These are then sampled and digitized at a sampling frequency of over 10,000 hertz. The digitized signals are broken into centisecond slices and spectrally analyzed by the system. This digital

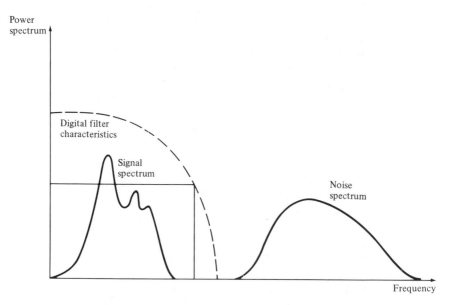

Figure 5. The power spectrum of signal and noise that facilitates separation of signal from noise via digital filtering.

spectrum analyzer requires a bank of digital bandpass filters, alongwith other functional units. For information on an advanced speech recognition system now under development at IBM Corporation's Thomas J. Watson Laboratories, see [14].

The preceding discussion centered around the various applications of one-dimensional digital filtering theory, required for example, in the processing of temporal signals. Spatial data processing and digital image processing require the tools of 2-D digital filter theory, while spatio-temporal signal processing might require higher than 2-D digital filter theory. Multidimensional digital filter theory is much more complex than its 1-D counterpart, but tremendous progress has been documented in the area over the last decade. The background of 1-D filter theory and the exposure to 2-D filtering principles in the last chapter of this book will certainly prepare the student for the challenges of multidimensional digital filtering.

References

1. Kahn, R. E. 1983. A new generation in computing. IEEE Spectrum, Nov.:36–41.

2. Lerner, E. J. 1983. Digital TV: makers bet on VLSI. IEEE Spectrum, Feb.:39–43.

3. Gallagher, E. F. 1977. Chipping in for digital telephones. IEEE Spectrum, Feb.:42–46.

4. Sondhi, M. M., and Berkley, D. A. 1980. Silencing echos on the telephone network. Proc. of IEEE, 68, Aug.:948–963.

5. Duttweiler, D. L. 1980. Bell's echo-killer chip. IEEE Spectrum, Oct.:34–37.

6. Blake, R. B., Bolling, A. C., and Farah, R. L. 1981. Voice-frequency transmission treatment for special-service telephone circuits. Bell System Tech. J. 60:1585–1619.

7. Boddie, J. R. 1981. Digital Signal Processor Overview: the device, support facilities and applications. Bell System Tech. J. 60:1431–1439.

8. Boddie, J. R., Sachs, N., and Tow, J. 1981. Receiver for TOUCH-TONE service. Bell System Tech. J. 60:1573–1579.

9. Zverev, A. I. 1968. Digital MTI radar filters. IEEE Trans. Audio and Electroacoustics, 3:422–432.

10. Della Carte, M., and Cerofolini, O. 1974. Application of a digital filter to biomedical signals. Med. Biol. Engg., 374–377.

11. Weaver, C. S., et al. 1968. Digital filtering with applications to electrocardiogram processing. IEEE Trans. Audio Electroacoustics 16:350–391.

12. Special Issue on Digital Signal Processing, 1975. Proc. IEEE, April.

13. Oppenheim, A. V., ed. 1978. Applications of Digital Signal Processing Prentice-Hall, Englewood Cliffs, NJ.

14. Reddy, R., and Zue, V. 1983. Recognizing continuous speech remains an elusive goal. IEEE Spectrum, Nov.:84–87.

Chapter 1
Linear Time-Invariant Digital Filters

1.1. Introduction

Digital filtering techniques are useful in the areas of digital telephony and communications, speech signal processing, seismic data processing, radar and sonar systems, facsimile and television image processing, space research, and so on. The chief advantages of digital over analog techniques are due to accuracy, stability, higher signal to noise ratio, and the decreasing cost of hardware and software implementation, spurred by the availability of low-cost, high-speed microprocessors and the advent of the very large-scale integration era, as well as greater flexibility and adaptivity in realizing filter design requirements. A disadvantage occurs from the fact that, though there is no drift or tolerance problem with digital filters, the errors due to finite arithmetic have to be attended to.

In this chapter, attention is restricted to the brief exposition of fundamentals required in the study of linear time-invariant (LTI) digital systems. This class of systems can be analyzed using powerful transform domain (Fourier transform, z-transform) techniques. One way of defining a LTI system is as follows: Let O be an operator on the set of one-dimensional discrete signals $\{x_i(k)\}$, which generates a set of corresponding outputs $\{y_i(k)\}$, where k is an integer variable and i belongs to an index set I. Thus,

$$O\{x_i(k)\} = \{y_i(k)\}. \tag{1.1}$$

O is a linear operator if and only if

$$O\{\mathsf{a}x_i(k)\} = \mathsf{a}\{y_i(k)\} \tag{1.2}$$

for all $i \in I$ and for all complex scalar constants a, and

$$O\sum_j \{x_j(k)\} = \sum_j \{y_j(k)\} \tag{1.3}$$

for all $j \in J$, where J may be any subset of the index set I. Both I and J may be infinite sets. O is a time-invariant operator if and only if

$$O\{x_i(k-r)\} = \{y_i(k-r)\} \tag{1.4}$$

for all $i \in I$ and an arbitrary but fixed integer r.

An LTI system is one that can be characterized by an operator that is both linear and time-invariant. An LTI system is characterizable by its unit sample response (or impulse response) $\{h(k)\} = O\{\delta(k)\}$, where the unit impulse sequence $\{\delta(k)\}$ is defined by the delta function $\delta(k) = 1$ for $k = 0$, and $\delta(k) = 0$, $k \neq 0$. The response to any input of the LTI system is obtainable as the discrete convolution of $\{h(k)\}$ with the input sequence.

This chapter is exclusively concerned with the fundamentals of linear time-invariant digital systems. The inputs to a digital system are, often, the sampled values, suitably quantized, of an analog signal. Section 1.2 is devoted, therefore, to the sampling theorem. Although several sampling theorems are available in the engineering literature, only the simplest case corresponding to that of a time function $y(t)$, whose spectrum (Fourier transform) $Y(\omega)$ is zero outside $-\omega_0 \leq \omega \leq \omega_0$, is considered here. The sampling process involves the multiplication of a continuous signal by a periodic train of delta functions, and the reconstruction of the original signal from its sampled values is regarded as a windowing operation on the frequency spectrum of the sampled impulses.

Discrete-time (and discrete-space) systems, which are LTI (linear shift-invariant [LSI]), are convenient to analyze using the theory of z-transforms, which converts a constant coefficient linear difference equation into an algebraic equation in a complex variable. Section 1.3 presents the rudiments of z-transform theory for readers who might not yet have been exposed to the topic. The z-transform of a sequence is defined as a power series (or polynomial if sequence is finite) in z^{-1} (definition as a power series in z is also used, especially by geophysicists), and causal sequences are emphasized. It should be noticed that the power series is just another way of representing the sequence, and z^{-k} is the position marker of element $r(k)$ in a sequence $(\ldots r(-2), r(-1), r(0), r(1), r(2), \ldots)$. Therefore, for noncausal sequences, $z^{\pm k}$ would be the power of z that one would associate with the element $r(\mp k)$ in the sequence. One-sided z-transform theory, presented here, extends quite naturally to the two-sided case.

Section 1.4 is concerned, first, with a discussion of the bounded-input bounded-output (BIBO) criterion for stability, followed by the description of several algebraic tests, which can, in principle, be implemented with 100% precision. In the literature, there is often some confusion about the precise implication of BIBO stability property. For linear systems, the BIBO stability property is equivalent to the uniform BIBO stability property. The condition to be satisfied by the system unit impulse response sequence for the system to be BIBO stable is proved. This condition leads to an equivalent characterization of the BIBO stability property in terms of the zero distribution of the denominator polynomial for any system describable by a rational transfer function. The test for zeros of a polynomial to be within the unit circle can be conveniently implemented by operating on the coefficients of the polynomial as described and illustrated.

In Section 1.5, the topics of discrete convolution and deconvolution are briefly described in relation to their role in the analysis of linear discrete-time systems. The computational complexity in their implementation will be considered in Chapter 2. The response of LTI digital filters is considered in Section 1.6. The objective, here, is briefly to review materials that the reader is expected to have encountered before. The sections on suggested readings and problems should further serve to reinforce this viewpoint.

1.2. Sampling Process

A continuous-time signal or an analog signal is a single-valued, albeit discontinuous, function of the independent continuous variable, time, and may take on any value in a continuous range. A discrete-time signal is defined by a function, which may take on a continuum of values at discrete instants of time. When the range of a discrete-time signal is quantized, the discrete-time signal is called a digital signal. The operation of sampling is an interfacing operation between continuous-time and discrete-time signals, which allows any member of a class (termed band-limited) of continuous signals to be reconstructed completely from its sampled values at a countable set of equally spaced sampling points, when the spacing is less than a predetermined value. Consider a band-limited signal $y(t)$, that is, the Fourier transform $Y(\omega)$ of $y(t)$, vanishes outside a finite frequency domain

$$Y(\omega) = 0, \qquad |\omega| > \omega_0 = 2\pi f_0. \tag{1.5}$$

For the reader's convenience, a Fourier transform pair is defined next.

Definition 1.1. The Fourier transform (FT) pair, $y(t) \leftrightarrow Y(\omega)$, is defined below, subject to the assumption of existence.

$$y(t) = \frac{1}{2\pi} \int_{-\infty}^{\infty} Y(\omega) e^{j\omega t} \, d\omega,$$

$$Y(\omega) = \int_{-\infty}^{\infty} y(t) e^{-j\omega t} \, dt.$$

The first equation is the inverse FT and the second equation is the FT. The result of sampling $y(t)$ at equispaced intervals (often referred to as first-order sampling), so that the sampling points are T seconds apart, is

$$\hat{y}(t) = y(t) \sum_{k=-\infty}^{\infty} \delta(t - kT) = \sum_{k=-\infty}^{\infty} y(kT)\delta(t - kT). \quad (1.6)$$

The Fourier transform (spectrum) of $\hat{y}(t)$, being the convolution of Fourier transforms of $y(t)$ and $\sum_k \delta(t - kT)$, is easily shown to equal[1]

$$\hat{Y}(\omega) \triangleq F[\hat{y}(t)] = \frac{1}{T} \sum_{k=-\infty}^{\infty} Y\left(\omega - \frac{2k\pi}{T}\right). \quad (1.7)$$

It should be noted that the spectrum of the sampled signal is $1/T$ times the spectrum of the original signal repeated at integer multiples of the sampling frequency, $f_s = 1/T$ Hz. As $y(t)$ has a band-limited spectrum with $f_0 = \omega_0/2\pi$ as the highest frequency, then, by selecting (Figure 1.1A)

$$f_s > 2f_0,$$

it is possible to obtain $Y(\omega)$ after low-pass filtration of the sampled signal. The possibility of recovering the original spectrum with $f_s = 2f_0$ is ensured if $Y(\omega)$ does not contain a δ-function at $f = f_0$. This lower bound of $2f_0$ Hz for the sampling rate is referred to as the Nyquist rate. The original spectrum, $Y(\omega)$, is recovered by multiplying $\hat{Y}(\omega)$ in (1.7) by a spectral window function (Figure 1.1B)

$$H(\omega) = \begin{cases} T & -\omega_0 \leq \omega \leq \omega_0, \\ 0 & \text{elsewhere} \end{cases} \quad (1.8)$$

whose inverse FT is the sinc function $(\sin \omega_0 t / \omega_0 t)$—the impulse response, $h(t)$, of an ideal low-pass filter. Therefore, in the time domain the convolution of $\hat{y}(t)$ in (1.6) and $h(t)$, just defined, should yield $y(t)$. Therefore, at the Nyquist rate of sampling,

$$\begin{aligned} y(t) &= \sum_{k=-\infty}^{\infty} y\left(\frac{k}{2f_0}\right) \frac{\sin \omega_0(t - kT)}{\omega_0(t - kT)} \\ &= \sum_{k=-\infty}^{\infty} y\left(\frac{k}{f_s}\right) \frac{\sin\left[(\pi/T)(t - kT)\right]}{(\pi/T)(t - kT)} \end{aligned} \quad (1.9)$$

If sampling is done at a frequency below the Nyquist rate, then a portion of the high-frequency information in $Y(\omega)$, the spectrum of the analog signal $y(t)$, is shifted into the lower frequencies of $\hat{Y}(\omega)$, the spectrum of the

---

[1](1.7) may also be derived after substituting the Fourier series representation

$$\sum_{k=-\infty}^{\infty} \delta(t - kT) = \frac{1}{T} \sum_{k=-\infty}^{\infty} e^{j2\pi kt/T}$$

of the periodic train of unit impulses in the first equality of (1.6) and then taking FT.

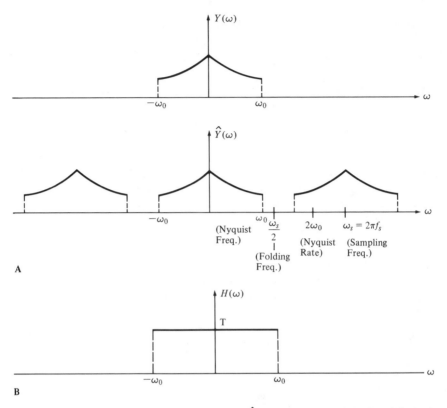

Figure 1.1. (A) Fourier transforms $Y(\omega)$ and $\hat{Y}(\omega)$, respectively, of a band-limited
signal and its sampled version. (B) The transfer function of an ideal
low-pass filter required to recover $Y(\omega)$ from $\hat{Y}(\omega)$.

sampled sequence. This effect is referred to as *aliasing* or the *folding* effect,
after a folding of the frequency axis. This folding is similar to the manner in
which a carpenter's scale folds itself. In general, the sampling should be
done at a rate greater than the Nyquist rate.

Sampling Theorem 1.1. *Let $y(t)$ be a band-limited signal having its highest
frequency content equal to f_0 Hz. Then $y(t)$ can be completely specified by
the sampled sequence $\{y(kT)\}$, provided $T < (1/2f_0)$, that is, $f_s = 1/T >
2f_0$.*

At this stage, several comments are in order. First, note that the sampling
period T, in general, is strictly less than $1/(2f_0)$. Otherwise, if T equaled
$1/(2f_0)$, a cosine wave of frequency ω_0 rad/sec or $f_0 = \omega_0/2\pi$ Hz could be
sampled every $1/(2f_0)$ sec at points where it takes the zero value (Figure

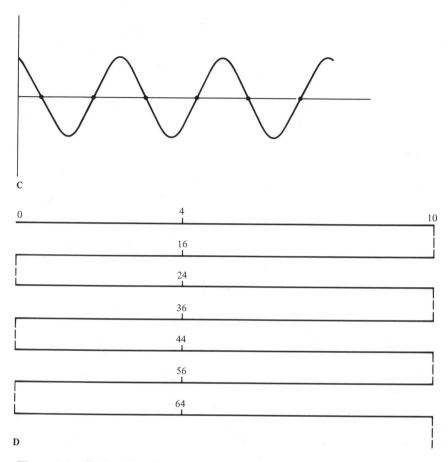

Figure 1.1. (C) Sampling frequency is two times the frequency of the sinusoid. (D) Foldover effect due to undersampling.

1.1C); however, it would not be possible to reconstruct the cosine wave via (1.9) from these samples. Note that

$$FT[\cos \omega_0 T] = \frac{1}{2} \int_{-\infty}^{\infty} e^{j(\omega + \omega_0)t} + e^{j(\omega - \omega_0)t} \, dt$$

$$= \pi [\delta(\omega + \omega_0) + \delta(\omega - \omega_0)],$$

and, as mentioned earlier, the Nyquist rate of sampling at $2f_0$ Hz is not, in general, permissible if the Fourier transform of the signal contains a δ-function at $f = f_0$. Second, if $T > 1/(2f_0)$, or, equivalently, $f_s < 2f_0$, aliasing occurs, as a result of which two frequencies, f_1, f_2, related by

$$f_2 = kf_s \pm f_1,$$

are indistinguishable from each other for any positive value of integer k due to the foldover of frequencies higher than $f_s/2$ into the range $0 \le f \le f_s/2$. This foldover effect is illustrated in Figure 1.1D, where $f_1 = 4$ Hz is indistinguishable from 16 Hz, 24 Hz, 36 Hz, 44 Hz, 56 Hz, etc., when the sampling frequency is 20 Hz.

The reconstruction formula in (1.9) can be construed as a scheme of interpolation; that is, the value of the function $y(t)$ between samples is determined by a weighted sum of its sampled values, which, however, are infinite in number. This follows from the fact that, in general, a time-limited function (i.e., a function of time that is zero outside a finite interval) cannot have a band-limited Fourier transform and vice-versa, as stated in the next paragraph.

In reconstructing $y(t)$ from its sampled values via (1.9), several difficulties are encountered. First, an infinite number of additions and multiplications are required, as a band-limited signal cannot be time-limited as well. Second, (1.9) implies that $y(t)$ can be constructed by passing $\{y(kT)\}$ through an ideal low-pass filter having an impulse response $\sin \omega_0 t / \omega_0 t$, which is physically unrealizable, as it is *not causal*.[2] Third, computation of $y(t)$ at any fixed $t = t_0$ via (1.9) requires the availability of all the samples $\{y(kT)\}$. Therefore, in practice, reconstruction of signals from its discrete-time samples are carried out via various types of sample and hold circuits, usually followed by low-pass filtration.

It should be borne in mind that the choice of sampling frequency, in practice, depends on factors like the speed of hardware used to implement operations in real time, the shape of the input spectrum yielding information about the percent of total energy that lies within a certain frequency range, and the error that is tolerable in the digital-analog (D/A) converter in addition to the bandwidth of the signal to be processed. As a rule of thumb, the sampling frequency may be chosen to be 2.5 times the highest frequency component of the signal, bearing in mind that the choice could also be influenced by the nature of the problem at hand and the experience of the system designer.

The final remark in this section is concerned with real world samplers, when the sampling function, $s(t)$ (Figure 1.2A), is a periodic train of rectangular pulses, say, of finite nonzero width b and unit height and period T. It can be verified that the Fourier series expansion for $s(t)$ is

$$s(t) = \frac{b}{T}\left[1 + 2 \sum_{k=1}^{\infty} \frac{\sin k\pi(b/T)}{k\pi(b/T)} \cos k \frac{2\pi}{T} t \right].$$

Therefore, if $y(t) \leftrightarrow Y(\omega)$ and $\hat{y}(t) \triangleq (y(t)s(t)) \leftrightarrow \hat{Y}(\omega)$ are Fourier trans-

---

[2] It is not zero for $t < 0$.

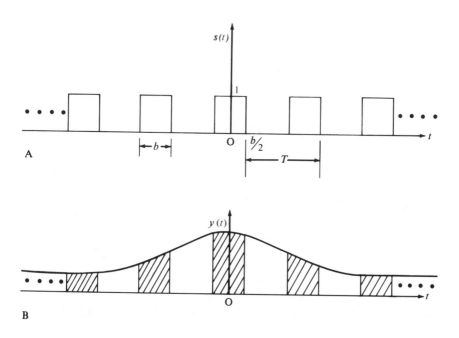

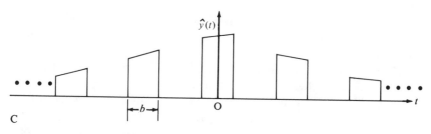

Figure 1.2. (A) A nonideal sampling function; (B) a signal for sampling; (C) the sampled signal.

form pairs, it can be checked that

$$\hat{Y}(\omega) = \frac{b}{T}\left[Y(\omega) + \sum_{k=1}^{\infty} \frac{\sin(k\pi b/T)}{k\pi b/T}\left\{Y\left(\omega - \frac{k2\pi}{T}\right) + Y\left(\omega + \frac{k2\pi}{T}\right)\right\}\right].$$

Plots of $y(t)$ vs. t, $\hat{y}(t)$ vs. t, $Y(\omega)$ vs. ω, and $\hat{Y}(\omega)$ vs. ω (when the sampling period is sufficiently small so that no overlapping of $Y(\omega)$ and $Y(\omega - (2\pi/T))$ or $Y(\omega + (2\pi/T))$ occur) are shown in Figures 1.2B through 1.2E, respectively. Thus $y(t)$ can, then, be reconstructed from $\hat{y}(t)$ by passing it through a low-pass filter.

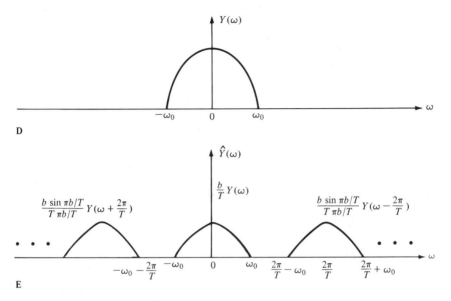

Figure 1.2. (D) Fourier transform of signal in (B); (E) Fourier transform of signal in (C).

When $y(t)$ is reconstructible from its samples, any filtering operation on $y(t)$ may be performed, without error, on its samples only and therefore may be implemented on a digital computer. It is usually possible to carry out multiple filtering operations. Let the sampling period $T = \frac{1}{50}$ sec, and suppose that one filtering operation takes 100 μsec. Then, within the sampling period T, as many as 200 filtering operations may be performed. This is similar to having 200 analog filters in cascade.

Advantages of digital filtering has resulted in a proliferation of a vast amount of literature on the subject, mainly in the past 15 years. This has necessitated the adoption of a common framework of terminologies and notations. The efforts made toward this goal in reference [23] are noteworthy. Conformity with accepted standards has been attempted in this text with regard to notations. The reader will be alerted in advance if possibilities of ambiguities in interpretation or representation exist. For example, in this section $\{y(kT)\}$ denotes a sequence obtained by uniformly sampling an analog waveform every T seconds. Sometimes the methods for obtaining sequences are independent of any time frame. To wit, a sequence could be generated from a difference equation with specified initial conditions. Then the sequence $(\ldots, y(-1), y(0), y(1), \ldots)$ could be described adequately via the notation $\{y(k)\}$. When the nature of derivation of a sequence is unimportant, it is convenient to adopt this notation, as has been done in the remainder of the chapter.

1.3. Elementary z-Transform Theory

Analogous to the use of Laplace transforms in solving linear ordinary differential equations with constant coefficients, z-transforms are widely used to analyze or solve linear difference equations with constant coefficients, which may be used to characterize the stimulus–response relationship of a class of discrete-time systems that are linear and time-invariant.

Definition 1.2. The z-transform, $F(z)$, of a sequence of numbers or a discrete-time series, $(\ldots, f(-2), f(-1), f(0), f(1), f(2), \ldots) \triangleq \{f(k)\}$, is defined here as

$$Z[\{f(k)\}] \triangleq F(z) \triangleq \sum_{k=-\infty}^{\infty} f(k)z^{-k} \qquad (1.10a)$$

for all values of the complex variable z, where the series converges.

It is mentioned that in geophysical applications a typical term in the power series for $F(z)$ is $f(k)z^k$ and not $f(k)z^{-k}$; but this is a matter of convention and results in no conceptual difference in the development and application of the theory of z-transforms. When $f(k) = 0$ for $k < 0$, the sequence $\{f(k)\}$ is a causal sequence, and the associated z-transform is the one-sided z-transform. The region of convergence of $F(z)$ in general is an annulus formed from the intersection of two concentric circles, centered at $z = 0$, and, for causal sequences, the region of convergence lies outside a circle of sufficiently large radius that encloses all the singularities of $F(z)$. The inverse z-transform relationship is given by

$$f(k) = \frac{1}{2\pi j} \oint_C F(z)z^{k-1}\,dz, \qquad (1.10b)$$

where C is any simple closed contour in the counterclockwise sense lying in the annulus of convergence and surrounding the inner circle. In case of causal sequences, C may be taken to be a circle of sufficiently large radius so that all singularities of $F(z)z^{k-1}$ are enclosed by it. For notational brevity,

$$\{f(k)\} \leftrightarrow F(z) \qquad (1.10c)$$

will be used to denote a z-transform pair. Causal sequences will be considered unless mentioned otherwise. Table 1.1 summarizes the various commonly encountered properties of z-transform pairs. In order to be able to solve a linear difference equation in $f(k)$ with constant coefficients, and given initial conditions, the following three steps need be implemented:

1. Transform the given difference equation with initial (or boundary) conditions into an algebraic equation using $Z[\{f(k)\}] \triangleq F(z)$.
2. Express $F(z)$ as a rational function (ratio of two polynomials) in z.
3. Apply the inverse z-transform to $F(z)$ in order to get $f(k)$.

Table 1.1. Common Properties of z-Transform of General Sequences

$$\{f(k)\} \leftrightarrow F(z)$$
$$\{g(k)\} \leftrightarrow G(z)$$

Property	Sequence	z-Transform (with appropriate region of convergence)
Linearity	$a\{f(k)\} + b\{g(k)\}$ for constants a, b	$aF(z) + bG(z)$
Translation or shift	$\{f(k + k_0)\}$ with zero initial condition	$z^{k_0} F(z)$
Differentiation of transform	$\{kf(k)\}$	$-z\dfrac{dF(z)}{dz}$
Initial value theorem	$\{f(k)\}$ with $(a)f(k) = 0, \quad k < 0$ $(b)f(k) = 0, \quad k > 0$	$(a)f(0) = \lim_{z \to \infty} F(z)$ $(b)f(0) = \lim_{z \to 0} F(z)$
Final value theorem	$\{f(k)\}$ $(1 - z^{-1})F(z)$ is analytic in $\lvert z \rvert \geq 1$)	$f(\infty) = \lim_{z \to 1}(1 - z^{-1})F(z)$
Time reversal	$\{f(-k)\}$	$F\left(\dfrac{1}{z}\right)$
Exponentiation	$\{a^k f(k)\}$ for constant a	$F(a^{-1}z)$
Parseval's relation	$\{f(k)g(k)\}$	$\dfrac{1}{2\pi j}\oint F(u)G\left(\dfrac{z}{u}\right)u^{-1}\,du$
Convolution	$\{f(k)\} * \{g(k)\}$	$F(z)G(z)$
One-sided z-transform for solving linear difference equations with constant coefficients	$\{f(k-m)\}, \quad m > 0$ $\{f(k+m)\}, \quad m > 0$	$z^{-m}\left[F(z) + \displaystyle\sum_{i=1}^{m} f(-i)z^{i}\right]$ $z^{m}\left[F(z) - \displaystyle\sum_{i=0}^{m-1} f(i)z^{-i}\right]$

As for all problems in this chapter and this book, $F(z)$ will be a rational function in z, other ways to invert $F(z)$, besides the one based on the use of (1.10b), are

1. via the partial fraction expansion method and
2. via straightforward division method.

The partial fraction expansion method is usually the most convenient to implement.

To motivate the presentation, an example is presented first. A spanning tree in a connected graph over n nodes is a subgraph containing $(n-1)$ edges and no loops.

Example 1.1. Let $t(k)$ denote the number of spanning trees in the graph of a $(k+1)$ node (including the datum node) doubly terminated ladder network. Then it can be shown that the following homogeneous constant coefficient linear difference equation holds, and $\{t(k)\}$ must be a causal sequence.

$$t(k+2)-3t(k+1)+t(k)=0, \quad \text{with } t(0)=1,\ t(1)=3. \quad (1.11)$$

It is required to find a closed form expression for $t(k)$.

For the sequence $\{t(k)\}$ define

$$Z[\{t(k)\}] \triangleq T(z),$$

where k ranges from 0 to ∞ in the definition for z-transform applied to this problem.

$$Z[\{t(k+1)\}] = \sum_{k=0}^{\infty} t(k+1)z^{-k} = z\sum_{k=0}^{\infty} t(k+1)z^{-(k+1)}$$

$$= z\sum_{k=1}^{\infty} t(k)z^{-k} = z\sum_{k=0}^{\infty} t(k)z^{-k} - zt(0)$$

$$= zT(z) - zt(0) = zT(z) - z.$$

Similarly (see also Table 1.1),

$$Z[\{t(k+2)\}] = z^2T(z) - zt(1) - z^2t(0) = z^2T(z) - 3z - z^2.$$

Therefore, on z-transforming (1.11), $[z^2 - 3z + 1]T(z) - z^2 = 0$.
Whence

$$T(z) = \frac{z^2}{z^2 - 3z + 1}. \qquad (1.12)$$

Expanding in partial fractions,

$$T(z) = \frac{3+\sqrt{5}}{2\sqrt{5}}\frac{z}{z-a} - \frac{3-\sqrt{5}}{2\sqrt{5}}\frac{z}{(z-b)}, \qquad (1.13)$$

where $a = (3+\sqrt{5})/2$ and $b = (3-\sqrt{5})/2$ are the roots of the denominator polynomial of $T(z)$ in (1.12). Applying the inverse z-transform to each term in (1.13) (Table 1.2 may be used to facilitate inversion),

$$t(k) = \frac{1}{2^{k+1}\sqrt{5}}\left[(3+\sqrt{5})^{k+1} - (3-\sqrt{5})^{k+1}\right].$$

Table 1.2. A Table of z-Transforms of Commonly Encountered Sequences

Sample function	$f(k)$	$F(z)$
Unit impulse, $\delta(k)$	$\begin{aligned} 1, & \quad k=0 \\ 0, & \quad k \neq 0 \end{aligned}$	1
Unit step, $u(k)$	$1, \quad k \geq 0$	$\dfrac{z}{z-1}$
Unit alternating	$(-1)^k, \quad k \geq 0$	$\dfrac{z}{z+1}$
Ramp	$k, \quad k \geq 0$	$\dfrac{z}{(z-1)^2}$
Exponential	$e^{-ak}, \quad k \geq 0$	$\dfrac{z}{z-e^{-a}}$
Power pulse	$a^k, \quad k \geq 0$	$\dfrac{z}{z-a}$
Ramp power-pulse	$ka^k, \quad k \geq 0$	$\dfrac{az}{(z-a)^2}$
Squared-ramp power-pulse	$k^2 a^k, \quad k \geq 0$	$\dfrac{az(z+a)}{(z-a)^3}$
Power-ramp power-pulse	$\dfrac{(k+1)(k+2)\cdots(k+m-1)a^k}{(m-1)!}$ $(k \geq 0, m > 1)$	$\dfrac{z^m}{(z-a)^m}$
Delayed unit impulse	$\delta(k-m), \quad m > 0$	z^{-m}
Sinusoid	$\sin \alpha k, \quad k \geq 0$	$\dfrac{z \sin \alpha}{z^2 - 2z \cos \alpha + 1}$
Cosinusoid	$\cos \alpha k, \quad k \geq 0$	$\dfrac{z(z - \cos \alpha)}{z^2 - 2z \cos \alpha + 1}$
Decaying cosinusoid	$e^{-\alpha k} \cos \beta k, \quad k \geq 0$	$\dfrac{z(z - e^{-\alpha} \cos \beta)}{z^2 - 2ze^{-\alpha} \cos \beta + e^{-2\alpha}}$

Since $t(k)$ must be an integer for each k and $(3-\sqrt{5})^{k+1}$ approaches 0 as k approaches ∞, for sufficiently large k,

$$t(k) = \left\lfloor \frac{1}{2^{k+1}\sqrt{5}}(3+\sqrt{5})^{k+1} \right\rfloor, \tag{1.14}$$

where $\lfloor x \rfloor$ denotes the largest integer smaller than or equal to x.

1.3.1. Partial Fraction Expansion Method

Case a: (Simple Poles). Let $F(z)$ be a rational function with simple poles at $z = p_1$, $z = p_2, \ldots, z = p_n$, and $p_i \neq 0$, $i = 1, 2, \ldots, n$.

$$F(z) = \frac{b_m z^m + \cdots + b_1 z + b_0}{(z - p_1)(z - p_2) \cdots (z - p_n)}, \quad m \leq n. \tag{1.15}$$

When $m \leq n$, the rational function $F(z)$ is called *proper*, and when $m < n$, it is called *strictly proper*. The objective is to expand $F(z)$ in the following form:

$$F(z) = k_0 + \frac{k_1 z}{z - p_1} + \frac{k_2 z}{z - p_2} + \cdots + \frac{k_n z}{z - p_n}. \tag{1.16}$$

Clearly,

$$k_0 = F(0) = (-1)^n \frac{b_0}{p_1 p_2 \cdots p_n} \tag{1.17a}$$

$$k_i = \left. \frac{z - p_i}{z} F(z) \right|_{z = p_i}, \quad i = 1, 2, \ldots, n. \tag{1.17b}$$

Using Table 1.2,

$$f(k) = k_0 \delta(k) + \sum_{i=1}^{n} k_i p_i^k, \quad k \geq 0. \tag{1.18}$$

The case when a pole (simple or multiple) is present at $z = 0$ will be considered following the discussion of Case b below.

Case b: (Multiple Poles). Let

$$F(z) = \frac{b_m z^m + b_{m-1} z^{m-1} + \cdots + b_1 z + b_0}{(z - p_1)^{n_1} (z - p_2)^{n_2} \cdots (z - p_r)^{n_r}},$$

where $p_i \neq 0$, $i = 1, 2, \ldots, r$, $n_1 + n_2 + \cdots + n_r = n$, and $m \leq n$. The objective is to expand $F(z)$ as

$$F(z) = k_0 + \sum_{i=1}^{r} \sum_{j=1}^{n_i} \frac{k_{ij} z^j}{(z - p_i)^j}. \tag{1.19}$$

Clearly,

$$k_0 = F(0) = (-1)^n \frac{b_0}{p_1^{n_1} p_2^{n_2} \cdots p_r^{n_r}} \tag{1.20a}$$

$$k_{in_i} = \left. \frac{(z - p_i)^{n_i}}{z^{n_i}} F(z) \right|_{z = p_i}, \quad i = 1, 2, \ldots, r. \tag{1.20b}$$

It is most convenient to solve for the remaining k_{ij}'s by specializing z to suitable values and forming a set of linear equations in an equal number of unknowns. The inverse z-transform of $F(z)$ in (1.19) can be obtained using the fact that

$$Z^{-1}\left[\frac{z^m}{(z-p)^m}\right] = \left\{\frac{(k+1)(k+2)\cdots(k+m-1)}{(m-1)!}p^k\right\}, \quad m>1.$$

Example 1.2. It is required to expand

$$F(z) = \frac{z^3 + 2z^2 + z + 1}{(z-2)^2(z+3)}, \tag{1.21a}$$

as in (1.19). Here, $p_1 = 2$, $p_2 = -3$ and $r = 2$, $n_1 = 2$, $n_2 = 1$, $n = 3$, $m = 3$. Therefore,

$$F(z) = k_0 + \frac{k_{11}z}{z-2} + \frac{k_{12}z^2}{(z-2)^2} + \frac{k_{21}z}{(z+3)}. \tag{1.21b}$$

Using (1.20a) and (1.20b),

$$k_0 = \frac{1}{12}, \quad k_{12} = \frac{19}{20}, \quad k_{21} = \frac{11}{75}. \tag{1.21c}$$

To find k_{11}, set z to any suitable value excluding 0, 2, and -3. Letting $z = 1$ and substituting (1.21c) in (1.21b), one gets

$$k_{11} = \frac{1}{12} + \frac{19}{20} + \frac{11}{300} - \frac{5}{4} = -\frac{9}{50}.$$

In case $F(z)$ has a pole of order n_0 at $p = 0$,

$$F(z) = \frac{B(z)}{z^{n_0}A(z)}$$

(where $B(z)$ and $A(z)$ are polynomials with $A(0) \neq 0$, $B(0) \neq 0$) may be expanded as

$$F(z) = k_{00} + \frac{k_{01}}{z} + \frac{k_{02}}{z^2} + \cdots + \frac{k_{0n_0}}{z^{n_0}} + \cdots, \tag{1.22}$$

where

$$k_{0n_0} = z^{n_0}F(z)\big|_{z=0}, \quad \text{but } k_{00} \neq F(0).$$

In fact $k_{00}, k_{01}, k_{02}, \ldots, k_{0(n_0-1)}$ may be computed by forming a set of linear equations and solving for the unknowns after using suitable specializations of z in (1.22).

Example 1.3. It is required to expand

$$F(z) = \frac{z+1}{z(z-1)} \tag{1.23a}$$

in the form

$$F(z) = k_{00} + \frac{k_{01}}{z} + \frac{k_{11}z}{z-1}.$$

Clearly,

$$k_{11} = \frac{z-1}{z} F(z)\Big|_{z=1} = 2$$

$$k_{01} = zF(z)\Big|_{z=0} = -1.$$

Finally, k_{00} is calculated after substituting any value (except $z = 0$ or $z = 1$), say $z = 2$, in

$$\frac{z+1}{z(z-1)} = k_{00} + \frac{2z}{z-1} - \frac{1}{z}.$$

Therefore,

$$k_{00} = \frac{3}{2} + \frac{1}{2} - 4 = -2.$$

The inverse z-transform (refer to Table 1.2)

$$Z^{-1}[F(z)] = -2\delta(k) - \delta(k-1) + 2u(k) \qquad (1.23b)$$

(where $u(k)$ is the unit sampled step) is obtainable after noting that

$$Z^{-1}[z^{-m}] = \delta(k-m). \qquad (1.23c)$$

1.3.2. Direct Division Method

The direct division method is very straightforward to apply. The denominator polynomial of $F(z)$ is used to divide the numerator polynomial (after writing each in descending powers of z), so that the quotient is a power series in z^{-1}. The coefficient of z^{-k} in this power series is $f(k)$. The inverse z-transform of $F(z)$ is, however, obtained as a power series and not in closed form. This method will be applied to the problem in Example 1.3.

Example 1.4. Obtain the first few terms of the sequence $\{f(k)\}$ associated with $F(z)$ in Example 1.3. The calculations run as follows:

$$
\begin{array}{r}
z^{-1} + 2z^{-2} + 2z^{-3} + 2z^{-4} + \cdots \\
z^2 - z \,\overline{\big)\, z + 1 } \\
z - 1 \\
\hline
2 \\
2 - 2z^{-1} \\
\hline
2z^{-1} \\
2z^{-1} - 2z^{-2} \\
\hline
2z^{-2} \\
2z^{-2} - 2z^{-3} \\
\hline
2z^{-3} \\
\vdots
\end{array}
$$

Thus, $f(0) = 0$, $f(1) = 1$, $f(2) = 2$, $f(3) = 2$, $f(4) = 2$, etc. These values agree with those obtainable from (1.23b).

1.3.3. Contour Integration Method

Usually, the contour integration method is quite cumbersome to apply for the computation of $f(k)$ via use of (1.10b). Cauchy's residue theorem in complex variable theory may be applied, and those who are familiar with the fundamentals of complex analysis need to study this subsection.

Fact 1.1 (Residue Theorem). *Let the function $F_1(z)$ be single-valued and analytic inside and on a simple closed curve C except at the singularities z_i, $i = 1, 2, \ldots, r$ inside C where the residues are given by a_{-i}, $i = 1, 2, \ldots, r$. Then*

$$\frac{1}{2\pi j} \oint_C F_1(z)\, dz = \sum_{i=1}^{r} a_{-i}$$

Example 1.5. Given

$$F(z) = \frac{z^2}{z^2 - 3z + 1},$$

it is required to compute $f(k)$ in (1.10b) by applying the residue theorem. $F(z)$ is expressible as

$$F(z) = \frac{z^2}{(z - z_1)(z - z_2)}, \tag{1.24}$$

where $z_1 + z_2 = 3$, $z_1 z_2 = 1$. The residues of $F_1(z) = F(z) z^{k-1}$ at $z = z_1$ and $z = z_2$, respectively, are

$$a_{-1} = \frac{z_1^{(k+1)}}{z_1 - z_2}, \quad a_{-2} = \frac{z_2^{k+1}}{z_2 - z_1}. \tag{1.25a}$$

From (1.24),

$$z_1 = \frac{3 + \sqrt{5}}{2}, \quad z_2 = \frac{3 - \sqrt{5}}{2}, \quad z_1 - z_2 = \sqrt{5}. \tag{1.25b}$$

Substituting (1.25b) in (1.24) and applying Fact 1.1 to compute (1.10b) using (1.25a),

$$f(k) = \frac{1}{2^{k+1}\sqrt{5}} \left[(3 + \sqrt{5})^{k+1} - (3 - \sqrt{5})^{k+1} \right].$$

1.3.4. Region of Convergence

The region of convergence is determined by the set of values of z for which the sequence $\{ f(k) z^{-k} \}$ is in l_1 (Theorem 1.2).

For a causal sequence (which is a type of right-sided sequence[3]), the region of convergence of its z-transform is the exterior of a circle. Similarly, for a left-sided sequence $\{f(k)\}$ with $f(k)=0$ for $k > k_0$, the region of convergence of its z-transform is the interior of a circle, except possibly $z=0$ if $k_0 > 0$. An *anticausal* sequence $\{f(k)\}$ is a left-sided sequence with $f(k)=0$, $k > 0$, and the region of convergence of its z-transform is the interior of a circle. A *two-sided sequence*, which can be viewed as the superposition of a causal and an anticausal sequence has a z-transform whose region of convergence is an annular region in the complex plane. To be able to get a sequence from its z-transform, the region of convergence must be specified, unless it is known a priori that only a particular class of sequence is of interest (i.e., causal sequences in the previous discussions).

Example 1.6. Given $F(z)=z/(z-2)$ with $|z| > 2$ as the region of convergence, the associated $f(k)$ is 2^k, $k \geq 0$. However, if $F(z)=z/(z-2)$ has the region of convergence specified as $|z| < 2$, the associated sequence is -2^k, $k < 0$, and 0 for $k \geq 0$.

Example 1.7. Given

$$F(z) = \frac{z^2 - z + 2}{(z-1)(z-2)(z-3)} \qquad 2 < |z| < 3,$$

it is required to find the two-sided sequence, $\{f(k)\}$.

Clearly, the factor $(z-3)$ in the denominator of $F(z)$ is associated with the anticausal portion (including part of $f(0)$) of the signal.

$$F(z) = \frac{4}{z-3} + \frac{-3z+2}{(z-1)(z-2)}$$

$$= \frac{4}{z-3} + 1 + \frac{z}{z-1} - \frac{2z}{z-2}.$$

Therefore,

$$f(k) = \begin{cases} -\dfrac{4}{3} 3^k & k < 0 \\ -\dfrac{4}{3} & k = 0 \\ -2^{k+1} + 1 & k > 0 \end{cases}$$

(Note that the inverse z-transform of $4/(z-3)$ is $(-4/3)3^k$, $k \leq 0$.)

In Example 1.7, it is clear that $\{f(k)\}$ is the superposition of an anticausal sequence and a causal sequence. The region of convergence for the anticausal part, $(-4/3)3^k$, for $k \leq 0$, whose z-transform is $4/(z-3)$, is

---

[3] For right-sided sequences, $\{f(k)\}, f(k) = 0, k < k_0$.

the region $|z| < 3$. The region of convergence for the causal part having a z-transform $(-3z + 2)/[(z - 1)(z - 2)]$ is the region $|z| > 2$, corresponding to the exterior of a circle centered at the origin in the z-plane with a radius equal to the largest magnitude pole at $z = 2$. The two-sided sequence, then, has a region of convergence bounded on the inside by the pole with the largest magnitude, which contributes to the causal part, and on the outside by the pole with the smallest magnitude, which contributes to the anticausal part. The reader is invited to verify that the two-sided sequence, associated with $F(z)$ in Example 1.7 when the region of convergence changes to $1 < |z| < 2$ is obtainable by expanding $F(z)$ in the following manner:

$$F(z) = -\frac{4}{z-2} + \frac{5z-7}{(z-1)(z-3)} = -\frac{4}{z-2} + \frac{4}{z-3} + \frac{1}{z-1}$$

$$= -\frac{4}{z-2} + \frac{4}{z-3} - 1 + \frac{z}{z-1}.$$

The required sequence, then, is

$$f(k) = \left[2^{k+1} - \frac{4}{3}3^k\right]u(-k-1) + \frac{2}{3}\delta(k) + u(k-1),$$

where $u(k) = 1$, $k = 0, 1, \ldots$, and $\delta(k)$ is the unit impulse. The sequences corresponding to regions of convergences specified, respectively, by $|z| < 1$ and $|z| > 3$, are obtained from the expansions

$$F(z) = \frac{4}{z-3} - \frac{4}{z-2} + \frac{1}{z-1}, \quad |z| < 1$$

and

$$F(z) = \frac{(4/3)z}{z-3} - \frac{2z}{z-2} + \frac{z}{z-1} - \frac{1}{3}, \quad |z| > 3.$$

1.4. Stability and Tests for Stability

A sequence $\{f(k)\}$ is said to be bounded or $\{f(k)\} \in l_\infty$ if there exists a finite positive constant K such that $|f(k)| < K$, for all k. The type of stability that is most popular in the analysis of digital filters is stability in the bounded-input bounded-output (BIBO) sense.

Definition 1.3. A linear time-invariant digital filter is BIBO stable if and only if for every bounded input sequence, the output sequence is also bounded.

Definition 1.4. A linear time-invariant digital filter is uniformly BIBO provided it is BIBO stable and there is a positive constant K such that $|y(k)| \le K|x(k)|$ for all k, where $\{x(k)\}$ is any bounded input sequence and $\{y(k)\}$ is the corresponding output sequence.

Fact 1.2 [1.8]. *For linear systems the notions of uniform BIBO stability and BIBO stability are equivalent.*

For a linear time-invariant digital filter the BIBO stability property can be characterized in terms of its impulse response sequence $\{h(n)\}$, where n instead of k has been chosen to be the integer variable.

1.4.1. Criteria for BIBO Stability

Theorem 1.2. *A linear time-invariant digital filter is BIBO stable if and only if $\sum_n |h(n)| < \infty$, that is, the impulse response sequence is absolutely summable, that is, $\{h(n)\} \in l_1$.*

PROOF:

If part: Let $\{x(n)\}$ be an arbitrary bounded input sequence so that $|x(n)| < K$, for all n and $0 < K < \infty$. The output sequence $\{y(n)\}$ is generated by convolution:

$$y(n) = \sum_k h(n-k)x(k).$$

Therefore, for any n,

$$|y(n)| \le \sum_k |h(n-k)||x(k)|$$

$$< K \sum_k |h(n-k)| < \infty.$$

Therefore,

$$\{y(n)\} \in l_\infty.$$

Only if part: Let $\sum_k |h(k)|$ be nonfinite; then $\sum_k^N |h(k)| \to \infty$ as $N \to \infty$. Choose the input sequence $\{x(n)\} \in l_\infty$ from

$$x_N(k) = \begin{cases} 1 & \text{if } h(N-k) \ge 0 \\ -1 & \text{if } h(N-k) < 0, \end{cases}$$

where $\{x(n)\} = \lim_{N \to \infty} \{x_N(n)\}$. Then

$$y_N(N) = \sum_k h(N-k)x_N(k)$$

$$= \sum_k |h(N-k)| \ge \sum_k^N |h(k)|.$$

Hence $y_N(N) \to \infty$, as $N \to \infty$. Therefore, the filter is not uniformly BIBO stable, and by Fact 1.2 it is also not BIBO stable—a contradiction. So $\{h(k)\}$ must be absolutely summable.

It should be noted that Theorem 1.2 holds for causal as well as noncausal filters. Also, any filter whose impulse response sequence has a finite number of nonzero terms is stable. Absolute summability of $\{h(n)\}$ also implies other properties, as summarized next.

Fact 1.3. *The impulse response sequence* $\{h(k)\}$ *of a BIBO stable filter, characterized by a rational transfer function, satisfies the following properties*:

1. $\lim\limits_{k \to \infty} |h(k)| = 0$

2. $\sum\limits_{k=0}^{\infty} |h(k)|^p < \infty$ *for any* $p \geq 1$

3. $\limsup\limits_{k \to \infty} \left(|h(k)| \right)^{1/k} < 1$

4. $|h(k)| \leq cr^k, 0 < c < \infty, 0 < r < 1$

As a matter of fact, the conditions are equivalent, and each is necessary and sufficient for BIBO stability of the filter.

It is possible to obtain a criterion for stability in the transform domain, and this criterion is convenient to test for via routine algebraic procedures. The transform domain stability criterion will be developed below for causal filters, characterized by rational transfer functions. It is recalled that every proper rational function $H(z)$ can be the transfer function of a causal linear time-invariant filter and vice versa. Let $H(z) = A(z)/B(z)$, where $A(z)$ and $B(z)$ are *relatively prime* polynomials in z, that is, they are devoid of any nonconstant common factor. Absence of a common factor in two polynomials can be checked by using the next theorem.

Theorem 1.3 [13]. *Polynomials*

$$A(z) = \sum_{k=0}^{m} a_k z^k$$

and

$$B(z) = \sum_{k=0}^{n} b_k z^k,$$

where a_m *and* b_n *are not both zero, are relatively prime if and only if the matrix*

$$R = \begin{bmatrix} a_m & a_{m-1} & a_{m-2} & \cdots & a_0 & 0 \cdots & 0 & 0 \\ 0 & a_m & a_{m-1} & \cdots & a_1 & a_0 \cdots & 0 & 0 \\ \cdot & \cdot & \cdot & \cdots & \cdot & \cdots & 0 & 0 \\ \cdot & \cdot & \cdot & \cdots & \cdot & \cdots & & \\ 0 & 0 & \cdot & a_m & a_{m-1} \cdot & \cdot & \cdots & a_1 & a_0 \\ 0 & 0 & \cdot & \cdot b_n & b_{n-1} & \cdots & b_1 & b_0 \\ \cdot & \cdot & \cdot & \cdots & \cdot & \cdots & & \\ \cdot & \cdot & \cdot & \cdots & \cdot & \cdots & & \\ 0 & b_n & b_{n-1} & \cdot b_2 \ b_1 & b_0 & \cdot\cdot & 0 & 0 \\ b_n & b_{n-1} & \cdot\cdot & \cdot b_1 \ b_0 \cdot & \cdot & \cdot\cdot & 0 & 0 \end{bmatrix} \begin{matrix} \\ \\ n \text{ rows} \\ \text{including } a_k\text{'s} \\ \\ \\ \\ \\ m \text{ rows} \\ \text{including } b_k\text{'s} \end{matrix}$$

of order $(m + n)$ *has a nonzero determinant.*

Example 1.8. The polynomials
$$A(z) = z^3 + 3z^2 + 3z + 1,$$
$$B(z) = z^2 + 5z + 6$$
are relatively prime because
$$R = \begin{bmatrix} 1 & 3 & 3 & 1 & 0 \\ 0 & 1 & 3 & 3 & 1 \\ 0 & 0 & 1 & 5 & 6 \\ 0 & 1 & 5 & 6 & 0 \\ 1 & 5 & 6 & 0 & 0 \end{bmatrix}$$
has a nonzero determinant.

Theorem 1.4. *The proper rational function $H(z) = A(z)/B(z)$ in reduced form characterizes a BIBO stable digital filter if and only if $B(z) \neq 0$, $|z| \geq 1$. The filter is understood to be causal.*

PROOF:

Only if part: Let the power series expansion about $z = \infty$ for the transfer function of the stable causal filter be
$$H(z) = \sum_{k=0}^{\infty} h(k) z^{-k}.$$
Then
$$|H(z)| \leq \sum_{k=0}^{\infty} |h(k)| \|z^{-k}|$$
$$\leq \sum_{k=0}^{\infty} |h(k)| < \infty, \quad |z| \geq 1.$$
Therefore, $B(z) \neq 0$, $|z| \geq 1$.

If part: Without any loss of generality, assume, for the sake of simplicity in presentation, that the poles of $H(z)$ at $z = b_i$, $i = 1, 2, \ldots m$ are distinct. Then $H(z)$ can be expanded as ($A(z)$, $B(z)$ have real coefficients),
$$H(z) = k_0 + \sum_{i=1}^{m} \frac{k_i z}{z - b_i},$$
where k_0 is a real constant and k_i, b_i, $i = 1, 2, \ldots, m$ are complex constants, each accompanied with its conjugate. If $H(z) \leftrightarrow \{h(k)\}$ is a z-transform pair, then
$$h(k) = k_0 \delta(k) + \sum_{i=1}^{m} k_i b_i^k.$$
Therefore,
$$\sum_{k=0}^{\infty} |h(k)| \leq |k_0| + \sum_{i=1}^{m} |k_i| \sum_{k=0}^{\infty} |b_i|^k.$$

Since $|b_i| < 1$, $i = 1,\ldots, m$, $|k_0|$, $|k_i|$ and m are each finite, it follows that $\{h(k)\} \in l_1$ and the filter is BIBO stable. The proof of the theorem is now complete. $\square$

1.4.2. Table Test for Stability

Polynomials that are devoid of zeros on or outside the unit circle will be referred to as *Schur polynomials*. Various tests exist to determine whether a polynomial (here only those having real coefficients are of interest) is a Schur polynomial or not. Let

$$B(z) = b_n z^n + b_{n-1} z^{n-1} + \cdots + b_1 z + b_0, \qquad (1.26)$$

be a polynomial with real coefficients, and let $b_0 \neq 0$, $b_n \neq 0$, $n \neq 0$. An algorithm, convenient to implement, for testing whether or not $B(z) \neq 0$, $|z| \geq 1$ is stated next. Without any loss in generality, b_n may be taken to be positive.

Fact 1.4. *From $B(z)$ in (1.26) form the polynomial $B_1(z)$ of degree $n - 1$.*

$$B_1(z) = \frac{1}{z}\left[b_n B(z) - b_0 z^n B\left(\frac{1}{z}\right)\right]. \qquad (1.27)$$

Then $B(z)$ is a Schur polynomial if and only if

1. $|b_0| < |b_n|$ *and*
2. $B_1(z)$ *is a Schur polynomial.*

It may be noted that the coefficients of increasing powers of z in $B_1(z)$ are the n determinants of 2×2 submatrices formed by the first column and each succeeding column in the matrix below.

$$
\begin{array}{cccccc}
b_n & b_{n-1} & \cdots & b_2 & b_1 & b_0 \\
b_0 & b_1 & \cdots & b_{n-2} & b_{n-1} & b_n.
\end{array}
$$

It is clear how the above matrix is formed from the coefficients of $B(z)$. In order to implement the algorithm each of the $(n - 2)$ polynomials $B_{i+1}(z)$ are generated from $B_i(z)$ for $i = 1, 2,\ldots, n - 2$ in the same manner as $B_1(z)$ is generated from $B(z)$. Then for $B(z)$ to be a Schur polynomial it is required that the absolute value of the constant term be less than the absolute value of the coefficient of the highest power in z for each of the polynomials in the set, $\{B(z), B_1(z),\ldots, B_{n-2}(z), B_{n-1}(z)\}$. A computer program that implements the stability test based on this method can be written, and this computer program, which also yields the coefficients of the generated polynomials required in the test, is given in Figure 1.3 (SCHUR1.FOR).

```
C       A FORTRAN PROGRAM (SCHUR1.FOR) TO TEST THE SCHUR PROPERTY
C       OF A GIVEN POLYNOMIAL, B(Z), BY IMPLEMENTING THE TABLE TEST
C          N=THE HIGHEST DEGREE OF THE INPUT POLYNOMIAL (<100)
C          B(I)=COEFFICIENTS OF THE INPUT POLYNOMIAL, B(Z)
C       POLYNOMIAL COEFFICIENTS ARE ORDERED AS IN THE TEXT.
C
        REAL B(0:99),B1(0:98)
        WRITE(6,10)
10      FORMAT(10X,'ENTER N, THE DEGREE OF INPUT POLYNOMIAL, B(Z)',/)
        READ(5,20) N
20      FORMAT(I)
        WRITE(6,30)
30      FORMAT(/,10X,'ENTER COEFFICIENTS, B(I), OF B(Z):I=0,1,...,N',/)
        READ(5,40)(B(I),I=0,N)
40      FORMAT(F)
        WRITE(6,50)
50      FORMAT(///,9X,
1       'THE POLYNOMIAL COEFFICIENTS IN ASCENDING POWERS OF Z ARE ;')
        WRITE(6,60)
60      FORMAT(//,10X,'B(I);')
        WRITE(6,70)(B(I),I=0,N)
70      FORMAT(13X,E13.5,2X,E13.5,2X,E13.5,2X,E13.5)
        J=1
100     CONTINUE
        R=B(0)/B(N)
        IF(ABS(R).GE.1.) GO TO 600
        NN=N
        IF(NN.EQ.1) GO TO 700
        DO 200 I=0,N-1
200     B1(I)=B(N)*B(I+1)-B(0)*B(N-I-1)
        IF(J.GT.9) GO TO 300
        WRITE(6,210) J
210     FORMAT(//,10X,'B',I1,'(I);')
        GO TO 400
300     WRITE(6,310) J
310     FORMAT(//,10X,'B',I2,'(I);')
400     WRITE(6,410) (B1(I),I=0,N-1)
410     FORMAT(13X,E13.5,2X,E13.5,2X,E13.5,2X,E13.5)
        B(N)=0.
        DO 500 I=0,N-1
500     B(I)=B1(I)
        N=N-1
        J=J+1
        GO TO 100
600     WRITE(6,610)
610     FORMAT(///,9X,'GIVEN POLYNOMIAL, B(Z), IS NOT OF SCHUR TYPE',/)
        GO TO 800
700     WRITE(6,710)
710     FORMAT(////,9X,'THE GIVEN POLYNOMIAL, B(Z), IS OF SCHUR TYPE',//)
800     STOP
        END
```

Figure 1.3. Listing of program SCHUR1.FOR to implement BIBO stability test.

Example 1.9. Let

$$B(z) = 12z^3 - 4z^2 - 3z + 1.$$

The following table gives $B_1(z), B_2(z)$. Note that the first two rows of the table are formed from the coefficients of decreasing powers of z in $B(z)$ and the reciprocal polynomial, $z^3 B(z^{-1})$. The third and fourth rows, similarly, give the coefficients of decreasing powers of z in $B_1(z)$ and $z^2 B(z^{-1})$:

12	-4	-3	1
1	-3	-4	12
143	-45	-32	
-32	-45	143	
19425	-7875		

From the third and fifth rows,

$$B_1(z) = 143z^2 - 45z - 32$$

$$B_2(z) = 19425z - 7875.$$

The absolute values of the ratios of the constant terms and the coefficients of the highest powers of z in $B(z)$, $B_1(z)$, and $B_2(z)$ are, respectively, $\frac{1}{12}$, $\frac{32}{143}$, and $\frac{105}{259}$. Therefore $B(z) \neq 0$, $|z| \geq 1$.

1.4.3. Schur-Cohn-Fujiwara Test for Stability

Another method to test a polynomial for zeros in $|z| \leq 1$ is based on the Schur-Cohn-Fujiwara criterion stated next.

Fact 1.5. *Associate with the polynomial $B(z)$ in (1.26) the $(n \times n)$ symmetric matrix $C = [c_{ij}]$, where c_{ij} for $i = 1, 2, \ldots, n$ and $j = 1, 2, \ldots, n$ is defined by*

$$c_{ij} = \sum_{k=1}^{\min(i,j)} \left(b_{n-i+k} b_{n-j+k} - b_{i-k} b_{j-k} \right).$$

Then, $B(z) \neq 0$, $|z| \geq 1$ if and only if C is a positive definite matrix, that is, the leading principal minors are positive.

A considerable amount of computation may be saved by noting that the matrix C exhibits symmetry not only about the main diagonal but also about the cross-diagonal.

Example 1.10. Let

$$B(z) = 24z^3 + 26z^2 + 9z + 1.$$

Then

$$C = \begin{bmatrix} 575 & \vdots & 615 & \vdots & 190 \\ 615 & 1170 & \vdots & 615 \\ 190 & 615 & 575 \end{bmatrix}.$$

It is seen that the determinant of C as well as the determinant of each of the submatrices enclosed within dotted lines are positive implying that $B(z)$ has all its zeros in $|z| \leq 1$. Also, note that only 4 of the 9 elements of C had actually to be computed.

1.4.4. An Alternate Test for Stability [9, 10, 22]

An alternate test for stability is the discrete counterpart of the fact that, in linear continuous systems theory, if a polynomial $P(s)$ with all coefficients of like sign is decomposed into the sum of an even polynomial $m(s)$ (i.e., $m(s) = m(-s)$ and an odd polynomial, $n(s)$ (i.e., $n(s) = -n(-s)$)) so that

$$P(s) = m(s) + n(s),$$

then $P(s) \neq 0$ in the closed right-half plane, defined by $\operatorname{Re} s \geq 0$ if and only if all the zeros of $m(s)$ and $n(s)$ occur on the imaginary axis, $\operatorname{Re} s = 0$, where those zeros are also simple and interlace. Schüssler's form of the test is stated.

Fact 1.6 [10]. *Let $B(z)$ be the polynomial in (1.26). Form the circularly symmetric and anticircularly symmetric components*

$$B_1(z) = \tfrac{1}{2}[B(z) + z^n B(z^{-1})]$$
$$B_2(z) = \tfrac{1}{2}[B(z) - z^n B(z^{-1})].$$

Then $B(z) \neq 0$, $|z| \geq 1$ if and only if $\left| \dfrac{b_0}{b_n} \right| < 1$, all the zeros of $B_1(z)$ and $B_2(z)$ are simple, are located on the unit circle, $|z| = 1$, and also separate each other on $|z| = 1$.

The example below serves only to illustrate the application of the test.

Example 1.11. Let

$$B(z) = 6z^2 + 5z + 1.$$

Then

$$B_1(z) = \tfrac{1}{2}[7z^2 + 10z + 7]$$
$$B_2(z) = \tfrac{1}{2}[5z^2 - 5].$$

The zeros of $B_1(z)$ are at $z = \dfrac{-5 \pm j\sqrt{24}}{7}$. The zeros of $B_2(z)$ are at $z = \pm 1$. All conditions in Fact 1.6 are satisfied, consistent with the fact that $B(z) \neq 0$, $|z| \geq 1$.

For an implementation of the test in Fact 1.6 based on a result in reference [11], see Problem 22.

1.4.5. Relationship Between Schur and Strictly Hurwitz Polynomials (Bilinear Transformation)

The counterpart of a Schur polynomial is the *strictly Hurwitz polynomial*, which is encountered in stability investigations of continuous systems. A strictly Hurwitz polynomial is required to be devoid of zeros in the closed right-half plane. It can be related to a Schur polynomial by a bilinear transformation. Thus, if $B(z)$ is a Schur polynomial of degree n, the polynomial [after replacing z by $(s+1)/(s-1)$]

$$D(s) \triangleq (s-1)^n B\left(\frac{s+1}{s-1}\right) \tag{1.28}$$

is a strictly Hurwitz polynomial, and similarly if $D(s)$ is a strictly Hurwitz polynomial of degree n, then (in $D(s)$ s is replaced by $(z+1)/(z-1)$)

$$B(z) \triangleq (z-1)^n D\left(\frac{z+1}{z-1}\right) \tag{1.29}$$

is a Schur polynomial. The mechanism of bilinear transformation, which relates continuous and discrete systems, is very useful in the analysis and design of discrete systems and is briefly discussed next. The bilinear transformation $z = (s+1)/(s-1)$ is a one-to-one onto mapping from the s-plane to the z-plane. It is simple to verify that this transformation maps the region $|z| < 1$ onto the region $\mathrm{Re}\, s < 0$, the circle $|z| = 1$ onto the boundary $\mathrm{Re}\, s = 0$, and the region $|z| > 1$ onto the region $\mathrm{Re}\, s > 0$. These properties of the bilinear transformation sketched in Figure 1.4 provide justification for the foregoing statements, summarized next for convenience.

Fact 1.7. *A polynomial $B(z)$ of degree n is devoid of zeros in the region $|z| \geq 1$ if and only if the polynomial*

$$D(s) = (s-1)^n B\left(\frac{s+1}{s-1}\right)$$

is devoid of zeros in the region $\mathrm{Re}\, s \geq 0$.

In order to generate the polynomial

$$D(s) = d_n s^n + d_{n-1} s^{n-1} + \cdots + d_1 s + d_0 \tag{1.30a}$$

from the polynomial

$$B(z) = b_n z^n + b_{n-1} z^{n-1} + \cdots + b_1 z + b_0 \tag{1.30b}$$

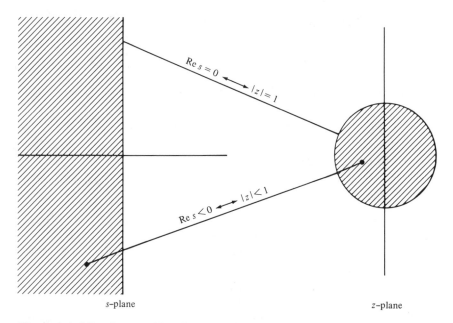

Figure 1.4. Mapping resulting from application of bilinear transformation $z = (s+1)/(s-1)$.

it is convenient computationally to relate the coefficient vector

$$\mathbf{d} \triangleq [d_n \; d_{n-1} \cdots d_1 \; d_0]^t \qquad (1.31a)$$

to the coefficient vector

$$\mathbf{b} \triangleq [b_n \; b_{n-1} \cdots b_1 \; b_0]^t \qquad (1.31b)$$

by a matrix, Q_n.

$$\mathbf{d} = Q_n \mathbf{b}. \qquad (1.32)$$

Clearly Q_n is a square matrix of order $(n+1)$. Let the elements of Q_n be denoted by q_{jk}, for $j = 0,1,\ldots,n$ and $k = 0,1,\ldots,n$, that is, q_{jk} is the element in the $(j+1)$th row and $(k+1)$th column of Q_n. Then (1.33) follows from (1.32) and (1.34) follows from (1.28), (1.30a), and (1.30b).

$$d_{n-j} = \sum_{k=0}^{n} q_{jk} b_{n-k}, \qquad j = 0,1,\ldots,n. \qquad (1.33)$$

$$D(s) \triangleq \sum_{k=0}^{n} d_{n-k} s^{n-k} = (s-1)^n \sum_{k=0}^{n} b_k \frac{(s+1)^k}{(s-1)^k}$$

$$= \sum_{k=0}^{n} b_{n-k} (1+s)^{n-k} (s-1)^k. \qquad (1.34)$$

After substituting (1.33) for the coefficients of $D(s)$, one gets

$$D(s) = \sum_{j=0}^{n} d_{n-j} s^{n-j} = \sum_{j=0}^{n} \left(\sum_{k=0}^{n} q_{jk} b_{n-k} \right) s^{n-j}. \qquad (1.35)$$

Comparing expressions in (1.34) and (1.35),

$$(s+1)^{n-k}(s-1)^{k} = \sum_{j=0}^{n} q_{jk} s^{n-j}. \qquad (1.36)$$

Clearly, from the preceding equation,

$$q_{0k} = 1, \quad q_{nk} = (-1)^{k}, \quad q_{j0} = \binom{n}{j}, \quad q_{jn} = (-1)^{j} \binom{n}{j}. \qquad (1.37)$$

Also, replacing k by $k-1$ in (1.36), one gets

$$(s+1)^{n-k+1}(s-1)^{k-1} = \sum_{j=0}^{n} q_{j(k-1)} s^{n-j}. \qquad (1.38)$$

From (1.36) and (1.38), (1.39) follows:

$$(s+1) \sum_{j=0}^{n} q_{jk} s^{n-j} = (s-1) \sum_{j=0}^{n} q_{j(k-1)} s^{n-j}. \qquad (1.39)$$

Using (1.37), since $q_{0k} = q_{0(k-1)} = 1$, the preceding equation simplifies to

$$\sum_{j=0}^{n-1} (q_{jk} + q_{(j+1)k}) s^{n-j} + q_{nk} s^{0}$$

$$= \sum_{j=0}^{n-1} (q_{(j+1)(k-1)} - q_{j(k-1)}) s^{n-j} - q_{n(k-1)} s^{0}$$

and again using (1.37), the previous equation further simplifies to

$$\sum_{j=0}^{n-1} (q_{(j+1)k} + q_{jk} - q_{(j+1)(k-1)} + q_{j(k-1)}) s^{n-j} = 0.$$

Therefore,

$$q_{(j+1)k} = q_{(j+1)(k-1)} - q_{j(k-1)} - q_{jk}. \qquad (1.40)$$

The above results lead to the following important theorem.

Theorem 1.5. *The matrix Q_n of order $(n+1)$ has the following interesting properties*:

1. *The first row is a row of 1's.*
2. *The elements of the first column are the binomial coefficients $\binom{n}{k}$, $k = 0, 1, \ldots, n$, arranged sequentially from top to bottom.*

3. *Any element q_{ij} in the $(i+1)$th row and $(j+1)$th column of the matrix Q_n is expressible as:*

$$q_{ij} = q_{i(j-1)} - q_{(i-1)(j-1)} - q_{(i-1)j},$$

for $i = 1, 2, \ldots, n$ and $j = 1, 2, \ldots, n$.

The above three properties can be used to generate the Q_n-matrix of any order very conveniently.

Example 1.12. The Q_5-matrix is

$$Q_5 = \begin{bmatrix} 1 & 1 & 1 & 1 & 1 & 1 \\ 5 & 3 & 1 & -1 & -3 & -5 \\ 10 & 2 & -2 & -2 & 2 & 10 \\ 10 & -2 & -2 & 2 & 2 & -10 \\ 5 & -3 & 1 & 1 & -3 & 5 \\ 1 & -1 & 1 & -1 & 1 & -1 \end{bmatrix}.$$

The Q_2-matrix is

$$Q_2 = \begin{bmatrix} 1 & 1 & 1 \\ 2 & 0 & -2 \\ 1 & -1 & 1 \end{bmatrix}.$$

Some additional general properties of Q_n worthy of note are summarized next.

Theorem 1.6. *The matrix Q_n of order $(n+1)$ has the following additional interesting properties, besides those summarized in Theorem 1.5:*

1. *The last row is a row of alternating 1's and -1's with 1 as its first element.*
2. *The last column consists of as elements the binomial coefficients of $(s+1)^n$ with signs alternated and the topmost element is 1.*
3. *$Q_n^2 = 2^n I$, where I is the identity matrix of appropriate order.*
4. *$[\det Q_n]^2 = 2^{n(n+1)}$.*
5. *The eigenvalues of Q_n are either $\pm 2^{n/2}$ and all the eigenvalues cannot be of the same sign.*

PROOF:

1. It follows from the second equation in (1.37).
2. This conclusion is verifiable by observing that the elements of the last column are the coefficients of $(s-1)^n$. See the last equality in (1.37).

3. $z = \dfrac{s+1}{s-1}$ implies that $s = \dfrac{z+1}{z-1}$. Then define a polynomial

$$F(z) = \sum_{k=0}^{n} f_k z^k \triangleq (z-1)^n D\left(\frac{z+1}{z-1}\right).$$ (1.41)

Then, combining (1.28) and (1.41),

$$F(z) = (z-1)^n \left(\frac{z+1}{z-1} - 1\right)^n B(z)$$

$$= 2^n B(z).$$ (1.42)

If

$$\mathbf{f} \triangleq [f_n \; f_{n-1} \cdots f_1 \; f_0]',$$ (1.43)

then from (1.42), (1.43), and (1.31b),

$$\mathbf{f} = 2^n \mathbf{b}.$$ (1.44)

Also, from (1.28), (1.32), and (1.41),

$$\mathbf{f} = Q_n \mathbf{d} = Q_n^2 \mathbf{b}.$$ (1.45)

From (1.44) and (1.45),

$$Q_n^2 = 2^n I.$$ (1.46)

Another derivation of this result follows Definition 1.5.

4. Follows from (1.46) by taking determinants of each side (it is possible to show using tedious calculation that

$$\det Q_n = (-1)^{[n(n+1)]/2} 2^{[n(n+1)]/2}).$$

5. This is easily verifiable and is left as an exercise.

Example 1.13.

1. Given $\mathbf{b} = [7 \quad 4 \quad -5 \quad 1]'$, it may be verified that

$$\mathbf{d} = [7 \quad 27 \quad 25 \quad -3]'$$

$$\mathbf{f} = [56 \quad 32 \quad -40 \quad 8]'.$$

2. Given $\mathbf{b} = [2 \quad 5 \quad -6 \quad 4 \quad 3]'$, it follows that

$$\mathbf{d} = [8 \quad -2 \quad 42 \quad -6 \quad -10]'$$

$$\mathbf{f} = [32 \quad 80 \quad -96 \quad 64 \quad 48]'.$$

3. Given $\mathbf{b} = [8 \quad 20 \quad 26 \quad 13 \quad 6 \quad 2]'$, then

$$\mathbf{d} = [75 \quad 85 \quad 74 \quad 6 \quad 11 \quad 5]'$$

$$\mathbf{f} = [256 \quad 640 \quad 832 \quad 416 \quad 192 \quad 64]'.$$

4. Given $\mathbf{b} = [3 \quad 7 \quad -2 \quad 5 \quad -4 \quad 7]'$, then

$$\mathbf{d} = [16 \quad 6 \quad 100 \quad -48 \quad 44 \quad -22]'.$$

A computer program that implements the calculations necessary to arrive at the results of the preceding example is given in Figure 1.5 (BT.FOR). The Q_n-matrix can be quite conveniently formed by using Theorem 1.5. It is also possible to calculate Q_{n+1} from Q_n as mentioned below.

Fact 1.8. *Given Q_n of order $(n+1)$, form Q_{na} by augmenting Q_n with a row of zeros. Then the matrix Q_{n+1} of order $(n+2)$ is formed by putting 1's in its first row, coefficients of $(s+1)^{n+1}$ in its first column, and if the remaining elements of Q_{n+1} are denoted by a matrix Q_{nb} of order $n+1$, then the $(i+1)$th row of Q_{nb} is obtained by subtracting the $(i+1)$th row of Q_{na} from the $(i+2)$th row of Q_{na}, $i = 0,1,\ldots, n$.*

Example 1.14. Given

$$Q_2 = \begin{bmatrix} 1 & 1 & 1 \\ 2 & 0 & -2 \\ 1 & -1 & 1 \end{bmatrix},$$

it follows that

$$Q_{2a} = \begin{bmatrix} 1 & 1 & 1 \\ 2 & 0 & -2 \\ 1 & -1 & 1 \\ 0 & 0 & 0 \end{bmatrix},$$

$$Q_{2b} = \begin{bmatrix} 1 & -1 & -3 \\ -1 & -1 & 3 \\ -1 & 1 & -1 \end{bmatrix},$$

whence

$$Q_3 = \begin{bmatrix} 1 & 1 & 1 & 1 \\ 3 & 1 & -1 & -3 \\ 3 & -1 & -1 & 3 \\ 1 & -1 & 1 & -1 \end{bmatrix}.$$

Furthermore,

$$Q_{3a} = \begin{bmatrix} 1 & 1 & 1 & 1 \\ 3 & 1 & -1 & -3 \\ 3 & -1 & -1 & 3 \\ 1 & -1 & 1 & -1 \\ 0 & 0 & 0 & 0 \end{bmatrix},$$

$$Q_{3b} = \begin{bmatrix} 2 & 0 & -2 & -4 \\ 0 & -2 & 0 & 6 \\ -2 & 0 & 2 & -4 \\ -1 & 1 & -1 & 1 \end{bmatrix}.$$

```
C       A FORTRAN PROGRAM (BT.FOR) TO FIND POLYNOMIAL COEFFICIENTS
C       AFTER APPLYING THE BILINEAR TRANSFORMATION
C          N1=NUMBER OF COEFFICIENTS OF THE INPUT POLYNOMIAL, B(Z), (<21)
C             =NUMBER OF COEFFICIENTS OF THE OUTPUT POLYNOMIAL, D(S)
C          B(I)=COEFFICIENTS OF THE INPUT POLYNOMIAL, B(Z)
C          D(I)=COEFFICIENTS OF THE OUTPUT POLYNOMIAL, D(S)
C       POLYNOMIAL COEFFICIENTS ARE ORDERED AS IN THE TEXT.
C
        DIMENSION B(20),D(20),IQN(20,20)
        WRITE(6,10)
10      FORMAT(2X,'ENTER N, THE DEGREE OF THE INPUT POLYNOMIAL')
        READ(5,20) N
20      FORMAT(I)
        N1=N+1
        WRITE(6,30)
30      FORMAT(/,2X,'ENTER POLY COEFF IN DESC POWER, ONE PER LINE')
        READ(5,*)(B(I),I=1,N1)
        WRITE(6,40)
40      FORMAT(/,2X,
     1  'THE Q(N) MATRIX, AND THE INPUT & OUTPUT COEFFICIENTS ARE :')
        DO 90 J=1,N1
90      IQN(1,J)=1
        N2=N/2+1
        DO 100 K=2,N2
100     IQN(K,1)=IQN(K-1,1)*(N-K+2)/(K-1)
        DO 110 K=1,N2
110     IQN(N1-K+1,1)=IQN(K,1)
        DO 200 I=2,N1
        DO 200 J=2,N1
200     IQN(I,J)=IQN(I,J-1)-IQN(I-1,J-1)-IQN(I-1,J)
        DO 210 I=1,N1
210     WRITE(6,220) (IQN(I,J),J=1,N1)
220     FORMAT(3X,10I7)
        DO 300 I=1,N1
        SUM=0
        DO 250 J=1,N1
        QN=FLOAT(IQN(I,J))
250     SUM=SUM+QN*B(J)
300     D(I)=SUM
        WRITE(6,*)(B(K),K=1,N1)
        WRITE(6,*)(D(K),K=1,N1)
        STOP
        END
```

Figure 1.5. Listing of program BT.FOR to compute coefficients after applying bilinear transformation.

Therefore,

$$Q_4 = \begin{bmatrix} 1 & 1 & 1 & 1 & 1 \\ 4 & 2 & 0 & -2 & -4 \\ 6 & 0 & -2 & 0 & 6 \\ 4 & -2 & 0 & 2 & -4 \\ 1 & -1 & 1 & -1 & 1 \end{bmatrix}.$$

In a private communication, Philippe Deslarte informed the author that the Q_n-matrix plays a very important role in algebraic coding theory, and its elements are related to a set of orthogonal polynomials called Krawtchouk polynomials, which are defined next.

Definition 1.5. A binary orthogonal polynomial in x of degree k and order n is defined as

$$P_k(x; n) \triangleq \sum_{j=0}^{k} (-2)^j \binom{x}{j} \binom{n-j}{k-j}, \quad k = 0, 1, \dots, n.$$

It is possible to show that the entry q_{ij} in the $(i+1)$th row and $(j+1)$th column, of Q_n for $i = 0, 1, \dots, n$ and $j = 0, 1, \dots, n$ is the value $P_i(j; n)$, which the Krawtchouk polynomial, $P_i(x; n)$, takes at $x = j$. It is easy to check, for example, that

$$P_0(x; 3) = 1$$
$$P_1(x; 3) = 3 - 2x$$
$$P_2(x; 3) = 3 - 6x + 2x^2$$
$$P_3(x; 3) = 1 - \frac{20}{3}x + 6x^2 - \frac{4}{3}x^3.$$

Clearly, the element q_{ij} in Q_3 is

$$q_{ij} = P_i(j; 3)$$

for $i = 0, 1, 2, 3$, and $j = 0, 1, 2, \dots, 3$. The Krawtchouk polynomials have numerous interesting properties (see reference [24], pp. 130–153), which can be applied directly to the elements of the Q_n-matrix. For example, (1.46) follows directly from the fact that

$$\sum_{i=0}^{n} P_r(i; n) P_i(s; n) = 2^n \delta(r - s),$$

where $\delta(r - s) = 1$, $r = s$ and $\delta(r - s) = 0$, $r \neq s$. Then, again, a recurrence between the elements of Q_n can be obtained in a form different from (1.40) by noting that for nonnegative integers, i and j,

$$(n - k) P_i(k + 1; n) = (n - 2i) P_i(k; n) - k P_i(k - 1; n).$$

Various other properties of the Q_n-matrix can be readily inferred from facts like

$$\sum_{i=0}^{n} P_i(k; n) = 2^n \delta(k).$$

$$\sum_{i=0}^{n} i P_i(k; n) = 2^{n-1}(n \delta(k) - \delta(k-1)).$$

$$\sum_{i=0}^{n} i^2 P_i(k; n) = 2^{n-2}\{n(n+1) \delta(k) - 2n \delta(k-1) + 2 \delta(k-2)\}.$$

1.4.6. Stability of Noncausal Filters

When digital filtering operations are done on the computer, it is not necessary that the filter be causal. The impulse response sequence could be anticausal or even two-sided. No matter what type of impulse response sequence occurs, the digital filter must have a stable implementation. Otherwise, numerical computations would very quickly generate very large numbers, which would lead to overflow. For BIBO stability, the region of convergence for the series expansion of the filter transfer function must contain the circle $|z| = 1$. The region of convergence of a sequence $\{h(k)\}$ is determined by the set of values of z for which $\{h(k)z^{-k}\}$ is absolutely summable, and therefore by Theorem 1.2 the BIBO stability condition is satisfied when the region of convergence for the filter unit impulse response sequence contains the circle $|z| = 1$. Therefore, the filters characterized by rational transfer functions whose unit impulse response sequences are anticausal, BIBO stability requires that all the poles be outside $|z| = 1$. Rational filters (i.e., filters characterized by rational transfer functions), whose impulse sequences are two-sided, might have poles in $|z| < 1$ and $|z| > 1$, but BIBO stability demands that the region of convergence contain $|z| = 1$, that is, no poles can occur on $|z| = 1$. Therefore, $H(z)$ in Example 1.7 can never characterize a stable filter, no matter how it is implemented. You will study recursive and nonrecursive implementations in Chapters 3 and 4, respectively. Here it is mentioned that for a digital filter to be implemented recursively, its unit impulse response sequence must be one-sided.

1.5. Discrete Convolution and Deconvolution

1.5.1. Linear Convolution

The reader is already familiar with the convolution summation that relates the output sequence $\{y(k)\}$ to the input sequence $\{x(k)\}$ and the impulse response sequence $\{h(k)\}$ of a linear time-invariant filter:

$$y(k) = \sum_{m=-\infty}^{\infty} h(k-m)x(m) = \sum_{m=-\infty}^{\infty} h(m)x(k-m). \quad (1.47)$$

The discrete linear convolution in (1.47) is usually denoted, for the sake of brevity, by

$$y(k) = h(k) * x(k). \quad (1.48)$$

For a causal filter, $h(k) = 0$ for $k < 0$, and therefore (1.47) becomes

$$y(k) = \sum_{m=-\infty}^{k} h(k-m)x(m) = \sum_{m=0}^{\infty} h(m)x(k-m). \quad (1.49)$$

In addition, if the input sequence is also causal, (1.49) further simplifies to

$$y(k) = \sum_{m=0}^{k} h(k-m)x(m) = \sum_{m=0}^{k} h(m)x(k-m). \quad (1.50)$$

It is easy to verify that if $\{h(k)\}$ and $\{x(k)\}$ are, respectively, of finite lengths M and N, the sequence $\{y(k)\}$ is of length $M + N - 1$. The linear convolution properties of discrete sequence are summarized next.

Fact 1.9. *Let* $\{h_1(k)\}, \{h_2(k)\}, \{h_3(k)\}$ *be arbitrary discrete-time sequences. Then*

$$\{h_1(k)\} * \{h_2(k)\} = \{h_2(k)\} * \{h_1(k)\}$$

$$(commutativity)$$

$$[\{h_1(k)\} * \{h_2(k)\}] * \{h_3(k)\} = \{h_1(k)\} * [\{h_2(k)\} * \{h_3(k)\}]$$

$$(associativity)$$

$$\{h_1(k)\} * [\{h_2(k)\} + \{h_3(k)\}] = \{h_1(k)\} * \{h_2(k)\}$$
$$+ \{h_1(k)\} * \{h_3(k)\}$$

$$(distributivity)$$

(The notation $\{h_1(k)\} * \{h_2(k)\}$ *has been used to denote the sequence obtained from the linear convolution of two sequences* $\{h_1(k)\}$ *and* $\{h_2(k)\}$.*)*

From Table 1.1, since the convolution of two sequences in the time domain is associated with the operation of multiplication of their transforms, the following important result is obvious.

Fact 1.10. *Let* $H_1(z)$ *and* $H_2(z)$ *be the formal power series associated with the causal sequences* $\{h_1(k)\}$ *and* $\{h_2(k)\}$, *that is,*

$$H_i(z) = \sum_{j=0}^{\infty} h_i(j)z^{-j}, \qquad for \ i = 1, 2.$$

Then the sequence obtained by convolving the two given sequences have as elements the coefficients of the formal power series $H_1(z)H_2(z)$.

From Fact 1.10, it follows that the computation of the convolution of two finite sequences is associated with the problem of multiplying two polynomials. A Fortran program, CONVOL.FOR, used for computing discrete linear convolutions, is shown in Figure 1.6A, and Figure 1.6B illustrates the linear convolution of two specified sequences.

Example 1.15. Let

$$h(k) = \left(\tfrac{1}{2}\right)^k, \quad k = 0,1,2,3.$$

Let

$$x(k) = k+1, \quad k = 0,1,2,\ldots,5.$$

Then, it can be verified that

$$y(0) = 1, \quad y(1) = 2.5, \quad y(2) = 4.25, \quad y(3) = 6.125, \quad y(4) = 8,$$
$$y(5) = 9.875, \quad y(6) = 4.75, \quad y(7) = 2.125, \quad y(8) = 0.75.$$

The reader is advised to verify the above calculations. Note that the output sequence is of length 9, as expected.

1.5.2. Deconvolution

Deconvolution or inverse filtering means the determination of the input from the measured output and the unit impulse response of a given system. The problem finds application in geophysical exploration, where inverse filtering is used to remove reverberations, and therefore permits the identification of reflections from depth, in restoring images that have been degraded by a linear spatial operation, in spectroscopy, in radioastronomy, and in other areas. When additive noise is present, the deconvolution becomes harder to implement, and it can be carried out satisfactorily to a certain extent, depending on the signal to noise ratio.

For simplicity, consider $\{h(k)\}$ and $\{x(k)\}$ to be causal sequences of lengths 4 and 2, respectively. Then (1.50) leads to the following matrix equation:

$$\begin{bmatrix} y(0) \\ y(1) \\ y(2) \\ y(3) \\ y(4) \end{bmatrix} = \begin{bmatrix} h(0) & & & & \\ h(1) & h(0) & & & \\ h(2) & h(1) & h(0) & & \\ h(3) & h(2) & h(1) & h(0) & \\ 0 & h(3) & h(2) & h(1) & h(0) \end{bmatrix} \begin{bmatrix} x(0) \\ x(1) \\ 0 \\ 0 \\ 0 \end{bmatrix}.$$

Note that the square matrix on the right-hand side of the above equation is

```
C          THIS PROGRAM LINEARLY CONVOLVES TWO SEQUENCES, {X(K)} AND
C          {H(K)}, TO YIELD THE OUTPUT SEQUENCE {Y(K)}.
           REAL X(0:100),H(0:100),Y(0:100)
           WRITE(6,10)
10         FORMAT(10X,'ENTER THE LENGTH M OF THE SEQUENCE H(I)',/)
           READ(5,30) M
           WRITE(6,20)
20         FORMAT(10X,'ENTER THE LENGTH N OF THE SEQUENCE X(I)',/)
           READ(5,30) N
30         FORMAT(I)
           WRITE(6,40)
40         FORMAT(/,10X,'ENTER THE SEQ. H(I); I=0,1,...,M-1',/)
           READ(5,60) (H(I),I=0,M-1)
           WRITE(6,50)
50         FORMAT(/,10X,'ENTER THE INPUT SEQ. X(I); I=0,1,...,N-1'//)
           READ(5,60) (X(I),I=0,N-1)
60         FORMAT(F)
           WRITE(6,70)
70         FORMAT(////,5X,'THE INPUT SEQUENCE H(I) IS:',//)
           WRITE(6,80) ((I,H(I)),I=0,M-1)
80         FORMAT(10X,'H(',I3,')=',2X,F)
           WRITE(6,90)
90         FORMAT(////,5X,'THE INPUT SEQUENCE X(I) IS:',//)
           WRITE(6,100) ((I,X(I)),I=0,N-1)
100        FORMAT(10X,'X(',I3,')=',2X,F)
           L=N+M-1
           CALL ZERO(H(M),H(L))
           CALL ZERO(X(N),X(L))
           WRITE(6,110)
110        FORMAT(////,10X,'THE OUTPUT SEQ. Y(K) OF THE CONVOLUTION IS:',//)
           DO 120 I=0,L-1
           ML=I
           CALL CONV(X,H,ML,Y)
120        WRITE(6,130) I,Y(I)
130        FORMAT(10X,'Y(',I3,')=',2X,F)
           STOP
           END
C
           SUBROUTINE CONV(H,U,K,Y)
           REAL H(0:100),U(0:100),Y(0:100)
           S=0.
           DO 100 I=0,K
100        S=S+H(I)*U(K-I)
           Y(K)=S
           RETURN
           END
```

Figure 1.6. (A) Listing of program CONVOL.FOR to compute linear convolution.

52

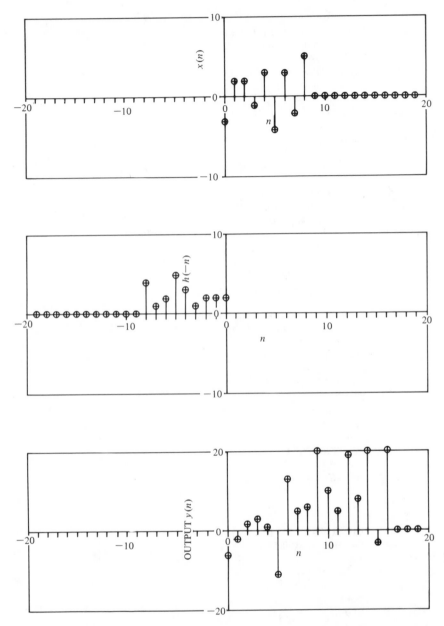

Figure 1.6. (B) Example showing the linear convolution of two finite sequences.

a lower triangular Toeplitz matrix, that is, the first column contains all necessary information for generation of the whole matrix. It is clear that $\{x(k)\}$ can be uniquely obtained from $\{y(k)\}$ recursively, provided $h(0) \neq 0$. In fact, $\{x(k)\}$ is, then, obtainable from known $\{y(k)\}$ and $\{h(k)\}$ by a series of forward substitutions. A computer program, DECONV.FOR, has been written for this purpose, as shown in Figure 1.7.

1.5.3. Circular Convolution

Consider two sequences

$$\{h_1(k)\} \triangleq [h_1(0), h_1(1), \ldots, h_1(N-2), h_1(N-1)],$$
$$\{h_2(k)\} \triangleq [h_2(0), h_2(1), \ldots, h_2(N-2), h_2(N-1)],$$

which are each of length N. The circular convolution of these two finite-length sequences is another sequence,

$$\{h_3(k)\} = \{h_1(k)\} \circledast \{h_2(k)\}, \tag{1.51a}$$

where

$$h_3(k) = \sum_{r=0}^{N-1} h_1(r) h_2\langle k-r \rangle = \sum_{r=0}^{N-1} h_1\langle k-r \rangle h_2(r), \tag{1.51b}$$

for $k = 0, 1, 2, \ldots, N-1$ and $\langle x \rangle$ denotes the number x computed modulo N. See Fact 2.5 for another method to compute circular convolution.

Example 1.16. Let

$$\{h_1(k)\} = (1, 2, 3, 12, 11)$$
$$\{h_2(k)\} = (3, 4, 5, 4, 14).$$

Then, $N = 5$ and

$$h_3(0) = 1 \times 3 + 2 \times 14 + 3 \times 4 + 12 \times 5 + 11 \times 4 = 147$$
$$h_3(1) = 1 \times 4 + 2 \times 3 + 3 \times 14 + 12 \times 4 + 11 \times 5 = 155$$
$$h_3(2) = 1 \times 5 + 2 \times 4 + 3 \times 3 + 12 \times 14 + 11 \times 4 = 234$$
$$h_3(3) = 1 \times 4 + 2 \times 5 + 3 \times 4 + 12 \times 3 + 11 \times 14 = 216$$
$$h_3(4) = 1 \times 14 + 2 \times 4 + 3 \times 5 + 12 \times 4 + 11 \times 3 = 118.$$

Note that the circular convolution sequence is obtained in Example 1.16 via computation of the matrix-vector product

$$\begin{bmatrix} 3 & 14 & 4 & 5 & 4 \\ 4 & 3 & 14 & 4 & 5 \\ 5 & 4 & 3 & 14 & 4 \\ 4 & 5 & 4 & 3 & 14 \\ 14 & 4 & 5 & 4 & 3 \end{bmatrix} \begin{bmatrix} 1 \\ 2 \\ 3 \\ 12 \\ 11 \end{bmatrix},$$

54

```
C
C       THIS PROGRAM YIELDS THE INPUT SEQUENCE {X(K)} GIVEN
C       THE OUTPUT SEQUENCE {Y(K)} AND THE IMPULSE RESPONSE
C       SEQUENCE {H(K)}.
C
        REAL X(0:100),H(0:100),Y(0:100)
        WRITE(6,10)
10      FORMAT(10X,'ENTER THE LENGTH L OF THE SEQUENCE Y(I)',/)
        READ(5,30) L
        WRITE(6,20)
20      FORMAT(10X,'ENTER THE LENGTH M OF THE SEQUENCE X(I)',/)
        READ(5,30) M
        JJ=L-M+1
        IF(JJ.LT.1) GO TO 170
30      FORMAT(I)
        WRITE(6,40)
40      FORMAT(/,10X,'ENTER THE SEQUENCE X(I); I=0,1,...,M-1',/)
        READ(5,60) (X(I),I=0,M-1)
        WRITE(6,50)
50      FORMAT(/,10X,'ENTER THE SEQUENCE Y(I); I=0,1,...,L-1'//)
        READ(5,60) (Y(I),I=0,L-1)
60      FORMAT(F)
        WRITE(6,70)
70      FORMAT(////,5X,'THE  SEQUENCE X(I) IS:',//)
        WRITE(6,80) ((I,X(I)),I=0,M-1)
80      FORMAT(10X,'X(',I3,')=',2X,F)
        WRITE(6,90)
90      FORMAT(///,5X,'THE SEQUENCE Y(I) IS:',//)
        WRITE(6,100) ((I,Y(I)),I=0,L-1)
100     FORMAT(10X,'Y(',I3,')=',2X,F)
        IF(JJ.GE.M) GO TO 110
        GO TO 120
110     CALL ZERO(X(M),X(JJ))
120     CONTINUE
        WRITE(6,130)
130     FORMAT(/////,10X,'THE SEQ. H(K) OF THE DECONVOLUTION IS:',//)
        H(0)=Y(0)/X(0)
        DO 140 I=1,JJ-1
        LS=I
        LI=I-1
        CALL ADD(H,X,LS,LI,SCONV)
        H(I)=(Y(I)-SCONV)/X(0)
140     CONTINUE
        DO 150 J=0,JJ-1
        WRITE(6,160) J,H(J)
150     CONTINUE
160     FORMAT(10X,'H(',I3,')=',2X,F)
        GO TO 190
170     WRITE(6,180)
```

Figure 1.7. (*Partial listing, continued.*)

```
180      FORMAT(//,10X,'THE LENGTH OF Y(I) MUST BE GREATER THAN
  #      THE LENGTH OF X(I)',/)
190      STOP
         END
C
C
C
         SUBROUTINE ADD(H,X,K,LSUM,SUM)
         REAL H(0:100),X(0:100)
         S=0.
         DO 100 I=0,LSUM
         S=S+H(I)*X(K-I)
100      CONTINUE
         SUM=S
         RETURN
         END
```

Figure 1.7. Listing of program DECONV.FOR to implement deconvolution.

where the matrix has the particular structure of a circulant matrix. Of course, from (1.51b) it is clear that the same result is obtained from the matrix-vector product

$$\begin{bmatrix} 1 & 11 & 12 & 3 & 2 \\ 2 & 1 & 11 & 12 & 3 \\ 3 & 2 & 1 & 11 & 12 \\ 12 & 3 & 2 & 1 & 11 \\ 11 & 12 & 3 & 2 & 1 \end{bmatrix} \begin{bmatrix} 3 \\ 4 \\ 5 \\ 4 \\ 14 \end{bmatrix}.$$

It should be noted that similar to the properties of linear convolution, summarized in Fact 1.9, circular convolution satisfies also the properties of commutativity, associativity, and distributivity. The linear convolution of two sequences of lengths M and N, respectively, may be computed by performing a circular convolution on two generated sequences, each of length $M + N - 1$, obtained by padding the sequence of length M with $N - 1$ zeros at the end and the sequence of length N with $M - 1$ zeros at the end. Fortran program CYCONV.FOR, which computes the cyclic or circular convolution of two sequences, is shown in Figure 1.8.

1.6. Response of Digital Filters

1.6.1. Zero-Input and Zero-Initial Condition Responses

Consider a time-invariant digital filter, whose input sequence $\{x(k)\}$ is related to its output sequence $\{y(k)\}$ by means of a constant coefficient linear difference equation,

$$\sum_{i=0}^{n} b_i y(k-i) = \sum_{i=0}^{m} a_i x(k-i). \qquad (1.52)$$

The transfer function $H(z)$, which is the ratio of the response transform

```
C
C         THIS PROGRAM OBTAINS THE CIRCULAR(CYCLIC) CONVOLUTION OF TWO
C         SEQUENCES, {H1(K)} AND {H2(K)}, TO YIELD THE OUTPUT SEQUENCE
C         {H3(K)}.
C
          REAL H1(0:100),H2(0:100),H3(0:100)
          WRITE(6,10)
10        FORMAT(10X,'ENTER THE LENGTH N OF THE SEQUENCES',/)
          READ(5,20) N
20        FORMAT(I)
          WRITE(6,30)
30        FORMAT(/,10X,'ENTER THE INPUT SEQ. H1(I); I=0,1,...,N-1',/)
          READ(5,50) (H1(I),I=0,N-1)
          WRITE(6,40)
40        FORMAT(/,10X,'ENTER THE SEQUENCE H2(I); I=0,1,...,N-1'//)
          READ(5,50) (H2(I),I=0,N-1)
50        FORMAT(F)
          WRITE(6,60)
60        FORMAT(////,5X,'THE INPUT SEQUENCES ARE:',//)
          WRITE(6,70) ((I,H1(I),I,H2(I)),I=0,N-1)
70        FORMAT(10X,'H1(',I3,')=',2X,F,5X,'H2(',I3,')=',2X,F)
          WRITE(6,80)
80        FORMAT(////,10X,'THE OUTPUT H3(I) OF CYCLIC CONVOL. IS:',///)
          DO 90 I=0,N-1
          II=I
          CALL CYCLIC(H2,H1,N,II,OUT)
          H3(I)=OUT
          WRITE(6,100) I,H3(I)
90        CONTINUE
100       FORMAT(10X,'H3(',I3,')=',2X,F)
          STOP
          END
C
          SUBROUTINE CYCLIC(H2,H1,N,I,YCONV)
          DIMENSION H2(0:100),H1(0:100)
          S=0.
          DO 200 K=0,N-1
          IF((I-K).LT.0) GO TO 100
          S=S+H2(K)*H1(I-K)
          GO TO 200
100       S=S+H2(K)*H1(N+I-K)
200       CONTINUE
          YCONV=S
          RETURN
          END
```

Figure 1.8. Listing of program CYCONV.FOR to compute circular (cyclic) convolution.

$Y(z)$ to the excitation transform $X(z)$ under zero initial conditions, is

$$H(z) = \frac{\displaystyle\sum_{i=0}^{m} a_i z^{-i}}{\displaystyle\sum_{i=0}^{n} b_i z^{-i}} \tag{1.53}$$

$$= \frac{z^{n-m} \displaystyle\sum_{i=0}^{m} a_{m-i} z^{i}}{\displaystyle\sum_{i=0}^{n} b_{n-i} z^{i}}. \tag{1.54}$$

The filter will be assumed to be BIBO stable. The response of the filter under zero-input conditions is the solution $y_h(k)$ of the homogeneous equation

$$\sum_{i=0}^{n} b_i y_h(k-i) = 0, \tag{1.55}$$

with appropriate initial conditions. The solution to (1.55) may be obtained using transform domain techniques, as illustrated in Example 1.1, or by writing directly the auxiliary equation

$$\sum_{i=0}^{n} b_i r^{n-i} = 0, \tag{1.56}$$

which is an algebraic equation of degree n. Denoting the roots, which could be simple or of multiple order, by $r_1, r_2, \ldots, r_{n_1}$ the solution $y_h(k)$ may be obtained as follows (if all roots are simple, then $n_1 = n$, and $n_1 < n$ when one or more multiple roots are present):

1. For each real unrepeated root r, write the solution
$$K_1 r^k,$$
where K_1 is an arbitrary constant.
2. If a real root r is repeated p times, write the solution
$$\left(K_1 + K_2 k + \cdots + K_p k^{p-1} \right) r^k,$$
where the K_i's are arbitrary constants.
3. For each pair of unrepeated complex conjugate roots with magnitude r and angle θ, write the solution
$$K_1 r^k \cos(k\theta + L_1),$$
where K_1 and L_1 are arbitrary constants.
4. If a pair of complex conjugate roots is repeated p times, write the solution
$$r^k \left[\sum_{i=1}^{p} k^{i-1} K_i \cos(k\theta + L_i) \right],$$
where the K_i's and L_i's are arbitrary constants.

The solution $y_h(k)$ is obtained by summing solutions of the above four types and then solving for the n constants by using the n specified initial conditions in the total solution $y(k)$ defined below in (1.57). The zero-state solution $y_s(k)$ may be obtained by taking the inverse z-transform of the product $H(z)X(z)$, where $H(z)$ is given in (1.53) and $X(z)$ is the z-transform of the input sequence. By virtue of linearity, the total solution $y(k)$ is

$$y(k) = y_h(k) + y_s(k). \qquad (1.57)$$

Example 1.17. It is required to solve for $y(k)$ for $k \geq 0$ in

$$6y(k) + y(k-1) - y(k-2) = 3^k, \quad k \geq 0,$$

with $y(-2) = 2$, $y(-1) = 1$, $x(k) = 3^k$ being the input.

For this problem, $n = 2$, $m = 0$ in (1.52), and the one-sided z-transform has to be used. Clearly, the auxiliary algebraic equation is

$$6r^2 + r - 1 = 0,$$

whose roots are $r_1 = -\frac{1}{2}$, $r_2 = \frac{1}{3}$. Therefore,

$$y_h(k) = K_1\left(-\tfrac{1}{2}\right)^k + K_2\left(\tfrac{1}{3}\right)^k.$$

$$Y_s(z) = \frac{z}{z-3} \frac{z^2}{(6z^2 + z - 1)} = X(z)H(z) \quad \left(\text{where } X(z) = \frac{z}{z-3}\right)$$

$$= \frac{(9/56)z}{z-3} + \frac{(1/70)z}{z + (1/2)} - \frac{(1/120)z}{z - (1/3)}.$$

$$y_s(k) = \frac{9}{56}(3)^k + \frac{1}{70}\left(-\frac{1}{2}\right)^k - \frac{1}{120}\left(\frac{1}{3}\right)^k, \quad k \geq 0.$$

The total solution can be written in the form,

$$y(k) = \frac{9}{56}3^k + K_{10}\left(-\frac{1}{2}\right)^k + K_{20}\left(\frac{1}{3}\right)^k, \quad k \geq 0,$$

where $K_{10} = K_1 + \frac{1}{70}$, $K_{20} = K_2 - \frac{1}{120}$. Actually, the above form could have been obtained without separately computing $y_h(k)$; but the objective was to determine the zero input and zero-state response, individually. Using the given initial condition

$$2 = \frac{1}{56} + 4K_{10} + 9K_{20},$$

$$1 = \frac{3}{56} - 2K_{10} + 3K_{20},$$

which, when solved, yield

$$K_{10} = -\frac{3}{35}, \qquad K_{20} = \frac{31}{120}.$$

Therefore,

$$y(k) = \frac{9}{56}3^k - \frac{3}{35}\left(-\frac{1}{2}\right)^k + \frac{31}{120}\left(\frac{1}{3}\right)^k, \quad k \geq 0. \qquad (1.58)$$

Alternate method. By applying z-transform to both sides of given equation,

$$(6 + z^{-1} - z^{-2})Y(z) = -y(-1) + y(-2) + z^{-1}y(-1) + \frac{z}{z-3}$$

$$= 1 + z^{-1} + \frac{z}{z-3}.$$

$$Y(z) = \frac{z(z+1)}{(3z-1)(2z+1)} + \frac{z^3}{(z-3)(3z-1)(2z+1)}$$

$$= \frac{(4/15)z}{z - \frac{1}{3}} - \frac{(1/10)z}{z + \frac{1}{2}} + \frac{(9/56)z}{z-3}$$

$$+ \frac{(1/70)z}{z + \frac{1}{2}} - \frac{1/120}{z - \frac{1}{3}}$$

$$= \frac{(31/120)z}{z - \frac{1}{3}} - \frac{(3/35)z}{z + \frac{1}{2}} + \frac{(9/56)z}{z-3}.$$

Therefore,

$$y(k) = \frac{9}{56}3^k - \frac{3}{35}\left(-\frac{1}{2}\right)^k + \frac{31}{120}\left(\frac{1}{3}\right)^k, \quad k \geq 0. \qquad (1.59)$$

As expected, (1.58) and (1.59) are identical.

In the total response $y(k)$, the modes excited by the poles of $H(z)$ are the *system modes* and decay with time when $H(z)$ is stable. Among the system modes present, one mode is *dominant* when its magnitude is much larger than the magnitude of each of the remaining system modes. It is possible, by proper choice of initial constants, to prevent some system modes from appearing in the output. The portion of the total response left after ignoring the system modes is usually referred to as the forced response or *steady-state response*. Similarly, the response associated with the stable system modes may be referred to as the *transient response*, since it decays toward zero with time.

1.6.2. Response to Sinusoidal Excitation

Consider a sampled complex exponential sequence with radian frequency ω,

$$\{x(k)\} = \{e^{jk\omega}\}. \qquad (1.60)$$

$\{x(k)\}$ could be causal or noncausal. Then the output of a linear time-invariant BIBO stable filter having unit impulse response $\{h(k)\}$ is

$$y(k) = \sum_{r=-\infty}^{\infty} h(r)e^{j(k-r)\omega}$$

$$= e^{jk\omega} \sum_{r=-\infty}^{\infty} h(r)e^{-jr\omega} \qquad (1.61)$$

$$\triangleq e^{jk\omega}H(e^{j\omega}),$$

where $H(z)$ is the z-transform of $\{h(k)\}$. The function $H(e^{j\omega})$, which is called the frequency response of the filter, describes how the magnitude and phase of the input complex exponential sequence is affected as it passes through the filter; since the filter is BIBO stable, the sequence $\{h(k)\}$ is of course absolutely convergent. $H(e^{j\omega})$ is a continuous function of ω, and it is also periodic with a period 2π. In terms of magnitude and phase $|H(j\omega)|$ and $\phi(j\omega)$ (also referred to as $|H(\omega)|$ and $\phi(\omega)$, respectively, without possibility of confusion),

$$H(e^{j\omega}) = |H(\omega)|e^{j\phi(\omega)}. \qquad (1.62a)$$

It is also important to note the logarithm of $H(e^{j\omega})$

$$\log H(e^{j\omega}) = \log|H(\omega)| + j\phi(\omega). \qquad (1.62b)$$

When $H(z)$ is representable as a rational function (see Fact 1.11 below), $|H(\omega)|$ is rational in ω^2; however, the phase function $\phi(\omega)$, is not, in general, rational. However, in that case, the phase function may be substituted by the group delay function, $\tau(\omega)$, defined as

$$\tau(\omega) = -\frac{d\phi(\omega)}{d\omega}. \qquad (1.63)$$

Fact 1.11. *The formal power series,* $\sum_{k=0}^{\infty} h(k)z^{-k}$ *associated with a causal sequence* $\{h(k)\}$ *of real or complex numbers has a rational representation (i.e., as a ratio of two polynomials in z) if and only if the infinite Hankel matrix*

$$
\begin{matrix}
h(1) & h(2) & h(3) & \cdots \\
h(2) & h(3) & h(4) & \cdots \\
h(3) & h(4) & h(5) & \cdots \\
\vdots & \vdots & \vdots & \cdots \\
\vdots & \vdots & \vdots & \cdots
\end{matrix}
$$

has a finite rank.

Example 1.18. It is verifiable that the causal infinite sequence $\{(\frac{1}{2})^k\}$, $k = 0, 1, 2, \dots$, has a rational representation.

However, the causal infinite sequences $\left\{\dfrac{1}{k}\right\}$ and $\left\{\dfrac{1}{k^2}\right\}$, $k = 0, 1, 2, \dots$ do not have rational representations. See Section 7.4.2 for a nontrivial application of the result stated in Fact 1.11.

1.6.3. Spectral Factorization

From (1.62a), the magnitude-squared function $|H(\omega)|^2$ is

$$|H(\omega)|^2 = H(e^{j\omega})H(e^{-j\omega}). \qquad (1.64)$$

On the unit circle $|z| = 1$, $e^{j\omega} = z$; therefore,

$$\left. |H(\omega)|^2 \right|_{e^{j\omega} = z} = H(z)H(z^{-1}). \tag{1.65}$$

Given the magnitude-squared function $|H(\omega)|^2$, it is often required to construct a stable real rational function $H(z)$ or even a real rational minimum phase function, which is constrained to have its numerator polynomial belong to the Schur class in addition to its denominator polynomial. The spectral factorization problem may be posed as follows: Given a "self-inversive polynomial" (with real coefficients)

$$P(z) = 2c_0 + \sum_{\substack{k = -n \\ k \neq 0}}^{n} c_k z^{-k}, \tag{1.66}$$

($P(z)$ may be obtainable from the numerator or denominator polynomials of $|H(\omega)|^2$ after replacing $e^{j\omega}$ by z; for a self-inversive polynomial, a zero at $z = z_0$ implies also a zero at $z = z_0^{-1}$), with $P(z) > 0$ on $|z| = 1$, it is required to find a polynomial $B(z)$ having real coefficients and belonging to the Schur class such that

$$P(z) = B(z)B(z^{-1}). \tag{1.67}$$

Sometimes the positivity property of $P(z)$ on $|z| = 1$ could be replaced by nonnegativity, in which case zeros of the spectral factor $B(z)$ on $|z| = 1$ would be permissible.

The function $P(z)$ in (1.66) is said to belong to the Caratheodory class when $P(z) > 0$ on $|z| = 1$ (or $P(z) \geq 0$ on $|z| = 1$, though here the nonnegativity case is not considered), and then it always admits a spectral factorization of the type $B(z)B(z^{-1})$ in (1.67) with $B(z) \neq 0$, $|z| \geq 1$. In fact, if the zeros of $P(z)$ are known, then those inside the circle are assigned to $B(z)$, so that the reciprocal zeros are assignable to $B(z^{-1})$. Bauer [20] gave an iterative procedure for determining the coefficients of $B(z)$. If $B(z)$ is written as

$$B(z) = \sum_{k=0}^{n} b_k z^k, \tag{1.68}$$

then at the mth iteration, the required coefficients b_k, $k = 1, 2, \ldots, n$, are estimated by $b_k^{(m)}$, $k = 1, 2, \ldots, n$, which in turn, are described by the Bauer recursion given next. For $m = 1, 2, 3, \ldots$,

$$b_0^{(m)} = \sqrt{2c_0 - \sum_{i=1}^{n} \left[b_i^{(m-i)} \right]^2}, \tag{1.69}$$

and for $k = 1, 2, \ldots, n$,

$$b_k^{(m)} = \left[c_k - \sum_{i=1}^{n} b_i^{(m-i)} b_{i+k}^{(m-i)} \right]. \tag{1.70}$$

The initial values for the recursion are given by

$$b_k^{(0)} = 0, \quad k = 0, 1, 2, \ldots, n.$$

It is quite easy to program the Bauer recursion described above, which leads directly to the coefficients of the spectral factor. Specifically,

$$b_k^{(m)} \to b_k$$

for $k = 0, 1, 2, \ldots, n$ as $m \to \infty$.

1.7. Summary and Suggested Readings

The sampling theorem for band-limited signals of finite energy has received extensive attention, as it forms the basis for the interchangeability of analog signals and discrete sequences. It is important to note from a practical standpoint that the recovery of a band-limited signal of finite energy from a knowledge of its samples taken at the Nyquist rate is stable in the sense that a small error in reading sample values produces only a correspondingly small error in the recovered signal [1, 2]. Sampling errors, however, do occur from a variety of causes. Foldover errors in the reconstructed frequency domain function, when the sampling rate is not sufficiently high, result in what is termed the *aliasing error*. Other common sampling errors are due to round-off of the samples, truncation of the sequence of the samples, jitter in the recording of the sampling instants, loss of a number of samples when these are stored in a magnetic tape, and the nonideal nature of physically realizable filters through which the sequence of samples have to be passed before the recovery of the band-limited analog signal is complete. More information on error analysis of sampling can be found in references [3] and [4].

Once a discrete sequence is obtained, it is convenient, for the purpose of analysis and synthesis of discrete-time systems, to associate with it a formal power series or a generating function in a single complex variable. The only text exclusively devoted to the mathematics of z-transform theory and its applications is the one by Jury [5], though many books of more recent origin contain sections on z-transform theory. The reader might want to refer to references [6] and [7] for additional information on the theory of double-sided z-transform.

Stability is a vital property to be satisfied in any system design. Various types of tests for BIBO stability are documented in references [5] and [13], and three algebraic stability tests are presented. Then the mechanism of bilinear transformation, which can be used to relate Schur polynomials to strictly Hurwitz polynomials, is discussed. It is of interest that the Q_n matrix occurring in the computations of the coefficients of a polynomial equation, resulting from a bilinear transformation of its variable, was first discussed by Unbehauen [14]. Transforming a polynomial of degree n by the Q_n-matrix requires $(n+1)^2$ multiplications and $n(n+1)$ additions. The structure of the Q_n-matrix, along with some clever programming techniques, may be used to reduce the number of multiplications by about a factor of half without changing the number of additions. For another algorithm, which compares favorably with the Q_n-matrix method, the reader may refer to reference [15], where also a program in BASIC has been presented to do the job. After application of the bilinear transformation on a specified polynomial, the known tests for strictly Hurwitz property [13] naturally become applicable to determine whether or not the originally specified polynomial belongs to the Schur class. In addition, there are a variety of numerical tests, which are fast but cannot be implemented with infinite precision. Techniques based on the use of fast Fourier transform (FFT), one of the topics to be covered in the next chapter, exist for numerical stability tests.

The literature on convolution, deconvolution, and inverse filtering is quite vast [16,17]. The important thing to note is that these problems are characterized by matrices, whose particular structures can be exploited to a considerable extent to reduce the computational chores (see Chapter 6).

The topic of Section 1.6 is adequately covered in most textbooks, whose contents might be viewed as prerequisites for a course or courses based on the subject matter of this book. The reader might wish to consult the material in reference [18]. Finally, reference [19] is a general mathematical reference concerned with difference equations and their solutions.

Problems

1. The formula for reconstruction of a band-limited analog signal $x(t)$ from its discrete sequence of samples $\{x(kT)\}$, as given by

$$x(t) = \sum_{k=-\infty}^{\infty} x(kT)h(t-kT),$$

    where

$$h(t) = \frac{\sin(\omega_s t/2)}{\omega_s t/2}, \quad \omega_s = \frac{2\pi}{T},$$

    may be viewed as the convolution summation of a sequence of pulses, $\{x(kT)\}$, with a low-pass analog filter impulse response, $h(t)$.
    a. Is the analog filter physically realizable?
    b. Determine the Fourier transform of the impulse response.
    c. Calculate $\int_{-\infty}^{\infty} h^2(t)\,dt$.

2. The input–output relationship for a first-order digital system is given to be
$$y(k) = y(k-1) + ax(k), \quad y(-1) = 0,$$
    where $x(k), y(k)$ are, respectively, the kth input and output samples, and a is a real scalar. Plot $y(k)$ vs. k, for $k = 0,1,\ldots,20$ when a unit impulse is applied as input for $a = 0.5$, $a = 1$, and $a = 2$.

3. The transfer function of a stable, causal digital system is given to be

$$H(z) = \frac{0.5z}{(z-0.5)(z-0.1)} .$$

Obtain the impulse response sequence $\{h(k)\}$, $k = 0,1,2,...$, by
a. partial fraction expansion;
b. direct division; and
c. contour integration.

4. The transfer function of a digital system is given to be

$$H(z) = \frac{0.5z}{z^2 - z + 0.5} .$$

For a sampled step input (sampling period is 1 second), obtain the output $y(k)$ in the form

$$y(k) = a + b(c)^k \cos(dk + e),$$

where a, b, c, d, and e are real constants to be calculated.

5. The transfer function of a digital system is

$$H(z) = \frac{0.1z(z-0.5)}{(z-0.1)(z-0.2)(z-0.9)} .$$

For a sampled step input (sampling period is 1 second), obtain
a. the steady-state response;
b. the transient response; and
c. a dominant mode, if any, in the transient response.

6. Consider the digital feedback system

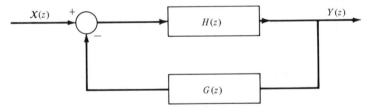

The open-loop system, characterized by $H(z) = \frac{0.01}{z-4}$ is unstable. It is required to place the poles of the feedback system at $z = 0.2$ and $z = 0.8$ using a
compensator characterized by $G(z) = \frac{a}{z-b}$, where a and b are real constants.
Calculate the values of a and b.

7. It is required to design a simple first-order stable digital filter whose transfer
function has the form

$$H(z) = \frac{a(z-b)}{(z-c)},$$

so that it will transmit the unit step input with gain one in the steady-state but will completely suppress the unit alternating input. Determine a set of values for the parameters a, b, c that will do the job.

8. In the feedback system of problem 6,

$$H(z) = \frac{6}{z-5}.$$

In order to obtain a fast responding system, determine the values of a and b so that the overall feedback system has a pole of multiplicity 2 (double-order pole) at $z = 0.5$.

9. Consider the difference equation relating the input sequence $\{x(k)\}$ to the output sequence $\{y(k)\}$:

$$y(k) = x(k) - 2x(k-1) + 3x(k-2) + 1.5y(k-1) - .75y(k-2).$$

a. Calculate the transfer function, $H(z) = Y(z)/X(z)$ (for transfer function calculations, remember to use zero initial conditions).
b. Write down the difference equation relating the input and output of the inverse system, characterized by a transfer function, $1/H(z)$.
c. Is the inverse system stable?
d. Determine the steady-state response of the original system to a unit step.

10. Which of the systems, described by their input–output difference equations, are linear time-invariant? $\{y(k)\}$ is the output sequence, and $\{x(k)\}$ denotes the input sequence.
a. $y(k) = x(k-2)y(k-1) + x(k-1)$.
b. $y(k) = 4x(k) + 2x(k-1) + 5k$.
c. $y(k) = x(k) + x(k-1) + 4x(k-2)$.
d. $y(k) = y(k-1) + x(k) - x(k-2)$.
e. $y(k) = x(k)\sin\left(\frac{3\pi}{7k} + \frac{\pi}{8}\right)$.

11. [1.21] Let $T(n, m, r)$ be the number of spanning trees for a multigraph wheel having n nodes on the rim in addition to a node at the center (called the hub), where the multiplicity of each rim edge is m and the multiplicity of each spoke is r. Then it is possible to show that

$$T(n, m, r) = (2m + r)T(n-1, m, r) - m^2 T(n-2, m, r) + 2rm^{n-1},$$

where

$$T(2, m, r) = 4mr + r^2$$

and

$$T(3, m, r) = (2m + r)^3 - 3m^2(2m + r) - 2m^3.$$

Obtain the solution for the above difference equation.

12. In a certain system the input and output signals have sampled values, patterned as shown below.

$$k:\ 0\ 1\ 2\ 3\ 4\ 5\ 6\ \cdots$$

$$\text{Input } x(k):\ 1\ 1\ 0\ 0\ 0\ 0\ 0\ \cdots$$

$$\text{Output } y(k):\ 1\ 0\ 1\ 0\ 1\ 0\ 1\ \cdots$$

What is the transfer function of the system?

13. Carry out the convolution of the following two causal sequences.

$$k:\ 0\ 1\ 2\ 3\ 4\ 5\ 6$$

$$x_1(k):\ 1\ 2\ 3\ 4\ 5\ 6\ 7$$

$$x_2(k):\ 6\ 5\ 4\ 3$$

Count the number of multiplications and additions you used to obtain the convolved output.

14. Evaluate the z-transform of the sequence

$$x(k) = \begin{cases} 0 & \text{if } k < 0 \\ 2^{-k}\cos k & \text{if } k \geq 0. \end{cases}$$

What is the region of convergence of this z-transform?

15. Determine the initial value $x(0)$ and the final value $x(+\infty)$ of a sequence $\{x(k)\}$ whose z-transform is

$$X(z) = \frac{12z}{z-1} - \frac{6z\sin 20°}{z^2 - z\cos 20° + 0.50}.$$

16. Consider the double differentiator $y = \dfrac{d^2x}{dt^2}$.
   a. Relate $y(kT)$ to $x(kT)$, $k = 0,1,2,\ldots$, with $T = 1$, by a difference equation approximating the above differentiator. Use the approximation

$$\frac{d}{dt}x(kT) \triangleq \frac{x(kT) - x[(k-1)T]}{T}$$

   and a corresponding approximation of the second derivative of the input x.
   b. Write a simple program for the realization of the approximate double differentiator in (a) by a programmable digital device (e.g., a microprocessor). Use FORTRAN or BASIC or an assembly language of your choice.

17. Consider an LTI, whose input $(x(k))$–output $(y(k))$ relationship is described by

$$y(k) = 0.5y(k-1) + 2x(k) - x(k-1),$$

with $y(-1) = 3$, $y(-k) = 0$, for $k > 1$, and $x(-k) = 0$, for $k > 0$.
   a. Find the unit impulse response sequence $\{h(k)\}$ for the system.
   b. Express $y(k)$ in terms of $x(k)$, for $k \geq 0$.

18. Determine whether or not the LTI systems characterized by transfer functions given below, are stable in the bounded-input bounded-output sense. Before applying stability tests, check the numerator and denominator polynomials for common factors.

a. $H(z) = \dfrac{z+2}{z^2 + \frac{3}{4}z + \frac{1}{8}}.$

b. $H(z) = \dfrac{z^2 + 2z + 1}{z^3 + 2z^2 + \frac{5}{4}z + \frac{1}{4}}.$

c. $H(z) = \dfrac{5z + 1}{20z^4 + 3z^2 + 8z + 10}.$

19. Given

$$H(s)H(-s) = \left. \frac{1 + \omega^2}{1 + \omega^8} \right|_{\omega^2 = -s^2},$$

obtain a transfer function $H(s)$ devoid of poles and zeros in $\operatorname{Re} s \geq 0$ (such transfer functions are called minimum phase). Plot the magnitude of the frequency response of the digital filter obtained by applying the bilinear transformation to $H(s)$ (i.e., replacing s by $\dfrac{z+1}{z-1}$).

20. Determine the unit impulse response sequence of an LTI causal digital system from the following observed causal input and output sequences of finite length.

k:	0	1	2	3	4	5	6	7	8	9	10	11	12
$x(k)$:	1	1	2	3	1	2	4	0	0	0	0	0	0.
$y(k)$:	1	3	6	9	10	9	9	10	9	1	-3	2	4

21. Show whether or not the coefficient vector

$$\mathbf{d}_1 \triangleq \begin{bmatrix} d_{1n} & d_{1(n-1)} & \cdots & d_{11} & d_{10} \end{bmatrix}'$$

of

$$D_1(s) = (s+1)^n B\left(\frac{s-1}{s+1}\right) = \sum_{k=0}^{n} d_{1k} s^k$$

is obtainable from the coefficient vector

$$\mathbf{b}_1 \triangleq \begin{bmatrix} b_0 & b_1 & \cdots & b_{(n-1)} & b_n \end{bmatrix}'$$

of

$$B(z) = b_0 + b_1 z + \cdots + b_n z^n$$

via computation of

$$\mathbf{d}_1 = Q_n \mathbf{b}_1,$$

where Q_n is the matrix in Section 1.4.5.

22. To implement the conditions in Fact 1.6 in order to test for Schur property of a polynomial $B(z)$, form the ratio

$$R(z) = \frac{B_1(z)}{B_2(z)}.$$

Then $B(z) \neq 0$, $|z| \geq 1$ if and only if the following continued fraction expansion is always possible with $K_i > 0$, $i = 0, 1, \ldots, n-1$.

$$R(z) = K_0 \frac{z+1}{z-1} + \cfrac{1}{K_1 \frac{z+1}{z-1} + \cfrac{1}{\ddots \atop{K_{n-2}\frac{z+1}{z-1} + \cfrac{1}{K_{n-1}(z+1)/(z-1)}}}}.$$

The functions $R_i(z)$ are defined recursively as

$$R_0(z) = R(z),$$

$$R_i(z) = K_i \frac{z+1}{z-1} + \frac{1}{R_{i+1}(z)}, \quad i = 0, 1, \ldots, n-2,$$

$$R_{n-1}(z) = K_{n-1} \frac{z+1}{z-1}.$$

The K_i's are calculated as

$$K_i = \tfrac{1}{2} \lim_{z \to 1} \{(z-1) R_i(z)\}, \quad i = 0, 1, \ldots, n-1.$$

Apply the above test on a polynomial

$$B(z) = 2z^3 + 2z^2 + z + 1.$$

See also reference [12].

23. A digital filter transfer function has a simple zero at $z = -1$ and three simple poles at $z = -\frac{1}{2}$, $z = -3$, and $z = -6$.
  a. Show that the unit impulse response-sequence of the filter must be two-sided, if the filter is to be stable in the BIBO sense.
  b. What is the region of convergence of the stable filter in (a)?
  c. What is a possible region of convergence, if the filter unit impulse response is two-sided, but the filter is implemented in an unstable manner?

24. Consider the z-transform

$$H(z) \triangleq \sum_{k=0}^{\infty} h(k) z^{-k} = -\log_e(1 + \tfrac{1}{3}z^{-1}), \quad |z| > \tfrac{1}{3}.$$

Calculate $h(0)$, $h(1)$, and $h(2)$.

25. Consider the input/output relationship of a causal digital filter:

$$y(k) - \tfrac{1}{2}y(k-1) = x(k) + \tfrac{1}{4}x(k-1).$$

Determine the steady-state response of the filter to

$$x(k) = 2 + \cos\left(\frac{\pi}{2}k + \frac{\pi}{4}\right), \quad k = 0, 1, 2, \ldots.$$

26. Two LTI digital systems, whose unit impulse response sequences are $\{h_1(k)\}$ and $\{h_2(k)\}$, are cascaded. Let

$$h_1(k) = \left(\tfrac{1}{2}\right)^k u(k),$$
$$h_2(k) = \delta(k) - \delta(k-3),$$

where $u(k)$ and $\delta(k)$ are, respectively, the unit step and unit impulse. Calculate the steady-state response of the cascaded system to an input

$$x(k) = 4\cos\left(\frac{2\pi}{3}k - \frac{\pi}{8}\right).$$

27. It has been seen in (1.61) that the frequency response, $H(e^{j\omega})$, of a digital filter is the z-transform of its unit impulse response sequence, $\{h(k)\}$, evaluated on $|z| = 1$. $H(e^{j\omega})$ is the Fourier transform of $\{h(k)\}$. Define a new sequence, $\{h_1(k)\}$, where

$$h_1(k) = \begin{cases} h\left(\dfrac{k}{2}\right) & k = 0, \pm 2, \pm 4, \ldots \\ 0 & \text{otherwise.} \end{cases}$$

Obtain the Fourier transform, $H_1(e^{j\omega})$, of $\{h_1(k)\}$ in terms of $H(e^{j\omega})$.

References

1. Linden, D. A. 1959. A discussion of sampling theorems. Proc IRE: pp. 1219–1226.
2. Landau, H. J. 1967. Sampling, data transmission and the Nyquist rate. Proc. IEEE, 10:1701–1706.
3. Papoulis, A. 1966. Error analysis in sampling theory. Proc. IEEE, 7:947–955.
4. Stickler, D. C. 1967. An upper bound on aliasing error. Proc. IEEE, 8:418–419.
5. Jury, E. I. 1973. Theory and Application of the z-Transform Method. Robert E. Krieger Publishing, New York.
6. Oppenheim, A. V., and Schafer, R. W. 1975. Theory and Application of Digital Signal Processing. Prentice-Hall, Englewood Cliffs, NJ.
7. Jong, M. T. 1982. Methods of Discrete Signal and System Analysis. McGraw-Hill, New York.
8. Desoer, C. A., and Thomasian, A. J. 1963. A note on zero-state stability of linear systems. Proceedings of the Allerton Conference on Circuit and System Theory, pp. 50–52.
9. Unbehauen, R. 1972. Determination of the transfer function of a digital filter from the real part of the frequency response. Arch. Elektr. Übertrag., 26:551–557.
10. Schüssler, H. W. 1976. A stability theorem for discrete systems. IEEE Trans. Acoust. Speech Signal Proc., 24:87–89.
11. Szczupak, J., Mitra, S. K., and Jury, E. I. 1977. Some new results on discrete system stability. IEEE Trans. Acoust. Speech Signal Proc., 25:101–102.
12. Steffen, P. 1977. An algorithm for testing stability of discrete systems. IEEE Trans. Acoust. Speech Signal Proc., 25:454–456.

13. Jury, E. I. 1974. Inners and Stability of Dynamic Systems. Wiley-Interscience, New York.

14. Unbehauen, R. 1964. Ein Beitrag zur Stabilitätsuntersuchung linearer Abtastsysteme. Regelungstechnik, 1:12–16.

15. Davies, A. C. 1974. Bilinear transformation of polynomials. IEEE Trans. Circuits Systems, 21:792–794.

16. Cuenod, M., and Durling, A. 1969. A Discrete-Time Approach for System Analysis. Academic Press, New York.

17. Sage, A. P., and Melsa, J. L. 1971. System Identification. Academic Press, New York.

18. Cadzow, J. A. 1973. Discrete-Time Systems. Prentice-Hall, Englewood Cliffs, NJ.

19. Goldberg, S. 1958. Introduction to Difference Equations. Wiley, New York.

20. Bauer, F. L. 1955. Ein direktes Iterationsverfahren zur Hurwitz Zerlegung eines Polynoms. Arch. Elektr. Übertrag., 9:285–290.

21. Bose, N. K., Feick, R., and Sun, F. K. 1973. General solution to the spanning tree enumeration problem in multigraph wheels. IEEE Trans. Circuit Theory, 20:69–70.

22. Nour Eldin, H. A. 1969. A new stability criterion for linear, stationary sampled-data systems. Sci. Electr., 15:35–66.

23. Rabiner, L. A. et al. 1972. Terminology in digital signal processing. IEEE Trans. Audio Electroacoustics, 20:322–337.

24. MacWilliams, F. J., and Sloane, N. J. A. 1977. The Theory of Error-Correcting Codes. North-Holland Publishing, Amsterdam.

Chapter 2
Transform Theory

2.1. Introduction

Section 1.5 in Chapter 1 showed that the output sequence of a linear shift-invariant (LSI) discrete system is obtainable by convolving the input sequence with the unit impulse response. The transform domain techniques are very powerful techniques for analysis and synthesis of LSI systems. This chapter provides the fundamentals of various such transform domain methods known to date, so that the reader can be equipped with the basics in this recently consolidated area of research. Since the subject matter of this chapter can be quite complicated and comprehensive, important principles are sometimes stated without proof, and carefully constructed examples often are relied upon to illustrate the application of those principles. The author feels that the readers to whom this book is primarily addressed, though not expected to become specialists in this area, would find the underlying theme informative and useful from both the theoretical and the application standpoint.

In Section 2.2, the steps leading to the derivation of the discrete Fourier transform (DFT), its properties, and its computation via the fast Fourier transform (FFT) algorithm are briefly presented. The reason for the brevity is that various special algorithms (like the radix-2 FFT, the radix-4 FFT, based on decimation-in-time and decimation-in-frequency approaches) have a common underlying philosophy. Moreover, discursive documentation is available in numerous textbooks, and most, if not all, prospective readers of this textbook are expected to have at least a passing familiarity with the subject.

Section 2.3 is devoted to the exposition of a fundamental algorithm and an important theorem, whose usage will be required in the remainder of the chapter. Euclid's algorithm is probably the oldest nontrivial algorithm. The division process, which supports this "divide-invert-divide" algorithm is stated for integers as well as polynomials in a single variable, whose coefficients belong to an arbitrary but fixed field. (The reader might refer to any elementary book on abstract algebra for definitions of fields, rings, and so on; the field of most interest to us here is the field of rational numbers.) Euclid's algorithm enables one to calculate the greatest common factor (unique up to units of the unique factorization domain) of any two elements of the set of integers or the set of polynomials. The role of generalized Euclid's algorithm to solve an equation of the type $ca + db = e$, for the unknowns c and d over the ring of integers, is expounded. The counterpart of this result, in the case of single variable polynomials whose coefficients are in an arbitrary but fixed field, is also required in some problems discussed in the text. The latter part of Section 2.3 is devoted to the exposition of the Chinese remainder theorem (CRT) over the ring of integers and also over the ring of single variable polynomials whose coefficients belong to any specified field.

The discrete Hilbert transform (DHT) has found applications in a variety of problems. Matrix forms of the 1-D DHT were proposed by some authors, and the scopes for applying DHT in digital filtering and the stabilization problem, without altering significantly the frequency response magnitude, have been noted in the literature (see Section 6.4). Scopes for applying the algebraic computational complexity theory to the computation of DHT exist and the advantages derived can be translated into reduced storage requirements, lesser overall computational time in implementation, and superior noise characteristics. In view of the possible impact of multiplicative computational complexity theory, to be discussed from Section 2.5 onward, on the various well-known uses of DHT, Section 2.4 is devoted to the discussion of this popular transform. Various forms of the impulse response operator for a 1-D DHT are derived and the scopes for exploiting the structure of the matrix in the matrix formulation of the DHT are noted.

In the earlier literature in digital signal processing, one of the most useful applications of the DFT has been in the development of fast algorithms for implementing convolution. Another important factor, which is of relatively recent origin, has been the recognition of the fact that linear convolution can be related to polynomial multiplication, whereas the computation of a circular convolution can be equated to the computation of a product of two polynomials modulo a third polynomial. This approach has led to the development of a theory of algorithms requiring the minimum multiplicative complexity, which is useful for computing convolutions and DFT's, when they are equated to circular convolutions. In Section 2.5, optimal algorithms are constructed to implement short-length circular convolutions. These optimal algorithms are also useful for developing suboptimal algorithms to compute longer length convolutions and DFT's.

Section 2.6 is concerned with number-theoretic transforms (NTT's). These transforms (defined over the ring of integers modulo numbers like Mersenne or Fermat numbers) are attractive for the implementation of convolution because it is possible to construct transforms, which are free of multiplications. Also, significant roundoff and quantization errors that occur in DFT's defined in the complex number field are avoided via use of number-theoretic techniques, even though these transforms require large wordlength and possibly complex hardware. Discussion is restricted to the two most widely investigated NTT's, namely, the Mersenne transform and the Fermat number transform.

Discrete polynomial transforms, defined in a ring of polynomials that can be used to compute polynomial circular convolutions without multiplications and round-off errors are briefly discussed in Section 2.7. Though evaluation of polynomial convolutions is usually equated to the computation of two-dimensional circular convolutions employed, for example, in digital processing of images, it is felt that the approach is interesting and potentially useful also in one-dimensional signal-processing applications.

In Section 2.8, additional comments are made on the results presented in the preceding sections.

2.2. Discrete-Time Fourier Series, Fourier Transform, and Discrete Fourier Transform

2.2.1. Preludes to DFT

In continuous-time system theory, the Fourier series arises when one studies the response of a linear time-invariant (LTI) system to a periodic input. Any periodic function of time $f(t)$, which is absolutely integrable over one period T and has at most a finite number of finite discontinuities, is expandable as

$$f(t) = \sum_{k=-\infty}^{\infty} c_k e^{jkt},$$

where the coefficients c_k are defined by the equation

$$c_k = \frac{1}{T} \int_{-T/2}^{T/2} f(t) e^{-jkt} \, dt.$$

Note that in the above expansion, the index of summation k runs through the set of all positive and negative integer values including zero. A discrete-time signal $x(r)$ is periodic if, for some positive value of N,

$$x(r) = x(r+N).$$

Any periodic discrete-time signal is expandable as a finite sum of weighted complex exponentials,

$$\lambda_i(r) = e^{j(2\pi/N)ir}, \quad i = 0, 1, \ldots, N-1.$$

The finiteness property follows from the fact that the complex exponentials

$\lambda_i(r)$ are periodic, with a period N, since

$$\lambda_i(r) = \lambda_i(r + N).$$

So, it is only necessary to consider a subset of N distinct elements of the set, $\{\lambda_k(r)\}$. Therefore,

$$x(r) = \sum_{k=0}^{N-1} d(k) e^{j(2\pi/N)kr},$$

where the index of summation, given above over $0 \leq k \leq N-1$, could be equivalently expressed over any period. Making use of the identity

$$\sum_{k=0}^{N-1} e^{j(2\pi/N)kr} = \begin{cases} N, & \text{if } r = 0, \pm N, \pm 2N, \ldots \\ 0, & \text{otherwise}, \end{cases}$$

the coefficients $d(k)$ in the preceding discrete-time Fourier series representation of $x(r)$ are given by

$$d(k) = \frac{1}{N} \sum_{r=0}^{N-1} x(r) e^{-j(2\pi/N)rk}, \quad k = 0, 1, \ldots, N-1.$$

Note that

$$d(k) = d(k + N).$$

The Fourier transform operator was defined in Section 1.2 for continuous-time aperiodic signals. The Fourier transform for discrete-time aperiodic signals can be obtained as a limiting case of the discrete Fourier series representation of a periodic signal, for which the aperiodic signal of finite duration is a restriction. In fact, if the discrete-time aperiodic sequence $\{x(r)\}$ is of finite duration, then it is straightforward to show that the following pair of equations form a discrete-time Fourier transform pair:

$$x(r) = \frac{1}{2\pi} \oint X(e^{j\omega}) e^{j\omega r} d\omega,$$

$$X(e^{j\omega}) = \sum_{r=-\infty}^{\infty} x(r) e^{-j\omega r}.$$

Note that when $\{x(r)\}$ is aperiodic, its *spectrum* $X(e^{j\omega})$ is a continuous function of frequency and is periodic; in the case when $\{x(r)\}$ is periodic, the corresponding distinct spectral coefficients are finite in number and the Fourier coefficients are periodic. The discrete-time Fourier series and Fourier transform have been elaborated upon in more elementary textbooks. (The exposition in reference [23] is discursive and appealing.) Here, we focus on the derivation and computation of the discrete Fourier transform (DFT) for finite-duration signals—a topic, though different from the discrete-time Fourier series and the discrete-time Fourier transform, that is closely related to both.

2.2.2. Discrete Fourier Transform Computation via Fast Fourier Transform

The Fourier transform pair in Definition 1.1 can also be written in the form,

$$y(t) = \int_{-\infty}^{\infty} Y(f) e^{j2\pi ft} \, df, \tag{2.1a}$$

$$Y(f) = \int_{-\infty}^{\infty} y(t) e^{-j2\pi ft} \, dt, \tag{2.1b}$$

where $f = \omega/2\pi$ is the frequency in hertz, and $Y(\omega) = Y(2\pi f)$ is written for brevity as $Y(f)$, without abuse of notation. If $y(t)$ is sampled every T sec, then at the sampling points rT, $r = 0, \pm 1, \pm 2, \ldots$, (2.1a) becomes

$$
\begin{aligned}
y(rT) &= \int_{-\infty}^{\infty} Y(f) e^{j2\pi frT} \, df \\
&= \sum_{k=-\infty}^{\infty} \int_{(k-\frac{1}{2})(1/T)}^{(k+\frac{1}{2})(1/T)} Y(f) e^{j2\pi frT} \, df \\
&= \int_{-1/2T}^{1/2T} Y_p(f) e^{j2\pi frT} \, df,
\end{aligned}
\tag{2.2a}
$$

where

$$Y_p(f) \triangleq \sum_{k=-\infty}^{\infty} Y\left(f + \frac{k}{T}\right). \tag{2.2b}$$

It can be checked that (2.2a) follows from the periodicity of the exponential function,

$$e^{j2\pi frT} = e^{j2\pi[f+(1/T)]rT}.$$

In (2.2b) $Y_p(f)$ is the "aliased" version of $Y(f)$, with the aliasing occurring relative to frequency of "folding," $(1/2T) \cdot Y_p(f)$ is periodic with a period $1/T$ and has a Fourier series expansion (under mild constraints almost always satisfied in real-world problems),

$$Y_p(f) = T \sum_{r=-\infty}^{\infty} y(rT) e^{-j2\pi frT}. \tag{2.3a}$$

Let $\{y(kT)\} = (y(0), y(T), \ldots, y((N-1)T)$ be a finite sequence of length N and duration NT. If the continuous function $Y_p(f)$ is sampled every $F = 1/NT$ Hz, then at the sampling points $f = kF$, $k = 0, \pm 1, \pm 2, \ldots$,

$$Y_p(kF) = \frac{1}{NF} \sum_{r=0}^{N-1} y(rT) e^{-j2\pi kr/N}. \tag{2.3b}$$

Thus $\{Y_p(kF)\}$ is a periodic sequence of period N (since $Y_p(kF) = Y_p[(k + N)F]$) so that (2.3b) need be computed only for $k = 0, 1, \ldots, N-1$. To obtain a relation for $y(rT)$ in terms of $Y_p(kF)$, we multiply both sides of (2.3b) by $e^{+j(2\pi/N)km}$, sum over all integers k in $0 \le k \le N-1$, interchange the orders of summation on the right-hand side, and make use of the identity

$$\sum_{k=0}^{N-1} e^{j(2\pi/N)k(m-r)} = \begin{cases} N & \text{if } r=m \\ 0 & \text{otherwise} \end{cases}$$

to get

$$y(mT) = \frac{1}{NT} \sum_{k=0}^{N-1} Y_p(kF) e^{j(2\pi km/N)}. \tag{2.3c}$$

If $T=1$, $NF=1$, and the following Definition follows as a logical consequence to (2.3b) and (2.3c).

Definition 2.1. The discrete Fourier transform (DFT) is a one-to-one mapping of any finite sequence $\{y(r)\}$, $r = 0,1,2,\ldots, N-1$ of N complex samples onto another sequence defined by

$$Y(k) = \sum_{r=0}^{N-1} y(r) w_N^{rk}, \tag{2.4}$$

where

$$w_N \triangleq e^{-j2\pi/N} = \cos\frac{2\pi}{N} - j\sin\frac{2\pi}{N}$$

for $k = 0,1,2,\ldots, N-1$. The inverse DFT (IDFT) is

$$y(r) = \frac{1}{N} \sum_{k=0}^{N-1} Y(k) w_N^{-rk}, \tag{2.5}$$

for $r = 0,1,\ldots, N-1$. $\{y(r)\}$ and $\{Y(k)\}$, each, is extendible to a periodic sequence, which in turn has a discrete-time Fourier series representation.

The fast Fourier transform (FFT) provides a rapid means for computing the DFT, which otherwise would have taken N^2 complex multiplications. FFT works especially well when N is highly composite. Let $N = N_1 N_2$, where N_1 and N_2 are two factors, not necessarily irreducible, of N. Define

$$r = N_1 r_2 + r_1 \quad \text{and} \quad k = N_2 k_2 + k_1, \tag{2.6a}$$

where

$$k_2 = 0,1,\ldots, N_1 - 1, \quad k_1 = 0,1,\ldots, N_2 - 1, \tag{2.6b}$$

$$r_2 = 0,1,\ldots, N_2 - 1, \quad r_1 = 0,1,\ldots, N_1 - 1. \tag{2.6c}$$

Also define

$$w_{N_1} = e^{-j2\pi/N_1}, \quad w_{N_2} = e^{-j2\pi/N_2}. \tag{2.6d}$$

Then, substituting (2.6a)–(2.6d) in (2.4),

$$Y(N_2 k_2 + k_1) = \sum_{r_1=0}^{N_1-1} w_{N_1}^{r_1 k_2} w_N^{r_1 k_1} \sum_{r_2=0}^{N_2-1} y(N_1 r_2 + r_1) w_{N_2}^{r_2 k_1}. \tag{2.7}$$

The computation of the above equation for all permissible values of k_1 and k_2 in (2.6b) requires the computation of N_1 DFT's of N_2 terms, plus N_2 DFT's of N_1 terms plus $N_1 N_2$ multiplication by the term $w_N^{r_1 k_1}$ (twiddle factor) in (2.7) for all permissible values of r_1, k_1. Therefore, in comparison to $N_1^2 N_2^2$ complex multiplications that would have been required if (2.7) were not used, its use reduces the number of complex multiplications to

$$N_1 N_2 (N_1 + N_2 + 1) = \sum_{i=1}^{2} N N_i + N. \tag{2.8}$$

Generalizing (2.8), it is not difficult to check that the number of complex multiplications sufficient for implementing the DFT in (2.4), when N factors as $N_1 N_2 \cdots N_r$, is

$$\left(\sum_{i=1}^{r} N N_i \right) + (r-1) N. \tag{2.9}$$

In the *radix-2* case, N is expressible as a power of 2. Then, if $N = 2^r$, (2.9) specializes to

$$r 2^{r+1} + (r-1) 2^r. \tag{2.10}$$

Note that a 2-point DFT (i.e., $N = 2$, in (2.4)),

$$Y(0) = y(0) + y(1),$$
$$Y(1) = y(0) - y(1),$$

requires no multiplication for its computation. Therefore, the summation term in (2.9) does not contribute to the multiplicative complexity in the radix-2 case, reducing therefore the number of complex multiplications in (2.10) to

$$(r-1) 2^r = N(\log_2 N - 1),$$

since $N = 2^r$. The various approaches for computing the FFT in this case are summarized next.

Radix 2-FFT (Decimation-in-Time). On considering the even and odd samples of $\{ y(k) \}$, (2.4) simplifies to

$$Y(k) = \sum_{i=0}^{(N/2)-1} \left[y(2i) + w_N^k y(2i+1) \right] w_N^{2ik}, \tag{2.11a}$$

$$Y\left(k + \frac{N}{2} \right) = \sum_{i=0}^{(N/2)-1} \left[y(2i) - w_N^k y(2i+1) \right] w_N^{2ik}, \tag{2.11b}$$

for $k = 0, 1, \ldots, (N/2) - 1$. The above basic step is repeated in $r = \log_2 N$ stages. In the $(i+1)$th stage, $0 \le i \le r - 2$, the problem of evaluating 2^i DFT's each of length 2^{r-i} is replaced by that of evaluating 2^{i+1} DFT's each of length 2^{r-i-1} at the cost of N additional additions and $N/2$ additional

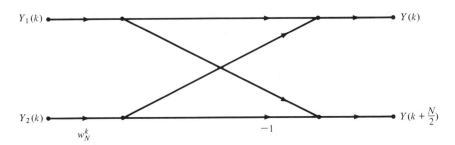

Figure 2.1. Flow graph for basic butterfly computation.

multiplications (in stage 1, these $N/2$ complex multiplications are due to the factor w_N^k in (2.11)). In (2.11) note that $w_N^{2ik} = w_{N/2}^{ik}$, so that after the first stage the original problem of computing an N-point DFT is replaced by the problem of computing two $N/2$-point DFT's. (Remember that $N = 2^r$, so that $N/2 = 2^{r-1}$ is an integer.) These two $N/2$-point DFT's are

$$Y_1(k) = \sum_{i=0}^{(N/2)-1} y(2i) w_{N/2}^{ik}, \qquad (2.12a)$$

$$Y_2(k) = \sum_{i=0}^{(N/2)-1} y(2i+1) w_{N/2}^{ik}. \qquad (2.12b)$$

Then

$$Y(k) = Y_1(k) + w_N^k Y_2(k), \quad 0 \le k \le \frac{N}{2} - 1, \qquad (2.12c)$$

$$Y\left(k + \frac{N}{2}\right) = Y_1(k) - w_N^k Y_2(k), \quad 0 \le k \le \frac{N}{2} - 1. \qquad (2.12d)$$

The flow graph illustrating the computation of $Y(k)$ and $Y\left(k + \frac{N}{2}\right)$ from $Y_1(k)$ and $Y_2(k)$ for any fixed k is shown in Figure 2.1, where the basic structure shown is referred to as a *butterfly*. A flow graph is a directed graph with numbers called node values associated with each node and weights associated with each branch. A branch directed from node j to node k contributes to the node value at node k by an amount equal to the branch weight times the node value at node j. The total node value at each node is the sum of the contributions of the node values by all branches entering the node. By convention, a branch with no number assigned to it will be understood to have a branch weight of unity. The decomposition procedure in the first stage is continued. After the $(r-1)$th stage, 2^{r-1} DFT's, each of length 2, are required to be computed. Each of these 2-point DFT's requires

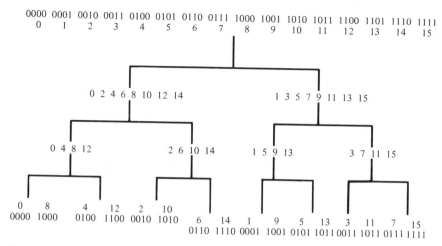

Figure 2.2. Decomposition of time sequence of length 16 for decimation-in-time algorithm.

no multiplication for its computation and the output of the rth stage gives the desired sample set, $\{Y(0)Y(1)\cdots Y(N-1)\}$. Summarizing, when $N=2^r$, the decimation-in-time approach requires the decomposition of the discrete-time sample set $\{y(0)y(1)\cdots y(N-1)\}$ into two subsets, one of which contains the even-indexed (including zero) samples and the second contains the odd-indexed samples. This decomposition procedure is repeated on each of these two subsets and is continued until the computation of 2^{r-1} DFT's, each of length 2, is required. Let $r=4$, so that $N=2^4=16$. The sample indices, $0,1,\ldots,15$ can be given their 4-bit binary representations, as shown at the top of Figure 2.2. Following the decomposition scheme described above, one obtains the indices for the output in *bit-reversed* order, as illustrated in Figure 2.2. The occurrence of the output sequence in bit-reversed order, when compared with the input sequence, becomes more evident after the algorithmic development of the DFT computational scheme (to be discussed next) is understood.

In the simple case, when $N=4=2^2$, implying that $r=2$, (2.11a) and (2.11b) can be rewritten as

$$Y(k) = \sum_{i=0}^{1} y(2i)w_2^{ik} + w_4^k \sum_{i=0}^{1} y(2i+1)w_2^{ik}, \quad k=0,1;$$

$$Y(k+2) = \sum_{i=0}^{1} y(2i)w_2^{ik} - w_4^k \sum_{i=0}^{1} y(2i+1)w_2^{ik}, \quad k=0,1.$$

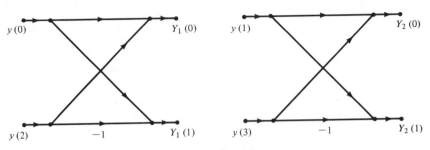

Figure 2.3. Two-point computations of even- and odd-indexed samples of a 4-point sequence.

The 2-point DFT's

$$Y_1(k) = \sum_{i=0}^{1} y(2i)w_2^{ik} \quad \text{and} \quad Y_2(k) = \sum_{i=0}^{1} y(2i+1)w_2^{ik}$$

for $k = 0, 1$, can be computed via the basic butterfly computation as shown in Figure 2.3, and then the 4-point DFT computational scheme is easily obtainable as illustrated in Figure 2.4. In Figure 2.4, the output sequence is indexed in the normal order, while the input sequence is in bit-reversed order relative to this normal sequential ordering. Also, the input and its computed DFT can be arranged, respectively, in normal and bit-reversed orders, as shown in Figure 2.5. The basic butterfly computation plays a dominant role in both Figures 2.4 and 2.5. A clear advantage of the butterfly in DFT computation is that computations are done *in-place*, that is, in Figure 2.1, the computed outputs $Y(k)$ and $Y(k+(N/2))$ can be stored in the same storage registers as the inputs $Y_1(k)$ and $Y_2(k)$, for use in subsequent calculations. However, note that data are not addressed sequentially, so that random access memory is desirable. When sequential addressing and storage of data is sought, rearrangement of the flow graph in Figure 2.4 is possible for DFT computation; however, in that case the butterfly computational structure is lost [8].

The flow graph in Figure 2.4 for $N = 4$ can be easily generalized for $N = 2^r$, $r > 2$. It is noticed that the total number of complex multipliers (including w_N^0, for notational convenience) sufficient for each stage is $N/2$. The distinct values of these multipliers for the stage where the two $N/2$-point DFT's occur are $w_N^0, w_N^1, \ldots, w_N^{(N/2)-1}$. The set of distinct values of the multipliers for the stage where the four $N/4$-point DFT's occur are $\{w_{N/2}^0, w_{N/2}^1, \ldots, w_{N/2}^{((N/4)-1)}\}$ or, equivalently, $\{w_N^0, w_N^2, \ldots, w_N^{(N/2)-2}\}$ (note that in this stage each multiplier value in the set occurs twice, which is consistent with the fact that there are $N/2$ multipliers per stage).

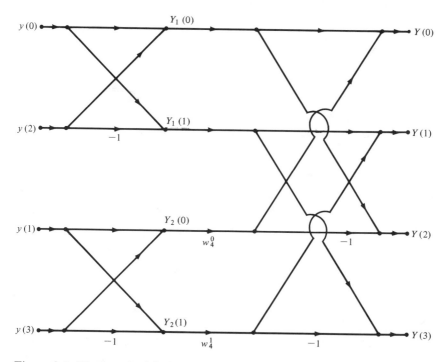

Figure 2.4. Flow graph of decimation-in-time decomposition for DFT computation of a 4-point sequence.

Radix 2-FFT (Decimation-in-Frequency). On considering first, the beginning $N/2$ samples of $\{y(k)\}$, and then the last $N/2$ samples of $\{y(k)\}$, (2.4) can be written as

$$Y(2k) = \sum_{i=0}^{(N/2)-1} \left[y(i) + y\left(i + \frac{N}{2}\right) \right] w_N^{2ik}, \qquad (2.13a)$$

$$Y(2k+1) = \sum_{i=0}^{(N/2)-1} \left[y(i) - y\left(i + \frac{N}{2}\right) \right] w_N^i w_N^{2ik}, \qquad (2.13b)$$

for $k = 0, 1, \ldots, (N/2)-1$. The above basic step is repeated $r = \log_2 N$ times. In the $(i+1)$th stage, $0 \le i \le r-2$, the problem of computing 2^i DFT's, each of length 2^{r-i}, is replaced by that of computing 2^{i+1} DFT's, each of length 2^{r-i-1}, at the cost of N additional additions and $N/2$ additional complex multiplications. In stage one the $N/2$ additional multiplications are due to the factor w_N^i in (2.13b). As in the decimation-in-time case, after the

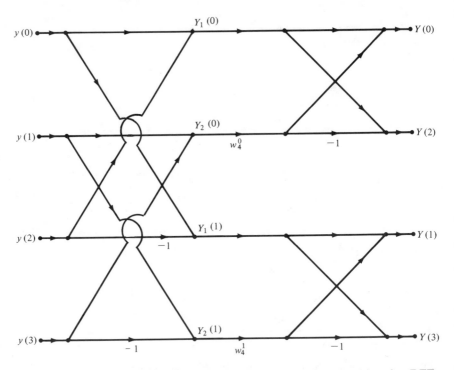

Figure 2.5. Alternate flow graph of decimation-in-time decomposition for DFT computation of a 4-point sequence.

$(r-1)$th stage, 2^{r-1} DFT's, each of length 2, are required to be computed. Again noting that $w_N^{2ik} = w_{N/2}^{ik}$, it is clear from (2.13a) and (2.13b) that the problem of computing an N-point DFT is replaced by the problem of computing two $N/2$-point DFT's. The two sequences, each of length $N/2$, whose DFT's are now under consideration are

$$\{y_1(i)\} = \left\{ y(i) + y\left(i + \frac{N}{2}\right) \right\}$$

and

$$\{y_2(i)\} = \left\{ \left[y(i) - y\left(i + \frac{N}{2}\right) \right] w_N^i \right\},$$

where $i = 0, 1, \ldots, (N/2) - 1$. The stage reached is illustrated in Figure 2.6A for the case when $N = 4$. Each of these $N/2$-point DFT's can be computed via two $N/4$-point DFT's and the iteration continued till 2^{r-1} 2-point

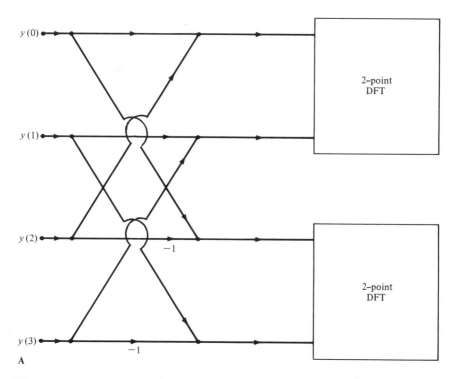

Figure 2.6. (A) Flow graph illustrating decimation-in-frequency algorithm, when $N = 4$, after the first stage of computations.

DFT's have to be computed, where $N = 2^r$. Each of the 2-point DFT's can be computed via use of the operation of addition only, as mentioned earlier. The complete flow graph, illustrating computation of a 4-point DFT via the decimation-in-frequency approach, is shown in Figure 2.6B. Note that the input sequence is indexed in normal order, while the output sequence is in bit-reversed order with respect to the input sequence. The complex multipliers total $N/2$ for each stage (provided w_N^0 is viewed as complex).

Considerable attention has been given to the development of special-purpose hardware for implementing FFT algorithms. Cost reduction resulting from this special-purpose hardware has been comparable to that achieved through the use of the technical devices discussed above [30]. Most dedicated hardware implementations use the decimation-in-time approach due to the popularity of multiplier-accumulator architecture over accumulator-multiplier architecture.

In digital signal-processing applications, a typical sampled sequence, whose DFT is sought, is usually a sequence of real numbers. It is possible to obtain the DFT of a real sequence, whose length is twice the computing

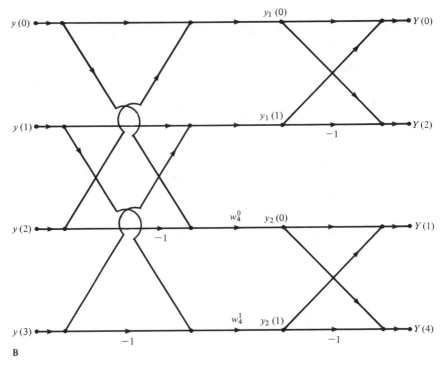

Figure 2.6. (B) Flow graph illustrating DFT computation, when $N = 4$, using the decimation-in-frequency approach.

capability available, using the following strategy. Let
$$\{ y(k) \} = (y(0),\, y(1),\ldots,\, y(2N-1))$$
be a real sequence of length $2N$. Write
$$z(k) = y(2k) + jy(2k+1), \quad k = 0,1,\ldots, N-1, \qquad (2.14)$$
so that $\{ z(k) \}$ is a complex sequence of length N. Compute the DFT,
$$Z(k) = \sum_{r=0}^{N-1} z(r)w_N^{rk} \quad (\text{for } k = 0,1,\ldots, N-1)$$
$$= \sum_{r=0}^{N-1} y(2r)w_N^{rk} + j \sum_{r=0}^{N-1} y(2r+1)w_N^{rk}$$
$$\triangleq Y_1(k) + jY_2(k), \qquad (2.15)$$
where the notations for DFT's $Y_1(k), Y_2(k)$ are self-explanatory. It can be verified that (note that $Y_i(k) = Y_i^*(N-k)$, for $i = 1,2$, since $\{ y(k) \}$ is a real sequence)
$$Y_1(k) = \tfrac{1}{2}[Z(k) + Z^*(N-k)], \quad k = 0,1,\ldots, N-1 \qquad (2.16a)$$
$$Y_2(k) = \frac{1}{2j}[Z(k) - Z^*(N-k)], \quad k = 0,1,\ldots, N-1 \qquad (2.16b)$$

The DFT, $\{Y(k)\}$, $k = 0,1,\ldots,2N-1$ of the original sequence $\{y(r)\}$ is written as in (2.11a) and (2.11b), where N is replaced by $2N$. Then, recognizing that $w_{2N}^{2rk} = w_N^{rk}$ and recalling what $Y_1(k)$ and $Y_2(k)$ represent, it follows from (2.15) that

$$Y(k) = Y_1(k) + w_{2N}^k Y_2(k), \quad k = 0,1,\ldots, N-1, \tag{2.17a}$$

$$Y(k) = Y_1(k-N) - w_{2N}^k Y_2(k-N), \quad k = N, N+1,\ldots, 2N-1. \tag{2.17b}$$

After substituting $Y_1(k)$ and $Y_2(k)$ from (2.16a) and (2.16b) in (2.17a) and (2.17b), the desired goal is reached.

Table 2.1 summarizes some of the useful properties of the DFT. It is emphasized that for derivation of the relevant properties, a finite length sequence $\{x(r)\}$ of length N is extended to a periodic sequence $\{\sum_{m=-\infty}^{\infty} x(r+mN)\}$, whenever necessary. For example, the symmetry property of DFT is derived as follows. Denote the DFT of sequence $\{y(-r)\}_{r=0}^{N-1}$ by $\{\hat{Y}(k)\}$, where

$$\hat{Y}(k) = \sum_{r=0}^{N-1} y(-r)w_N^{rk}, \quad k = 0,1,\ldots, N-1.$$

Visualize $\{y(-r)\}$ as the truncated version of a periodic sequence of period N. Then

$$\hat{Y}(k) = \sum_{r=0}^{N-1} y(N-r)w_N^{rk}$$

$$= \sum_{m=1}^{N} y(m)w_N^{(N-m)k}.$$

Since $w_N^{Nk} = 1$ and $y(0) = y(N)$, the preceding equation simplifies to

$$\hat{Y}(k) = \sum_{m=0}^{N-1} y(m)w_N^{-mk}$$

$$= Y(-k),$$

where the DFT of $\{y(r)\}$ is given by

$$Y(k) = \sum_{m=0}^{N-1} y(m)w_N^{mk}, \quad k = 0,1,\ldots, N-1.$$

Also, $Y(N-k) = Y(-k)$ since $w_N^{Nm} = 1$.

The other properties in Table 2.1 might also be conveniently verified. In particular, it is emphasized that the DFT is quite useful in the computation of convolutions and correlation. The linear convolution of two sequences, $\{h(k)\}$ and $\{x(k)\}$, of lengths n and m, respectively, is another sequence, $\{y(k)\}$, of length $n + m - 1$. In order to implement the radix-2 FFT for

Table 2.1. Summary of Properties of DFT

$$\{x(r)\} \overset{\text{DFT}}{\leftrightarrow} \{X(k)\}, \{y(r)\} \overset{\text{DFT}}{\leftrightarrow} \{Y(k)\}$$

$\{x(r)\}, \{y(r)\}$ are sequences of length N

	Property	Sequence	DFT
1.	Linearity	$a\{x(r)\} + b\{y(r)\}$ a and b are complex constants	$a\{X(k)\} + b\{Y(k)\}$
2.	Time translation	$\{y(r+k_0)\}$	$\{W_N^{-k_0 k} Y(k)\}$
3.	Frequency translation	$\{w_N^{k_0 r} y(r)\}$	$\{Y(k+k_0)\}$
4.	Symmetry	$\{y(-r)\}$	$\{Y(-k)\} = \{Y(N-k)\}$
5.	Circular convolution	$\left\{ \sum_{n=0}^{N-1} x(n) y\langle r-n\rangle \right\}$	$\{X(k)Y(k)\}$
6.	Circular correlation	$\left\{ \sum_{k=0}^{N-1} x^*(k) y\langle k+r\rangle \right\}$	$\{X^*(k)Y(k)\}$
7.	Term-by-term product	$\{x(r)y(r)\}$	$\left\{ \dfrac{1}{N} \sum_{i=0}^{N-1} X(i) Y\langle k-i\rangle \right\}$
8.	Periodicity	$\{y(r)\} = \{y(N+r)\}, \quad \forall r$	$\{Y(k)\} = \{Y(N+k)\}, \quad \forall k$
9.	Sequence complex conjugation	$\{y^*(r)\}$	$\{Y^*(-k)\} = \{Y^*(N-k)\}$
10.	Parseval's theorem	$\sum_{r=0}^{N-1} |y(r)|^2$	$= \dfrac{1}{N} \sum_{k=0}^{N-1} |Y(k)|^2$
11.	Complex conjugation of DFT	$\left\{ \dfrac{Y^*(r)}{N} \right\}$	$\{y^*(k)\}$
12.	Stretching ($K > 0$ and integer)	$\hat{x}(k) = x\left(\dfrac{r}{K}\right), r = iK$ $i = 0,1,\dots, N-1$	$\dfrac{X(k)}{K}, \ k = 0,1,\dots, NK-1$
13.	Sampling ($K > 0$, integer and $K|N$)	$\hat{x}(k) = x(rK),$ $r = 0,1,\dots, \dfrac{N}{K} - 1$ $\left(\hat{x}(k) \text{ has period } \dfrac{N}{K}\right)$	$\hat{X}(k) = \sum_{i=0}^{K-1} X\left(k + \dfrac{iN}{K}\right)$

Note: For derivation of the relevant properties of DFT, a finite length sequence $\{x(r)\}$ of length N is extended to a periodic sequence $\left\{ \sum_{m=-\infty}^{\infty} x(r+mN) \right\}$ (denoted by $\{x\langle r\rangle\}$), whenever necessary, where $\langle r\rangle = r \bmod N$. Then, time translation is a circular shift.

computing the sequence $\{y(k)\}$, the following steps are necessary:

Step 1. Select the minimum integer N such that

$$N \geq n + m - 1 \text{ and } N = 2^k \text{ for some positive integer } k.$$

Step 2. Form sequences $\{\hat{h}(k)\}$ and $\{\hat{x}(k)\}$, each of length N, by augmenting with the appropriate number of zeros, the sequences $\{h(k)\}$ and $\{x(k)\}$, respectively.

Step 3. Compute the DFT's $\{\hat{H}(k)\}$ and $\{\hat{X}(k)\}$, of $\{\hat{h}(k)\}$ and $\{\hat{x}(k)\}$, respectively, by using the radix-2 FFT.

Step 4. Compute the product sequence

$$\{\hat{Y}(k)\} \triangleq \{\hat{H}(k)\hat{X}(k)\}.$$

Step 5. Compute

$$\text{IDFT}\{\hat{Y}(k)\},$$

by again using the radix-2 FFT (from (2.4) and (2.5) determine what simple modifications in the DFT computation algorithms facilitates the computation of IDFT via FFT).

The first $(n + m - 1)$ terms of $[\text{IDFT}\{\hat{Y}(k)\}]$ yield the linearly convolved sequence, $\{y(k)\}$. The cross-correlation $R_{hx}(k)$ of two periodic sequences, $\{h(k)\}$ and $\{x(k)\}$, each of period N, is

$$R_{hx}(k) \triangleq \frac{1}{N} \sum_{j=0}^{N-1} h(j)x\langle j-k \rangle, \quad k = 0,1,\ldots, N-1.$$

The autocorrelation $R_{xx}(k)$, of the sequence $\{x(k)\}$, is

$$R_{xx}(k) = \frac{1}{N} \sum_{j=0}^{N-1} x(j)x\langle j-k \rangle, \quad k = 0,1,\ldots, N-1.$$

As in the case of convolution, the DFT via use of FFT provides an efficient means for computing the sequences $\{R_{hx}(k)\}$ and $\{R_{xx}(k)\}$. The DFT's of $\{R_{hx}(k)\}$ and $\{R_{xx}(k)\}$ are, respectively, the cross-power spectrum and autopower spectrum (or, simply, power spectrum) sequences.

The DFT is also useful in approximating the Fourier transform of a continuous-time signal, $y(t)$, which for obvious practical reasons, is assumed to be time-limited. This is interpreted to imply that for some finite T_1, $y(t)$ is zero outside the interval $0 \leq t \leq T_1$. Before applying DFT computations, $y(t)$ has to be sampled to yield $\hat{y}(t)$ as in (1.6). It is understood that there is a finite number, N, of samples, so that $T_1 = NT$, where T is the sampling period. From (1.7) and (2.3b), it is clear that the DFT, $\{\hat{Y}(k)\}$, of the sampled sequence is

$$\hat{Y}(k) = \frac{1}{T}\left[\sum_{i=-\infty}^{\infty} Y\left(\omega - \frac{2i\pi}{T}\right)\right]\Bigg|_{\omega = 2\pi k/NT},$$

for $k = 0, 1, \ldots, N - 1$. The preceding expression can be rewritten as

$$\hat{Y}(k) = \frac{1}{T} Y\left(\frac{2\pi k}{NT} \right) + \sum_{\substack{i = -\infty \\ i \neq 0}}^{\infty} Y\left[2\pi \frac{(k - iN)}{NT} \right].$$

Since $y(t)$ is time-limited, $Y(\omega)$ cannot be band-limited, and therefore the second term in the preceding equation is responsible for the aliasing error inherent in the use of DFT for approximating the spectrum of a continuous-time signal. This aliasing error may be reduced by increasing the number, N, of samples, that is, by decreasing the sampling period, T. The second source of error in the use of DFT for approximating the spectrum of an analog signal is due to the phenomenon of leakage, which occurs because of the practical necessity of observing the signal over a finite time interval. This requirement may be viewed as a type of "windowing" of the signal. Design of proper windows (of the types used in the design of FIR filters in Section 4.3) is required to reduce the undesirable effects of leakage. The final source of error is due to what is commonly called the picket-fence effect, resulting from lack of information about the continuous spectrum between the records at a discrete set of points. One way to reduce the picket-fence effect is by adding zero samples to the original sequence and calculating the DFT over a larger set of points occurring in this new sequence.

2.3. Euclid's Algorithm and the Chinese Remainder Theorem

Since the subsequent sections in this chapter will be devoted to the exposition of recently developed techniques which reduce the multiplicative complexity of computation of the discrete Hilbert transform (DHT), DFT, and convolutions and correlations, this section will familiarize the reader with the necessary mathematical prerequisites. The most fundamental tools are presented, while some of the specialized results will be introduced in later sections as needed.

2.3.1. Division Theorem over Integers

Theorem 2.1. *If b is positive and a is any integer, there is a unique pair of integers q and r such that the conditions*

$$a = bq + r, \quad 0 \leq r < b \tag{2.18}$$

hold; q is called the quotient and r is the remainder or residue of a modulo b.

The proof for the preceding theorem can be found in almost any book in elementary number theory, such as reference [3]. When $r = 0$ in (2.18), b and q are divisors of a. This fact is denoted by the symbol $b|a$ and $q|a$.

Definition 2.2. The positive integer a is prime when it has no divisors other than 1 and a.

Theorem 2.2. *Given any two integers a and b, not both zero, there is an unique integer d, called the greatest common divisor (gcd) with the following properties*:

$$\text{(i)} \quad d|a \text{ and } d|b;$$

$$\text{(ii)} \quad \text{if } d_1|a \text{ and } d_1|b, \text{ then } d_1|d;$$

$$\text{(iii)} \quad d > 0.$$

Note that condition (iii) in the above theorem imposes uniqueness, since otherwise, $-d$ could also be a gcd, that is, then gcd is unique up to units ± 1. For proof of Theorem 2.2, see reference [3].

The gcd of two specified positive integers a and b, denoted by $d = (a, b) = (b, a)$ may be determined by the Euclid's division algorithm, which is based on the repeated application of Theorem 2.1. Without any loss of generality, the integers a and b are assumed positive, since

$$(a, b) = (-a, b) = (a, -b) = (-a, -b)$$

and when $a = 0$, $d = (0, b) = |b|$ while when $b = 0$, $d = (a, 0) = |a|$. Euclid's division algorithm (also referred to as the Euclidean algorithm or division algorithm) for positive integers a and b runs as follows:

$$
\begin{aligned}
a &= bq_1 + r_1 & 0 &< r_1 < b \\
b &= r_1 q_2 + r_2 & 0 &< r_2 < r_1 \\
r_1 &= r_2 q_3 + r_3 & 0 &< r_3 < r_2 \\
&\;\;\vdots & &\;\;\vdots \\
r_{k-3} &= r_{k-2} q_{k-1} + r_{k-1} & 0 &< r_{k-1} < r_{k-2} \\
r_{k-2} &= r_{k-1} q_k + r_k & 0 &< r_k < r_{k-1} \\
r_{k-1} &= r_k q_{k+1} &
\end{aligned}
\tag{2.19}
$$

The sequence of steps in (2.19) is a divide–invert–divide sequence and can be expressed as a continued fraction expansion (CFE):

$$
\frac{a}{b} = q_1 + \cfrac{1}{q_2 + \cfrac{1}{q_3 + \cfrac{1}{q_4 + \cdots + \cfrac{1}{q_k + \cfrac{1}{q_{k+1}}}}}}
$$

The above CFE is usually denoted for brevity by

$$\frac{a}{b} = q_1 + \frac{1}{q_2 +} \frac{1}{q_3 +} \cdots \frac{1}{q_{k+1}}. \qquad (2.20)$$

Another notation for CFE is brought out in Problem 8 of Chapter 3. The number of steps in (2.19) is finite, since the r_i's form a strictly decreasing sequence of positive integers, and therefore the continued fraction expansion in (2.20) is also finite.

Theorem 2.3. *The greatest common divisor of two positive integers a and b is r_k, where r_k is obtained in (2.19) via the application of the Euclidean algorithm.*

PROOF: Denoting by (a, b) the required greatest common divisor, it is clear that

$$
\begin{aligned}
(a, b) &= (a - bq_1, b) \\
&= (b, r_1) = (b - r_1 q_2, r_1) \\
&= (r_1, r_2) = (r_1 - r_2 q_3, r_2) \\
&= \cdots\cdots\cdots\cdots\cdots \\
&= (r_{k-1}, r_k) = (r_{k-1} - r_k q_{k+1}, r_k) \\
&= (r_k, 0) \\
&= r_k.
\end{aligned}
$$

Fact 2.1. *It is simple to establish from the division algorithm that there exist integers c and d such that*

$$r_k = ca + db.$$

In Fact 2.1, it is often necessary to calculate c and d along with r_k for specified a, b. There are various procedures for doing this; the algorithm given below is quite convenient for the purpose. The proof may be found in reference [4].

Algorithm (Extended Euclid)

Step 1. Set $(a_1, a_2, a_3) \leftarrow (1, 0, a)$, $(b_1, b_2, b_3) \leftarrow (0, 1, b)$

Step 2. If $b_3 = 0$, *go to Step* 4

Step 3. Set $q \leftarrow \left\lfloor \dfrac{a_3}{b_3} \right\rfloor$, *where* $\lfloor x \rfloor$ *denotes the integer part of x. Then, set*

$$
\begin{aligned}
(e_1, e_2, e_3) &\leftarrow (a_1, a_2, a_3) - (b_1, b_2, b_3)q \\
(a_1, a_2, a_3) &\leftarrow (b_1, b_2, b_3) \\
(b_1, b_2, b_3) &\leftarrow (e_1, e_2, e_3)
\end{aligned}
$$

   go to Step 2

Step 4. Set $c \leftarrow a_1$, $d \leftarrow a_2$, $r_k \leftarrow a_3$
       Stop.

The implementation of the above algorithm is illustrated in the next example. It should be noted that c and d in Fact 2.1 are not unique.

Example 2.1. It is required to calculate r_k and a set of $\{c, d\}$ satisfying the equation in Fact 2.1 when $a = 202$ and $b = 508$ are specified. The preceding algorithm is implemented as follows:

q	a_1	a_2	a_3	b_1	b_2	b_3
—	1	0	202	0	1	508
0	0	1	508	1	0	202
2	1	0	202	-2	1	104
1	-2	1	104	3	-1	98
1	3	-1	98	-5	2	6
16	-5	2	6	83	-33	2
3	83	-33	2	-254	101	0

Therefore, $c = 83$, $d = -33$, and $(a, b) = 2$. It may be checked that
$$83 \times 202 + (-33)508 = 2.$$

Fact 2.2 is a simple consequence of Fact 2.1.

Fact 2.2. *Two positive integers a and b are relatively prime if and only if there exist integers c and d such that*
$$ca + db = 1.$$

EUCLID.FOR in Figure 2.7A implements the result in Fact 2.2.

2.3.2. Division Theorem over Polynomials

Consider the set of polynomials in a single complex variable, whose coefficients belong to an arbitrary but fixed field like the field of real numbers or the finite field Z_q whose elements are $(0, 1, \ldots, q-1)$ with q prime. The next theorem is the polynomial counterpart of Theorem 2.1.

Theorem 2.4. *For two arbitrary polynomials $a(z)$ and $b(z)$ (whose coefficients are in an arbitrary but fixed field), $b(z) \not\equiv 0$, one can find the quotient and remainder polynomials $q(z)$ and $r(z)$, respectively, in*
$$a(z) = q(z)b(z) + r(z), \quad \delta[r(z)] < \delta[b(z)], \qquad (2.21)$$
where $\delta[r(z)]$ denotes the degree of the polynomial $r(z)$. The preceding representation is unique.

```
      C
            SUBROUTINE EUCLID(A,B,C,D)
      C
      C     GIVEN TWO MUTUALLY PRIME POSITIVE NUMBERS A AND B,
      C     THIS SUBROUTINE OBTAINS TWO INTEGERS C AND D SUCH THAT
      C     C*A + D*B = 1 IS SATISFIED.
      C     A AND B ARE INPUT INTEGERS, C AND D ARE OUTPUT INTEGERS.
      C
            INTEGER A,B,C,D,A1,A2,A3,B1,B2,B3,E1,E2,E3,Q
            A1=1
            A2=0
            A3=A
            B1=0
            B2=1
            B3=B
      100   Q=A3/B3
            E3=A3-Q*B3
            E2=A2-Q*B2
            E1=A1-Q*B1
            IF(E3.EQ.1) GO TO 200
            A1=B1
            A2=B2
            A3=B3
            B1=E1
            B2=E2
            B3=E3
            GO TO 100
      200   C=E1
            D=E2
            RETURN
            END
```

Figure 2.7. (A) Listing of program EUCLID.FOR.

All the results in this section hold because of the validity of the division process in (2.21). Therefore, as in the case of integers, a mathematical structure paralleling the contents of the preceding section can be developed in the case of polynomials over a field.

Definition 2.3. A polynomial whose highest coefficient is 1, the multiplicative identity in the field of coefficients, is monic and *the greatest common divisor* of two polynomials is the unique monic common divisor of highest degree.

The greatest common divisor $d(z)$ of two polynomials $a(z)$ and $b(z)$ can be calculated via the repeated application of the division process in (2.21). In fact, the algorithm is of the divide-invert-divide type and is, essentially, the polynomial counterpart of (2.19). Naturally, a continued fraction expansion development, similar to (2.20) is possible, and this expansion is finite because the sequence of remainder polynomials is a sequence of strictly decreasing degrees. More precisely, the steps of successive divisions via repeated application of (2.21) can be summarized by [similar to (2.19)],

$$f_{i-2}(z) = q_i(z)f_{i-1}(z) + f_i(z), \quad i = 1, 2, 3, \ldots, \qquad (2.22)$$

where $f_{-1}(z) \triangleq a(z)$, $f_0(z) \triangleq b(z)$. If $f_{k+1}(z) \equiv 0$, then $f_k(z)$ is a gcd of $a(z)$ and $b(z)$. After division of $f_k(z)$ by its leading coefficient, the gcd of $a(z)$ and $b(z)$ is obtained as a monic polynomial. The next example illustrates the steps which are the polynomial counterparts of those in (2.19) that were necessary to compute the gcd of two integers.

Example 2.2. Evaluate the greatest common divisor of polynomials:

$$b(z) = 2z^3 + 11z^2 + 17z + 6,$$

$$a(z) = 2z^2 + 9z + 9,$$

$$\frac{b(z)}{a(z)} = z + 1 + \frac{-(z+3)}{2z^2 + 9z + 9}$$

$$= z + 1 + \frac{1}{(2z^2 + 9z + 9)/-(z+3)}$$

$$= z + 1 + \frac{1}{-(2z+3)}.$$

The last divisor which yields a zero remainder is $z + 3$, and this monic polynomial is the required greatest common divisor.

Fact 2.3. *The gcd, $f(z)$, of two polynomials $a(z)$ and $b(z)$ can be expressed as*

$$f(z) = c(z)a(z) + d(z)b(z), \qquad (2.23)$$

where $c(z)$ and $d(z)$ are also polynomials. $c(z)$ and $d(z)$ are unique when the restrictions

$$\delta[c(z)] < \delta[b(z)],$$
$$\delta[d(z)] < \delta[a(z)] \qquad (2.24)$$

are imposed.

From the preceding fact the following result follows.

Fact 2.4. *Polynomials $a(z)$ and $b(z)$ are relatively prime provided there are polynomials $c(z)$ and $d(z)$ such that*

$$c(z)a(z) + d(z)b(z) = 1 \qquad (2.25)$$

and $c(z)$ *and* $d(z)$ *are uniquely determinable provided the degree constraints in Fact* 2.3 *are again imposed here.*

The polynomials $c(z)$ and $d(z)$ in (2.25) can be determined by solving a set of linear equations. In order to illustrate the general technique, consider the case of two relatively prime polynomials $a(z)$ and $b(z)$, each of degree 3:

$$a(z) = a_0 + a_1 z + a_2 z^2 + a_3 z^3,$$
$$b(z) = b_0 + b_1 z + b_2 z^2 + b_3 z^3. \qquad (2.26a)$$

For $c(z)$ and $d(z)$ to be unique, they must be of the form

$$c(z) = c_0 + c_1 z + c_2 z^2,$$
$$d(z) = d_0 + d_1 z + d_2 z^2, \qquad (2.26b)$$

where the degree constraints in (2.24) have been imposed. On substituting (2.26) in (2.25) and equating coefficients of various powers of z, the following system of equations is obtained:

$$\begin{bmatrix} a_0 & 0 & 0 & b_0 & 0 & 0 \\ a_1 & a_0 & 0 & b_1 & b_0 & 0 \\ a_2 & a_1 & a_0 & b_2 & b_1 & b_0 \\ a_3 & a_2 & a_1 & b_3 & b_2 & b_1 \\ 0 & a_3 & a_2 & 0 & b_3 & b_2 \\ 0 & 0 & a_3 & 0 & 0 & b_3 \end{bmatrix} \begin{bmatrix} c_0 \\ c_1 \\ c_2 \\ d_0 \\ d_1 \\ d_2 \end{bmatrix} = \begin{bmatrix} 1 \\ 0 \\ 0 \\ 0 \\ 0 \\ 0 \end{bmatrix}. \qquad (2.27)$$

The unknowns c_0, c_1, c_2, d_0, d_1, and d_2 in (2.27) can be uniquely determined since the coefficient matrix is nonsingular [this follows from the fact that this coefficient matrix, after appropriate permutation of rows and columns, may be made equivalent to the resultant matrix (see Chap. 1) of $a(z)$ and $b(z)$, and this resultant matrix must have a nonzero determinant because $a(z)$ and $b(z)$ are relatively prime polynomials]. For specified relatively prime polynomials $a(z)$ and $b(z)$ of arbitrary degrees, the polynomials $c(z)$ and $d(z)$ satisfying (2.24) and (2.25) can be uniquely determined, after writing down a set of linear equations, which are obtainable from a routine and natural generalization of (2.27).

Example 2.3. Let

$$a(z) = 5z + 3,$$
$$b(z) = 2z^2 + 4z + 1.$$

Evidently, $a(z)$ and $b(z)$ are relatively prime. It is necessary to uniquely determine $c(z)$ and $d(z)$ in (2.25). Applying (2.24), $c(z)$ and $d(z)$ are of the form

$$c(z) = c_0 + c_1 z,$$
$$d(z) = d_0.$$

On equating coefficients of powers of z in (2.25) applicable for this problem, the counterpart of (2.27) is obtained:

$$\begin{bmatrix} 3 & 0 & 1 \\ 5 & 3 & 4 \\ 0 & 5 & 2 \end{bmatrix} \begin{bmatrix} c_0 \\ c_1 \\ d_0 \end{bmatrix} = \begin{bmatrix} 1 \\ 0 \\ 0 \end{bmatrix}.$$

Therefore, after solving for c_0, c_1 and d_0,

$$c(z) = \tfrac{14}{17} + \tfrac{10}{17}z, \quad d(z) = -\tfrac{25}{17}$$

2.3.3. Chinese Remainder Theorem over Integers

In (2.18) the fact that r is the residue of integer a modulo integer b is sometimes written as a linear congruence,

$$a \equiv r \quad \mod b,$$

where b is the modulus. The Chinese remainder theorem (CRT) enables one to construct an integer from its specified residues with respect to a sufficiently large number of modulii which are mutually prime.

Theorem 2.5. *If* $(m_i, m_j) = 1$ *for* $1 \le i < j \le k$, *then the system of linear congruences,*

$$x \equiv r_j \quad \mod m_j, \quad j = 1, 2, \ldots, k$$

with specified r_j *and* m_j *has a unique solution,*

$$x = \left[\sum_{j=1}^{k} r_j s_j \left(\prod_{\substack{i=1 \\ i \neq j}}^{k} m_i \right) \right] \mod \left(\prod_{i=1}^{k} m_i \right), \qquad (2.28)$$

where s_j *is specified by*

$$s_j \left(\prod_{\substack{i=1 \\ i \neq j}}^{k} m_i \right) \equiv 1 \quad \mod m_j \qquad (2.29)$$

provided that

$$\prod_{i=1}^{k} m_i > x.$$

PROOF: Since $\left(\prod\limits_{\substack{i=1 \\ i \neq j}}^{k} m_i \right)$ and m_j are relatively prime integers, there must

exist integers s_j and t_j such that

$$s_j \left(\prod_{\substack{i=1 \\ i \neq j}}^{k} m_i \right) + t_j m_j = 1, \quad j = 1, 2, \ldots, k.$$

Actually, s_j and t_j may be constructed by applying Euclid's extended algorithm, discussed in Section 2.2.1. Clearly, then

$$s_j \left(\prod_{\substack{i=1 \\ i \neq j}}^{k} m_i \right) \equiv 1 \quad \mod m_j,$$

and (2.28), then, satisfies the specified system of linear congruences. Uniqueness of solution, modulo $(\prod_{i=1}^{k} m_i)$, is easy to establish. □

Figure 2.7B shows the listing for a program CHINES.FOR, written in FORTRAN, which can be used for implementing the CRT over integers. Hardware implementation of the CRT is also possible, and this is useful in FIR filter design. The CRT can also be conveniently used to reconstruct a negative integer from its specified residues with respect to a sufficiently large number of modulii. Let $m \triangleq \{m_1 \ m_2 \ldots m_k\}$ be a set of relatively prime positive integers comprising the modulii set. Define,

$$M \triangleq \prod_{i=1}^{k} m_i \quad \text{and} \quad w \triangleq 1/2(M - 1).$$

Then, any integer $i \in [-w, w]$ can be uniquely coded in a *residue number system* (RNS) as a sequence of residues, $(r_1 \ r_2 \ldots r_k)$, defined by,

$$r_j = \begin{cases} |i| \quad \mod m_j, & i \in [0, w] \\ m_j - |i| \quad \mod m_j, & i \in [-w, 0], \end{cases}$$

for $j = 1, 2, \ldots, k$. Note that the r_j's are nonnegative. The integer i can then be reconstructed by the CRT as follows. Obtain the integer x in (2.28). If x fall in the interval $[0, w]$, then $i = x$ is positive; if x falls in the interval $[w + 1, M]$, then $i = x - M$ is negative. Refer to Problem 22 at the end of Chapter 4 for possible use of the preceding scheme in FIR filter implementation. The next two examples illustrate the use of CRT, expounded in this section.

```
C
C          GIVEN THE PRIMES M(1),M(2),...,M(K) AND THE INTEGERS
C          R(1),R(2),...,R(K) SUCH THAT X=R(I)MOD M(I) FOR I=1,..,K
C          HAS A UNIQUE SOLUTION MODULO M(1)*M(2)*...*M(K)
C          THIS PROGRAM PROVIDES S(I),T(I) I=1,..,K FROM THE EQUATION
C          S(I)*(M(1)*...*M(I-1)*M(I+1)*..*M(K))+T(I)*M(I)=1  I=1,..,K
C          AND ALSO GIVES THE FINAL SOLUTION X.
C
C
C          *******************************************************************
C
           INTEGER M(50),R(50),S(50),R1,R2
           WRITE(6,110)
110        FORMAT(/,10X,'ENTER THE NUMBER K OF PRIMES NOW',//)
           READ(5,120) NPRIM
120        FORMAT(I)
           WRITE(6,130)
130        FORMAT(/,10X,'ENTER M(I) AND R(I) I=1,..,K A PAIR PER LINE',//)
           NN=0
           DO 140 I=1,NPRIM
           READ(5,150) M(I),R(I)
           IF(M(I).EQ.1) GO TO 260
140        CONTINUE
150        FORMAT(2I)
           DO 180 I=1,NPRIM
           R1=1
           DO 160 L=1,NPRIM
160        R1=R1*M(L)
           R1=R1/M(I)
           R2=M(I)
           IF(R2.GT.R1) GO TO 170
           CALL EUCLID(R1,R2,S(I),LL)
           WRITE(6,190) I,S(I),I,LL
           GO TO 180
170        CALL EUCLID(R2,R1,LL,S(I))
           WRITE(6,190) I,S(I),I,LL
180        CONTINUE
190        FORMAT(10X,'S(',I2,')=',1X,I,5X,'T(',I2,')=',1X,I)
           NTOT=1
           DO 200 J=1,NPRIM
200        NTOT=NTOT*M(J)
           NS=0
           DO 210 I=1,NPRIM
           NS=NS+R(I)*NTOT/M(I)*S(I)
210        CONTINUE
220        IF(NS.GT.0) GO TO 230
           NS=NTOT+NS
           GO TO 220
230        IF(NS.LT.NTOT) GO TO 240
```

Figure 2.7B. (*Partial listing, continued.*)

```
          NS=NS-NTOT
          GO TO 230
240       WRITE(6,250) NS
250       FORMAT(////,10X,'THE SOLUTION IS X=',2X,I)
          GO TO 280
260       WRITE(6,270)
270       FORMAT(10X,'M(I)=1 FOR SOME I; PLEASE LEAVE IT OUT',//)
280       STOP
          END
```

Figure 2.7. (B) Listing of program CHINES.FOR. Must be used with EUCLID.FOR
    in Figure 2.7(A).

Example 2.4. The number k of primes $= 5$. It is specified that [in the
program CHINES.FOR, $m(i) \leftarrow m_i$ and $r(i) \leftarrow r_i$]

$$
\begin{aligned}
m_1 &= 3, & r_1 &= 2, \\
m_2 &= 5, & r_2 &= 4, \\
m_3 &= 7, & r_3 &= 6, \\
m_4 &= 11, & r_4 &= 10, \\
m_5 &= 13, & r_5 &= 12.
\end{aligned}
$$

On executing CHINES.FOR [in the program, $b(i) \leftarrow t_i$ and $a(i) \leftarrow s_i$],

$$
\begin{aligned}
t_1 &= -1668, & s_1 &= 1, \\
t_2 &= -1201, & s_2 &= 2, \\
t_3 &= 613, & s_3 &= -2, \\
t_4 &= -124, & s_4 &= 1, \\
t_5 &= -533, & s_5 &= 6, \\
\end{aligned}
$$

$$
x = 15{,}014 \quad \mathrm{mod}\left(\prod_{i=1}^{5} m_i \right).
$$

Example 2.5. The number k of primes $= 8$. It is specified that

$$
\begin{aligned}
m_1 &= 2, & r_1 &= 1, \\
m_2 &= 3, & r_2 &= 0, \\
m_3 &= 5, & r_3 &= 1, \\
m_4 &= 11, & r_4 &= 2, \\
m_5 &= 13, & r_5 &= 1, \\
m_6 &= 17, & r_6 &= 3, \\
m_7 &= 19, & r_7 &= 4, \\
m_8 &= 23, & r_8 &= 5.
\end{aligned}
$$

On executing CHINES.FOR

$$t_1 = -7,967,602, \qquad s_1 = 1,$$
$$t_2 = 3,541,157, \qquad s_2 = -1,$$
$$t_3 = 2,549,633, \qquad s_3 = -2,$$
$$t_4 = -1,316,959, \qquad s_4 = 5,$$
$$t_5 = 565,747, \qquad s_5 = -3,$$
$$t_6 = 441,113, \qquad s_6 = -4,$$
$$t_7 = -264,851, \qquad s_7 = 3,$$
$$t_8 = -120,493, \qquad s_8 = 2,$$

$$x = 4,423,641 \quad \mathrm{mod}\left(\prod_{i=1}^{8} m_i \right).$$

It is readily verified that

$$s_j \left(\prod_{\substack{i=1 \\ i \neq j}}^{k} m_i \right) + t_j m_j = 1, \quad j = 1, 2, \ldots, 8.$$

2.3.4. Chinese Remainder Theorem over the Ring of Univariate Polynomials with Coefficients in a Field

Prime integers were defined in Definition 2.1. Irreducible polynomials are defined next.

Definition 2.4. A polynomial $a(z)$ with coefficients in an arbitrary but fixed field F is said to be irreducible over the field F provided that the only divisors of $a(z)$ have degree either zero or a degree equal to the degree of $a(z)$. The divisors and quotients are also polynomials (of degree greater than or equal to zero) with coefficients in F.

Example 2.6. Over the coefficient field of real numbers, the polynomial, $z^2 + 4$, is irreducible. Over the coefficient field of complex numbers the polynomial, $z^2 + 4$, factorizable as $(z + j2)(z - j2)$ is reducible.

When the polynomial coefficients belong to a finite field consisting of elements $0, 1$, the polynomial, $z^2 + 1$, factorizable as $(z + j)(z - j)$ is reducible. Note that the elements 0 and 1 form a field with addition and multiplication operations defined modulo 2.

The polynomial division process in (2.21) is representable as

$$a(z) \equiv r(z) \quad \mathrm{mod}\, b(z).$$

Theorem 2.6. *Let* $(p_1(z), p_2(z), \ldots, p_k(z))$ *be a set of mutually prime polynomials, i.e., the monic gcd of* $(p_i(z), p_j(z))$ *is* 1, $1 \leq i < j \leq k$. *The system of congruences,*

$$a(z) \equiv r_j(z) \mod p_j(z), \quad j = 1, 2, \ldots, k,$$

has a unique solution

$$a(z) = \left[\sum_{j=1}^{k} r_j(z) s_j(z) \prod_{\substack{i=1 \\ i \neq j}}^{k} p_i(z) \right] \mod \left(\prod_{j=1}^{k} p_j(z) \right),$$

where

$$\left[s_j(z) \prod_{\substack{i=1 \\ i \neq j}}^{k} p_i(z) \right] \equiv 1 \mod p_j(z)$$

provided that

$$\sum_{j=1}^{k} \delta[p_j(z)] > \delta[a(z)]$$

$\{\delta[a(z)]$ *is the degree of* $a(z)\}$.

PROOF: Note that the polynomials

$$\prod_{\substack{i=1 \\ i \neq j}}^{k} p_i(z) \quad \text{and} \quad p_j(z)$$

are relatively prime. Therefore, applying Fact 2.4, there must exist polynomials $s_j(z)$ and $t_j(z)$ such that

$$s_j(z) \left(\prod_{\substack{i=1 \\ i \neq j}}^{k} p_i(z) \right) + t_j(z) p_j(z) = 1. \tag{2.30}$$

Polynomials $s_j(z)$ and $t_j(z)$ may be constructed in the manner discussed in Section 2.3.2. The rest of the proof is easy to complete.

Example 2.7. Given that

$$p_1(z) = z + 1, \quad p_2(z) = z + 2, \quad p_3(z) = z + 3$$

and that the polynomial $a(z)$ is of degree 2 with

$$a(z) \equiv 1 \mod p_1(z),$$
$$a(z) \equiv -1 \mod p_2(z),$$
$$a(z) \equiv -1 \mod p_3(z).$$

Note that $(p_1(z), p_2(z), p_3(z))$ forms a mutually prime set of polynomials. Calculate

$$v_1(z) \triangleq p_2(z)p_3(z) = z^2 + 5z + 6,$$

$$v_2(z) \triangleq p_1(z)p_3(z) = z^2 + 4z + 3,$$

$$v_3(z) \triangleq p_1(z)p_2(z) = z^2 + 3z + 2.$$

The polynomials $s_j(z)$ and $t_j(z)$ required to satisfy

$$s_j(z)v_j(z) + t_j(z)p_j(z) = 1, \quad j = 1, 2, 3,$$

are chosen to be of the form

$$s_j(z) = s_{0j},$$

$$t_j(z) = t_{0j} + t_{1j}z$$

so that the degree constraints in (2.24) are satisfied. The unknowns are computed by setting up systems of linear equations as in Example 2.3.

$$\begin{bmatrix} 1 & 0 & 6 \\ 1 & 1 & 5 \\ 0 & 1 & 1 \end{bmatrix} \begin{bmatrix} t_{01} \\ t_{11} \\ s_{01} \end{bmatrix} = \begin{bmatrix} 1 \\ 0 \\ 0 \end{bmatrix}, \quad \begin{bmatrix} t_{01} \\ t_{11} \\ s_{01} \end{bmatrix} = \begin{bmatrix} -2 \\ -\frac{1}{2} \\ \frac{1}{2} \end{bmatrix},$$

$$\begin{bmatrix} 2 & 0 & 3 \\ 1 & 2 & 4 \\ 0 & 1 & 1 \end{bmatrix} \begin{bmatrix} t_{02} \\ t_{12} \\ s_{02} \end{bmatrix} = \begin{bmatrix} 1 \\ 0 \\ 0 \end{bmatrix}, \quad \begin{bmatrix} t_{02} \\ t_{12} \\ s_{02} \end{bmatrix} = \begin{bmatrix} 2 \\ 1 \\ -1 \end{bmatrix},$$

$$\begin{bmatrix} 3 & 0 & 2 \\ 1 & 3 & 3 \\ 0 & 1 & 1 \end{bmatrix} \begin{bmatrix} t_{03} \\ t_{13} \\ s_{03} \end{bmatrix} = \begin{bmatrix} 1 \\ 0 \\ 0 \end{bmatrix}, \quad \begin{bmatrix} t_{03} \\ t_{13} \\ s_{03} \end{bmatrix} = \begin{bmatrix} 0 \\ -\frac{1}{2} \\ \frac{1}{2} \end{bmatrix}.$$

Therefore,

$$a(z) = \sum_{j=1}^{3} r_j(z)s_j(z)v_j(z)$$

$$= \tfrac{1}{2}(z^2 + 5z + 6) + (z^2 + 4z + 3) - \tfrac{1}{2}(z^2 + 3z + 2)$$

$$= z^2 + 5z + 5.$$

Check:

$$a(z) \mod(z+1) = a(-1) = 1 = r_1,$$

$$a(z) \mod(z+2) = a(-2) = -1 = r_2,$$

$$a(z) \mod(z+3) = a(-3) = -1 = r_3.$$

Note that, in the division process,

$$a(z) = q(z)b(z) + r(z)$$

if the divisor $b(z)$ is of the form $b_0 + b_1 z$, then the remainder $r(z)$ computable as $r(z) = a$ $(z = -b_0/b_1)$, is a constant.

2.3.5. Use of CRT in IDFT (DFT) Computation

Let $(y(0), y(1), \ldots, y(N-1))$ be a sequence of length N whose DFT is another sequence $(Y(0), Y(1), \ldots, Y(N-1))$ of length N. $Y(k)$ is defined in (2.4). Clearly, the DFT computation is equivalent to the evaluation of a polynomial,

$$\hat{Y}(z) = \sum_{r=0}^{N-1} y(r)z^r$$

at the points w_N^k, $k = 0, 1, \ldots, N-1$, on the unit circle, $|z| = 1$. The numbers w_N^k, $k = 0, 1, \ldots, N-1$, happen to be the roots of polynomial $(z^N - 1)$. Therefore, it can be concluded that

$$Y(k) = \left[\hat{Y}(z) \quad \mathrm{mod}(z^N - 1) \right] \quad \mathrm{mod}\left(z - w_N^k \right)$$

for $k = 0, 1, \ldots, N-1$. If the polynomial $(z^N - 1)$ is factorable into m relatively prime factors,

$$(z^N - 1) = \prod_{i=1}^{m} f_i(z),$$

where

$$f_i(z) = \prod_{j=1}^{N_i} \left(z - w_N^{k_{i,j}} \right)$$

and $\sum_{i+1}^{m} N_i = N$, then the calculation of a single DFT of size N may be replaced by computation of several smaller DFT's of size N_i. Typically, then

$$Y(k_{i,j}) = \left[\hat{Y}(z) \quad \mathrm{mod}\, f_i(z) \right] \quad \mathrm{mod}\left(z - w_N^{k_{i,j}} \right)$$

for $j = 1, 2, \ldots, N_i$ and $i = 1, 2, \ldots, m$. The IDFT computation in (2.5) requires the finding of $(y(0), y(1), \ldots, y(N-1))$ from the specified $\{Y(k)\}$ or equivalently from $\{Y(k_{i,j})\}$. For each fixed i, given $Y(k_{i,j})$ for $j = 1, 2, \ldots, N_i$, it is possible to derive the polynomial $Y(z) \bmod f_i(z)$ by applying the CRT for polynomials (see Theorem 2.6). Finally, these m intermediate polynomials, $Y(z) \bmod f_i(z)$ for $i = 1, 2, \ldots, m$, may be used to determine $Y(z)$ and hence $(y(0), y(1), \ldots, y(N-1))$ via another application of Theorem 2.6. Note that an analogous procedure can be used to compute the DFT.

2.4. Discrete Hilbert Transform

It is well known in continuous system theory that the real and imaginary parts, x and y, of an analytic function of a complex variable are, except for an additive constant, explicitly related to each other. Furthermore, under the assumption that the function vanishes at infinity, one can obtain the pair of relations [5, pp. 330–349]

$$x(t) = -\frac{1}{\pi} \int_{-\infty}^{\infty} \frac{y(\tau)}{t - \tau} \, d\tau, \tag{2.31a}$$

$$y(t) = \frac{1}{\pi} \int_{-\infty}^{\infty} \frac{x(\tau)}{t - \tau} \, d\tau. \tag{2.32b}$$

The second relation defines the Hilbert transform of $x(t)$ while the first relation defines the inverse Hilbert transform and $x(t) \overset{HT}{\leftrightarrow} y(t)$ forms a Hilbert transform pair. The function $y(t) = H[x(t)]$, where H denotes the Hilbert transform operator may be viewed as the convolution of $x(t)$ and $1/\pi t$

$$y(t) = H[x(t)] = (1/\pi t) * x(t). \tag{2.32}$$

The Fourier transform of $1/\pi t$ is,

$$F\left[\frac{1}{\pi t}\right] = \frac{1}{\pi} \int_{-\infty}^{\infty} \frac{e^{-j\omega t}}{t} \, dt$$

$$= \frac{1}{\pi} \int_{-\infty}^{\infty} \frac{\cos \omega t - j \sin \omega t}{t} \, dt$$

$$= \frac{2}{\pi} \int_{0}^{\infty} \frac{-j \sin \omega t}{t} \, dt.$$

Using the fact that [6, pp. 103–105]

$$\frac{2}{\pi} \int_{0}^{\infty} \frac{\sin \omega t}{t} \, dt = \begin{cases} 1 & \text{if } \omega > 0, \\ -1 & \text{if } \omega < 0, \end{cases}$$

$$H(j\omega) \triangleq F\left[\frac{1}{\pi t}\right] = \begin{cases} -j & \text{if } \omega > 0, \\ 0 & \text{if } \omega = 0, \\ j & \text{if } \omega < 0, \end{cases} \tag{2.33}$$

or, more compactly,

$$H(j\omega) = -j \operatorname{sgn} \omega \tag{2.34a}$$

where the signum function

$$\operatorname{sgn} \omega = \begin{cases} 1 & \text{if } \omega > 0, \\ 0 & \text{if } \omega = 0, \\ -1 & \text{if } \omega < 0. \end{cases} \tag{2.34b}$$

Hilbert transformers have been applied to diverse problems originating in, for example, radar moving target indicators, stabilization of unstable recursive filters without appreciable change in the magnitude of frequency response, measurement of voice fundamental frequency, calculation of phase of minimum phase signals from amplitude, and generation of single sideband amplitude modulation and quadrature modulation filters. Possibilities for applications in various problems of digital signal processing along with the scope for using effectively some of the results of algebraic computational complexity theory when implementing the discrete Hilbert transform (DHT) motivate us to include this topic in this chapter.

In continuous system theory, it is often convenient to represent a real signal in terms of a complex signal which is an analytic function of time. For example, the analytic signal $w(t)$ of a real bandlimited signal $x(t)$ has been defined to be [24]

$$w(t) = x(t) + jy(t),$$

where $y(t) = H[x(t)]$ is the Hilbert transform of $x(t)$. For discrete-time signals, the notion of analyticity is meaningless. However, a complex representation of real discrete-time signals is possible in a manner analogous to analytic signals. Let $\{x(k)\}$ be a real sequence, whose Fourier transform is $X(e^{j\omega})$. Then, it is possible to construct a complex sequence whose kth element is

$$w(k) = x(k) + jy(k)$$

and whose Fourier transform is

$$W(e^{j\omega}) = \begin{cases} 2X(e^{j\omega}), & 0 < \omega < \pi, \\ 0, & \pi < \omega < 2\pi. \end{cases}$$

It is simple to show (see Problem 20 at the end of Chapter 4) that to construct $\{w(k)\}$ with its Fourier transform satisfying the conditions in the preceding equation the sequence $\{y(k)\}$ should be generated by passing $\{x(k)\}$ through a linear time-invariant system whose frequency response, $H(e^{j\omega})$ is described by

$$H(e^{j\omega}) = \begin{cases} -j, & 0 < \omega < \pi, \\ +j, & \pi < \omega < 2\pi. \end{cases}$$

Following comparison with the continuous-time counterpart in (2.34a) and (2.34b), it is meaningful to name the system responsible for generating $\{y(k)\}$ from $\{x(k)\}$ an ideal discrete time Hilbert transformer. In Section 2.1, it was seen that the DFT is defined for a sequence of finite length, and it may be viewed as an approximation to the Fourier transform. For the ideal discrete-time Hilbert transformer we associate an approximant, which is the appropriate counterpart of the DFT, and call this the discrete Hilbert transform (DHT).

Definition 2.5. The DHT of a sequence $\{x(k)\}$, $k = 0,1,\ldots, N-1$, is another sequence $\{y(k)\}$, $k = 0,1,\ldots, N-1$, defined by

$$y(k) = \text{IDFT}[H(k)X(k)],$$

where $\{X(k)\}$ is the DFT of $\{x(k)\}$ and $\{H(k)\}$ is the periodic discrete representation of $-j\,\text{sgn}\,\omega$ in (2.34a). Specifically, for N even,

$$H(k) = \begin{cases} -j, & k = 1,2,\ldots, N/2-1, \\ 0, & k = 0, N/2, \\ j, & k = N/2+1,\ldots, N-1, \end{cases} \quad (2.35a)$$

and for N odd,

$$H(k) = \begin{cases} -j, & k = 1,2,\ldots,(N-1)/2, \\ 0, & k = 0, \\ j, & k = (N+1)/2,\ldots, N-1. \end{cases} \quad (2.35b)$$

From the preceding definition, it follows that

$$y(k) = \frac{1}{N} \sum_{r=0}^{N-1} H(r) X(r) w_N^{-kr}, \quad w_N \triangleq e^{-j2\pi/N},$$

$$= \frac{1}{N} \sum_{r=0}^{N-1} H(r) \sum_{s=0}^{N-1} x(s) w_N^{(s-k)r}$$

$$= \frac{1}{N} \sum_{s=0}^{N-1} x(s) \sum_{r=0}^{N-1} H(r) w_N^{(s-k)r}. \quad (2.36)$$

(2.36) will be simplified after substituting $H(k)$ in (2.35a) and (2.35b), for the N even and N odd cases, respectively.

N even: Substituting (2.35a) in (2.36), and simplifying, we obtain

$$y(k) = \frac{1}{N} \sum_{s=0}^{N-1} x(s) j \left[\sum_{r=N/2+1}^{N-1} w_N^{(s-k)r} - \sum_{r=1}^{(N-2)/2} w_N^{(s-k)r} \right] \quad (2.37a)$$

$$= \frac{1}{N}(-1) \sum_{s=0}^{N-1} jx(s) \left[1-(-1)^{s-k}\right] \sum_{r=1}^{N/2-1} w_N^{(s-k)r}. \quad (2.37b)$$

Now,

$$\sum_{r=1}^{N/2-1} w_N^{(s-k)r} = \frac{w_N^{(s-k)}\left[1 - w_N^{(s-k)(N/2-1)}\right]}{1 - w_N^{(s-k)}}$$

$$= \begin{cases} -1, & \text{if } |k-s| \text{ is even,} \\ j\cot(k-s)\pi/N, & \text{if } |k-s| \text{ is odd.} \end{cases} \quad (2.38)$$

Substituting (2.38) in (2.37),

$$y(k) = \begin{cases} 0, & \text{if } |k-s| \text{ is even,} \\ \dfrac{2}{N} \displaystyle\sum_{s=0}^{N-1} x(s)\cot(k-s)\dfrac{\pi}{N}, & \text{if } |k-s| \text{ is odd} \end{cases} \tag{2.39}$$

or

$$y(k) = \frac{2}{N} \sum_{r=0,2,4,\ldots}^{N-2} x(r)\cot(k-r)\frac{\pi}{N}, \quad \text{for } k \text{ odd,} \tag{2.40a}$$

and

$$y(k) = \frac{2}{N} \sum_{r=1,3,5,\ldots}^{N-1} x(r)\cot(k-r)\frac{\pi}{N}, \quad \text{for } k \text{ even.} \tag{2.40b}$$

Another representation for $y(k)$, called the sine form representation, is obtainable by rewriting $y(k)$ in (2.37a) in the form

$$\begin{aligned} y(k) &= \frac{1}{N}(-1)\sum_{s=0}^{N-1} jx(s)\left[\sum_{r=1}^{N/2-1} \left(w_N^{(s-k)r} - w_N^{-(s-k)r} \right) \right] \\ &= \frac{2}{N}\sum_{s=0}^{N-1} x(s) \sum_{r=1}^{N/2-1} \sin\frac{(k-s)r2\pi}{N}. \end{aligned} \tag{2.41}$$

N odd: Substituting (2.35b) in (2.36), it may be verified in a manner analogous to the N even case that

$$\begin{aligned} y(k) &= \frac{1}{N}\sum_{s=0}^{N-1} x(s)\left\{ \cot\frac{(k-s)\pi}{N} - \frac{(-1)^{k-s}}{\sin[(k-s)\pi/N]} \right\}, \quad \text{if } k \neq s, \\ &= 0, \quad \text{if } k = s. \end{aligned} \tag{2.42}$$

The sine form representation in this case is

$$y(k) = \frac{2}{N}\sum_{s=0}^{N-1} x(s) \sum_{r=1}^{(N-1)/2} \sin\frac{r(k-s)2\pi}{N}. \tag{2.43}$$

It may be noted that the above representation of $y(k)$ for N odd is identical to the representation of $y(k)$ for N even in (2.41) with $(N/2-1)$ replaced by $(N-1)/2$ in the second summation of (2.43).

Equations (2.41) and (2.43) can be combined into one equation,

$$y(k) = \sum_{s=0}^{N-1} x(s)h(k-s), \tag{2.44a}$$

where

$$h(k) = \frac{2}{N}\sum_{r=1}^{M} \sin\frac{2\pi rk}{N}, \quad \begin{cases} M \triangleq N/2-1 & \text{for } N \text{ even} \\ M \triangleq (N-1)/2 & \text{for } N \text{ odd} \end{cases}$$

$$\tag{2.44b}$$

for $k = 0, 1, \ldots, N-1$. Define

$$h(k-s) \triangleq h_{ks}, \qquad (2.45)$$

$k = 0, 1, \ldots, N-1$, and $s = 0, 1, \ldots, N-1$. Then, the following properties of h_{ks} are readily verifiable:

i. $h_{ks} = -h_{sk}$.
ii. $h_{ks} = h_{\langle k+1 \rangle \langle s+1 \rangle}$, where $\langle k \rangle \triangleq k \bmod N$.
iii. $h_{ks} = 0$, $k = s$.
iv. $h_{0 \langle s \rangle} = -h_{0 \langle N-s \rangle}$ for N even.

It is convenient to write (2.44a) for $k = 0, 1, \ldots, N-1$ as a matrix vector product

$$\begin{bmatrix} y(0) \\ y(1) \\ \vdots \\ y(N-1) \end{bmatrix} = \begin{bmatrix} h_{ks} \end{bmatrix} \begin{bmatrix} x(0) \\ x(1) \\ \vdots \\ x(N-1) \end{bmatrix}, \qquad (2.46)$$

where the matrix $H \triangleq [h_{ks}]$, for $k = 0, 1, \ldots, N-1$, and $s = 0, 1, \ldots, N-1$. The computation of $y(0), y(1), \ldots, y(N-1)$ in (2.46) is facilitated by the use of the special properties of $[h_{ks}]$. In fact, the use of the Chinese remainder theorem on the structure of H can be used to reduce the multiplicative complexity in calculating $y(0), y(1), \ldots, y(N-1)$ in comparison to that required via direct multiplication. The next example substantiates this fact.

Example 2.8. Let $N = 5$. Then, the reader can easily verify that

$$H = \begin{bmatrix} 0 & h_0 & h_1 & -h_1 & -h_0 \\ -h_0 & 0 & h_0 & h_1 & -h_1 \\ -h_1 & -h_0 & 0 & h_0 & h_1 \\ h_1 & -h_1 & -h_0 & 0 & h_0 \\ h_0 & h_1 & -h_1 & -h_0 & 0 \end{bmatrix}, \qquad (2.47)$$

where, for the sake of brevity, define $h_{01} \triangleq h_0$ and $h_{02} \triangleq h_1$. The elements along the main diagonal of H are zero, and all the remaining elements in the matrix are obtainable in terms of h_0 and h_1. Note that H is a Toeplitz matrix (where the elements along any fixed diagonal are the same) of a special type. The Toeplitz property of H holds, in general, for any N. For the $H = [h_{ks}]$ in (2.47), $y(0), y(1), \ldots, y(4)$, in (2.46) with $N = 5$, can obviously be computed with 20 multiplications. (This number is much more than actually required.) Dutta Roy [7] showed that the number of multiplications can be reduced by a factor of 2 after rewriting the matrix vector product in (2.46), for the case in this example, as

$$
\begin{bmatrix} y(0) \\ y(1) \\ \vdots \\ y(4) \end{bmatrix} = \begin{bmatrix} x(1)-x(4) & x(2)-x(3) \\ x(2)-x(0) & x(3)-x(4) \\ x(3)-x(1) & x(4)-x(0) \\ x(4)-x(2) & x(0)-x(1) \\ x(0)-x(3) & x(1)-x(2) \end{bmatrix} \begin{bmatrix} h_0 \\ h_1 \end{bmatrix}. \tag{2.48}
$$

Two multiplications, then, suffice to compute each $x(i)$, $i = 0,1,\ldots,4$. Bart Rice pointed out (at the NSF sponsored Conference on Algebraic Computational Complexity, held at Pittsburgh in 1978) that the number of multiplications can be reduced to eight by noting that

$$
y(0)+y(1)+y(2)+y(3)+y(4) = 0 \tag{2.49}
$$

since each column of H in (2.47) adds to zero. He also produced an algorithm for computing the output points using only seven multiplications. To wit, define

$$
m_1 = h_1\big(x(1)-x(0)\big),
$$
$$
m_2 = h_1\big(x(3)-x(4)\big),
$$
$$
m_3 = h_0\big(x(2)-x(0)\big),
$$
$$
m_4 = (h_0 + h_1)\big(x(3)-x(1)\big),
$$
$$
m_5 = h_0\big(x(4)-x(2)\big),
$$
$$
m_6 = h_0\big(x(0)-x(3)\big),
$$
$$
m_7 = h_1\big(x(0)-x(2)\big).
$$

Then, from (2.49)

$$
y(0) = -\big(y(1)+y(2)+y(3)+y(4)\big),
$$

where

$$
y(1) = m_2 + m_3,
$$
$$
y(2) = m_1 - m_2 + m_4,
$$
$$
y(3) = m_5 - m_1,
$$
$$
y(4) = m_1 + m_6 + m_7.
$$

It will be shown in the next section that six multiplications are necessary and sufficient to compute the output points, $y(0), y(1),\ldots, y(4)$.

2.5. Winograd's Multiplicative Complexity Theory

In Section 2.2, it was seen that when the number of samples, N, is a power of 2, the FFT algorithms make feasible the computation of DFT using about $(N/2)\log_2 N$ complex multiplications and $N\log_2 N$ additions. One complex multiplication can obviously be done with four real multiplications. However, on a special purpose computer the time required to carry out a multiplication could be significantly more than the time required to implement an addition. The question, then, arises as to whether the overall computational cost in implementing an algorithm using the operations of addition and multiplication may be reduced by trading off some multiplications for some extra additions. In principle, it is possible to carry out a complex multiplication using three real multiplications and five real additions (i.e., addition of real numbers) instead of four real multiplications and two real additions (see Problem 8 for one possible way of achieving this). Another possible way of reaching this goal is discussed next. Given real variables x_1, x_2, x_3, and x_4, it is required to calculate $x_1 x_3 - x_2 x_4$, $x_2 x_3 + x_1 x_4$, which are, respectively, the real and imaginary parts of the product $(x_1 + jx_2)(x_3 + jx_4)$, using less than four real multiplications. The real and imaginary parts are seen to be obtainable by multiplying the two polynomials in z with indeterminate coefficients, namely, $(x_1 + zx_2)$ and $(x_3 + zx_4)$ and then reducing the product modulo the polynomial $z^2 + 1$, which is the minimal degree polynomial with real coefficients which has j as a root. The coefficients of the reduced polynomial of first degree in z are the required real and imaginary parts sought. Clearly, the reduction modulo $(z^2 + 1)$ requires no multiplication, and, therefore, it becomes necessary to compute the coefficients z_0, z_1, and z_2 of the product

$$(x_1 + zx_2)(x_3 + zx_4) \equiv z_0 + z_1 z + z_2 z^2. \qquad (2.50a)$$

Therefore, after reduction modulo $(z^2 + 1)$ the preceding identity becomes

$$(x_1 x_3 - x_2 x_4) + z(x_2 x_3 + x_1 x_4) = (z_0 - z_2) + z_1 z.$$

Select three distinct values of z; suppose that the chosen values are $0, 1, -1$. Then, substituting each in the preceding identity in (2.50a), one gets

$$\begin{bmatrix} 1 & 0 & 0 \\ 1 & 1 & 1 \\ 1 & -1 & 1 \end{bmatrix} \begin{bmatrix} z_0 \\ z_1 \\ z_2 \end{bmatrix} = \begin{bmatrix} x_1 x_3 \\ (x_1 + x_2)(x_3 + x_4) \\ (x_1 - x_2)(x_3 - x_4) \end{bmatrix}. \qquad (2.50b)$$

Therefore, after solving for z_0, z_1, and z_2,

$$z_1 = x_2 x_3 + x_1 x_4 = \tfrac{1}{2}(x_1 + x_2)(x_3 + x_4) - \tfrac{1}{2}(x_1 - x_2)(x_3 - x_4),$$

$$z_0 - z_2 = x_1 x_3 - x_2 x_4 = 2x_1 x_3 - \tfrac{1}{2}(x_1 + x_2)(x_3 + x_4)$$
$$- \tfrac{1}{2}(x_1 - x_2)(x_3 - x_4).$$

It is important to note that on ignoring the multiplications by the constants $\tfrac{1}{2}$ and 2, the three real multiplications sufficient to implement a complex multiplication, by the procedure described above, are

$$x_1 x_3, \quad (x_1 + x_2)(x_3 + x_4), \quad \text{and} \quad (x_1 - x_2)(x_3 - x_4).$$

Since these three products are linearly independent, three is also the minimum number of real multiplications necessary for the purpose.

Minimal multiplication algorithms are not unique. Another alternate algorithm for implementing a complex multiplication (besides the one suggested in Problem 8) is

$$m_0 \triangleq x_1(x_3 + x_4),$$
$$m_1 \triangleq x_3(x_1 - x_2),$$
$$m_2 \triangleq x_4(x_1 + x_2),$$
$$z_1 = x_2x_3 + x_1x_4 = m_0 - m_1,$$
$$z_0 - z_2 = x_1x_3 - x_2x_4 = m_0 - m_2.$$

The preceding discussion brings to focus the basic assumptions in the multiplicative complexity theory, to be discussed in this section, for the computation of DFT and the related problems of convolution and correlation. The inputs to the algorithms constructed are variables and constants in a field. Multiplication by the constant elements from the specified field (usually the field of rational or real numbers) will not be counted as multiplications. Also, though in practice some control over the number of additions may be necessary, no such constraint will be placed here. The main justification for these assumptions is the mathematical ease and theoretical completeness that result from it. Nevertheless, it is satisfying to know that these assumptions do make considerable sense, at least in certain applications.

2.5.1. Minimal Algorithms for Computing Linear Convolution

It is desired to compute the linear convolution of length $m + n + 1$ of two specified finite sequences,

$$\{x_k\} \triangleq (x_0 \quad x_1 \quad \cdots \quad x_m), \tag{2.51a}$$

$$\{h_k\} \triangleq (h_0 \quad h_1 \quad \cdots \quad h_n). \tag{2.51b}$$

The convolved sequence $\{y_k\}$ has a typical element

$$y_k = \sum_{r=0}^{k} x_r h_{k-r}, \quad k = 0, 1, \ldots, m + n, \tag{2.52}$$

where $x_r = 0$ for $r > m$ and $h_r = 0$ for $r > n$. Associate with sequences $\{x_k\}$ and $\{y_k\}$, respectively, polynomials

$$X(z) = \sum_{k=0}^{m} x_k z^k, \tag{2.53a}$$

$$H(z) = \sum_{k=0}^{n} h_k z^k. \tag{2.53b}$$

It is readily verifiable that the elements of $\{y_k\}$ in (2.52) are the coefficients of the polynomial

$$Y(z) \triangleq X(z)H(z).$$

The objective, then, is to compute the coefficients of $Y(z)$ with the minimum number of multiplications, excluding multiplications by constants in the chosen field, say, of rational numbers. For any set of $m + n + 1$ distinct values, $z_0, z_1, \ldots, z_{m+n}$ of z in the field, Lagrange's interpolation formula enables one to construct $Y(z)$ uniquely from its sampled values $Y(z_0), Y(z_1), \ldots, Y(z_{m+n})$

$$Y(z) = \sum_{k=0}^{m+n} Y(z_k) \left[\prod_{r \neq k} (z - z_r) \Big/ \prod_{r \neq k} (z_k - z_r) \right]. \qquad (2.54)$$

Note that, based on the premise that one is free to ignore multiplications by field elements, the only multiplications required to obtain $Y(z)$ in the preceding equation are the $m + n + 1$ multiplications:

$$Y(z_k) = X(z_k)H(z_k) = \left(\sum_{r=0}^{m} x_r z_k^r \right) \cdot \left(\sum_{r=0}^{n} h_r z_k^r \right), \qquad (2.55)$$

for $k = 0, 1, \ldots, m + n$. Of course, due to the assumptions made, no multiplications are required to compute either $X(z_k)$ or $H(z_k)$, while the coefficients of the polynomial

$$W_k(z) \triangleq \left[\prod_{r \neq k} (z - z_r) \Big/ \prod_{r \neq k} (z_k - z_r) \right] \qquad (2.56)$$

may be precomputed from the chosen values $z_0, z_1, \ldots, z_{m+n}$ in the field of rational numbers.

Example 2.9. The product

$$Y(z) = (x_0 + x_1 z)(h_0 + h_1 z)$$

is desired. Choose three distinct values of z, say $z_0 = 0$, $z_1 = 1$, and $z_2 = -1$, in the field of rational numbers. These are the most convenient values to choose, for obvious reasons. Then,

$$W_0(z) = 1 - z^2,$$
$$W_1(z) = \tfrac{1}{2}(z^2 + z),$$
$$W_2(z) = \tfrac{1}{2}(z^2 - z)$$

and

$$Y(z_0) = x_0 h_0 \triangleq m_1,$$
$$Y(z_1) = (x_0 + x_1) \cdot (h_0 + h_1) \triangleq m_2,$$
$$Y(z_2) = (x_0 - x_1) \cdot (h_0 - h_1) \triangleq m_3.$$

Therefore, on using (2.54),

$$Y(z) \triangleq y_0 + y_1 z + y_2 z^2$$

$$= m_1(1 - z^2) + \tfrac{1}{2}m_2(z + z^2) + \tfrac{1}{2}m_3(z^2 - z)$$

$$= m_1 + \tfrac{1}{2}(m_2 - m_3)z + (\tfrac{1}{2}m_3 + \tfrac{1}{2}m_2 - m_1)z^2.$$

Therefore,

$$y_0 = m_1 = x_0 h_0,$$

$$y_1 = \tfrac{1}{2}(m_2 - m_3) = \tfrac{1}{2}(x_0 + x_1)(h_0 + h_1) - \tfrac{1}{2}(x_0 - x_1)(h_0 - h_1),$$

$$y_2 = \left[\tfrac{1}{2}(m_2 + m_3) - m_1\right] = \tfrac{1}{2}(x_0 + x_1)(h_0 + h_1)$$

$$+ \tfrac{1}{2}(x_0 - x_1)(h_0 - h_1) - x_0 h_0.$$

The result is, naturally, in agreement with (2.50) after the minor notational changes are taken into consideration.

The fact that $(m + n + 1)$ multiplications are necessary to compute $Y(z)$ follows from the linear independence of the y_k's in (2.52). The result discussed so far in this section is summarized in the theorem below.

Theorem 2.7. *The minimum number of multiplications required to compute the coefficients of the polynomial product, $(\sum_{k=0}^{m} x_k z^k)(\sum_{k=0}^{n} h_k z^k)$ is $m + n + 1$, when the multiplications by elements of the chosen field to which the polynomial coefficients belong are ignored.*

Winograd [10] showed that all algorithms of minimal multiplicative complexity for computing the product of two polynomials (or the linear convolution of two finite sequences) fall under either of two general classes. The underlying field should be large enough, and in our case this field, unless otherwise specified, *is the field of rational numbers*, which contains an infinite number of elements. For minimal algorithms in the first class, choose $(m + n + 1)$ distinct elements, z_k, $k = 0, 1, 2, \ldots, m + n$, in the field (the size of the field allows this to be possible) and form the polynomial

$$M_1(z) = \prod_{k=0}^{m+n} (z - z_k), \qquad (2.57)$$

which is of degree $m + n + 1$. Then, the product $X(z)H(z)$, where $X(z)$ and $H(z)$ are polynomials of degrees m and n, respectively, is computed using the following identity:

$$Y(z) = X(z)H(z) \equiv X(z)H(z) \mod M_1(z). \qquad (2.58)$$

In $M_1(z)$, the $m + n + 1$ linear factors form a mutually relatively prime set (since the z_k's in (2.57) are distinct). The $m + n + 1$ multiplications required are

$$Y(z_k) = X(z_k)H(z_k) = X(z)H(z) \mod(z - z_k), \qquad (2.59)$$

$k = 0, 1, \ldots, m + n$. The polynomial $Y(z)$ can, then, be constructed from $Y(z_k)$'s by using the Chinese remainder theorem. No additional multiplications, other than multiplications by elements of the field, are required. The algorithm described in this section, using Lagrange's interpolation formula, belongs to this class.

The second class of minimal algorithm for computing the polynomial product $X(z)H(z) = (\sum_{k=0}^{m} x_k z^k)(\sum_{k=0}^{n} h_k z^k)$ is based on the following identity:

$$Y(z) = X(z)H(z) \equiv X(z)H(z) \mod M_2(z) + x_m h_n M_2(z),$$

$$(2.60)$$

where the polynomial

$$M_2(z) = \prod_{k=1}^{m+n} (z - z_k) \qquad (2.61)$$

is formed from $m + n$ distinct elements, z_k, $k = 1, 2, \ldots, m + n$, in the field. The $m + n + 1$ multiplications, in this case, are

$$Y(z_k) = X(z_k)H(z_k), \quad k = 1, 2, \ldots, m + n,$$

and

$$x_m h_n.$$

Again, the desired product $Y(z)$ may be computed from the $Y(z_k)$'s and $x_m h_n$ using the Chinese remainder theorem without additional multiplications except for multiplications by field elements. The choices of $M_1(z)$ and $M_2(z)$ as products of linear factors often lead to large constants even when m and n are small. To combat this problem, Winograd suggested the use of irreducible factors, which are not necessarily linear. Example 2.11 will illustrate how computations modulo an irreducible factor, which is not linear, may be carried out.

2.5.2. Multiplicative Complexity for Computing Cyclic (Circular) Convolutions

The cyclic convolution of two sequences, each of length n,

$$\{x_k\} \triangleq (x_0 \quad x_1 \quad \cdots \quad x_{n-1}), \qquad (2.62a)$$

$$\{h_k\} \triangleq (h_0 \quad h_1 \quad \cdots \quad h_{n-1}), \qquad (2.62b)$$

can be expressed in terms of polynomial multiplication and subsequent computation of the polynomial product modulo a third polynomial, $z^n - 1$.

Fact 2.5. *The cyclic convolution of two sequences $\{x_k\}$ and $\{h_k\}$, each of length n as in (2.62) is a sequence $\{y_k\}$, where $y_k = \sum_{r=0}^{n} x_{\langle k-r \rangle} h_r =$*

$\sum_{r=0}^{n} h_{\langle k-r \rangle} x_r, \quad k = 0, 1, \ldots, n-1, \quad \langle k-r \rangle \triangleq (k-r) \mod n,$ *and this sequence is obtainable by calculating the coefficients of the polynomial*

$$\left(\sum_{k=0}^{n-1} x_k z^k \right) \left(\sum_{k=0}^{n-1} h_k z^k \right) \mod (z^n - 1).$$

Example 2.10. Let $n = 4$. Then

$$\left(\sum_{k=0}^{3} x_k z^k \right) \left(\sum_{k=0}^{3} h_k z^k \right) = \sum_{k=0}^{6} \hat{y}_k z^k,$$

where

$$\hat{y}_k = \sum_{r=0}^{k} x_r h_{k-r}, \quad k = 0, 1, \ldots, 6, \qquad (2.63)$$

with $x_r = 0, \quad r > 3, \quad h_r = 0, \quad r > 3$. Now,

$$\hat{y}_4 z^4 \mod (z^4 - 1) = \hat{y}_4,$$

$$\hat{y}_5 z^5 \mod (z^4 - 1) = \hat{y}_5 z,$$

$$\hat{y}_6 z^6 \mod (z^4 - 1) = \hat{y}_6 z^2.$$

Therefore,

$$\sum_{k=0}^{6} \hat{y}_k z^k \mod (z^4 - 1) = (\hat{y}_0 + \hat{y}_4) + (\hat{y}_1 + \hat{y}_5)z + (\hat{y}_2 + \hat{y}_6)z^2 + \hat{y}_3 z^3.$$

Using (2.63), the elements of the cyclically convolved sequence are

$$y_0 = \hat{y}_0 + \hat{y}_4 = (x_0 h_0) + (x_1 h_3 + x_2 h_2 + x_3 h_1),$$

$$y_1 = \hat{y}_1 + \hat{y}_5 = (x_0 h_1 + x_1 h_0) + (x_2 h_3 + x_3 h_2),$$

$$y_2 = \hat{y}_2 + \hat{y}_6 = (x_0 h_2 + x_1 h_1 + x_2 h_0) + (x_3 h_3),$$

$$y_3 = \hat{y}_3 = (x_0 h_3 + x_1 h_2 + x_2 h_1 + x_3 h_0).$$

The above set of equations can obviously be written in the form,

$$\begin{bmatrix} y_0 \\ y_1 \\ y_2 \\ y_3 \end{bmatrix} = \begin{bmatrix} h_0 & h_3 & h_2 & h_1 \\ h_1 & h_0 & h_3 & h_2 \\ h_2 & h_1 & h_0 & h_3 \\ h_3 & h_2 & h_1 & h_0 \end{bmatrix} \begin{bmatrix} x_0 \\ x_1 \\ x_2 \\ x_3 \end{bmatrix},$$

where the square matrix on the right-hand side is a circulant (defined in Chap. 6). Another characterization in terms of a circulant involving the elements of x_k is

$$
\begin{bmatrix} y_0 \\ y_1 \\ y_2 \\ y_3 \end{bmatrix} = \begin{bmatrix} x_0 & x_3 & x_2 & x_1 \\ x_1 & x_0 & x_3 & x_2 \\ x_2 & x_1 & x_0 & x_3 \\ x_3 & x_2 & x_1 & x_0 \end{bmatrix} \begin{bmatrix} h_0 \\ h_1 \\ h_2 \\ h_3 \end{bmatrix}.
$$

The polynomial, $z^n - 1$, that occurs in the computation of the cyclic convolution of two sequences, each of length n, is factorable over the field of rationals into a product of k irreducible factors, $C_{d_r}(z)$, $r = 1, 2, \ldots, k$, where k is the number of divisors of n including $d_1 = 1$ and $d_k = n$. Each of these factors belongs to a class of polynomials called cyclotomic polynomials, defined next.

Definition 2.6. A complex number $z = e^{j\alpha}$ is called an rth ($r \geq 1$) root of unity if $z^r - 1 = 0$ and a primitive rth root of unity if, in addition, $z^s - 1 \neq 0$ for $1 \leq s < r$. The rth cyclotomic polynomial $C_r(z)$ is the monic polynomial whose zeros are the distinct primitive rth roots of unity. The number of such distinct primitive rth roots of unity equals $\phi(r)$, where $\phi(r)$, *Euler's totient function*, is the number of positive integers less than r including 1, which are relatively prime to r. [See Problem 17 for more on $\phi(r)$, also $\phi(1) \triangleq 1$.]

The cyclotomic polynomials have several remarkable properties. First, all their coefficients are integers. Second, for $r < 105$, the coefficients of $C_r(z)$ are either 0, 1, or -1. Furthermore, since a cyclotomic polynomial is irreducible over rationals, computation of a polynomial product modulo $(z^n - 1)$ may, first, be done modulo each cyclotomic polynomial which is a factor of $z^n - 1$ and then the desired product modulo $(z^n - 1)$ may be constructed by the Chinese remainder theorem. Based on these types of arguments, Winograd arrived at an important theorem, stated next, without proof. Examples 2.12 and 2.13 will illustrate the ideas in the general proof.

Theorem 2.8. *The minimum number of multiplications required to cyclically convolve two sequences, each of length n, provided that multiplications by constants in the chosen field of rational numbers are not counted, is $2n - k$, where k is the number of positive divisors of n including 1 and n.*

It is further mentioned that Theorem 2.8 is a special case of the minimal multiplicative complexity theory developed by Winograd for computing the product of two polynomials modulo a third polynomial. If this third polynomial, say $M(z)$, factors into a product of k irreducible factors over the field of rationals, then the minimum number of multiplications required to compute the product of two polynomials, each of degree $(n - 1)$, modulo

$M(z)$ is $(2n - k)$, provided multiplications by constants in the chosen field of rationals are not counted.

Before proceeding any further, an example will be supplied to illustrate the computation of the product of two polynomials modulo a third polynomial. This example will also involve computations modulo an irreducible polynomial, which is not of degree 1.

Example 2.11. The output points, $y(0), y(1),\ldots, y(4)$ in Example 2.8 are obtained from the matrix vector product [remember that notations $y(k)$ and y_k are used interchangeably]:

$$\begin{bmatrix} 0 & h_0 & h_1 & -h_1 & -h_0 \\ -h_0 & 0 & h_0 & h_1 & -h_1 \\ -h_1 & -h_0 & 0 & h_0 & h_1 \\ h_1 & -h_1 & -h_0 & 0 & h_0 \\ h_0 & h_1 & -h_1 & -h_0 & 0 \end{bmatrix} \begin{bmatrix} x(0) \\ x(1) \\ x(2) \\ x(3) \\ x(4) \end{bmatrix} = \begin{bmatrix} y(0) \\ y(1) \\ y(2) \\ y(3) \\ y(4) \end{bmatrix}.$$

Clearly,

$$\hat{h}(z)\hat{x}(z) \equiv y(0) + y(1)z^4 + y(2)z^3 + y(3)z^2 + y(4)z \quad \mathrm{mod}(z^5 - 1)$$

(this interesting observation and subsequent manipulations were suggested by Bart Rice), where

$$\hat{h}(z) \triangleq \left(h_0 z + h_1 z^2 - h_1 z^3 - h_0 z^4 \right),$$

$$\hat{x}(z) \triangleq \left(x(0) + x(1)z^4 + x(2)z^3 + x(3)z^2 + x(4)z \right).$$

Note that

$$z^5 - 1 = (z - 1)(z^4 + z^3 + z^2 + z + 1).$$

By direct substitution of $z = 1$ in $\hat{h}(z)$,

$$\hat{h}(z)\hat{x}(z) \quad \mathrm{mod}(z - 1) = 0$$

Therefore,

$$y(0) + y(1) + y(2) + y(3) + y(4) = 0$$

in agreement with (2.49). Also, from the first identity following the matrix equation in this example,

$$\hat{h}(z)\hat{x}(z) \quad \mathrm{mod}(z^4 + z^3 + z^2 + z + 1)$$

$$= y(0) - y(1) + (y(4) - y(1))z + (y(3) - y(1))z^2 + (y(2) - y(1))z^3.$$

Noting that

$$h_0 z + h_1 z^2 - h_1 z^3 - h_0 z^4 = z(1 - z)\left[h_0 + (h_0 + h_1)z + h_0 z^2 \right],$$

it is simple to verify that

$$\hat{h}(z)\hat{x}(z) \equiv \hat{h}_1(z)\hat{x}_1(z) \quad \mathrm{mod}(z^4 + z^3 + z^2 + z + 1),$$

where

$$\hat{h}_i(z) = \left[h_0 + (h_0 + h_1)z + h_0 z^2 \right],$$

$$\hat{x}_1(z) = \left[(x(1) - 2x(2) + x(3)) + z(x(0) - x(1) - x(2) + x(3)) \right.$$
$$\left. + z^2(-x(0) - x(2) + x(3) + x(4)) + z^3(-x(2) + 2x(3) - x(4)) \right].$$

In order to compute $\hat{h}_1(z)\hat{x}_1(z) \triangleq w(z)$, it may be rewritten as (the next identity is very similar in structure to the identity used for the construction of the second class of minimal algorithms for computing polynomial products as discussed in Section 2.5.1):

$$w(z) = \sum_{k=0}^{5} w_k z^k = v(z) + h_0(-x(2) + 2x(3) - x(4))(z^5 - z),$$

where

$$v(z) = \sum_{k=0}^{4} v_k z^k = w(z) \mod(z^5 - z).$$

The computation of $w(z)$ using the minimum number of multiplications is crucial to the solution of the problem.

Noting that

$$(z^5 - z) = z(z-1)(z+1)(z^2+1),$$

$$v(z) \mod z = v_0 = h_0(x(1) - 2x(2) + x(3)) \triangleq m_1,$$

$$v(z) \mod(z-1) = \sum_{k=0}^{4} v_k = 15(h_0 + \tfrac{1}{3}h_1)(x(3) - x(2)) \triangleq m_2,$$

$$v(z) \mod(z+1) = \sum_{k=0}^{4} (-1)^k v_k$$
$$= (h_0 - h_1)\{-x(2) - x(3) + 2[x(4) + x(1) - x(0)]\}.$$

Similarly, computing $\mod(z^2+1)$,

$$v_0 - v_2 + v_4 = (h_0 + h_1)(-x(0) + x(1) + x(3) - x(4)) \triangleq m_4,$$

$$v_1 - v_3 = (h_0 + h_1)(x(0) + x(1) - x(2) - x(4)) \triangleq m_5.$$

Define

$$h_0(-x(2) + 2x(3) - x(4)) \triangleq m_6.$$

Then,

$$\begin{bmatrix} w_0 \\ w_1 \\ w_2 \\ w_3 \\ w_4 \\ w_5 \end{bmatrix} = \begin{bmatrix} v_0 \\ v_1 - m_6 \\ v_2 \\ v_3 \\ v_4 \\ m_6 \end{bmatrix} = M_1 \begin{bmatrix} v_0 \\ v_1 \\ v_2 \\ v_3 \\ v_4 \\ m_6 \end{bmatrix},$$

where

$$M_1 = \begin{bmatrix} 1 & 0 & 0 & 0 & 0 & 0 \\ 0 & 1 & 0 & 0 & 0 & -1 \\ 0 & 0 & 1 & 0 & 0 & 0 \\ 0 & 0 & 0 & 1 & 0 & 0 \\ 0 & 0 & 0 & 0 & 1 & 0 \\ 0 & 0 & 0 & 0 & 0 & 1 \end{bmatrix}.$$

Also,

$$M_2 \begin{bmatrix} v_0 \\ v_1 \\ v_2 \\ v_3 \\ v_4 \\ m_6 \end{bmatrix} = \begin{bmatrix} m_1 \\ m_2 \\ m_3 \\ m_4 \\ m_5 \\ m_6 \end{bmatrix}, \quad M_2 = \begin{bmatrix} 1 & 0 & 0 & 0 & 0 & 0 \\ 1 & 1 & 1 & 1 & 1 & 0 \\ 1 & -1 & 1 & -1 & 1 & 0 \\ 1 & 0 & -1 & 0 & 1 & 0 \\ 0 & 1 & 0 & -1 & 0 & 0 \\ 0 & 0 & 0 & 0 & 0 & 1 \end{bmatrix},$$

$$\begin{bmatrix} \sum_{k=0}^{4} y(k) \\ y(0) - y(1) \\ y(4) - y(1) \\ y(3) - y(1) \\ y(2) - y(1) \end{bmatrix} = M_3 \begin{bmatrix} y(0) \\ y(1) \\ y(2) \\ y(3) \\ y(4) \end{bmatrix}, \quad M_3 = \begin{bmatrix} 1 & 1 & 1 & 1 & 1 \\ 1 & -1 & 0 & 0 & 0 \\ 0 & -1 & 0 & 0 & 1 \\ 0 & -1 & 0 & 1 & 0 \\ 0 & -1 & 1 & 0 & 0 \end{bmatrix}.$$

Also,

$$M_3 \begin{bmatrix} y(0) \\ y(1) \\ y(2) \\ y(3) \\ y(4) \end{bmatrix} = M_4 \begin{bmatrix} w(0) \\ w(1) \\ w(2) \\ w(3) \\ w(4) \\ w(5) \end{bmatrix}, \quad M_4 = \begin{bmatrix} 0 & 0 & 0 & 0 & 0 & 0 \\ 1 & 0 & 0 & 0 & -1 & 1 \\ 0 & 1 & 0 & 0 & -1 & 0 \\ 0 & 0 & 1 & 0 & -1 & 0 \\ 0 & 0 & 0 & 1 & -1 & 0 \end{bmatrix}.$$

From the previous equations, it follows that

$$\begin{bmatrix} y(0) \\ y(1) \\ y(2) \\ y(3) \\ y(4) \end{bmatrix} = \begin{bmatrix} 1 & -\frac{1}{5} & 0 & 0 & 0 & 1 \\ -1 & \frac{1}{20} & \frac{1}{4} & \frac{1}{2} & 0 & 0 \\ 0 & \frac{1}{20} & -\frac{1}{4} & 0 & -\frac{1}{2} & 0 \\ 0 & \frac{1}{20} & \frac{1}{4} & -\frac{1}{2} & 0 & 0 \\ 0 & \frac{1}{20} & -\frac{1}{4} & 0 & \frac{1}{2} & -1 \end{bmatrix} \begin{bmatrix} m_1 \\ m_2 \\ m_3 \\ m_4 \\ m_5 \\ m_6 \end{bmatrix},$$

where the matrix in the preceding equation is obtained by computing the matrix product

$$M_3^{-1} M_4 M_1 M_2^{-1}.$$

Ignoring the multiplications by the constants in the field of rationals, the number of multiplications sufficient to do the job is six. The necessity part follows from a trivial application of the polynomial multiplicative complexity theorem. Refer to comments following Theorem 2.8.

Example 2.12. It is required to compute the circular convolution of two sequences:

$$(h_0 \quad h_1 \quad h_2) \quad \text{and} \quad (x_0 \quad x_1 \quad x_2).$$

Define

$$H(z) \triangleq h_0 + h_1 z + h_2 z^2,$$
$$X(z) \triangleq x_0 + x_1 z + x_2 z^2.$$

It is then required to compute

$$Y(z) \triangleq H(z) X(z) \quad \mathrm{mod}\,(z^3 - 1).$$

Note that

$$(z^3 - 1) = (z - 1)(z^2 + z + 1),$$

where

$$M_1(z) \triangleq (z - 1), \quad M_2(z) \triangleq z^2 + z + 1$$

are irreducible.

$$H_1(z) \triangleq H(z) \quad \mathrm{mod}\,(z - 1) = (h_0 + h_1 + h_2),$$
$$X_1(z) \triangleq X(z) \quad \mathrm{mod}\,(z - 1) = (x_0 + x_1 + x_2),$$
$$H_2(z) \triangleq H(z) \quad \mathrm{mod}\,(z^2 + z + 1) = (h_0 - h_2) + (h_1 - h_2)z,$$
$$X_2(z) \triangleq X(z) \quad \mathrm{mod}\,(z^2 + z + 1) = (x_0 - x_2) + (x_1 - x_2)z.$$

Define

$$(h_0 + h_1 + h_2)(x_0 + x_1 + x_2) \triangleq m_1,$$
$$(h_0 - h_2)(x_0 - x_2) \triangleq m_2,$$
$$[(h_1 - h_2) + (h_0 - h_2)][(x_1 - x_2) + (x_0 - x_2)] \triangleq m_3, \tag{2.64}$$
$$(h_1 - h_2)(x_1 - x_2) \triangleq m_4.$$

Then,

$$Y_1(z) \triangleq H_1(z) X_1(z) \quad \mathrm{mod}\,(z - 1) = m_1,$$
$$Y_2(z) \triangleq H_2(z) X_2(z) \quad \mathrm{mod}\,(z^2 + z + 1)$$
$$= (m_2 - m_4) + (m_3 - m_2 - 2m_4)z.$$

Before applying CRT, compute $S_i(z)$, $i = 1, 2$, in

$$S_1(z) M_2(z) + S_2(z) M_1(z) = 1.$$

For uniqueness of solution, impose the degree constraints:

$$\delta[S_i(z)] < \delta[M_i(z)], \quad i = 1, 2,$$

where $\delta[A(z)]$ denotes the degree of polynomial $A(z)$. It may be verified that

$$S_1(z) = \tfrac{1}{3}, \quad S_2(z) = -\tfrac{1}{3}z - \tfrac{2}{3}.$$

Applying the CRT for polynomials,

$$
\begin{aligned}
Y(z) &\triangleq y_0 + y_1 z + y_2 z^2 \\
&= \tfrac{1}{3}(z^2 + z + 1)m_1 + \left(-\tfrac{1}{3}z^2 - \tfrac{1}{3}z + \tfrac{2}{3}\right) \\
&\quad \times \left[(m_2 - m_4) + (m_3 - m_2 - 2m_4)z\right]\left[\bmod\left(z^3 - 1\right)\right] \\
&= \tfrac{1}{3}(z^2 + z + 1)m_1 + \tfrac{2}{3}(m_2 - m_4) - \tfrac{1}{3}(m_2 - m_4)z \\
&\quad + \tfrac{2}{3}(m_3 - m_2 - 2m_4)z - \tfrac{1}{3}(m_3 - m_2 - 2m_4)z^2 \\
&\quad - \tfrac{1}{3}(m_2 - m_4)z^2 - \tfrac{1}{3}(m_3 - m_2 - 2m_4).
\end{aligned}
$$

Comparing coefficients of respective powers of z,

$$
\begin{aligned}
y_0 &= \frac{m_1 + 3m_2 - m_3}{3}, \\
y_1 &= \frac{m_1 - 3m_2 + 2m_3 - 3m_4}{3}, \\
y_2 &= \frac{m_1 - m_3 + 3m_4}{3}.
\end{aligned}
\tag{2.65}
$$

The number of multiplications required is four, in agreement with the implication of Theorem 2.8.

Example 2.13. It is required to compute the cyclic convolution of two real sequences

$$(h_0 \quad h_1 \quad h_2 \quad h_3) \quad \text{and} \quad (x_0 \quad x_1 \quad x_2 \quad x_3)$$

by constructing an algorithm of minimal multiplicative complexity. Define

$$
\begin{aligned}
H(z) &\triangleq h_0 + h_1 z + h_2 z^2 + h_3 z^3, \\
X(z) &\triangleq x_0 + x_1 z + x_2 z^2 + x_3 z^3.
\end{aligned}
$$

The objective is to compute

$$Y(z) = H(z)X(z) \quad \bmod\left(z^4 - 1\right),$$

using the minimum number of multiplications. Over the field of rational numbers, $(z^4 - 1)$ factors into the product of three cyclotomic polynomials:

$$(z^4 - 1) = (z - 1)(z + 1)(z^2 + 1),$$
$$H_1(z) \triangleq H(z) \mod (z - 1) = h_0 + h_1 + h_2 + h_3,$$
$$X_1(z) \triangleq X(z) \mod (z - 1) = x_0 + x_1 + x_2 + x_3,$$
$$H_2(z) \triangleq H(z) \mod (z + 1) = h_0 - h_1 + h_2 - h_3,$$
$$X_2(z) \triangleq X(z) \mod (z + 1) = x_0 - x_1 + x_2 - x_3,$$
$$H_3(z) \triangleq H(z) \mod (z^2 + 1) = h_0 - h_2 + (h_1 - h_3)z,$$
$$X_3(z) \triangleq X(z) \mod (z^2 + 1) = x_0 - x_2 + (x_1 - x_3)z.$$

The products $H_i(z)X_i(z) \mod M_i(z)$, $i = 1, 2, 3$, have to be computed where $M_1(z) = (z - 1)$, $M_2(z) = (z + 1)$, $M_3(z) = (z^2 + 1)$. Define

$$(h_0 + h_1 + h_2 + h_3)(x_0 + x_1 + x_2 + x_3) \triangleq m_1,$$
$$(h_0 - h_1 + h_2 - h_3)(x_0 - x_1 + x_2 - x_3) \triangleq m_2,$$
$$(h_0 - h_2)(x_0 - x_2) \triangleq m_3, \qquad (2.66)$$
$$[(h_0 - h_2) + (h_1 - h_3)][(x_0 - x_2) + (x_1 - x_3)] \triangleq m_4,$$
$$(h_1 - h_3)(x_1 - x_3) \triangleq m_5.$$

Then,

$$Y_1(z) \triangleq H_1(z)X_1(z) \mod (z - 1) = m_1,$$
$$Y_2(z) \triangleq H_2(z)X_2(z) \mod (z + 1) = m_2,$$
$$Y_3(z) \triangleq H_3(z)X_3(z) \mod (z^2 + 1) = m_3 - m_5 + (m_4 - m_3 - m_5)z.$$

Before applying the CRT, it is required to compute via Euclid's division algorithm or otherwise the polynomial $S_i(z)$, $i = 1, 2, 3$, in

$$S_1(z)M_2(z)M_3(z) + T_1(z)M_1(z) = 1,$$
$$S_2(z)M_1(z)M_3(z) + T_2(z)M_2(z) = 1,$$
$$S_3(z)M_1(z)M_2(z) + T_3(z)M_3(z) = 1.$$

To solve for $S_1(z)$ and $T_1(z)$ uniquely (up to constants) the degree constraints below are imposed:

$$\delta[T_1(z)] < \delta[M_2(z)M_3(z)],$$
$$\delta[S_1(z)] < \delta[M_1(z)],$$

where $\delta[A(z)]$ denotes the degree of polynomial $A(z)$. Since

$$M_2(z)M_3(z) = z^3 + z^2 + z + 1,$$
$$M_1(z) = z - 1,$$

let

$$S_1(z) = a^1,$$

$$T_1(z) = b^1 z^2 + c^1 z + d^1,$$

where a^1, b^1, c^1, and d^1 are constants. Solving for a^1, one gets

$$S_1(z) = a^1 = \tfrac{1}{4}.$$

Similarly, it may be verified that

$$S_2(z) = -\tfrac{1}{4},$$

$$S_3(z) = -\tfrac{1}{2}.$$

Applying the CRT for polynomials,

$$\begin{aligned}
Y(z) &= Y_1(z)S_1(z)M_2(z)M_3(z) + Y_2(z)S_2(z)M_1(z)M_3(z) \\
&\quad + Y_3(z)S_3(z)M_1(z)M_2(z) \\
&= \tfrac{1}{4}(z^3 + z^2 + z + 1)m_1 - \tfrac{1}{4}(z^3 - z^2 + z - 1)m_2 \\
&\quad - \tfrac{1}{2}(z^2 - 1)\left[(m_3 - m_5) + (m_4 - m_3 - m_5)z\right].
\end{aligned}$$

If the cyclically convolved sequence is $(y_0 \ y_1 \ y_2 \ y_3)$,

$$Y(z) = y_0 + y_1 z + y_2 z^2 + y_3 z^3.$$

Therefore,

$$\begin{aligned}
y_0 &= \frac{m_1 + m_2}{4} + \frac{m_3 - m_5}{2}, \\
y_1 &= \frac{m_1 - m_2}{4} + \frac{m_4 - m_3 - m_5}{2}, \\
y_2 &= \frac{m_1 + m_2}{4} - \frac{m_3 - m_5}{2}, \\
y_3 &= \frac{m_1 - m_2}{4} - \frac{m_4 - m_3 - m_5}{2}.
\end{aligned} \tag{2.67}$$

The minimal algorithm constructed requires $(2 \times 4) - 3 = 5$ multiplications, consistent with the result stated in Theorem 2.8.

It is of educational value to note that a minimal multiplication algorithm for computation of a circular convolution of two sequences, each of length n, can be written in the form [11]

$$\mathbf{y} = C\left[(A\mathbf{h}) \odot (B\mathbf{x})\right], \tag{2.68}$$

where $\mathbf{y} = (y_0 \ y_1 \ \cdots \ y_{n-1})^t$ contains the elements of the convolved sequence, while

$$\mathbf{h} \triangleq (h_0 \ \ h_1 \ \ \cdots \ \ h_{n-1})^t,$$

$$\mathbf{x} \triangleq (x_0 \ \ x_1 \ \ \cdots \ \ x_{n-1})^t$$

contain the elements of the two input sequences. A, B, and C are matrices of orders $M \times n$, $M \times n$, and $n \times M$, respectively, where $M = 2n - k$ and k is the number of cyclotomic polynomials whose product is $z^n - 1$. Also, the symbol $\odot$ denotes the vector formed by taking the element by element products of the $(M \times 1)$ vectors $A\mathbf{h}$ and $B\mathbf{x}$. The reader should be able to understand how the matrices A, B, and C are formed after studying the illustrative examples.

Example 2.14. Construct matrices A, B, and C which yield the algorithm for computation of the three-point convolution in Example 2.12. From (2.64)

$$A = \begin{bmatrix} 1 & 1 & 1 \\ 1 & 0 & -1 \\ 1 & 1 & -2 \\ 0 & 1 & -1 \end{bmatrix}, \quad B = \begin{bmatrix} 1 & 1 & 1 \\ 1 & 0 & -1 \\ 1 & 1 & -2 \\ 0 & 1 & -1 \end{bmatrix}.$$

From (2.65), and the defined values of m_i, $i = 1, 2, 3, 4$, in (2.64)

$$C = \frac{1}{3} \begin{bmatrix} 1 & 3 & -1 & 0 \\ 1 & -3 & 2 & -3 \\ 1 & 0 & -1 & 3 \end{bmatrix}.$$

By redefining terms, the factor $\frac{1}{3}$ in the matrix C could be absorbed in either matrix A or matrix B.

Example 2.15. Construct matrices A, B, and C which yield the algorithm for computation of the four-point convolution in Example 2.13. From (2.66)

$$A = B = \begin{bmatrix} 1 & 1 & 1 & 1 \\ 1 & -1 & 1 & -1 \\ 1 & 0 & -1 & 0 \\ 1 & 1 & -1 & -1 \\ 1 & 0 & -1 & 0 \end{bmatrix}.$$

Finally, from (2.66) and (2.67)

$$C = \frac{1}{4} \begin{bmatrix} 1 & 1 & 0 & 2 & -2 \\ 1 & -1 & -2 & 2 & -2 \\ 1 & 1 & -2 & 0 & 2 \\ 1 & -1 & 2 & -2 & 2 \end{bmatrix}.$$

A few comments are in order. The preceding two examples show how algorithms can be constructed to implement cyclic convolutions of lengths 3 and 4 using the minimum number of multiplications. Of course, multiplications involving the elements of the chosen field (in this case, the field of rational numbers) are not counted. The reader can attempt to construct a minimal algorithm on sequences of length 5 (see Problem 13). This problem is computationally more tedious, but the manipulations done in Example 2.11 should be helpful for this purpose. The number of additions in the minimal algorithm might turn out to be high, and an algorithm of a subminimal nature with more multiplications than the theoretical minimum might be useful to develop so that the number of additions are reduced. For sequences of length 5, for example, the reader might want to see how effectively the number of additions are controlled when the number of multiplications are allowed to increase from 8 [the theoretical minimum = $(2 \times 5) - 2$, according to the Theorem 2.8] to, say, 10. For long sequences the minimal algorithms can become quite complicated. In those cases it is expedient to use optimal algorithms for short convolutions to get a suboptimal algorithm for long convolutions, provided the length n of the long sequence is factorable into a product of small mutually prime factors. This strategy, of course, will not work when n is a large prime. The artifice underlying the strategy, when it works, is a one-dimensional (1-D) to a multidimensional index mapping. Since this index mapping is also very important in the computation of discrete Fourier transform, it will be summarized in the theorem given next.

Theorem 2.9. *Let the integer n be factorable into a product of k mutually prime integers*:

$$n = n_1 n_2 \ldots n_k.$$

The multidimensional index mapping is defined by

$$i_j \equiv i \quad \mathrm{mod}\, n_j, \quad 0 \le i_j < n_j, \quad j = 1, 2, \ldots, k$$

for each i which is an element in the set $(0, 1, 2, \ldots, n-1)$. Let

$$(n/n_j)s_j \equiv 1 \quad \mathrm{mod}\, n_j, \quad j = 1, 2, \ldots, k$$

[s_j is possible to construct by Fact 2.2 and the extended Euclid's algorithm in Section 2.3.1 since (n/n_j) and n_j are relatively prime]. The inverse mapping is uniquely defined by

$$i \equiv \sum_{j=1}^{k} \left(\frac{n}{n_j} \right) i_j s_j \quad \mathrm{mod}\, n. \qquad (2.69)$$

The proof of the preceding theorem follows easily from the CRT over integers. It is interesting to note [1, p. 11] that (2.69) may be replaced by

$$i \equiv \sum_{j=1}^{k} \left(\frac{n}{n_j} \right) i_j \quad \mathrm{mod}\, n. \qquad (2.70)$$

Both (2.69) and (2.70) will span the complete set of values for i in $(0, 1, 2, \ldots, n-1)$, although in a different order. Note, however, that (2.70) does not require the computation of s_j's.

The ideas behind the computation of a long circular convolution from optimal algorithms available for short convolutions will be illustrated for the 2-factor case. Let the integer n be factorable as a product, $n_1 n_2$, where n_1 and n_2 are relatively prime integers. In order to find the circular convolution $\{y_k\}$ of two sequences $\{x_k\}$ and $\{h_k\}$, defined in (2.62), it is first necessary to obtain the one-to-one index mapping

$$i \rightarrow (i_1, i_2), \qquad (2.71)$$

where i_1 and i_2 are defined by the congruence relation in Theorem 2.9 with $k = 2$ in the theorem. Thus, one-dimensional (1-D) vectors $\mathbf{y}$, $\mathbf{x}$, and $\mathbf{h}$, occurring in (2.68) may be mapped onto the respective 2-D arrays, Y, X, and H. Then the convolution in Fact 2.5 can be written as

$$y_{i_1, i_2} = \sum_{j_2 = 0}^{n_2 - 1} \sum_{j_1 = 0}^{n_1 - 1} h_{\langle i_1 - j_1 \rangle, \langle i_2 - j_2 \rangle} x_{j_1, j_2}. \qquad (2.72)$$

The inner summation may be viewed as a circular convolution of column $\langle i_2 - j_2 \rangle$ of the array H with column j_2 of the array X, for each fixed i_2 and j_2. Each of these n_1-point convolutions are computable via the minimal algorithm, assumed to be available. This is followed by a suitable readjustment of terms which then allows the implementation of the minimal algorithm, again assumed available, to implement n_2-point convolutions. Let the counterparts of matrices A, B, and C for n-point convolutions in (2.68) be associated with n_1-point and n_2-point convolutions. Denote these sets of matrices by $\{A_1, B_1, C_1\}$ and $\{A_2, B_2, C_2\}$, respectively. Agarwal and Cooley [12] obtained an expression for computing long convolutions in terms of short convolutions. For the 2-factor case under consideration, their results are summarized in the next theorem. For the sake of brevity, the notations introduced in the preceding discussion will not be repeated.

Theorem 2.10. *The n-point circular convolution of two known sequences* x_k *and* h_k *is computable from*

$$y = C_1 \textcircled{r} C_2 \textcircled{c} \left[\left(A_2 \textcircled{r} \left(A_1 \textcircled{c} H \right) \right) \boxdot \left(B_2 \textcircled{r} \left(B_1 \textcircled{c} X \right) \right) \right],$$

where matrices A_2, B_2, *and* C_2 *are of order* $M_2 \times n_2$, $M_2 \times n_2$, *and* $n_2 \times M_2$ *while matrices* A_1, B_1, *and* C_1 *are of order* $M_1 \times n_1$, $M_1 \times n_1$, *and* $n_1 \times M_1$ *[these being the counterparts of matrices* A, B, *and* C *in (2.68)].* M_1 *and* M_2 *are, respectively, the minimum number of multiplications required to compute* n_1- *and* n_2-*point circular convolutions where* n_1 *and* n_2 *are relatively prime integers in the factorization* $n = n_1 n_2$. *Furthermore,* $A_1 \textcircled{c} H$ *denotes the computation of the transform* A_1 *of columns of* H *while* $A_2 \textcircled{r} (A_1 \textcircled{c} H)$ *denotes the computation of the transform* A_2 *of the rows of the result of* $A_1 \textcircled{c} H$. $A \boxdot B$ *is the matrix obtained by taking the element by element product of matrices* A *and* B.

The next example will illustrate the application of the preceding theorem. In order to be able to use the above theorem, the reader should find this example very helpful.

Example 2.16. It is required to apply Theorem 2.10 to the computation of a 6-point circular convolution.

Here $n = 6 = 3 \times 2$. Therefore, $n_1 = 2$ and $n_2 = 3$. The index mapping in (2.71) with $i_1 = i \mod 2$ and $i_2 = i \mod 3$ yields in this case

$$
\begin{aligned}
0 &\to (0,0), \quad 3 \to (1,0), \\
1 &\to (1,1), \quad 4 \to (0,1), \\
2 &\to (0,2), \quad 5 \to (1,2).
\end{aligned}
\tag{2.73}
$$

Then, if

$$
\begin{aligned}
\mathbf{h} &= (h_0 \quad h_1 \quad h_2 \quad h_3 \quad h_4 \quad h_5)^t, \\
\mathbf{x} &= (x_0 \quad x_1 \quad x_2 \quad x_3 \quad x_4 \quad x_5)^t,
\end{aligned}
\tag{2.74}
$$

the one-to-one mapping in (2.73) yields the 2-D arrays H and U

$$
H = \begin{bmatrix} h_0 & h_4 & h_2 \\ h_3 & h_1 & h_5 \end{bmatrix}, \quad X = \begin{bmatrix} x_0 & x_4 & x_2 \\ x_3 & x_1 & x_5 \end{bmatrix}.
\tag{2.75}
$$

On applying the algorithm of Problem 15 (the factor of $\frac{1}{2}$ is absorbed in matrix A_2 and not C_2),

$$
A_1 = \begin{bmatrix} \frac{1}{2} & \frac{1}{2} \\ \frac{1}{2} & -\frac{1}{2} \end{bmatrix}, \quad B_1 = \begin{bmatrix} 1 & 1 \\ 1 & -1 \end{bmatrix}, \quad C_1 = \begin{bmatrix} 1 & 1 \\ 1 & -1 \end{bmatrix}.
$$

On using the matrices in Problem 16 at the end of this chapter to compute a three-point circular convolution,

$$
A_2 = \frac{1}{3} \begin{bmatrix} 1 & 1 & 1 \\ 3 & 0 & -3 \\ 0 & 3 & -3 \\ 1 & 1 & -2 \end{bmatrix}, \quad B_2 = \begin{bmatrix} 1 & 1 & 1 \\ 1 & 0 & -1 \\ 0 & 1 & -1 \\ 1 & 1 & -2 \end{bmatrix},
$$

$$
C_2 = \begin{bmatrix} 1 & 1 & 0 & -1 \\ 1 & -1 & -1 & 2 \\ 1 & 0 & 1 & -1 \end{bmatrix}.
$$

Only for the sake of brevity in exposition the column matrices $\mathbf{h}$ and $\mathbf{x}$ are specialized to

$$
\begin{aligned}
\mathbf{h} &= (1 \quad -1 \quad \tfrac{1}{2} \quad 0 \quad 2 \quad -\tfrac{1}{2})^t, \\
\mathbf{x} &= (1 \quad 2 \quad -1 \quad -2 \quad 0 \quad 1)^t.
\end{aligned}
\tag{2.76}
$$

The calculations below are shown with respect to these specializations; however, the reader could easily do parallel calculations when the elements of **h** and **x** are indeterminates as in (2.74). On substituting the specializations from (2.76) in (2.75),

$$H = \begin{bmatrix} 1 & 2 & \frac{1}{2} \\ 0 & -1 & -\frac{1}{2} \end{bmatrix}, \quad X = \begin{bmatrix} 1 & 0 & -1 \\ -2 & 2 & 1 \end{bmatrix}. \tag{2.77}$$

Then the operation of transforming the columns of H is denoted by

$$A_1 \text{ⓒ} H = \frac{1}{2} \begin{bmatrix} 1 & 1 & 0 \\ 1 & 3 & 1 \end{bmatrix}.$$

The operation of transforming the rows of $A_1 \text{ⓒ} H$ is denoted by $A_2 \text{ⓡ} A_1 \text{ⓒ} H$, and in this case,

$$A_2 \text{ⓡ} A_1 \text{ⓒ} H = \frac{1}{6} \begin{bmatrix} 1 & 1 & 1 \\ 3 & 0 & -3 \\ 0 & 3 & -3 \\ 1 & 1 & -2 \end{bmatrix} \begin{bmatrix} 1 & 1 \\ 1 & 3 \\ 0 & 1 \end{bmatrix} = \frac{1}{6} \begin{bmatrix} 2 & 5 \\ 3 & 0 \\ 3 & 6 \\ 2 & 2 \end{bmatrix}.$$

Similarly,

$$B_1 \text{ⓒ} X = \begin{bmatrix} -1 & 2 & 0 \\ 3 & -2 & -2 \end{bmatrix},$$

$$B_2 \text{ⓡ} B_1 \text{ⓒ} X = \begin{bmatrix} 1 & 1 & 1 \\ 1 & 0 & -1 \\ 0 & 1 & -1 \\ 1 & 1 & -2 \end{bmatrix} \begin{bmatrix} -1 & 3 \\ 2 & -2 \\ 0 & -2 \end{bmatrix} = \begin{bmatrix} 1 & -1 \\ -1 & 5 \\ 2 & 0 \\ 1 & 5 \end{bmatrix}.$$

Then,

$$\left[A_2 \text{ⓡ} \left(A_1 \text{ⓒ} H \right) \right] \boxed{\cdot} \left[B_2 \text{ⓡ} \left(B_1 \text{ⓒ} X \right) \right] = \begin{bmatrix} \frac{1}{3} & -\frac{5}{6} \\ -\frac{1}{2} & 0 \\ 1 & 0 \\ \frac{1}{3} & \frac{5}{3} \end{bmatrix}.$$

Subsequently,

$$C_2 \text{ⓒ} \left[\left(A_2 \text{ⓡ} \left(A_1 \text{ⓒ} H \right) \right) \boxed{\cdot} \left(B_2 \text{ⓡ} \left(B_1 \text{ⓒ} X \right) \right) \right] = \begin{bmatrix} -\frac{1}{2} & -\frac{5}{2} \\ \frac{1}{2} & \frac{5}{2} \\ 1 & -\frac{5}{2} \end{bmatrix}$$

and

$$C_1 \text{ⓡ} C_2 \text{ⓒ} \left[\left(A_2 \text{ⓡ} \left(A_1 \text{ⓒ} H \right) \right) \boxed{\cdot} \left(B_2 \text{ⓡ} \left(B_1 \text{ⓒ} X \right) \right) \right]$$

$$= \begin{bmatrix} -3 & 3 & -\frac{3}{2} \\ 2 & -2 & \frac{3}{2} \end{bmatrix}.$$

Thus,

$$Y = \begin{bmatrix} y_0 & y_4 & y_2 \\ y_3 & y_1 & y_5 \end{bmatrix} = \begin{bmatrix} -3 & 3 & -\frac{3}{2} \\ 2 & -2 & \frac{3}{2} \end{bmatrix}$$

Therefore, the convolved output sequence is

$$\{y_k\} = (y_0 \quad y_1 \quad y_2 \quad y_3 \quad y_4 \quad y_5)$$
$$= (-3 \quad -2 \quad -\tfrac{3}{2} \quad 2 \quad 3 \quad \tfrac{3}{2}).$$

The reader can verify by straightforward calculation that the output sequences arrived at via application of Theorem 2.10 is, indeed, correct. Note that the number of multiplications required if the elements of $\{h_k\}$ and $\{x_k\}$ were indeterminates is equal to the product of the minimal number of multiplications necessary to compute a 3-point circular convolution and a 2-point circular convolution.

2.5.3. Cyclic (Circular) Convolution and DFT

Fast algorithms developed for the computation of cyclic convolution are also useful in the computation of DFT. This is based on an important concept which allows the conversion of a DFT to a circular convolution, when the length of the DFT satisfies certain conditions. Rader [9] was the first to show that when the length of the transform is a prime number, the N-point DFT contains an $(N-1)$-point circular convolution. This interesting observation follows from the following mathematical fact.

Fact 2.6. *If N is a positive prime integer, then the set of elements $(0,1,\ldots, N-1)$ forms a field, and there exists an integer g for which the sequence $\{g^r \mod N\}_{r=1}^{N-1}$ is a permutation of the sequence $(1,2,\ldots,N-1)$. Also, $g^{N-1} \mod N = 1$, g is a primitive root of N, and there are $\phi(N-1)$ such roots.*

The fact that when N is prime the set of elements, $Z_N \triangleq (0,1,\ldots, N-1)$ forms a finite field is well known. It is the existence of a "primitive root" g, not necessarily unique, that is of interest in the present context.

Example 2.17. Consider the set of integers,

$$Z_7 = (0,1,2,\ldots,5,6).$$

For this case $g = 3$ is a primitive root and

$$\{g^r \mod 7\}_{r=1}^{6} = (3,2,6,4,5,1).$$

Fact 2.6 is the special case of a result in number theory that every finite field has a primitive root, an element whose powers generate all the elements of the field except zero. Before proving Rader's result relating the N-point DFT to a circular convolution, another well-known result in number theory, referred to as Euler's theorem, is stated without proof.

Theorem 2.11 (Euler). *Given two relatively prime integers m and n (> 0),*

$$m^{\phi(n)} \equiv 1 \mod n,$$

where $\phi(n)$ is the Euler totient function.

From Euler's theorem it follows that the solution to the equation

$$mx \equiv 1 \quad \mathrm{mod}\, n, \qquad (2.78a)$$

when m and n are relatively prime positive integers is

$$x = \left[m^{\phi(n)-1} \right] \quad \mathrm{mod}\, n. \qquad (2.78b)$$

Theorem 2.12 (Rader [9]). *For a prime N, an N-point DFT is essentially a circular convolution.*

PROOF: Let

$$\{ X(m) \} \leftrightarrow \{ x(n) \}$$

be an N-point DFT pair. Then,

$$X(0) = \sum_{i=0}^{N-1} x(i), \qquad (2.79a)$$

$$X(m) = \sum_{i=1}^{N-1} x(i) w_N^{im} + x(0), \quad m = 1, 2, \ldots, N-1, \qquad (2.79b)$$

where $w_N = e^{-j2\pi/N}$. By Fact 2.6, there exists a primitive root g that allows the index mapping

$$i \rightarrow \left(g^i \quad \mathrm{mod}\, N \right) \triangleq \langle g^i \rangle,$$
$$k \rightarrow \left(g^k \quad \mathrm{mod}\, N \right) \triangleq \langle g^k \rangle. \qquad (2.80)$$

Then, from (2.79b) and (2.80)

$$X\left[\langle g^k \rangle \right] = \sum_{i=0}^{N-2} x\left[\langle g^i \rangle \right] w_N^{\langle g^{(i+k)} \rangle} + x(0), \qquad (2.81a)$$

$k = 0, 1, \ldots, N-2$, since the relation between the new index k and the old index m in (2.79b) is described by

$$m = g^k \quad \mathrm{mod}\, N,$$

$$m = 1, 2, \ldots, N-1 \quad \text{and} \quad k = 0, 1, 2, \ldots, N-2,$$

where

$$g^k \neq 1, \quad 0 < k < N-1,$$

$$g^{N-1} = g^0 = 1$$

Equation (2.81a) is a circular correlation, and it can be easily converted into a circular convolution by defining $\langle g^{-i} \rangle$ as $\langle g^{\phi(N)-i} \rangle$ and making the substitution

$$i \rightarrow \langle g^{-i} \rangle$$

instead of

$$i \rightarrow \langle g^i \rangle$$

in (2.80), so that

$$X\left[\langle g^k\rangle\right] - x(0) = \sum_{i=0}^{N-2} x\left[\langle g^{-i}\rangle\right] w_N^{\langle g^{k-i}\rangle}, \quad k = 0,1,\ldots, N-2,$$

(2.81b)

since $\langle g^{N-1}\rangle = \langle g^0\rangle$. Note that $X(0)$ in (2.79a) may be computed separately, while $X[\langle g^k\rangle]$ is easily obtainable, after $X[\langle g^k\rangle] - x(0)$, $k = 0,1,\ldots, N-2$, has been computed as an $(N-1)$-point circular convolution.

Example 2.18. Let $N = 5$, so that

$$X(0) = \sum_{i=0}^{4} x(i),$$

$$X(k) = \sum_{i=0}^{4} x(i)w_5^{ki}, \quad k = 1,2,3,4.$$

For the sake of brevity, redefine in this problem

$$w_5 = e^{-j2\pi/5} \triangleq w.$$

The primitive roots (see Fact 2.6) nonunique in this case are 2 and 3. Take $g = 2$. Then, from (2.81b),

$$X\left[\langle 2^k\rangle\right] - x(0) \triangleq \hat{X}\left[\langle 2^k\rangle\right] = \sum_{i=0}^{3} x\left[\langle 2^{-i}\rangle\right] w^{\langle 2^{(k-i)}\rangle}$$

for $k = 0,1,2,3$. Here $\langle 2^k\rangle = 2^k \mod 5$. Noting that $\phi(5) = 5 - 1 = 4$, it follows that [from (2.78b)]

$$\langle 2^{-1}\rangle = \langle 2^{4-1}\rangle = \langle 2^3\rangle = 3,$$

$$\langle 2^{-2}\rangle = \langle 2^{4-2}\rangle = \langle 2^2\rangle = 4,$$

$$\langle 2^{-3}\rangle = \langle 2^{4-3}\rangle = 2^1 = 2.$$

Using the fact that $\langle 2^0\rangle = 1$, $\langle 2^1\rangle = 2$, $\langle 2^2\rangle = 4$, $\langle 2^3\rangle = 3$, it is easy to verify that

$$\begin{bmatrix} \hat{X}(1) \\ \hat{X}(2) \\ \hat{X}(4) \\ \hat{X}(3) \end{bmatrix} = \begin{bmatrix} w^1 & w^3 & w^4 & w^2 \\ w^2 & w^1 & w^3 & w^4 \\ w^4 & w^2 & w^1 & w^3 \\ w^3 & w^4 & w^2 & w^1 \end{bmatrix} \begin{bmatrix} x(1) \\ x(3) \\ x(4) \\ x(2) \end{bmatrix}.$$

(2.82)

So far, the computation of an N-point DFT has been equated to the computation of a circular convolution, when N is a prime. This is possible because every prime has primitive roots [$\phi(N-1)$ in number] and each such primitive root g generates a sequence, $(\langle g^1 \rangle, \langle g^2 \rangle, \langle g^3 \rangle, \ldots, \langle g^{N-1} \rangle)$ (where $\langle g^r \rangle \equiv g^r \mod N$), which is a permutation of the sequence $(1, 2, \ldots, N-1)$. [Remember that this sequence is of length $\phi(N)$ when g is a primitive root and $\phi(N) = N - 1$ for N prime.] Primitive roots exist not only for primes but also for certain other classes of integers as stated next.

Fact 2.7. *The numbers having primitive roots are* 2, 4, n^r, *and* $2n^r$, *where r is any positive integer and n is a positive odd prime* (*i.e., any positive prime excluding* 2). *If the positive integer N has a primitive root, it has $\phi[\phi(N)]$ of those.*

Winograd showed that the DFT may also be equated to a convolution when the number N of discrete sample points is an odd prime power, i.e., $N = n^r$ for a prime $n \neq 2$. Specifically, he showed that this problem may be equated to the problem of computing a circular convolution of length $n^{r-1}(n-1)$ in conjunction with the problem of computing two DFT's, each of length n^{r-1}. The circular convolution of length $n^{r-1}(n-1)$ is done following the removal of all integers which contain a factor n from the index set $(1, 2, \ldots, n^r - 1)$; the set of remaining integers has $n^{r-1}(n-1)$ elements. This set of integers is isomorphic to the set $(0, 1, 2, \ldots, n^r - n^{r-1} - 1)$, which forms an additive group under the operation of addition modulo the integer $(n^r - n^{r-1})$. These $(n^r - n^{r-1})$ elements in the set of remaining integers, referred to above (after eliminating the integers which contain a factor n), are a permutation of the set generated by the powers g^k of a primitive root g, known to exist, of $N = n^r$, for $k = 0, 1, 2, \ldots, n^r - n^{r-1} - 1$, when taken modulo $(n^r - n^{r-1})$. This permutation makes feasible a circular convolution of length $(n^r - n^{r-1})$. Similarly, two DFT's each of length n^{r-1} can be equated to the computation of two $n^{r-2}(n-1)$-point circular convolutions and four DFT's, each of length $n^{r-3}(n-1)$. Repeating the procedure on these four DFT's, one can compute a n^r-point DFT for odd prime n by computing several circular convolutions as summarized next.

Fact 2.8. *The computation of an n^r-point DFT, where n is a positive odd prime and r an arbitrary but fixed positive integer, may be equated to the computation of 2^k circular convolutions, each of length $n^{r-k-1}(n-1)$, for $k = 0, 1, 2, \ldots, r - 1$.*

It should be noted that Fact 2.8 requires the use of minimal algorithms (from the standpoint of multiplicative complexity) developed for computing circular convolutions of shorter lengths than the original DFT.

Example 2.19. Let $N = 9 = 3^2$, so that $n = 3$ and $r = 2$. From Problem 21 at the end of this chapter, it can be concluded that $(z^6 - 1)$ is factorable as a product of four cyclotomic polynomials, each of which must be irreducible over the field of rationals. (Note that 1, 2, 3, and 6 are the positive integers less than or equal to 6 which divide 6). On applying Theorem 2.8, it is clear that the minimum number of multiplications required to compute a 6-point circular convolution is $(2 \times 6 - 4) = 8$. Similarly, $(z^2 - 1)$ is factorable as a product of two cyclotomic polynomials, and the number of multiplications necessary to compute a 2-point circular convolution is $(2 \times 2) - 2 = 2$. According to Fact 2.8, a 3^2-point DFT may be computed by computing one 6-point circular convolution and two 2-point circular convolutions. Therefore, $(8 + 2 \times 2) = 12$ multiplications will suffice to compute a 9-point DFT.

From the above discussion it becomes clear how optimal algorithms for short circular convolutions may be used to compute long DFT's whose lengths are of the form n^r. Long DFT's may also be computed by using a 1-D to k-D index mapping. The idea is to convert a 1-D length $N = n_1 n_2 \ldots n_k$ transform into a k-D transform which requires the computation of k shorter length n_r-point transforms, for $r = 1, 2, \ldots, k$. One way to implement the desired conversion is by an index mapping formula as in Theorem 2.9. Consider, for brevity the $k = 2$ case; i.e., let $N = n_1 n_2$, where n_1 and n_2 are relatively prime integers. Let

$$\{ X(k) \} \leftrightarrow \{ x(i) \}$$

be a DFT pair, so that

$$X(k) = \sum_{i=0}^{N-1} x(i) w_N^{ik}, \quad k = 0, 1, \ldots, N-1, \quad w_N \triangleq e^{-j2\pi/N}.$$

The objective is to establish an one-to-one onto map of the input index i into a pair of indices (i_1, i_2) and another similar type of map of the output index k into a pair of indices (k_1, k_2). For the output index, define

$$k_1 \triangleq k \mod n_1, \quad k_2 \triangleq k \mod n_2 \tag{2.83}$$

so that by CRT

$$
\begin{aligned}
k &= (n_2 s_1 k_1 + n_1 s_2 k_2) \mod N \\
&= (t_1 k_1 + t_2 k_2) \mod N,
\end{aligned}
\tag{2.84}
$$

where

$$t_1 \triangleq n_2 s_1, \quad t_2 \triangleq n_1 s_2,$$

$$t_1 \equiv 1 \mod n_1, \quad t_1 \equiv 0 \mod n_2,$$

$$t_2 \equiv 0 \mod n_1, \quad t_2 \equiv 1 \mod n_2. \tag{2.85}$$

Clearly, (2.84) is similar in structure to (2.69). The input index i may be mapped as above or use may be made of the following fact that leads to an inverse mapping similar in form to (2.70) as opposed to (2.69).

Fact 2.9. *The set of r distinct integers, $\{a_i\} = (0, 1, \ldots, r-1)$, may be uniquely permuted to the set of integers $\{b_i\}$ under the map*

$$b_i = ca_i \mod r$$

if and only if the integer c is chosen to be relatively prime to r.

Example 2.20. Let

$$\{a_i\} \triangleq (0, 1, 2, \ldots, 8, 9).$$

Then,

$$\{3a_i \mod 10\} = (0, 3, 6, 9, 2, 5, 8, 1, 4, 7)$$

is a permuted version of $\{a_i\}$ since 3 is relatively prime to 10, while

$$\{5a_i \mod 10\} = (0, 5, 0, 5, 0, 5, 0, 5, 0, 5)$$

is not a permuted version of $\{a_i\}$ because 5 and 10 are not mutually prime.

Since n_1 and n_2 are mutually prime in $N = n_1 n_2$, $c_1 \triangleq n_2 \mod n_1$ is relatively prime to n_1 while $c_2 \triangleq n_1 \mod n_2$ is relatively prime to n_2. For the index i, define

$$i_1 \triangleq c_1 \hat{i}_1 \mod n_1, \tag{2.86a}$$

$$i_2 \triangleq c_2 \hat{i}_2 \mod n_2, \tag{2.86b}$$

$$\hat{i}_1 \triangleq \hat{i} \mod n_1, \tag{2.86c}$$

$$\hat{i}_2 \triangleq \hat{i} \mod n_2, \tag{2.86d}$$

($\{\hat{i}\} \triangleq (0, 1, 2, \ldots, N-1)$ and $\{i\}$ below is a permuted version of $\{\hat{i}\}$) so that an element i in $\{i\}$,

$$i = \left(n_2 \hat{i}_1 + n_1 \hat{i}_2\right) \mod N, \tag{2.87}$$

is uniquely constructed from (i_1, i_2) [or $(\hat{i}_1, \hat{i}_2)$] specified in (2.86a) and (2.86b). Then, using (2.84) and (2.87),

$$ik = \left(t_1 k_1 + t_2 k_2\right)\left(n_2 \hat{i}_1 + n_1 \hat{i}_2\right) \mod N. \tag{2.88}$$

Letting $\langle x \rangle \triangleq x \mod N$, $w_{n_1} = e^{-j2\pi/n_1}$, $w_{n_2} = e^{-j2\pi/n_2}$, consistent with earlier notations, (2.88) leads to

$$w_N^{ik} = w_{n_1}^{\langle t_1 k_1 \hat{i}_1 \rangle} w_{n_1}^{\langle t_2 k_2 \hat{i}_1 \rangle} w_{n_2}^{\langle t_1 k_1 \hat{i}_2 \rangle} w_{n_2}^{\langle t_2 k_2 \hat{i}_2 \rangle}. \tag{2.89}$$

Recalling the definitions for t_1 and t_2, (2.89) can be simplified to

$$w_N^{ik} = w_{n_1}^{k_1 \hat{i}_1} w_{n_2}^{k_2 \hat{i}_2}. \tag{2.90}$$

Therefore, the 1-D DFT, $\{X(k)\}$, may be computed after computing

$$X(k_1, k_2) = \sum_{i_1=0}^{n_1-1} \sum_{i_2=0}^{n_2-1} x(i_1, i_2) w_{n_1}^{k_1 i_1} w_{n_2}^{k_2 i_2},$$

$$k_1 = 0, 1, \ldots, n_1 - 1, \quad k_2 = 0, 1, \ldots, n_2 - 1.$$

Example 2.21. Let $N = 15 = 5 \times 3$, so that $n_1 = 5$ and $n_2 = 3$ are relatively prime. The mapping between the output index k and indices (k_1, k_2), as defined by (2.83) and (2.84), yields the following isomorphism:

k	(k_1, k_2)	k	(k_1, k_2)
0	(0,0)	8	(3,2)
1	(1,1)	9	(4,0)
2	(2,2)	10	(0,1)
3	(3,0)	11	(1,2)
4	(4,1)	12	(2,0)
5	(0,2)	13	(3,1)
6	(1,0)	14	(4,2)
7	(2,1)		

The mapping between the input index i in the sequence $\{i\}$ (or $\hat{i}$ in the sequence $\{\hat{i}\}$, a permuted version of $\{i\}$) and indices $(\hat{i}_1, \hat{i}_2)$ as well as (i_1, i_2) defined in (2.86) and (2.87) leads to the following isomorphisms (note that $c_1 = 3 \mod 5 = 3$, $c_2 = 5 \mod 3 = 2$).

i	$(\hat{i}_1, \hat{i}_2)$	(i_1, i_2)	i
0	(0,0)	(0,0)	0
1	(1,1)	(3,2)	8
2	(2,2)	(1,1)	1
3	(3,0)	(4,0)	9
4	(4,1)	(2,2)	2
5	(0,2)	(0,1)	10
6	(1,0)	(3,0)	3
7	(2,1)	(1,2)	11
8	(3,2)	(4,1)	4
9	(4,0)	(2,0)	12
10	(0,1)	(0,2)	5
11	(1,2)	(3,1)	13
12	(2,0)	(1,0)	6
13	(3,1)	(4,2)	14
14	(4,2)	(2,1)	7

Note that the column under $(\hat{i}_1, \hat{i}_2)$ is generated from a mapping identical to that used between k and (k_1, k_2).

2.5.4. Systolic Arrays for Hardware Implementation

The advent of the very large scale integrated (VLSI) circuit era motivated research into the design of algorithms suitable for direct hardware implementation. Systolic systems provide a unified model of parallel VLSI computation, pipelining, and interconnection structures in order to ensure high throughput and speedy response. The nucleus of a systolic system is comprised of processors, which receive, compute, and propagate multiple streams of data throughout the system. Systolic arrays which have only nearest-neighbor connections are very attractive for VLSI because of insignificant propagation delay. The problem of an N-point DFT computation in (2.4) can transform to a problem of matrix-vector multiplication,

$$W\mathbf{y} = \mathbf{Y},$$

where $\mathbf{y} = [y(0)\ y(1)\ \dots\ y(N-1)]'$, $\mathbf{Y} = [Y(0)\ Y(1)\ \dots\ Y(N-1)]'$, and W is an $N \times N$ matrix having, in its ith row and jth column, the entry $w_N^{(i-1)(j-1)}$, $i = 1, 2, \dots, N$ and $j = 1, 2, \dots, N$. The problem of computing the linear convolution in (2.52) can also be translated to a problem of matrix-vector multiplication. It has been shown [27] that, using a linearly connected systolic array of N processors, both the convolution of two N-point sequences and the N-point DFT can be computed in only $O(N)$ units of time rather than $O(N \log N)$ units of time as required by the sequential FFT algorithm.

2.6. Number Theoretic Transforms

A number theoretic transform (NTT) has the same structure as the DFT but with the complex exponential $w_N = e^{-j2\pi/N}$ replaced by an integer g_N and with all operations performed modulo an integer q, which may be prime or composite. However, very little is gained by working with a composite q. To keep the discussion within reasonable bounds and at the same time convey to the reader the essential ingredients of NTT techniques, the presentation is centered around the Mersenne and Fermat primes as choices for q. An NTT pair, $\{X(k) \triangleq X_k\} \leftrightarrow \{x(k) \triangleq x_k\}$, where each sequence (i.e., $\{X_k\}_{k=0}^{N-1}$, $\{x\}_{k=0}^{N-1}$) is a sequence of integers of length N, is related by

$$X_k = \sum_{r=0}^{N-1} x_r g_N^{kr} \mod q, \quad k = 0, 1, \dots, N-1, \tag{2.91a}$$

$$x_r = N^{-1} \sum_{k=0}^{N-1} X_k g_N^{-kr} \mod q, \quad r = 0, 1, \dots, N-1, \tag{2.91b}$$

where

$$NN^{-1} \equiv 1 \mod q. \tag{2.91c}$$

[Since q is a prime in our discussion here, N has an "inverse" $N^{-1} \mod q$ as shown in (2.91c); for the same reason g_N, which must be an integer that is a root of 1 of order $N \mod q$, i.e., $g_N^N \equiv 1 \mod q$, must have an inverse, $g_N^{-1} \mod q$, which justifies the occurrence of negative powers in (2.91b).] Clearly, $g_N^{-1} \mod q = g_N^{N-1} \mod q$. The Mersenne transform is considered next.

2.6.1. Mersenne Transforms (MT)

Definition 2.7. The numbers $M_N = 2^N - 1$, where N is any positive integer, are referred to as Mersenne numbers. A prime M_N is referred to as a Mersenne prime.

Some examples of Mersenne primes are M_2, M_3, M_5, M_7, M_{13}, M_{17}, M_{19}, M_{31}, M_{61}, M_{67}, M_{89}, M_{107}, M_{127}, and M_{257}. Very recently (*Time Magazine*, Feb. 13, 1984, p. 47) scientists at Sandia Laboratories in Albuquerque, New Mexico, targetted the last unfactored number in a famous list compiled by the 17th century French mathematician Martin Mersenne. Using about 32 h of computer time, they showed that M_{251} (a 69-digit number) is not prime and has three basic factors. In this section, the following notations will be used, for the sake of brevity:

$$((x)) \triangleq x \quad \mathrm{mod}\, M_N, \tag{2.92a}$$

$$\langle x \rangle \triangleq x \quad \mathrm{mod}\, N. \tag{2.92b}$$

In working with Mersenne transforms all exponents and indices are taken modulo N and all operations are performed modulo M_N. From the definition of a Mersenne prime, it is clear that

$$2^N \equiv 1 \quad \mathrm{mod}\, M_N \tag{2.93}$$

and for $x > 0$,

$$((2^x)) \equiv ((2^{\langle x \rangle})). \tag{2.94}$$

(2.94) implies that the following definition is meaningful:

$$(2^{-x}) \quad \mathrm{mod}\, M_N \equiv 2^{\langle -x \rangle} \quad \mathrm{mod}\, M_N. \tag{2.95}$$

Definition 2.8. Let $\{x_k\}_{k=0}^{N-1}$ be a sequence of N integers. Let $\{X_k\}_{k=0}^{N-1}$ be the sequence of Mersenne transform. Then,

$$X_k = \left[\sum_{r=0}^{N-1} x_r 2^{rk} \right] \quad \mathrm{mod}\, M_N, \quad k = 0, 1, \ldots, N-1,$$

and the inverse Mersenne transform (IMT) is

$$x_r = \left[\left(M_N - \frac{(M_N - 1)}{N} \right) \sum_{k=0}^{N-1} X_k 2^{-rk} \right] \quad \mathrm{mod}\, M_N,$$

$$r = 0, 1, \ldots, N-1,$$

where M_N is a Mersenne prime. $\{X_k\} \leftrightarrow \{x_k\}$ will denote an N-point MT pair.

Note that in the preceding Definition, $(M_N - 1)/N$ is an integer since (apply Theorem 2.11, noting that $\phi(N) = N - 1$ for prime N)

$$2^{N-1} \equiv 1 \quad \mathrm{mod}\, N.$$

Therefore, defining

$$N^{-1} \triangleq M_N - (M_N - 1)/N, \tag{2.96}$$

it is clear that

$$NN^{-1} \equiv 1 \quad \mathrm{mod}\, M_N. \tag{2.97}$$

Thus, Definition 2.8 conforms with Equation (2.91) (q is replaced by M_N). Next, several general properties of sequences in MT pairs are stated and proved.

Theorem 2.13. *Let* $\{X_k\} \leftrightarrow \{x_k\}$ *be an MT pair, where each sequence is of length* N. *If* $x_k = x_{\langle -k \rangle}$, *then* $X_k = X_{\langle -k \rangle}$.

PROOF:

$$X_{\langle -k \rangle} = \left(\sum_{r=0}^{N-1} x_r 2^{r\langle -k \rangle} \right) \quad \mathrm{mod}\, M_N.$$

Using (2.95),

$$\left((2^{r\langle -k \rangle}) \right) = \left((2^{r(-k)}) \right) = \left((2^{(-r)k}) \right) = \left((2^{\langle -r \rangle k}) \right).$$

Therefore,

$$X_{\langle -k \rangle} = \left(\sum_{r=0}^{N-1} x_r 2^{\langle -r \rangle k} \right) \quad \mathrm{mod}\, M_N.$$

Since $x_k = x_{\langle -k \rangle}$,

$$X_{\langle -k \rangle} = \left(\sum_{r=0}^{N-1} x_{\langle -r \rangle} 2^{\langle -r \rangle k} \right) \quad \mathrm{mod}\, M_N$$

$$= \left(\sum_{r=0}^{N-1} x_r 2^{rk} \right) \quad \mathrm{mod}\, M_N$$

$$= X_k. \quad \square$$

Theorem 2.14. *Let* $\{X_k\} \leftrightarrow \{x_k\}$ *be an N-point MT pair. If* $x_k = -x_{\langle -k \rangle}$, *then* $X_k = -X_{\langle -k \rangle}$.

The proof of Theorem 2.14 runs parallel to the proof of Theorem 2.13 and is omitted, for the same of brevity.

Theorem 2.15. *Let* $\{X_k\} \leftrightarrow \{x_k\}$ *be an N-point MT pair. Then,*

$$\sum_{r=0}^{N-1} x_r^2 \equiv N^{-1} \sum_{k=0}^{N-1} X_k X_{\langle -k \rangle} \quad \mathrm{mod}\, M_N,$$

where N^{-1} *is defined in* (2.96).

PROOF: From Definition 2.8,

$$\sum_{r=0}^{N-1} x_r^2 \equiv \left[\sum_{r=0}^{N-1} \left(N^{-1} \sum_{k=0}^{N-1} X_k 2^{rk} \right)^2 \right] \mod M_N$$

$$\equiv \left[\sum_{k=0}^{N-1} \sum_{i=0}^{N-1} (N^{-1})^2 X_k X_i \sum_{r=0}^{N-1} 2^{(i+k)r} \right] \mod M_N.$$

Note that, working mod M_N

$$\sum_{r=0}^{N-1} 2^{(i+k)r} = \begin{cases} N, & \langle i+k \rangle = 0, \\ 0, & \langle i+k \rangle \neq 0. \end{cases}$$

Using (2.97) in the above equations,

$$\sum_{r=0}^{N-1} x_r^2 = N^{-1} \sum_{k=0}^{N-1} X_k X_{\langle -k \rangle} \mod M_N.$$

Combining the results of Theorems 2.13 and 2.15, one arrives at the following result.

Fact 2.10. *Let* $\{X_k\} \leftrightarrow \{x_k\}$ *be an N-point MT pair. If* $x_k = x_{\langle -k \rangle}$, *then*

$$\sum_{k=0}^{N-1} x_k^2 \equiv N^{-1} \sum_{k=0}^{N-1} X_k^2 \mod M_N.$$

Example 2.22. Let $N = 3$. $M_3 = 2^3 - 1 = 7$. Let $\{x_k\}_{k=0}^2 = (1,2,3)$. Then,

$$X_k = \left(\sum_{r=0}^{2} x_r 2^{rk} \right) \mod 7.$$

Therefore,

$$X_0 = (1+2+3) \mod 7 = 6,$$

$$X_1 = (1+4+12) \mod 7 = 3,$$

$$X_2 = (1+8+48) \mod 7 = 1,$$

$$N^{-1} = M_3 - \frac{(M_3 - 1)}{3} = 7 - \frac{6}{3} = 5,$$

$$\sum_{k=0}^{2} x_k^2 = 1^2 + 2^2 + 3^2 = 14,$$

$$N^{-1} \sum_{k=0}^{2} X_k X_{\langle -k \rangle} = \left(X_0^2 + X_1 X_2 + X_2 X_1 \right) N^{-1}$$

$$= (36 + 3 + 3)5 = (42)(5)$$

$$= 210.$$

As expected,

$$14 \equiv 210 \quad \mathrm{mod}\ 7.$$

Theorem 2.16. *Let*

$$\{X_k\} \leftrightarrow \{x_k\},$$
$$\{Y_k\} \leftrightarrow \{y_k\}$$

be two N-point MT pairs. Define

$$Z_k \triangleq X_k Y_k \quad \mathrm{mod}\ M_N. \tag{2.98}$$

If $\{Z_k\} \leftrightarrow \{z_k\}$ *denotes another N-point MT pair, then*

$$z_k = \left(\sum_{r=0}^{N-1} x_r y_{\langle k-r \rangle} \right) \quad \mathrm{mod}\ M_N$$

$$= \left(\sum_{r=0}^{N-1} x_{\langle k-r \rangle} y_r \right) \quad \mathrm{mod}\ M_N.$$

The proof of the above theorem is left as an exercise (see Problem 26). Theorem 2.16 shows the possibility for implementing a cyclic convolution of two sequences via Mersenne transform techniques. An MT can be computed using only additions and multiplications by powers of 2 (multiplications by powers of 2 mod M_N can be replaced by rotational shifts). It is simple to verify that an N-point MT,

$$X_k = \left(\left(\sum_{r=0}^{N-1} x_r 2^{rk} \right) \right), \quad k = 0, 1, \ldots, N-1,$$

requires $N(N-1)$ additions and $(N-1)^2$ shifts (note that no shifts are required in the computation of X_0). For computing an N-point circular convolution by using (2.98) in Theorem 2.16 it can be easily verified that $3N(N-1)$ additions, $3(N-1)^2$ shifts and $2N$ multiplications are sufficient (N multiplications are required to compute the product in (2.98) for $k = 0, 1, \ldots, N-1$ and the remaining N multiplications are due to N^{-1} in the IMT).

2.6.2. Fermat Transform (FT)

Let

$$N \triangleq 2^{m+1}, \tag{2.99}$$

$$M_N \triangleq 2^N - 1. \tag{2.100}$$

Definition 2.9. For any positive integer m, a Fermat number F_m is defined as

$$F_m \triangleq 2^{2^m} + 1. \tag{2.101}$$

It is verifiable that F_1, F_2, and F_3 are primes. F_4 is known to be a prime.

From (2.99), (2.100), and (2.101),

$$
\begin{aligned}
M_N &= 2^N - 1 = 2^{2^{m+1}} - 1 \\
&= (2^{2^m})^2 - 1 \\
&= (2^{2^m} - 1)(2^{2^m} + 1) \\
&= (F_m - 2)F_m.
\end{aligned}
\tag{2.102}
$$

Equation (2.102) shows a relationship between Mersenne and Fermat numbers. For a integer x, then

$$
x \quad \mathrm{mod}\ F_m = (x \quad \mathrm{mod}\ M_N) \quad \mathrm{mod}\ F_m.
\tag{2.103}
$$

For the sake of brevity, the following notations will be used in this section:

$$
((x)) \triangleq x \quad \mathrm{mod}\ F_m,
\tag{2.104a}
$$

$$
\langle x \rangle \triangleq x \quad \mathrm{mod}\ N.
\tag{2.104b}
$$

The developments below will be based on the above preliminaries.

Definition 2.10. Let $\{x_k\}_{k=0}^{N-1}$ be a sequence of N integers. Let $\{X_k\}_{k=0}^{N-1}$ denote the sequence of the Fermat number transform (FNT). Then,

$$
X_k = \left(\sum_{r=0}^{N-1} x_r 2^{rk} \right) \quad \mathrm{mod}\ F_m, \quad k = 0, 1, \ldots, N-1,
$$

and the inverse Fermat number transform (IFNT) is

$$
x_r = \left(2^{N-m-1} \sum_{k=0}^{N-1} X_k 2^{\langle -rk \rangle} \right) \quad \mathrm{mod}\ F_m,
$$

for $r = 0, 1, \ldots, N-1$, where F_m is a Fermat prime. $\{X_k\} \leftrightarrow \{x_k\}$ will denote an N-point FNT pair. (The symbol FNT instead of FT has been used to avoid confusion with Fourier transform.)

Note that on defining

$$
N^{-1} \triangleq 2^{N-m-1}
\tag{2.105}
$$

and using (2.99),

$$
\begin{aligned}
NN^{-1} &= 2^N \\
&= 2^{2^{m+1}}.
\end{aligned}
$$

Using (2.103) in the preceding equation,

$$
NN^{-1} \equiv 1 \quad \mathrm{mod}\ F_m.
\tag{2.106}
$$

Thus, Definition 2.10 yields a compatible transform pair. Note that 2 is an Nth order root of 1 $\quad \mathrm{mod}\ F_m$, where $N = 2^{m+1}$ since

$$
2^N = 2^{2^{m+1}} = (2^{2^m})^2,
$$

which in turn implies that

$$2^N \equiv (-1)^2 \equiv 1 \quad \bmod F_m.$$

The following result will be needed in order to prove the theorem concerning the evaluation of the cyclic convolution via Fermat transform techniques.

Theorem 2.17. *With reference to the notations introduced,*

$$\left(N^{-1} \sum_{k=0}^{N-1} 2^{\langle nk \rangle} \right) \bmod F_m = \begin{cases} 1, & \langle n \rangle = 0, \\ 0, & \langle n \rangle \neq 0. \end{cases}$$

PROOF:

$\langle n \rangle = 0$ *case:*

$$\left(N^{-1} \sum_{k=0}^{N-1} 2^{\langle nk \rangle} \right) \bmod F_m = N^{-1} N \quad \bmod F_m = 1 \quad \bmod F_m = 1$$

[following the use of (2.106)].

$\langle n \rangle \neq 0$ *case:* Applying (2.103),

$$2^{\langle nk \rangle} \quad \bmod F_m = \left(2^{\langle nk \rangle} \quad \bmod M_N \right) \quad \bmod F_m = 2^{nk} \quad \bmod F_m.$$

$$(2.107)$$

Therefore,

$$\left(N^{-1} \sum_{k=0}^{N-1} 2^{\langle nk \rangle} \right) \bmod F_m = \left(N^{-1} \sum_{k=0}^{N-1} 2^{nk} \right) \bmod F_m$$

$$= \left[N^{-1} \left(\frac{1 - 2^{nN}}{1 - 2^n} \right) \right] \bmod F_m$$

$$= \left[N^{-1} \left(\frac{1 - 2^{nN}}{1 - 2^n} \right) \bmod M_N \right] \bmod F_m$$

$$= 0 \quad \text{for } \langle n \rangle \neq 0,$$

since

$$2^{nN} \equiv 1 \quad \bmod M_N.$$

The proof of the theorem is now complete.

The main theorem in this section is now stated and proved.

Theorem 2.18. *Let*

$$\{ X_k \} \leftrightarrow \{ x_k \},$$
$$\{ Y_k \} \leftrightarrow \{ y_k \}$$

denote two N-point FNT pairs. Define

$$Z_k = X_k Y_k \quad \mathrm{mod}\ F_m, \qquad\qquad (2.108)$$

and let $\{Z_k\} \leftrightarrow \{z_k\}$ *be another N-point FNT pair. Then,*

$$z_k = \left(\sum_{r=0}^{N-1} x_r y_{\langle k-r \rangle} \right) \quad \mathrm{mod}\ F_m.$$

PROOF: Since

$$2^N \quad \mathrm{mod}\ F_m = \left(2^N \quad \mathrm{mod}\ M_N \right) \quad \mathrm{mod}\ F_m = 1,$$

it follows that

$$2^x \equiv 2^{\langle x \rangle} \quad \mathrm{mod}\ F_m.$$

Then,

$$z_k = \left[N^{-1} \sum_{r=0}^{N-1} \left(\sum_{i=0}^{N-1} x_i 2^{ir} \right) \left(\sum_{t=0}^{N-1} y_t 2^{tr} \right) 2^{\langle -rk \rangle} \right] \quad \mathrm{mod}\ F_m$$

$$= \left(\sum_{i=0}^{N-1} \sum_{t=0}^{N-1} x_i y_t N^{-1} \sum_{r=0}^{N-1} 2^{\langle (i+t-k)r \rangle} \right) \quad \mathrm{mod}\ F_m$$

$$= \left(\sum_{i=0}^{N-1} x_i y_{\langle k-i \rangle} \right) \quad \mathrm{mod}\ F_m$$

$$= \left(\sum_{i=0}^{N-1} x_{\langle k-i \rangle} y_i \right) \quad \mathrm{mod}\ F_m.$$

The last two steps follow from application of Theorem 2.17. The proof is now complete. $\square$

Like Theorem 2.16 for the Mersenne transform, Theorem 2.19 for the Fermat transform suggests that the transform of the cyclic convolution of two sequences is equal to the product of their transforms. The FNT has been found to be very attractive from the standpoint of hardware and is very suitable for digital filtering computations because no multiplications are required and also FFT-type algorithms apply.

2.7. Polynomial Transforms

Quite recently, Nussbaumer introduced several multiplication free polynomial transform methods to support efficient implementation of 2-D convolution and 2-D DFT [1]. Since it is possible to rapidly execute 1-D convolution by multidimensional techniques [22], polynomial transforms are also useful in the context of 1-D digital signal processing. Therefore, for the sake of completeness, polynomial transform methods are introduced. However, in order that brevity is simultaneously attended to, the presentation is geared towards the implementation of 2-D convolution only.

Definition 2.11. Let $\{x_k(z)\}_{k=0}^{N-1}$ denote a sequence of N polynomials in the complex variable z, whose coefficients belong to an arbitrary but fixed field. The polynomial transform (PT) $\{X_k(z)\}_{k=0}^{N-1}$, is another sequence of N polynomials defined by

$$X_k(z) = \sum_{r=0}^{N-1} x_r(z)[g(z)]^{rk} \mod m(z)$$

for $k = 0, 1, \ldots, N-1$. Polynomials $g(z)$ and $m(z)$ must be chosen so that they satisfy the following conditions:

a. $g^N(z) \equiv 1 \mod m(z)$;   (2.109a)
b. there must exist N^{-1} and $[g(z)]^{-1}$ so that

$$NN^{-1} \equiv 1 \mod m(z),$$

$$g(z)[g(z)]^{-1} \equiv 1 \mod m(z); \qquad (2.109b)$$

c. $S \triangleq \sum\limits_{r=0}^{N-1} [g(z)]^{rk} \mod m(z) \equiv \begin{cases} 0, & k \neq 0 \mod N, \\ N, & k = 0 \mod N. \end{cases}$ (2.109c)

The inverse polynomial transform (IPT) is defined by

$$x_r(z) = N^{-1} \sum_{k=0}^{N-1} X_k(z)[g(z)]^{-kr} \mod m(z),$$

$r = 0, 1, \ldots, N-1$. $\{X_k(z)\} \leftrightarrow \{x_k(z)\}$ will denote an N-polynomial PT pair.

Example 2.23. $g(z) = z$ and $m(z) = z^N - 1$ will define an N-polynomial PT pair because, in this case,

$$g^N(z) = z^N \equiv 1 \mod m(z),$$

$$N^{-1} = \frac{1}{N} \Rightarrow NN^{-1} \equiv 1 \mod m(z),$$

$$[g(z)]^{-1} = z^{N-1} \Rightarrow [g(z)][g(z)]^{-1} \equiv 1 \mod m(z),$$

and

$$S = \sum_{r=0}^{N-1} z^{rk} \mod m(z)$$

$$= \frac{z^{kN} - 1}{z^k - 1} \mod m(z),$$

implying that

$$S = \begin{cases} 0, & k \neq 0 \mod N, \\ N, & k = 0 \mod N. \end{cases}$$

[For $k = 0 \mod N$, $S = 1 + 1 + \cdots + 1 = N$, since the 1's are repeated N times. For $k \neq 0 \mod N$, $(z^N)^k \equiv 1 \mod m(z)$, implying that $z^{kN} - 1 \equiv 0 \mod m(z)$, while $z^k - 1 \not\equiv 0 \mod m(z)$.]

Example 2.24. $g(z) = z$ and $m(z) = (z^N - 1)/(z - 1)$, where N is an odd positive prime will define an N-polynomial PT pair because of the following:

$$m(z) = \frac{z^N - 1}{z - 1} = z^{N-1} + z^{N-2} + \cdots + z + 1,$$

$$[g(z)]^N = z^N = zm(z) - (z^{N-1} + \cdots + z^2 + z).$$

Therefore,

$$[g(z)]^N \equiv 1 \mod m(z).$$

N^{-1}, obviously, exists such that $NN^{-1} \equiv 1 \mod m(z)$. Also,

$$[g(z)]^{-1} \triangleq -(z^{N-2} + \cdots + z + 1)$$

satisfies

$$g(z)[g(z)]^{-1} \equiv 1 \mod m(z).$$

The expression for S is

$$S = \frac{z^{kN} - 1}{z^k - 1} \mod m(z)$$

and, using arguments analogous to those in the previous problem,

$$S \equiv \begin{cases} 0, & k \neq 0 \mod N, \\ N, & k = 0 \mod N. \end{cases}$$

Next the feasibility for computing a 2-D circular convolution (defined next) using the PT approach will be considered.

Definition 2.12. Let $\{h_{k_1, k_2}\}$ and $\{x_{k_1, k_2}\}$ denote two arrays each of size $(N \times N)$. The indices k_1 and k_2 each range over the index set $(0, 1, \ldots, N - 1)$. The 2-D circular convolution of $\{h_{k_1, k_2}\}$ and $\{x_{k_1, k_2}\}$ is another $(N \times N)$ array, $\{y_{k_1, k_2}\}$, defined by

$$y_{k_1, k_2} = \sum_{r_1 = 0}^{N-1} \sum_{r_2 = 0}^{N-1} x_{r_1, r_2} h_{\langle k_1 - r_1 \rangle, \langle k_2 - r_2 \rangle}, \qquad (2.110)$$

$$k_1 = 0, 1, \ldots, N - 1, \quad k_2 = 0, 1, \ldots, N - 1,$$

$$\text{and} \quad \langle k - r \rangle \triangleq (k - r) \mod N.$$

Associate a polynomial $h_{k_2}(z)$ with the $(k_2 + 1)$th column of array $\{h_{k_1, k_2}\}$:

$$h_{k_2}(z) \triangleq \sum_{k_1 = 0}^{N-1} h_{k_1, k_2} z^{k_1}, \quad k_2 = 0, 1, \ldots, N - 1. \qquad (2.111)$$

The element h_{k_1, k_2} is located in the $(k_1 + 1)$th row, starting from the top and $(k_2 + 1)$th column of array $\{h_{k_1, k_2}\}$. Similarly, associate another polynomial, $x_{k_2}(z)$ with the $(k_2 + 1)$th column of array $\{x_{k_1, k_2}\}$

$$x_{k_2}(z) \triangleq \sum_{k_1 = 0}^{N-1} x_{k_1, k_2} z^{k_1}, \quad k_2 = 0, 1, \ldots, N-1. \qquad (2.112)$$

Let $\{X_{k_2}(z)\} \leftrightarrow \{x_{k_2}(z)\}$ and $\{H_{k_2}(z)\} \leftrightarrow \{h_{k_2}(z)\}$ each be an N-polynomial PT pair defined with $m(z) = z^N - 1$ and $g(z) = z$ as in Example 2.23. Then,

$$X_{k_2}(z) H_{k_2}(z) \equiv \left[\sum_{k_1 = 0}^{N-1} x_{k_1}(z) z^{k_1 k_2} \right] \left[\sum_{k_1 = 0}^{N-1} h_{k_1}(z) z^{k_1 k_2} \right] \quad \mathrm{mod}\, (z^N - 1)$$

$$\equiv \sum_{k_1 = 0}^{N-1} \sum_{r_1 = 0}^{N-1} x_{k_1}(z) h_{r_1}(z) z^{(r_1 + k_1) k_2} \quad \mathrm{mod}\, (z^N - 1). \qquad (2.113)$$

Define $\{Y_{k_2}(z) \triangleq X_{k_2}(z) H_{k_2}(z) \ \mathrm{mod}\, (z^N - 1)\} \leftrightarrow \{y_{k_2}(z)\}$ to be another N-polynomial PT pair. Then,

$$y_{k_2}(z) = N^{-1} \sum_{i_1 = 0}^{N-1} X_{i_1}(z) H_{i_1}(z) z^{-i_1 k_2} \quad \mathrm{mod}\, (z^N - 1). \qquad (2.114)$$

Substituting (2.113) in (2.114), one gets

$$y_{k_2}(z) = \sum_{k_1 = 0}^{N-1} \sum_{r_1 = 0}^{N-1} x_{k_1}(z) h_{r_1}(z) N^{-1} \sum_{i_1 = 0}^{N-1} z^{(k_1 + r_1 - k_2) i_1} \quad \mathrm{mod}\, (z^N - 1). \qquad (2.115)$$

Recall that (2.109c) applied to this case yields

$$N^{-1} \sum_{i_1 = 0}^{N-1} z^{(k_1 + r_1 - k_2) i} = \begin{cases} 0, & k_1 + r_1 - k_2 \neq 0 \quad \mathrm{mod}\, N, \\ 1, & k_1 + r_1 - k_2 = 0 \quad \mathrm{mod}\, N. \end{cases} \qquad (2.116)$$

Substituting (2.116) in (2.115), one gets

$$y_{k_2}(z) = \sum_{k_1 = 0}^{N-1} x_{k_1}(z) h_{\langle k_2 - k_1 \rangle}(z) \quad \mathrm{mod}\, (z^N - 1). \qquad (2.117)$$

After substituting (2.111) and (2.112) in (2.117),

$$y_{k_2}(z) = \sum_{k_1 = 0}^{N-1} \sum_{r_1 = 0}^{N-1} \sum_{j_1 = 0}^{N-1} x_{r_1, k_1} h_{j_1, \langle k_2 - k_1 \rangle} z^{(r_1 + j_1)} \quad \mathrm{mod}\, (z^N - 1).$$

Let $(r_1 + j_1) \equiv i_1 \pmod{N}$, $i_1 = 0, 1, \ldots, N-1$. Then, the preceding equation simplifies to

$$y_{k_2}(z) = \sum_{k_1=0}^{N-1} \sum_{r_1=0}^{N-1} \sum_{i_1=0}^{N-1} x_{r_1,k_1} h_{\langle i_1-r_1 \rangle, \langle k_2-k_1 \rangle} z^{i_1}$$

$$= \sum_{i_1=0}^{N-1} \left(\sum_{r_1=0}^{N-1} \sum_{k_1=0}^{N-1} x_{r_1,k_1} h_{\langle i_1-r_1 \rangle, \langle k_2-k_1 \rangle} \right) z^{i_1}. \tag{2.118}$$

Substituting (2.110) in (2.118), one gets

$$y_{k_2}(z) = \sum_{i_1=0}^{N-1} y_{i_1,k_2} z^{i_1}. \tag{2.119}$$

Thus, $y_{k_2}(z)$ is a polynomial associated with the k_2th column of the convolved array, $\{y_{k_1,k_2}\}$, $k_2 = 0,1,\ldots,N-1$.

The computation of (2.117) when N is an odd positive prime will be discussed. In this case, the polynomial, $(z^N - 1)$, is factorable as a product of two cyclotomic polynomials, which are, therefore, irreducible over the field of rational numbers:

$$(z^N - 1) = (z-1)(z^{N-1} + z^{N-2} + \cdots + z + 1). \tag{2.120}$$

The objective is to calculate (2.117) via use of the Chinese remainder theorem (CRT) for polynomials (Theorem 2.6) after calculating

$$\hat{y}_{k_2}(z) \triangleq \sum_{k_1=0}^{N-1} x_{k_1}(z) h_{\langle k_2-k_1 \rangle}(z), \tag{2.121a}$$

modulo each of the irreducible polynomials,

$$p_1(z) \triangleq (z-1), \tag{2.121b}$$

$$p_2(z) \triangleq (z^{N-1} + z^{N-2} + \cdots + z + 1). \tag{2.121c}$$

Define the residues

$$r_{1k_2}(z) \triangleq \hat{y}_{k_2}(z) \mod p_1(z) = \sum_{k_1=0}^{N-1} x_{k_1}(1) h_{\langle k_2-k_1 \rangle}(1),$$

$$\tag{2.122a}$$

$$r_{2k_2}(z) \triangleq \hat{y}_{k_2}(z) \mod p_2(z) = \hat{y}_{k_2}(z) \mod \frac{(z^N-1)}{(z-1)}.$$

$$\tag{2.122b}$$

Then, by Theorem 2.6,

$$y_{k_2}(z) = \left[r_{1k_2}(z) p_2(z) s_1(z) + r_{2k_2}(z) p_1(z) s_2(z) \right] \mod (z^N - 1),$$

where the polynomials $s_1(z)$ and $s_2(z)$ are chosen to satisfy

$$s_1(z) p_2(z) + s_2(z) p_1(z) = 1 \tag{2.123}$$

or

$$s_1(z)p_2(z) \equiv 1 \mod p_1(z),$$
$$s_2(z)p_1(z) \equiv 1 \mod p_2(z).$$

It should be noted that polynomials $s_1(z)$ and $s_2(z)$ satisfying (2.123) must exist since the polynomials $p_1(z)$ and $p_2(z)$ are relatively prime (Fact 2.4). On dividing $p_2(z)$ by $p_1(z)$,

$$p_2(z) = \left[z^{N-2} + 2z^{N-3} + \cdots + (N-2)z + (N-1) \right] p_1(z) + N$$

or

$$1 = \frac{1}{N} p_2(z) - \frac{1}{N} \left[z^{N-2} + 2z^{N-3} + \cdots + (N-2)z + (N-1) \right] p_1(z).$$

On comparing the previous equation, term-by-term, with (2.123),

$$s_1(z) = \frac{1}{N},$$

$$s_2(z) = -\frac{1}{N} \left[z^{N-2} + 2z^{N-3} + 3z^{N-2} + \cdots + (N-2)z + (N-1) \right].$$

The various steps in the algorithm for implementing a 2-D digital convolution by PT method will be illustrated by the nontrivial example, given next.

Example 2.25. It is required to compute the 2-D circular convolution of the (3×3) arrays,

$$\{ x_{k_1, k_2} \} = \begin{bmatrix} 1 & 3 & 2 \\ 2 & 2 & 1 \\ 1 & 4 & 5 \end{bmatrix}$$

and

$$\{ h_{k_1, k_2} \} = \begin{bmatrix} 2 & 6 & 7 \\ 1 & 4 & 1 \\ 1 & 3 & 2 \end{bmatrix}.$$

Step 1: Associate polynomials $x_{k_2}(z)$ and $h_{k_2}(z)$ with the $(k_2 + 1)$th columns of $\{ x_{k_1, k_2} \}$ and $\{ h_{k_1, k_2} \}$, respectively, for $k_2 = 0, 1, \ldots, N-1$ (here $N = 3$).

$$x_0(z) = z^2 + 2z + 1, \quad h_0(z) = z^2 + z + 2,$$
$$x_1(z) = 4z^2 + 2z + 3, \quad h_1(z) = 3z^2 + 4z + 6,$$
$$x_2(z) = 5z^2 + z + 2, \quad h_2(z) = 2z^2 + z + 7.$$

Step 2: Take $m(z) = z^2 + z + 1 \triangleq p_2(z)$, $g(z) = z$. Compute
$$x_{2k_2}(z) \triangleq x_{k_2}(z) \mod p_2(z), \quad k_2 = 0, 1, \ldots, N-1,$$
$$h_{2k_2}(z) \triangleq h_{k_2}(z) \mod p_2(z), \quad k_2 = 0, 1, \ldots, N-1.$$

Here

$$x_{20}(z) = z, \qquad h_{20}(z) = 1,$$
$$x_{21}(z) = -2z - 1, \quad h_{21}(z) = z + 3,$$
$$x_{22}(z) = -4z - 3, \quad h_{22}(z) = -z + 5.$$

Step 3: Compute the polynomial transforms, $\{X_{2k_2}(z)\}$ and $\{H_{2k_2}(z)\}$, where $\{X_{2k_2}(z)\} \leftrightarrow \{x_{2k_2}(z)\}$ and $\{H_{2k_2}(z)\} \leftrightarrow \{h_{2k_2}(z)\}$ are N-polynomial PT pairs. Here

$$X_{2k_2}(z) = \sum_{k_1=0}^{2} x_{2k_1}(z) z^{k_1 k_2} \quad \mathrm{mod}\, (z^2 + z + 1).$$

Therefore,

$$X_{20}(z) = -5z - 4,$$
$$X_{21}(z) = 5z + 1,$$
$$X_{22}(z) = 3z + 3.$$

Similarly,

$$H_{2k_2}(z) = \sum_{k_1=0}^{2} h_{2k_1}(z) z^{k_1 k_2} \quad \mathrm{mod}\, (z^2 + z + 1).$$

Therefore,

$$H_{20}(z) = 9,$$
$$H_{21}(z) = -3z - 6,$$
$$H_{22}(z) = 3z.$$

Step 4: Form

$$R_{2k_2}(z) \triangleq H_{2k_2}(z) X_{2k_2}(z) \quad \mathrm{mod}\, p_2(z), \quad k_2 = 0, 1, \ldots, N-1.$$

In this case,

$$R_{20}(z) = -45z - 36,$$
$$R_{21}(z) = -18z + 9,$$
$$R_{22}(z) = -9.$$

Step 5: Form the N-polynomial IPT of $\{R_{2k_2}(z)\}$ to get $\{r_{2k_2}(z)\}$, where $\{R_{2k_2}(z)\} \leftrightarrow \{r_{2k_2}(z)\}$ is an N-polynomial PT pair. Here,

$$r_{2k_2}(z) = \tfrac{1}{3} \sum_{k_1=0}^{2} R_{2k_1}(z) z^{-k_1 k_2} \quad \mathrm{mod}\, (z^2 + z + 1), \quad k_2 = 0, 1, 2.$$

From Example 2.24, it is clear that, for this example, $[g(z)]^{-1} = -(z+1)$, where $g(z) = z$ and $g(z)[g(z)]^{-1} \equiv 1 \quad \mathrm{mod}\, (z^2 + z$

$+1$). Therefore, simple algebraic manipulations yield

$$r_{20}(z) = -21z - 12,$$
$$r_{21}(z) = -21z - 21,$$
$$r_{22}(z) = -3z - 3.$$

Step 6: Determine residues:

$$r_{1k_2}(z) = \left[\sum_{k_1=0}^{N-1} x_{k_1}(z) h_{\langle k_2-k_1 \rangle}(z) \right] \mod (z-1).$$

Here,

$$r_{1k_2}(z) = \sum_{k_1=0}^{2} x_{k_1}(1) h_{\langle k_2-k_1 \rangle}(1), \quad k_2 = 0,1,2.$$

Therefore,

$$r_{10} = 210,$$
$$r_{11} = 168,$$
$$r_{12} = 189.$$

Step 7: Compute

$$s_1(z) = \frac{1}{N},$$

$$s_2(z) = -\frac{1}{N} \left[z^{N-2} + 2z^{N-3} + 3z^{N-2} + \cdots + (N-2)z + (N-1) \right].$$

Here,

$$s_1(z) = \tfrac{1}{3},$$
$$s_2(z) = -\tfrac{1}{3}(z+2).$$

Step 8: Apply CRT to obtain

$$y_{k_2}(z) = \left[r_{1k_2}(z)p_2(z)s_1(z) + r_{2k_2}(z)p_1(z)s_2(z) \right] \mod (z^3 - 1).$$

Therefore, after simple algebraic manipulations,

$$y_0(z) = 69 + 60z + 81z^2,$$
$$y_1(z) = 49 + 49z + 70z^2,$$
$$y_2(z) = 62 + 62z + 65z^2.$$

Step 9: Denoting

$$y_{k_2}(z) = \sum_{k_1=0}^{N-1} y_{k_1 k_2} z^{k_1}, \quad k_2 = 0,1,2,$$

write down the output array of size $N \times N$, $\{ y_{k_1, k_2} \}$. Here,

$$\{ y_{k_1, k_2} \} = \begin{bmatrix} 69 & 49 & 62 \\ 60 & 49 & 62 \\ 81 & 70 & 65 \end{bmatrix}.$$

2.8. Summary and Suggested Readings

Recent textbooks, where the past and current developments of transform theory are documented are references [1], [25], [28] and [29]. The topic has been treated at a slightly more elementary but concise level in reference [2], which also contains a collection of papers on various aspects of the subject published before 1979.

Next, the reader is alerted to sources that contain additional material on specific topics covered in this chapter. A vast number of books and survey papers discuss the DFT and its computation via FFT algorithms. A detailed coverage of this topic can be found, for example, in reference [8].

The importance of number theory in digital signal processing has been evident in this chapter. Though no specialized knowledge of number theory is expected of the reader for comprehension of the subject matter at the level presented, it might be useful to know that various books on number theory, like [3], are readily available for consultation. For clarity, the exposition of Euclid's algorithm and the CRT has been done separately over the ring of integers and the ring of polynomials, whose coefficients belong to any arbitrary but fixed field (the field of rational numbers is of prime interest in this chapter). A reader versed in algebra will recognize that the division process in Euclid's algorithm exists in any principal ideal ring, and, if the principal ideal ring is restricted to be an Euclidean domain, the division process can actually be constructed. The CRT is also known to hold in any principal ideal ring.

Rational minimum phase functions in the complex variable z are devoid of poles and zeroes in $|z| \geq 1$. For such functions, the logarithm of the magnitude of the frequency response is uniquely related to the phase through the Hilbert transform. For a thorough discussion on discrete Hilbert transforms, the reader is referred to Chapter 7 in reference [8]. In Section 2.4, discussion is centered around the matrix formulation of the DHT. Exploitation of the structure of a matrix (like Toeplitz, Hankel, circulant, centrosymmetric, persymmetric; the reader should refer to Chapter 6 for relevant definitions and detailed analysis of some typical problems in digital filter theory, where particular matrix structures characterizing corresponding problems provide links to particular classes of orthogonal polynomials, which in turn support recursive computational capabilities in the solution of those problems) is crucial to the development of fast algorithms.

The results in Section 2.5 are an outgrowth of relatively recent developments in the theory of algebraic computational complexity. The problems of linear convolution and circular (or cyclic) convolution can be related, respectively, to polynomial multiplication and computation of a polynomial product modulo a third polynomial. Furthermore, the indices of the DFT can often be rearranged so as to obtain circular convolution (when the number of points is a prime, this possibility was pointed out by Rader [9], and the case when the number of points is the power of a prime has been considered by Winograd [10]). Therefore, fast algorithms for implementing polynomial multiplications (modulo a third polynomial, when necessary) lead to efficient algorithms for computing convolutions and DFTs. The reader will find in the two papers, [11] and [12], both of which have been included in reference [2], additional illustrative examples on the subject besides the ones worked out in the book. Several comments are now in order. First, the fact that the minimum multiplicative complexity theory required to compute a polynomial product modulo a third polynomial (where the polynomial coefficients belong to an arbitrary but fixed field) ignore multiplications by fixed elements of the field is emphasized. Second, minimal algorithms for computing the products just referred to become increasingly difficult to construct with the increase in the degrees of the polynomial. This, in turn, implies that in order to compute long-term convolutions or DFT's of long transform lengths, one might have to be satisfied with suboptimal algorithms. This possibility need not be a serious disadvantage, if it is a disadvantage at all, since minimal algorithms from the standpoint of the number of multiplications could involve a very large number of additions, so that the gain achieved from the reduction in the number of multiplications might be offset by the loss due to the increased number of additions. In fact, recent technological breakthroughs have led to the development of general-purpose computers where the time to implement an addition can no more be considered to be insignificant compared to the time required to implement a multiplication. This technological innovation implies that algorithms need to be developed, where an overall cost factor based on both the number of additions and multiplications is optimized. The establishing of effective guidelines to make feasible a tradeoff between the number of multiplications and additions (by allowing a larger number of multiplications than the minimum suggested from computational complexity theory so that the number of additions become controllable in a desired manner), in the development of fast computational algorithms for implementation on the latest as well as future general purpose computers remains a challenging open problem. Though some attention has been directed to this problem, the guidelines set are, at best, heuristic. For an authoritative account of the results until about 1971 in the area covering schemes to speed up multiplication in a broader context (like matrix multiplication of two n-bit numbers) than required in this text, the reader might like to consult reference [4].

Number-theoretic transforms, briefly discussed in Section 2.6, had, probably, its formal origin in a paper by Pollard [13] (and also in the work of Knuth [4]). He showed that transforms analogous to the DFT can be defined in a finite field of q^n elements, where q is a prime number and n is a positive integer. These transforms have the cyclic convolution property as well as other properties, most of which are similar to the properties of a

DFT. Pollard also derived conditions for defining transforms with the cyclic convolution property in a finite ring of integers. At about the same time, Nicholson [14] developed an algebraic transform theory over any ring and showed that fast FFT-type algorithm can be used to compute these transforms. Rader [15] defined transforms modulo a Mersenne prime or a Fermat number. (Schonage and Strassen [16], in a paper published in German, also defined transforms modulo a Fermat number.) The chief advantage of these transforms is that multiplications by powers of 2 are performed by simple bit rotations. The disadvantages are the large wordlength required and the difficulties that could be faced in computing a number modulo a Mersenne prime or a Fermat number. Some researchers [17] claim that computation of convolution in the ring of integers modulo a Fermat number is best suited for implementation on a digital computer. For hardware realization of a Fermat number transform, see [18] and [19].

Polynomial transforms, which can be viewed as DFT's defined in rings of polynomials, are especially useful in computing multidimensional convolutions and DFT's. The advantages of polynomial transforms over other techniques discussed can be spelled out in terms of reduced computational complexity, fast FFT-like transform computation structure, reduced computation noise, and real arithmetic, in addition to other attractive features. Since the topic is of very recent origin and of a relatively advanced nature, the reader is introduced to the bare essentials of the theory in Section 2.6. The nontrivial example to compute a 2-D circular convolution of $N \times N$ arrays, when N is prime, should be of help in understanding the philosophy underlying the polynomial transform approach. The interested reader can refer to Chapters 6 and 7 of reference [1] for many additional details. Reference [20] should also be of interest. For another approach to compute 2-D convolutions, see reference [21].

Problems

1. Consider the DFT of a sequence $\{ y(k) \} = \{ y(0), y(1), \dots, y(N-1) \}$ of length N, where N is an even positive integer:

$$Y(k) = \sum_{r=0}^{N-1} y(r) w_N^{rk}, \quad k = 0, 1, \dots, N-1, \quad w_N = e^{-j2\pi/N}.$$

Using the transformation,

$$k = k_2 \frac{N}{2} + k_1 \quad \text{and} \quad r = 2r_2 + r_1,$$

where $k_1 = 0, 1, \dots, N/2 - 1$, $k_2 = 0, 1$, $r_1 = 0, 1$, and $r_2 = 0, 1, \dots, N/2 - 1$ obtain the expressions in (2.11).

2. Given two real finite sequences, $\{ y(r) \}$ and $\{ x(r) \}$ each of length N (where $r = 0, 1, \dots, N-1$) show that the DFT's of both can be calculated by com-

puting the DFT of the complex sequence $\{x(r)+jy(r)\}$ of length N. Note that for a real sequence $\{y(r)\}$ having a DFT $\{Y(k)\}$, $\operatorname{Re} Y(k) = \operatorname{Re} Y(-k)$ and $\operatorname{Im} Y(k) = -\operatorname{Im} Y(-k)$.

3. Prove the expression for the DFT of the term-by-term product of two sequences, as given in Table 2.1. The DFT is a circular convolution.

4. Obtain the DFT for each case below:
   a. $\{y(k)\} = \{\sin^2(\pi k/N), \quad k = 0,1,2,\ldots,N-1.$
   b. $\{y(k)\} = \{c\}$, constant c, and $k = 0,1,\ldots,N-1.$

5. Verify the validity of the following equality in DFT computation.
$$Y(k) \triangleq \sum_{r=0}^{N-1} y(r)w_N^{rk} = w_N^{-k^2/2} \sum_{r=0}^{N-1} x(r)h(k+r),$$
where $x(r) = w_N^{-r^2/2}y(r)$, $h(r) = w_N^{r^2/2}$. Use it to your advantage.

6. Justify whether or not a 4-point ($N = 4$) DFT can be implemented with only additions and subtractions.

7. **a.** Show that the number of complex multiplications and complex additions sufficient to calculate an N-point DFT (where $N = 2^r$ with r a positive integer) via the decimation in time approach are, respectively, $\frac{1}{2}N\log_2 N$ and $N\log_2 N$.
   b. Comment on the number of complex multiplications and complex additions sufficient to compute the N-point DFT in part (a) via the decimation in frequency approach.

8. It is stated that a complex multiplication can be implemented with three real multiplications and five real additions. Specifically, if x_1, x_2, x_3, and x_4 are real numbers, to compute the product.
$$(x_1 + jx_2)(x_3 + jx_4) = x_1 x_3 - x_2 x_4 + j(x_2 x_3 + x_1 x_4),$$
four real multiplications and two real additions are sufficient. However, the following equation demonstrates the feasibility of computing the product with one less multiplication (and three more additions):
$$\begin{bmatrix} x_1 x_3 - x_2 x_4 \\ x_2 x_3 + x_1 x_4 \end{bmatrix} = \begin{bmatrix} 1 & -1 & 0 \\ -1 & -1 & 1 \end{bmatrix} \begin{bmatrix} x_1 x_3 \\ x_2 x_4 \\ (x_1 + x_2)(x_3 + x_4) \end{bmatrix}.$$
It is given that a certain computer implements a real multiplication and a real addition (or subtraction) in 2 and 0.1 μsec, respectively. Estimate the time required to implement a 1024-point DFT.

9. Establish the "periodicity property" (see Table 2.1) for a DFT pair by noting that
$$w_N^{rk} = w_N^{(r+N)k} = w_N^{r(k+N)}.$$
Then, verify properties 4 and 9 in Table 2.1.

10. Use properties 4 and 9 in Table 2.1 to prove the following:

   a. $\{x(r)\}$ is a real and an even periodic sequence if and only if $\{X(k)\}$ is a real and an even periodic sequence.

   b. $\{x(r)\}$ is a real and an odd periodic sequence if and only if $\{X(k)\}$ is pure imaginary and an odd periodic sequence.

   c. $\{x(r)\}$ is a pure imaginary and an odd periodic sequence if and only if $\{X(k)\}$ is a real and an odd periodic sequence.

   d. $\{x(r)\}$ is a pure imaginary and an even periodic sequence if and only if $\{X(k)\}$ is a pure imaginary and an even periodic sequence.

11. In order to compute a DFT of length N, where N is prime, one can use the map

$$n = \alpha^m \mod N, \quad n = 1,2,\ldots,N-1 \quad \text{and} \quad m = 0,1,\ldots,N-2.$$

   a. For what value or values of α does the above map provide an isomorphism between a multiplicative group, $\{1,2,\ldots,N-1\} \mod N$, and an additive group, $\{0,1,\ldots,N-2\} \mod (N-1)$, when $N = 7$?

   b. In the case when $N = 7$, derive the counterpart of the relation in (2.82) for DFT pairs that was obtained in Section 2.5.3.

12. It can be seen that when x_1, x_2, y_1, and y_2 are real scalars, the terms $x_1 y_1 + x_2 y_2$ and $x_2 y_1 + x_1 y_2$ can be computed using only two real multiplications (ignoring multiplications by elements of the field of rational numbers) on using the identity

$$\begin{bmatrix} x_1 y_1 + x_2 y_2 \\ x_2 y_1 + x_1 y_2 \end{bmatrix} = \begin{bmatrix} \frac{1}{2}(x_1 + x_2)(y_1 + y_2) + \frac{1}{2}(x_1 - x_2)(y_1 - y_2) \\ \frac{1}{2}(x_1 + x_2)(y_1 + y_2) - \frac{1}{2}(x_1 - x_2)(y_1 - y_2) \end{bmatrix}.$$

Extend this fact to show that the product

$$\begin{bmatrix} X_1 & X_2 \\ X_2 & X_1 \end{bmatrix} \begin{bmatrix} y_1 \\ y_2 \end{bmatrix},$$

where X_1, X_2 are each $(n \times 1)$ column matrices with real entries and y_1, y_2 are real scalars, can be computed using $2n$ multiplications.

13. Construct an algorithm to compute

$$\left(\sum_{k=0}^{4} y_k z^k \right) \left(\sum_{k=0}^{4} x_k z^k \right) \mod (z^5 - 1),$$

using the minimum number of multiplications.

   If the algorithm you constructed requires quite a few additions, investigate the possibility of reducing the number of additions by sacrificing the minimal multiplicative property of the algorithm you initially designed.

14. Take $n > 1$.

   a. Show that $\rho = e^{j2\pi/n}$ is a primitive nth root of unity.

   b. Show that the nth cyclotomic polynomial, defined as the monic polynomial whose zeros are the distinct primitive nth roots of unity, must be of the form

$$C_n(z) = \prod_{\substack{1 \le k < n \\ (k,n)=1}} (z - \rho^k),$$

where $(k, n) = 1$ denotes that k is relatively prime to n.

c. Show that

$$z^n - 1 = \prod_{k \mid n} C_k(z),$$

where $k \mid n$ denotes that k divides n (the set, $\{k\}$, includes 1 and n).

d. From b conclude that the degree of $C_n(z)$ is $\phi(n)$, where ϕ is the Euler's totient function and $\phi(n)$, therefore, is the number of integers less than or equal to n, which are relatively prime to n.

e. From d and c conclude that for n prime

$$C_n(z) = z^{n-1} + z^{n-2} + \cdots + z + 1.$$

15. Show constructively that a minimal multiplication algorithm for computing

$$(h_0 + h_1 z)(x_0 + x_1 z) \mod (z^2 - 1) \triangleq y_0 + y_1 z$$

is

$$m_1 \triangleq \tfrac{1}{2}(h_0 + h_1)(x_0 + x_1),$$

$$m_2 \triangleq \tfrac{1}{2}(h_0 - h_1)(x_0 - x_1),$$

$$\begin{bmatrix} y_0 \\ y_1 \end{bmatrix} = \begin{bmatrix} m_1 + m_2 \\ m_1 - m_2 \end{bmatrix} = \begin{bmatrix} 1 & 1 \\ 1 & -1 \end{bmatrix} \begin{bmatrix} m_1 \\ m_2 \end{bmatrix}.$$

Find matrices A, B, and C of (2.68) for the above algorithm.

16. Show that an alternate set of matrices $\{A, B, C\}$, besides the one in Example 2.14, that can be used to implement an algorithm to compute a 3-point circular convolution is

$$A = \frac{1}{3} \begin{bmatrix} 1 & 1 & 1 \\ 3 & 0 & -3 \\ 0 & 3 & -3 \\ 1 & 1 & -2 \end{bmatrix}, \quad B = \begin{bmatrix} 1 & 1 & 1 \\ 1 & 0 & -1 \\ 0 & 1 & -1 \\ 1 & 1 & -2 \end{bmatrix},$$

$$C = \begin{bmatrix} 1 & 1 & 0 & -1 \\ 1 & -1 & -1 & 2 \\ 1 & 0 & 1 & -1 \end{bmatrix}.$$

17. For integers $n \ge 2$, the Euler totient function, $\phi(n)$, is the number of positive integers, $<$ (or $\le$ for obvious reasons) n, which are relatively prime to n. This description also gives $\phi(1) = 1$.

a. Show that $\phi(n) = n - 1$, when n is prime.

b. Show that $\phi(n^m) = n^{m-1}(n-1)$, when n is prime and integer $m \ge 1$. (*Hint:* note that integers $\le n^m$ which are not relatively prime to a prime n must be multiples of n; there are n^{m-1} such integers). .

18. Given that m and n are relatively prime positive integers, show that the solution of the congruence

$$mx \equiv 1 \quad \mathrm{mod}\, n$$

is

$$x = m^{\phi(n)-1},$$

where $\phi(n)$ is the Euler totient function.

19. In Example 2.18, choose the primitive root to be $g = 3$ and not $g = 2$. Show that the matrix W in

$$\begin{bmatrix} \hat{X}(1) \\ \hat{X}(3) \\ \hat{X}(4) \\ \hat{X}(2) \end{bmatrix} = W \begin{bmatrix} x(1) \\ x(2) \\ x(4) \\ x(3) \end{bmatrix}$$

is the transpose of the square matrix in Equation (2.82).

20. The rth cyclotomic polynomial $C_r(z)$ (consult Definition 2.6) is definable as

$$C_r(z) = \prod_{\substack{(k,r)=1 \\ 1 \le k < r}} \left(z - e^{-j2\pi k/r}\right),$$

where the integer k is relatively prime to r. Make a list of the first 25 cyclotomic polynomials, simplifying each polynomial in the list to have integer coefficients.

21. It is known that the polynomial $z^N - 1$ may be factorized as a product of cyclotomic polynomials:

$$z^N - 1 = \prod_{d_r \mid N} C_{d_r}(z),$$

where the k positive integers d_r divide N and include $d_1 = 1$, $d_k = N$. Find k for each the following cases: (a) $N = 3$, (b) $N = 6$, (c) $N = 9$, (d) $N = 12$, (e) $N = 15$, (f) $N = 18$.

22. Find the number of multiplications required to compute DFT's of the following specified lengths by applying the result stated in Fact 2.8: (a) 27, (b) 81, and (c) 25.

23. Show that each of the computations (polynomial product modulo a third polynomial) described below can be implemented via algorithms requiring only three multiplications:
a. $[(h_0 + h_1 z)(x_0 + x_1 z)] \quad \mathrm{mod}\,(z^2 + z + 1)$.
b. $[(h_0 + h_1 z)(x_0 + x_1 z)] \quad \mathrm{mod}\,(z^2 - z + 1)$.

24. Prove Theorem 2.14.

25. Let $\{X_k\} \leftrightarrow \{x_k\}$ be an N-point Mersenne transform pair. Prove that

$$\sum_{k=0}^{N-1} X_k^2 \equiv N \sum_{r=0}^{N-1} x_r x_{\langle -r \rangle} \quad \text{mod } M_N,$$

where $M_N = 2^N - 1$ is a Mersenne prime and $\langle r \rangle = r \mod N$.

26. Prove Theorem 2.16.

27. Take two (2×2) arrays, each of whose entries is a distinct literal coefficient:
   a. Compute the circular convolution of the two arrays using Definition 2.12.
   b. Check the answer you obtained in part a via the polynomial transform approach.

28. An algorithm to compute the circular convolution of two real sequences is given. Show how you will use this algorithm to compute the circular convolution of two complex sequences.

References

1. Nussbaumer, H. J. 1981. *Fast Fourier Transform and Convolution Algorithms.* Springer-Verlag, Berlin, Heidelberg.
2. McClellan, J. H., and Rader, C. M. 1979. *Number Theory in Digital Signal Processing.* Prentice-Hall, Englewood Cliffs, NJ.
3. LeVeque, W. J. 1977. *Fundamentals of Number Theory.* Addison-Wesley, Reading, MA.
4. Knuth, D. E. 1969. *The Art of Computer Programming, Vol. 2: Seminumerical Algorithms.* Addison-Wesley, Reading, MA.
5. Guillemin, E. A. 1949. *Mathematics of Circuit Theory.* Wiley, New York.
6. Titchmarsh, E. C. 1939. *The Theory of Functions*, 2nd ed. Oxford University Press, London.
7. Dutta Roy, S. C. 1976. Alternative matrix formulation of the discrete Hilbert transform. Proc. IEEE, 64:1435.
8. Oppenheim, A. V., and Schafer, R. W. 1975. *Digital Signal Processing*, Prentice-Hall, Englewood Cliffs, NJ.
9. Rader, C. M. 1968. Discrete Fourier transforms when the number of data samples is prime. Proc. IEEE, 56:1107–1108.
10. Winograd, S. 1978. On computing the discrete Fourier transform. Math. Comput., 32:175–199.
11. Kolba, D. P., and Parks, T. W. 1977. A prime factor FFT algorithm using high-speed convolution. IEEE Trans. Acoust. Speech Signal Proc., 25:282–294.
12. Agarwal, R. C., and Cooley, J. W. 1977. New algorithms for digital convolution. IEEE Trans. Acoust. Speech Signal Proc., 25:392–410.
13. Pollard, J. M. 1971. The Fast Fourier transform in a finite field. Math. Comput., 25:365–374.
14. Nicholson, P. J. 1971. Algebraic theory of finite Fourier transforms. J. Comput. Syst. Sci., 7:281–282.

15. Rader, C. M. 1972. Discrete convolution via Mersenne transform. IEEE Trans. Comput., 21:1269–1273.
16. Schonage, A., and Strassen, V. 1971. Fast multiplication of large numbers. Comput., 7:281–292.
17. Agarwal, R. C., and Burrus, C. S. 1974. Fast convolution using Fermat number transforms with applications to digital filtering. IEEE Trans. Acoust. Speech Signal Proc., 22:168–178.
18. McClellan, J. H. 1976. Hardware realization of a Fermat number transform. IEEE Trans. Acoust. Speech Signal Proc., 24:216–225.
19. Leibowitz, L. M. 1976. A simplified binary arithmetic for the Fermat number transform. IEEE Trans. Acoust. Speech Signal Proc., 5:356–359.
20. Kriz, T. A. 1982. Reduced data reorder complexity properties of polynomial transform 2-D convolution and Fourier transform methods, IBM J. Res. Dev., 26:708–713.
21. Martens, J. B. 1982. Fast polynomial transforms for two-dimensional convolution. IEEE Trans. Acoust. Speech Signal Proc., 30:1007–1010.
22. Agarwal, R. C., and Burrus, C. S. 1974. Fast one-dimensional digital convolution by multidimensional techniques. IEEE Trans. Acoust. Speech Signal Proc., 23:179–188.
23. Oppenheim, A. V., and Willsky, A. S., with Young, I. T. 1983. *Signals and Systems*. Prentice-Hall, Englewood Cliffs, NJ.
24. Schroeder, M. R., Flanagan, J. L., and Lundry, E. A. 1967. Bandwidth compression of speech by analytic signal rooting. Proc. IEEE, 55:296–401.
25. Elliott, D. F., and Rao, K. R. 1982. *Fast Transforms: Algorithms, Analyses and Applications*. Academic, New York.
26. Martens, J. B. 1984. Recursive cyclotomic factorization—a new algorithm for calculating the discrete Fourier transform. IEEE Trans. Acoust. Speech Signal Proc., 32:750–761.
27. Leiserson, C. E. 1983. *Area-Efficient VLSI Computation*. MIT Press, Cambridge, MA (ACM Doctoral Dissertation Award 1982).
28. Burrus, C. S., and Parks, T. W. 1985. *DFT/FFT and Convolution Algorithms*. John Wiley & Sons Inc., Somerset, NJ.
29. Blahut, R. E. 1984. *Fast Algorithms for Digital Signal Processing*. Addison-Wesley, Reading, MA.
30. Bergland, G. D. 1969. Fast Fourier Transform Hardware Implementations—An Overview. IEEE Trans. Audio Electroacoustics, 17:104–108.

Chapter 3
Infinite Impulse Response Digital Filter Design

3.1. Introduction

In this chapter, some techniques for designing linear time-invariant digital filters characterizable, in general, by a rational transfer function, will be discussed. The unit impulse response sequences characterizing such types of filters are infinite sequences, and therefore these filters are also referred to as infinite impulse response (IIR) filters. Usually IIR filters require less hardware and can perform a filtering task with greater speed than those filters that are characterizable by finite unit impulse response sequences (discussed in Chapter 4). In a filter implemented recursively, the output sample $y(k)$ is a linear combination of the present and past input samples in the sequence $\{x(k)\}$ and also the past output samples.[1] Specifically,

$$y(k) = \sum_{j=0}^{m} a_j x(k-j) - \sum_{j=1}^{n} b_j y(k-j), \qquad (3.1)$$

where the a_j's and b_j's are suitable real constants. Therefore, IIR filters have both feedback and feedforward paths. The transforms $Y(z)$ and $X(z)$ of $\{y(k)\}$ and $\{x(k)\}$, respectively, are related by [on z-transforming (3.1) under zero initial conditions]:

$$Y(z) = \frac{\sum_{j=0}^{m} a_j z^{-j}}{1 + \sum_{j=1}^{n} b_j z^{-j}} X(z). \qquad (3.2)$$

The possibility of using already computed output samples in the computation of the present output sample can lead to significant increase in the speed with which the filter can be implemented. For example, consider the transfer function

$$H(z) = \frac{Y(z)}{X(z)} = 1 + az^{-1} + a^2 z^{-2} + \cdots + a^{10} z^{-10} \qquad (3.3)$$

---

[1] IIR filters are sometimes called recursive filters, because they are, usually, implemented recursively.

whose implementation without feedback involves the difference equation

$$y(k) = x(k) + ax(k-1) + a^2 x(k-2) + \cdots + a^{10} x(k-10). \quad (3.4)$$

However, if the polynomial in (3.3) is viewed as a rational function

$$H(z) = \frac{1 - a^{11} z^{-11}}{1 - az^{-1}}$$

with nonrelatively prime numerator and denominator, a recursive implementation is possible as shown next:

$$y(k) = ay(k-1) + x(k) - a^{11} x(k-11). \quad (3.5)$$

The number of arithmetic operations (multiplications and additions in this case) to implement (3.4) will be far greater than that required to implement (3.5).

Disadvantages usually accompany advantages. An IIR filter is useless unless it is stable. Therefore, stability checks must accompany all design and if a filter is found to be unstable, satisfactory stabilization schemes should be provided. In many applications, where linearity of the phase characteristic is sought, an IIR filter is usually impractical. Moreover, sustained parasitic oscillations are possible in such filters.

3.2. IIR Filter Design via Bilinear Transformation

A popular method for designing digital IIR filters is based on the design of an analog filter prototype followed by the use of bilinear transformation, described in Section 1.5.5. In handbooks various types of analog filters that have specific characteristics in the pass, stop, and transition bands are documented. Here, attention is directed toward how the digital filter realizations may be obtained from some of those analog filter specifications. The bilinear transform to be used will be of the form

$$s = k\frac{z-1}{z+1} = k\frac{1-z^{-1}}{1+z^{-1}}, \quad k > 0, \quad (3.6a)$$

or

$$z^{-1} = \frac{1 - s/k}{1 + s/k}, \quad k > 0, \quad (3.6b)$$

where k is a positive real constant to be determined as follows. The transformation in (3.6b) maps,

$$\text{Re}\, s = 0 \quad \text{to} \quad |z| = 1,$$
$$\text{Re}\, s > 0 \quad \text{to} \quad |z| > 1,$$
$$\text{Re}\, s < 0 \quad \text{to} \quad |z| < 1.$$

On $\text{Re}\, s = 0$, $s = j\Omega$ and, on $|z| = 1$, $z = e^{j\omega T}$, where T is the sampling period for the digital filter. Clearly then, via (3.6), if $s = \sigma + j\Omega$ (for brevity,

ω will be referred to as the digital frequency)

$$\Omega \rightarrow k \tan \frac{\omega T}{2}$$

so that the imaginary axis in the s-plane is mapped into the regions $(2n-1)\pi/T < \omega < (2n+1)\pi/T$ for each integer valued n. If the cutoff frequency of the analog filter is normalized to 1 rad/sec, then for the digital filter to have a desired cutoff frequency of ω_c, it is necessary that

$$k = \frac{1}{\tan(\omega_c T/2)} = \cot \frac{\omega_c T}{2} \tag{3.7}$$

For a cutoff frequency ω_c of the digital filter, the analog filter must have a cutoff frequency of $k \tan(\omega_c T/2)$ and the choice of k in (3.7) enables one to conveniently use the tables available for normalized analog lowpass filter design. It is also mentioned that the shape of the plot of $\tan(\omega T/2)$ vs. ω is responsible for what is known as *warping* of the frequency scale. Consequently, only in the low-frequency range (where $k \tan(\omega T/2) \simeq k\omega T/2$), the digital filter has a frequency response very close to an analog filter.

3.2.1. Digital Butterworth Filter

It will be proved that a normalized analog lowpass Butterworth filter of order n has a magnitude-squared response:

$$|H_a(j\Omega)|^2 = \frac{1}{1 + \Omega^{2n}}, \tag{3.8}$$

where n is the order of the filter, the cutoff frequency has been normalized to 1 rad/sec and the $dc(\Omega = 0)$ gain magnitude has been normalized to 1 [i.e., $|H_a(0)| = 1$]. Consider the problem of approximating the ideal lowpass magnitude-squared frequency response characteristic,

$$|\hat{H}_a(j\Omega)|^2 = K, \quad |\Omega| < \Omega_c,$$
$$= 0, \quad |\Omega| > \Omega_c,$$

by a rational transfer function of the form

$$|H_a(j\Omega)|^2 = \frac{K\left[1 + a_1\Omega^2 + a_2\Omega^4 + \cdots + a_{n-1}\Omega^{2n-2}\right]}{1 + b_1\Omega^2 + b_2\Omega^4 + \cdots + b_n\Omega^{2n}}, \quad b_n > 0$$

Note that $|H_a(j\Omega)|^2$ must be an even function of Ω and its denominator degree must be higher than its numerator degree since it is a lowpass magnitude-squared frequency response function. For maximal flatness at the origin, $\Omega = 0$, the first $(2n-1)$ derivatives with respect to Ω of $|H_a(j\Omega)|^2$ or, equivalently, of

$$|H_a(j\Omega)|^2 - K = \frac{K\left[\sum_{i=1}^{n-1}(a_i - b_i)\Omega^{2i} - b_n\Omega^{2n}\right]}{1 + \sum_{i=1}^{n}b_i\Omega^{2i}}$$

must be zero. For this to be possible, the power series expansion about $\Omega = 0$ of the expression in the preceding equation should be of the form

$$|H_a(j\Omega)|^2 - K = \sum_{i=2n}^{\infty} g_i \Omega^i.$$

In order to satisfy the constraint of maximal flatness at $\Omega = 0$, it then becomes clear that

$$a_i = b_i, \quad i = 1, 2, \ldots, n-1.$$

If the condition of maximal flatness is also imposed at $\Omega = \infty$ on the function $|H_a(j\Omega)|^2$, rewritten as

$$|H_a(j\Omega)|^2 = \frac{K\left[\Omega^{-2n} + \sum_{i=1}^{n-1} a_{n-i} \Omega^{-2i}\right]}{\Omega^{-2n} + \sum_{i=0}^{n-1} b_{n-i} \Omega^{-2i}}$$

then

$$a_i = 0, \quad i = 1, 2, \ldots, n-1.$$

Therefore, for maximal flatness at $\Omega = 0$ and at $\Omega = \infty$, it is required that $|H_a(j\Omega)|^2$ be of the form $K/(1 + b_n \Omega^{2n})$ and following frequency scaling so that the half-power frequency is at $\Omega = 1$ (and then magnitude scale to make $K = 1$), the filter magnitude-squared frequency response function reduces to the expression given in (3.8). It is emphasized that the filter whose magnitude-squared response is given by (3.8) is maximally flat not only at $\Omega = 0$ (i.e., the derivatives of orders 1 to $2n-1$ of $|H_a(j\Omega)|^2$ are zero at $\Omega = 0$), but also at $\Omega = \infty$. In the complex frequency plane, (3.8) analytically continues to

$$H_a(s)H_a(-s) = \frac{1}{1 + (-s^2)^n}. \tag{3.9}$$

On applying the transformation in (3.6a) to (3.9), one gets

$$H(z)H(z^{-1}) = \frac{1}{1 + \left\{-\left[k(1 - z^{-1})/(1 + z^{-1})\right]^2\right\}^n}, \tag{3.10}$$

where $H(z)$ is defined to be the transfer function of the digital filter. The magnitude-squared response of the digital filter, on the use of (3.7) in (3.10), then has the form

$$|H(e^{j\omega T})|^2 = \frac{1}{1 + \left\{\left[\tan(\omega T/2)\right]\left[\cot(\omega_c T/2)\right]\right\}^{2n}}. \tag{3.11}$$

The following fact, stated without proof, justifies the validity of the maximally flat property of the digital filter, whose magnitude-squared characteristic is derived from a maximally flat analog filter characteristic by using the transformation in (3.6), which is an analytic transformation or mapping.

Fact 3.1. *The maximally flat property of the magnitude-squared lowpass function, $|H_a(j\Omega)|^2$ at $\Omega = 0$ is invariant at a point or points in the ω-plane to which $\Omega = 0$ is mapped under an analytic mapping, $\phi(\cdot)$, of the type*

$$\Omega \overset{\phi(\cdot)}{\to} \omega.$$

It is also mentioned that the above transformation has a natural extension to the two (see Chapter 7) or more dimensional cases.

The poles of $H(z)H(z^{-1})$ in (3.10) can be explicitly obtained after some algebraic manipulation. In the s-plane, with s related to z as in (3.6), these poles are at the zeros of [consistent with (3.9)],

$$(-1)^n s^{2n} + 1 = 0. \tag{3.12}$$

Let the zeros of (3.12) be at $s = s_r$, $r = 0, 1, 2, \ldots, 2n - 1$. Then, for odd n,

$$s_r = e^{jr\pi/n} \tag{3.13a}$$

and for even n

$$s_r = e^{j(2r+1)\pi/2n}. \tag{3.13b}$$

As expected, the poles in the s-plane lie on a unit circle. Let the location of the poles in the z-plane be given by $z_r = x_r + jy_r$, $r = 0, 1, 2, \ldots, 2n - 1$, where x_r and y_r are, respectively, the real and imaginary parts of z_r. Then, by straightforward manipulations on (3.6),

$$x_r = \frac{1 - 1/k^2}{(1 - \operatorname{Re} s_r/k)^2 + (\operatorname{Im} s_r/k)^2} \tag{3.14a}$$

$$y_r = \frac{2 \operatorname{Im} s_r/k}{(1 - \operatorname{Re} s_r/k)^2 + (\operatorname{Im} s_r/k)^2} \tag{3.14b}$$

and after substituting the value of k from (3.7) and values of $\operatorname{Re} s_r, \operatorname{Im} s_r$ [obtainable from 3.13; for odd n, $\operatorname{Re} s_r = \cos(r\pi/n)$, $\operatorname{Im} s_r = \sin(r\pi/n)$] one gets the following:
odd n case

$$x_r = \frac{1 - \tan^2(\omega_c T/2)}{1 + \tan^2(\omega_c T/2) - 2\tan(\omega_c T/2)\cos(r\pi/n)},$$

$$y_r = \frac{2\tan(\omega_c T/2)\sin(r\pi/n)}{1 + \tan^2(\omega_c T/2) - 2\tan(\omega_c T/2)\cos(r\pi/n)} \tag{3.15}$$

164

```
C
C          FORTRAN PROGRAM (BUTTER.FOR) TO OBTAIN THE POLES OF THE
C          STABLE BUTTERWORTH DIGITAL LOW-PASS FILTER, AS WELL AS
C          THE DENOMINATOR POLYNOMIAL, FROM WHICH THE TRANSFER
C          FUNCTION CAN BE FORMED
C
           REAL X(200),Y(200),U(200),V(200)
           PI=3.141592654
C
           WRITE(6,10)
   10      FORMAT(//,12X,'ENTER N, THE NUMBER OF POLES OF H(Z)')
           READ(5,20) N
           WRITE(1,20) N
   20      FORMAT(I)
           M=2*N
           RN=FLOAT(N)
           WRITE(6,30)
   30      FORMAT(//,12X,'ENTER WC, THE CUTOFF FREQUENCY IN RAD/SEC')
           READ(5,40) WC
   40      FORMAT(F)
           WRITE(6,50)
   50      FORMAT(//,12X,'ENTER T, THE SAMPLING PERIOD IN SEC')
           READ(5,40) T
C
           A=TAN(WC*T/2.)
           L0=MOD(N,2)
           IF(L0.EQ.1) GO TO 200
C
           DO 110 I=1,M
           X(I)=A*COS((2.*FLOAT(I-1)+1.)*PI/(2.*RN))
           Y(I)=A*SIN((2.*FLOAT(I-1)+1.)*PI/(2.*RN))
  110      CONTINUE
           GO TO 300
C
  200      DO 210 I=1,M
           X(I)=A*COS(FLOAT(I-1)*PI/RN)
           Y(I)=A*SIN(FLOAT(I-1)*PI/RN)
  210      CONTINUE
C
  300      WRITE(6,320)
           DO 310 I=1,M
           U(I)=(1.-X(I)**2-Y(I)**2)/((1.-X(I))**2+Y(I)**2)
           V(I)=2.*Y(I)/((1.-X(I))**2+Y(I)**2)
           K=I-1
           WRITE(6,330) K,U(I),V(I)
  310      CONTINUE
  320      FORMAT(//,10X,'POLE LOCATIONS OF H(Z)*H(1/Z) IN 1/Z-PLANE',
      1    //,14X,'K;',5X,'REAL PART',11X,'IMAG PART',//)
```

Figure 3.1. (*Partial listing, continued.*)

```
330     FORMAT(12X,I3,3X,E17.8,3X,E17.8)
C
        CALL ZERO(X(1),X(N+1))
        CALL ZERO(Y(1),Y(N+1))
        L=0
        DO 400 I=1,M
        II=I
        J=1
410     IF((X(J).NE.U(I)).OR.(ABS(Y(J)).NE.ABS(V(I)))) GO TO 420
        IF(J.EQ.II) GO TO 400
        J=J+1
        GO TO 410
420     L=L+1
        X(L)=U(I)
        Y(L)=V(I)
400     CONTINUE
C
        L=0
        DO 430 I=1,N+1
        R=X(I)**2+Y(I)**2
        IF(R.LE.1.) GO TO 430
        L=L+1
        X(L)=X(I)/R
        Y(L)=-Y(I)/R
430     CONTINUE
C
        IF(L0.EQ.0) GO TO 500
        MAX=N/2+1
        DO 440 I=1,MAX
        IF(ABS(Y(I)).GT.10.E-6) GO TO 440
        INDEX=I
440     CONTINUE
        RX=X(INDEX)
        DO 450 I=INDEX,MAX-1
        X(I)=X(I+1)
        Y(I)=Y(I+1)
450     CONTINUE
C
500     CONTINUE
        NN=N/2
        DO 510 I=1,NN
        U(I)=-2.*X(I)
        V(I)=X(I)**2+Y(I)**2
510     CONTINUE
        CALL REALCO(NN,X,Y,U,V)
        IF(L0.EQ.0) GO TO 530
        IF(NN.NE.1) GO TO 529
        Y(2)=X(2)
```

Figure 3.1. (*Partial listing, continued.*)

```
        Y(3)=X(3)
529     CONTINUE
        Y(1)=1.
        DO 520 I=2,N+1
        X(I)=X(I)-Y(I-1)*RX
520     CONTINUE
530     CONTINUE
        WRITE(6,540)
540     FORMAT(//,10X,'COEFF OF STABLE DENOMINATOR IN Z, NOT IN 1/Z',
   1    /,14X,'STARTING FROM THE HIGHEST DEGREE COEFFICIENT',//)
        WRITE(6,550) ((I,X(I)),I=1,N+1)
550     FORMAT(12X,I3,';',3X,E17.8)
        STOP
        END
C
C
        SUBROUTINE REALCO(NN,X,Y,U,V)
        REAL X(200),Y(200),U(200),V(200)
        CALL ZERO(X(1),X(200))
        CALL ZERO(Y(1),Y(200))
        X(1)=1.
        X(2)=U(1)
        X(3)=V(1)
        IF(NN.EQ.1) RETURN
        M=2
100     MM=2*M+1
        DO 110 I=2,MM
        II=I
        IF(II.EQ.2) GO TO 111
        Y(I)=X(I)+X(I-1)*U(M)+X(I-2)*V(M)
        GO TO 110
111     Y(I)=X(I)+U(M)
110     CONTINUE
        DO 120 I=2,MM
        X(I)=Y(I)
120     CONTINUE
        IF(M.EQ.NN) RETURN
        M=M+1
        GO TO 100
        END
```

Figure 3.1. Listing of program BUTTER.FOR for use in digital Butterworth filter design.

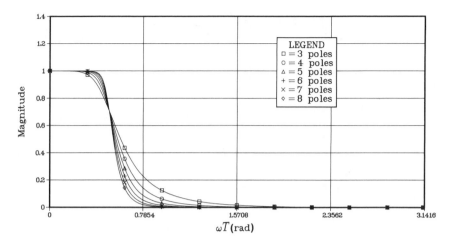

Figure 3.2. Typical plots of the magnitude of the frequency response for various orders n of a digital Butterworth filter; $\omega_c = 1000$ rad/sec, $T = 0.5$ msec.

for $r = 0, 1, 2, \ldots, 2n - 1$;
even n case

$$x_r = \frac{1 - \tan^2(\omega_c T/2)}{1 + \tan^2(\omega_c T/2) - 2\tan(\omega_c T/2)\cos[(2r + 1)/2n]\pi}$$

$$y_r = \frac{2\tan(\omega_c T/2)\sin[(2r + 1)/2n]\pi}{1 + \tan^2(\omega_c T/2) - 2\tan(\omega_c T/2)\cos[(2r + 1)/2n]\pi} \tag{3.16}$$

for $r = 0, 1, 2, \ldots, 2n - 1$.

The computer program BUTTER.FOR (listing provided in Figure 3.1) can be used to obtain the location of the poles for any specified n, ω_c, and T. The poles of the stable filter transfer function $H(z)$ are the n poles located within the unit circle. BUTTER.FOR also forms the polynomial [the denominator of $H(z)$] whose roots coincide with the location of these n poles within the unit circle. From (3.10), it is clear that the numerator of $H(z)$ must have a zero of multiplicity n at $z = -1$. Before illustrating the design of Butterworth filters, the reader's attention is directed to the shape of the magnitude of frequency responses of such filters for different values of n in Figure 3.2 (in these plots, the cutoff frequency ω_c, which plays an identical role in both the continuous and digital filter designs, has been taken as 1000 rad/sec and the sampling period T is 0.5 msec corresponding to a sampling frequency of 2 kHz). Also, the plots are normalized so that $|H(1)| = 1$, which implies that at $\omega = 0$ the magnitude of the gain is unity. It is clear from the plots that the response falls off faster as n increases. In fact, the order n of the filter is normally chosen so that a certain minimum attenuation in decibels is obtained at a desired frequency, referred to as the transition frequency f_t (or transition angular frequency $\omega_t = 2\pi f_t$). It is also clear from the figure that monotonic behavior is displayed in both the pass and stop bands.

Example 3.1. It is required to design a lowpass digital Butterworth filter, with a maximally flat magnitude characteristic around $\omega = 0$, so that the cutoff frequency is at $f_c = \omega_c/2\pi = 2.25$ kHz. At the transition frequency $f_t = 2.5$ kHz, a minimum of 9 db attenuation is sought. The specified sampling rate is 9 kHz.

From the above data

$$T = 1/(9 \times 10^3) = \tfrac{1}{9} \text{ msec},$$

$$\tan(\omega_c T/2) = \tan 0.25\pi = 1,$$

$$\tan(\omega_t T/2) = \tan(2.5\pi/9) = \tan 50°,$$

$$10 \log\left[1 + (\tan 50°)^{2n}\right] \geq 9.$$

The minimum value of integer n that will satisfy the above inequality is $n = 6$. Using BUTTER.FOR, the poles of $H(z)H(z^{-1})$ are located as follows:

x_r	y_r
0.42364937E − 05	0.75957542E + 01
0.50239807E − 06	0.24142136E + 01
0.19099392E − 06	0.13032254E + 01
0.11541480E − 06	0.76732700E + 00
0.88380107E − 07	0.41421356E + 00
0.72007320E − 07	0.13165250E + 00
0.72954785E − 07	− 0.13165250E + 00
0.88380108E − 07	− 0.41421358E + 00
0.11541480E − 06	− 0.76732700E + 00
0.20104623E − 06	− 0.13032254E + 01
0.48967917E − 06	− 0.24142137E + 01
0.41544967E − 05	− 0.75957539E + 01

From (3.16), you would expect the x_r's to be zero in this problem, provided calculations were carried out with infinite precision. Also, the coefficients of the stable denominator polynomial (stability may be checked via the use of SCHUR1.FOR, whose listing was provided in Figure 1.3), $D(z)$, of the

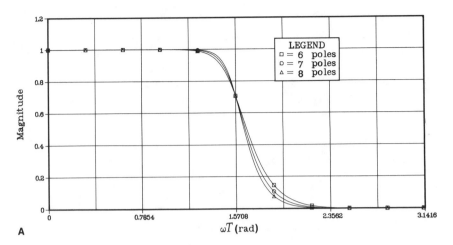

A

Figure 3.3A. Plot of magnitude of frequency response of Butterworth filter when $n = 6, 7, 8$.

filter, written in descending powers of z are

$$0.10000000E+01, \quad -0.54416368E-06, \quad 0.77769595E+01,$$
$$-0.25864464E-06, \quad 0.11419942E+00, \quad -0.17263729E-07,$$
$$0.17509259E-02.$$

$$(3.17a)$$

The numerator polynomial of the filter when $n = 6$ is

$$N(z) = (z+1)^6. \tag{3.17b}$$

Letting $H(1) = KN(1)/D(1) = 1$,

$$K = D(1)/N(1) = 0.138951716. \tag{3.17c}$$

The desired filter transfer function is

$$H(z) = KN(z)/D(z), \tag{3.17d}$$

where $D(z)$, $N(z)$, and K are as in (3.17a), (3.17b), and (3.17c). The magnitude of the frequency response of the digital filter is plotted in Figure 3.3A and in db in Figure 3.3B while the phase response is given in Figure 3.3C (both plots are vs. ωT in radians). In case a different attenuation is desired at the transition frequency, the order n of the filter has to be changed. Table 3.1 may be helpful for that purpose. In Figures 3.3A–3.3C, the magnitude and phase response plots done for $n = 6$ are given also for $n = 7, 8$. Figures 3.3D–3.3F are plots of the magnitude of the frequency responses as ratios between 0 and 1, in decibels, and also the phase

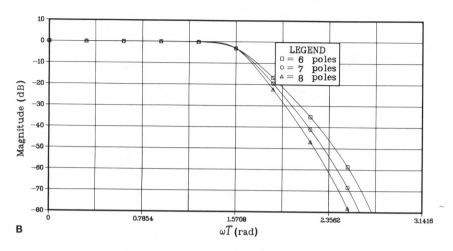

B

Figure 3.3B. Plot of magnitude of frequency response in db of Butterworth filter when $n = 6, 7, 8$.

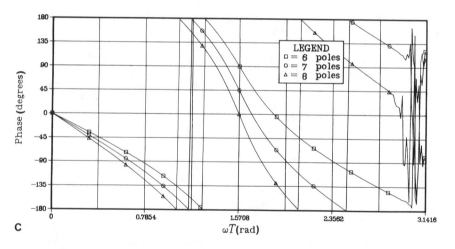

C

Figure 3.3C. Plot of phase response of Butterworth filter when $n = 6, 7, 8$.

responses when $n = 14, 20, 27$. Use of BUTTER.FOR was made in the design of all these filters.

3.2.2. Digital Chebyshev Filter

The magnitude of the frequency response of Chebyshev filters is known to have an equiripple behavior, in either the passband or the stopband, and the ripple is such that the maximum error in the band where the ripple occurs is

Table 3.1. Order n versus Minimum
Desired Attenuation at $\omega_t T =$
1.74532925 Rad for Example 3.1

Order n of filter	Attenuation at ω_t, $-20\log|H(e^{j\omega_t})|$
3	-5.871451933 db
4	-7.049255571 db
5	-8.311758668 db
6	-9.64165569 db
7	-11.02352682 db
8	-12.44452071 db

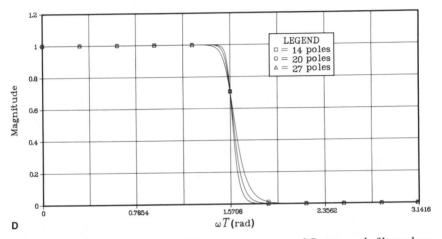

D

Figure 3.3D. Plot of magnitude of frequency response of Butterworth filter when $n = 14, 20, 27$.

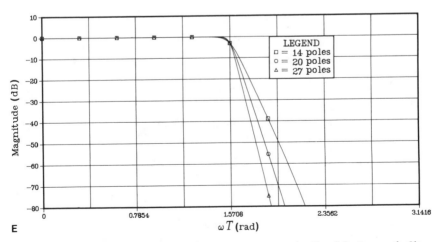

E

Figure 3.3E. Plot of magnitude of frequency response in db of Butterworth filter when $n = 14, 20, 27$.

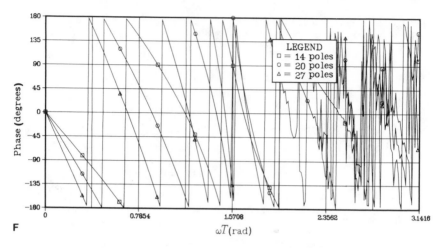

F

Figure 3.3F. Plot of phase response of Butterworth filter when $n = 14, 20, 27$.

minimized. The response in the band, free of ripples, is monotonic. Here, the design of lowpass Chebyshev filters where the passband contains ripples will be considered. The Chebyshev polynomial (in the complex variable x) of the first kind, of order n, which plays a major role in the design of this type of filters, is defined as

$$T_n(x) = \cos(n \cos^{-1} x) = \cos n\theta, \quad \theta \triangleq \cos^{-1} x.$$

Clearly, when $x \geq 1$, $T_n(x) = \cosh n \cosh^{-1} x$, and, therefore, $T_n(x) \neq 0$, $|x| \geq 1$. It is easy to verify that $T_n(x) \neq 0$, when x is strictly complex. Therefore, the only zeros of $T_n(x)$ are in $-1 < x < 1$. The sequence $\{T_n(x)\}$ is known to satisfy the three term recurrence relation [see Problem 4(a)]:

$$T_{n+1}(x) - 2x T_n(x) + T_{n-1}(x) = 0,$$

with $T_0(x) = 1$, $T_1(x) = x$. The listing for the computer program CHEGE.FOR which uses this recurrence relation to generate the Chebyshev polynomials of the first kind is given in Figure 3.4. For the convenience of the reader, the coefficients of $T_n(x)$, for $n = 1, 2, \ldots, 20$, are tabulated in Table 3.2; these coefficients are ordered according to increasing powers of x in $T_n(x)$. The sequence of polynomials $\{T_n(x)\}$ satisfies the following important properties:

a. $\{T_n(x)\}$ forms an orthogonal set over $-1 \leq x \leq 1$; it is readily verifiable that

$$\int_{-1}^{1} \frac{1}{\sqrt{1 - x^2}} T_n(x) T_m(x) \, dx = \begin{cases} 0, & m \neq n, \\ \pi/2, & m = n \neq 0, \\ \pi, & m = n = 0. \end{cases}$$

b. $T_n(x) = -T_n(-x)$, for odd n,
   $T_n(x) = T_n(-x)$, for even n.
c. $T_n(1) = 1$, for each n.
d. $T_n(-1) = 1$, for even n, $T_n(0) = (-1)^{n/2}$, for even n,
   $T_n(-1) = -1$, for odd n, $T_n(0) = 0$, for odd n.
e. The range of $T_n(x)$ is between -1 and $+1$, for $-1 \le x \le 1$, and within this range $T_n(x)$ varies with an equal ripple as x varies.
f. The zeros of $T_n(x)$ lie in $-1 < x < 1$.

```
C
C        A FORTRAN PROGRAM (CHEGE.FOR) TO COMPUTE THE COEFFICIENTS
C        OF THE CHEBYSHEV POLYNOMIALS OF THE FIRST KIND & ORDER N.
C
         INTEGER IT(21,20)
C
         WRITE(6,110)
110      FORMAT(2X,'ENTER THE ORDER N')
         READ(5,120) N
120      FORMAT(I)
C
         IT(2,1)=1
         IT(1,2)=-1
         IT(3,2)=2
         DO 200 J=3,N
         DO 200 I=1,J+1
200      IT(I,J)=2*IT(I-1,J-1)-IT(I,J-2)
C
         WRITE(6,300) N
300      FORMAT(//,10X,46HCOEFFICIENTS OF CHEBYSHEV POLYNOMIALS, TI(X)'S,
1        /,24X,'FROM T1(X) TO T',I2,'(X)',///,10X,
2        'STARTING FROM THE LOWEST DEGREE COEFFICIENT OF EACH POLY.',/)
C
         DO 500 J=1,N
         N=J
         IF(N.GT.9) GO TO 400
         WRITE(6,510) N,(IT(I,J),I=1,N+1)
         GO TO 500
400      WRITE(6,520) N,(IT(I,J),I=1,N+1)
500      CONTINUE
510      FORMAT(/,10X,'T',I1,'(X) ;',5(I9,','),/,17X,
1        5(I9,','),/,17X,5(I9,','),/,17X,5(I9,','),/,17X,I9,',')
520      FORMAT(/,10X,'T',I2,'(X);',5(I9,','),/,17X,5(I9,','),
1        /,17X,5(I9,','),/,17X,5(I9,','),/,17X,I9,',')
         STOP
         END
```

Figure 3.4. Listing of program CHEGE.FOR to compute the Chebyshev polynomials of the first kind.

Table 3.2. Coefficients of Chebyshev Polynomials $T_n(x)$ of the First Kind, of Order n, in Ascending Powers of Variable x

n	Coefficients of $T_n(x)$ in ascending powers of x; t_{ij} is the element in the ith row, $i = 1,2,3,\ldots$ and $(j+1)$th column, $j = 0,1,2,\ldots$
1	$0, 1$
2	$-1, 0, 2$
3	$0, -3, 0, 4$
4	$1, 0, -8, 0, 8$
5	$0, 5, 0, -20, 0, 16$
6	$-1, 0, 18, 0, -48, 0, 32$
7	$0, -7, 0, 56, 0, -112, 0, 64$
8	$1, 0, -32, 0, 160, 0, -256, 0, 128$
9	$0, 9, 0, -120, 0, 432, 0, -576, 0, 256$
10	$-1, 0, 50, 0, -400, 0, 1120, 0, -1280, 0, 512$
11	$0-11, 0, 220, 0, -1232, 0, 2816, 0, -2816, 0, 1024$
12	$1, 0, -72, 0, 840, 0, -3584, 0, 6912, 0, -6144, 0, 2048$
13	$0, 13, 0, -364, 0, 2912, 0, -9984, 0, 16640, 0, -13,312, 0, 4096$
14	$-1, 0, 98, 0, -1568, 0, 9408, 0, -26,880, 0, 39,424, 0, -28,672, 0, 8192$
15	$0, -15, 0, 560, 0, -6048, 0, 28,800, 0, -70,400, 0, 92,160, 0, -61,440, 0, 16,384$
16	$1, 0, -128, 0, 2688, 0, -21,504, 0, 84,480, 0, -180,224, 0, 212,992, 0, -131,072, 0, 32,768$
17	$0, 17, 0, -816, 0, 11,424, 0, -71,808, 0, 239,360, 0, -452,608, 0, 487,424, 0, -278,528, 0, 65,536$
18	$-1, 0, 162, 0, -4320, 0, 44,352, 0, -228,096, 0, 658,944, 0, -1,118,208, 0, 1,105,920, 0, -589,824, 0, 131,072$
19	$0, -19, 0, 1140, 0, -20,064, 0, 160,512, 0, -695,552, 0, 1,770,496, 0, -2,723,840, 0, 2,490,368, 0, -1,245,184, 0, 262,144$
20	$1, 0, -200, 0, 6600, 0, -84,480, 0, 549,120, 0, -2,050,048, 0, 4,659,200, 0, -6,553,600, 0, 5,570,560, 0, -2,621,440, 0, 524,288$

Note: If $T_n(x) = \sum_{j=0}^{n} t_{nj} x^j$, then $|t_{ij}| = 2|t_{(i-1)(j-1)}| + |t_{(i-2)j}|$ for $i = 3,4,\ldots$, and $j = 0,1,2,\ldots$, with $t_{i(-1)} \triangleq 0$, $i = 1,2,\ldots$, and $t_{ij} = 0$ for $i < j$. Also $t_{ii} = 2t_{(i-1)(i-1)}$, $i = 2,3,4,\ldots$ [21].

Figure 3.5A–I shows the plots of $T_n(x)$ vs. x, for $n = 2,\ldots,10$, and the reader can verify that the properties stated above are indeed satisfied. From the magnitude-squared response function [obtained by substituting $s = j\Omega$ in (3.18)],

$$H_a(s)H_a(-s) = \frac{1}{1 + \varepsilon^2 T_n^2(s/j)} \tag{3.18}$$

of an analog Chebyshev filter, the counterpart of (3.11) can be obtained via use of (3.6) and (3.7). In (3.18), ε^2 is the *ripple factor*. The *ripple amplitude* δ is given by

$$\delta = 1 - 1/\sqrt{1 + \varepsilon^2}. \tag{3.19}$$

It is readily verifiable by implementing the operations suggested above that

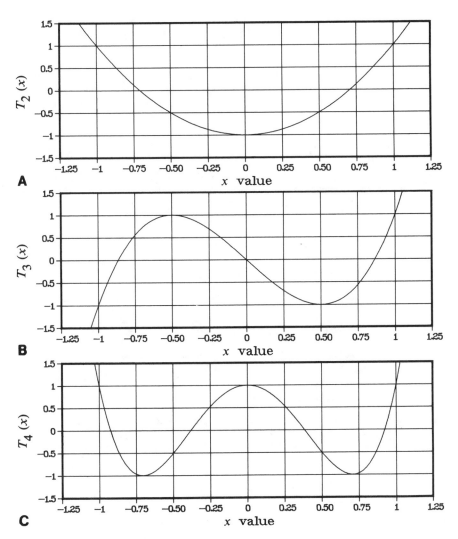

Figure 3.5. A to I. $T_p(x)$ versus x for $n = 2, 3, 4, 5, 6, 7, 8, 9, 10$.

the magnitude-squared response function of a digital Chebyshev lowpass filter, with an equiripple behavior in the passband and a monotonic decay in the stopband is given by[2]

$$\left| H\!\left(e^{j\omega T}\right)\right|^2 = \frac{1}{1 + \varepsilon^2 T_n^2\!\left[k \tan(\omega T/2)\right]}, \tag{3.20}$$

---

[2] In (3.11) and (3.20), $\tan(\omega T/2)$, is, justifiably, referred to as the lowpass frequency function (see Section 3.5.1); in this case, in (3.6a), $k = \cot(\omega_p T/2)$.

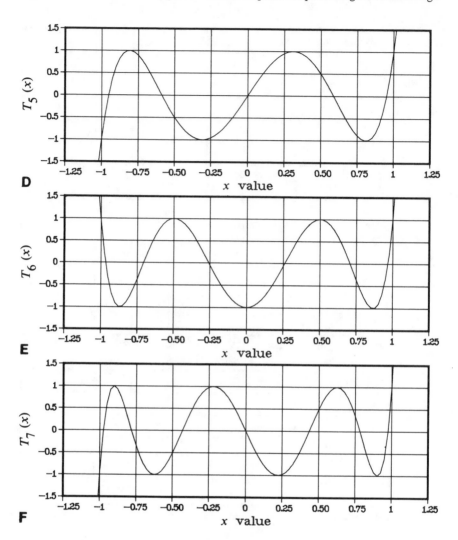

Figure 3.5. (*Continued.*)

where $k = \cot(\omega_p T/2)$ and ω_p is the frequency in rad/sec at the end of the equiripple passband. (3.20) is obtained from (3.18) via the use of (3.6a). Since $T_n(1) = 1$, for each n, the above equation suggests that

$$\left| H(e^{j\omega_p T}) \right|^2 = \frac{1}{1 + \varepsilon^2}.$$

The half-power frequency in rad/sec for the analog Chebyshev filter characterized in (3.18) is $\cosh[(1/n)\cosh^{-1}(1/\varepsilon)]$. Plots of $|H(e^{j\omega T})|$ vs. ωT, for

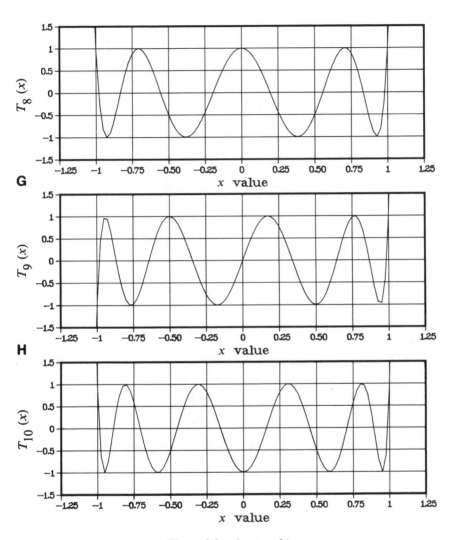

Figure 3.5. (*Continued*.)

chosen values of ε, ω_p, and T are shown in Figure 3.6, and these plots depict the shape of the Chebyshev lowpass digital filter magnitude response.

To obtain the poles of $H(z)$, the Chebyshev digital filter transfer function obtainable from $H_a(s)$ in (3.18) after applying the bilinear transformation of (3.6), consider first the zeros of

$$1 + \varepsilon^2 T_n^2(s/j) = 0 \qquad (3.21)$$

in the s-plane. (3.21) is obtained by equating to zero the denominator of

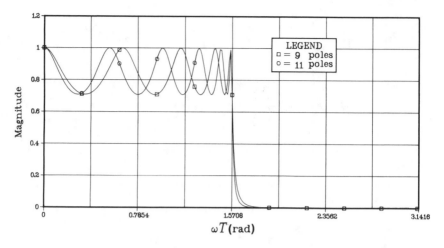

Figure 3.6. Plots of magnitude of frequency response of Chebyshev filters.

(3.18). Let

$$\cos w = \cos(u + jv) = s/j, \tag{3.22}$$

where $w = u + jv$ is another complex variable. Then, from the definition of $T_n(x)$,

$$T_n(s/j) = \cos(n\cos^{-1}(s/j)) = \cos nw = \cos n(u + jv) \tag{3.23}$$
$$= \cos nu \cosh nv - j\sin nu \sinh nv.$$

From (3.21), $T_n(s/j) = \pm j/\varepsilon$ and using (3.23), it follows that

$$\cos nu \cosh nv = 0, \tag{3.24a}$$

$$-\sin nu \sinh nv = \pm 1/\varepsilon. \tag{3.24b}$$

(3.24a) holds only if u takes on values

$$u = u_r = (2r+1)\pi/2n, \quad r = 0,1,\ldots,2n-1. \tag{3.25}$$

Substituting (3.25) in (3.24b), one gets (it is convenient to consider v positive and show the sign explicitly in the values of the roots)

$$v = \frac{1}{n}\sinh^{-1}\frac{1}{\varepsilon}. \tag{3.26}$$

Note that the right-hand side of (3.26) is independent of r, and therefore the subscript r is not used in v.

From (3.26), it follows that

$$(e^{nv})^2 - \frac{2}{\varepsilon}e^{nv} - 1 = 0 \tag{3.27}$$

after using the formula

$$\sinh x = \frac{e^x - e^{-x}}{2}.$$

Solving (3.27), one gets (note that e^v cannot be negative for real v)

$$e^v = \left[\frac{1}{\varepsilon} + \left(\frac{1}{\varepsilon^2} + 1\right)^{1/2}\right]^{1/n}.\qquad(3.28)$$

Therefore,

$$\cosh v = \tfrac{1}{2}\left\{\left[\frac{1}{\varepsilon} + \left(\frac{1}{\varepsilon^2} + 1\right)^{1/2}\right]^{1/n} + \left[\frac{1}{\varepsilon} + \left(\frac{1}{\varepsilon^2} + 1\right)^{1/2}\right]^{-1/n}\right\},$$

$$(3.29a)$$

$$\sinh v = \tfrac{1}{2}\left\{\left[\frac{1}{\varepsilon} + \left(\frac{1}{\varepsilon^2} + 1\right)^{1/2}\right]^{1/n} - \left[\frac{1}{\varepsilon} + \left(\frac{1}{\varepsilon^2} + 1\right)^{1/2}\right]^{-1/n}\right\}.$$

$$(3.29b)$$

Let the zeros of (3.21) be at $s = s_r = \sigma_r + j\Omega_r$, $r = 0, 1, \ldots, 2n - 1$, and let $s_r = j\cos(u_r + jv_r) = j\cos(u_r + jv)$ from (3.22). Then

$$s_r = \sin u_r \sinh v + j\cos u_r \cosh v = \sigma_r + j\Omega_r,\qquad(3.30)$$

whence

$$\left(\frac{\sigma_r}{\sinh v}\right)^2 + \left(\frac{\Omega_r}{\cosh v}\right)^2 = 1.\qquad(3.31)$$

Therefore, the poles of a Chebyshev filter lie on an ellipse in the s-plane. The ellipse is tangent to the circles of radii $\cosh v$ and $\sinh v$, respectively. These circles may be referred to as the Butterworth circles. The major and minor semiaxes of the ellipse are, respectively, $\cosh v$ and $\sinh v$, and the foci are at $s = +j$ and $s = -j$, which imply that the semiaxis, whose length is $\cosh v$, lies on the imaginary axis in the s-plane.[3] From (3.30), σ_r and Ω_r can be parametrized as

$$\sigma_r = (\sinh v)\cos\theta_r, \quad \Omega_r = (\cosh v)\sin\theta_r,$$

where θ_r is the angle which the line joining $(0,0)$ and the Butterworth filter pole given in (3.13a) or (3.13b) makes with the positive direction of the

---

[3] The nth-order Chebyshev filter poles are obtainable as points of intersection of, respectively, the horizontal and vertical lines through the nth-order Butterworth filter poles on circles of radii $\cosh v$ and $\sinh v$.

abscissa in the s-plane. The locations of the poles in the z-plane are obtainable via the use of (3.6b). Noting that $k = \cot(\omega_p T/2)$ and $\theta_r = r\pi/n$, $r = 0, 1, \ldots, 2n - 1$ *for odd* n, the z-plane pole locations can be verified to be given by

$$x_r = \frac{1 - \tan^2(\omega_p T/2)\left[\sinh^2 v \cos^2(r\pi/n) + \cosh^2 v \sin^2(r\pi/n)\right]}{\left[1 - \sinh v \tan(\omega_p T/2) \cos(r\pi/n)\right]^2 + \left[\cosh v \tan(\omega_p T/2) \sin(r\pi/n)\right]^2},$$

$$y_r = \frac{2 \cosh v \tan(\omega_p T/2) \sin(r\pi/n)}{\left[1 - \sinh v \tan(\omega_p T/2) \cos(r\pi/n)\right]^2 + \left[\cosh v \tan(\omega_p T/2) \sin(r\pi/n)\right]^2}$$

for odd n and $r = 0, 1, 2, \ldots, (2n - 1)$. For the case of even n, $r\pi/n$ in the above expressions should be replaced by $(2r + 1)\pi/2n$. See Figure 3.7 for listing of CHEBYS.FOR which yields the poles and also forms the stable denominator polynomial of a Chebyshev filter being designed.

Example 3.2. Consider design of a lowpass digital Chebyshev filter with the ripple band edge frequency at 2.25 kHz, where the loss is specified to be 3 db. The filter is, furthermore, required to have at least 10 db attenuation at 2.5 kHz. Also, the transfer function $H(z)$ evaluated at $z = 1$ is to be normalized to unity. The sampling frequency may be taken to be 9 kHz. From (3.20),

$$\left|H(e^{j\omega_p T})\right|^2 = \frac{1}{1 + \varepsilon^2 T_n^2(1)}.$$

Since $T_n(1) = 1$ for all n, and the 3 db loss occurs at ω_p rad/sec,

$$10\log\left[1 + \varepsilon^2 T_n^2(1)\right] = 3,$$

which when solved for ε yields

$$\varepsilon = \varepsilon_0 = 0.9976283451.$$

Actually, if ω_p is the half-power frequency, ε should be 1. At $f_1 = 2.5$ kHz, $\omega_1 = 2\pi \times 2.5$ krad/sec;

$$\frac{\omega_1 T}{2} = \frac{5\pi}{2 \times 9} = \frac{100°}{2} = 50°.$$

Therefore,

$$10\log\left[1 + \varepsilon_0^2 T_n^2(\tan 50°)\right] \geq 10.$$

The minimum value of n which satisfies the above inequality for the value of ε_0, already determined, is

$$n = 3.$$

To obtain the value of n given above, you may consult Table 3.3. Using

```
C       FORTRAN PROGRAM (CHEBYS.FOR) TO OBTAIN THE POLES OF THE
C       STABLE CHEBYSHEV DIGITAL LOW-PASS FILTER, AS WELL AS
C       THE DENOMINATOR POLYNOMIAL, FROM WHICH THE TRANSFER
C       FUNCTION CAN BE OBTAINED
C
        REAL X(200),Y(200),U(200),V(200)
        PI=3.141592654
C
        WRITE(6,10)
  10    FORMAT(//,12X,'ENTER N, THE NUMBER OF POLES OF H(Z)')
        READ(5,20) N
        WRITE(3,20) N
  20    FORMAT(I)
        M=2*N
        RN=FLOAT(N)
        WRITE(6,30)
  30    FORMAT(//12X,'ENTER WP,',/,12X,
   1    'THE FREQUENCY IN RAD/SEC AT THE END OF THE RIPPLE BAND')
        READ(5,40) WP
  40    FORMAT(F)
        WRITE(6,50)
  50    FORMAT(//,12X,'ENTER T, THE SAMPLING PERIOD IN SEC')
        READ(5,40) T
        WRITE(6,60)
  60    FORMAT(//,12X,'ENTER E,THE RIPPLE FACTOR, 0<E<1')
        READ(5,40) E
C
        A=TAN(WP*T/2.)
        D1=1./E**2+1.
        D2=1./E
        D3=1./RN
        BB=.5*((SQRT(D1)+D2)**D3-(SQRT(D1)+D2)**(-D3))
        CC=.5*((SQRT(D1)+D2)**D3+(SQRT(D1)+D2)**(-D3))
        L0=MOD(N,2)
        IF(L0.EQ.1) GO TO 200
C
        DO 110 I=1,M
        X(I)=A*BB*COS((2.*FLOAT(I-1)+1.)*PI/(2.*RN))
        Y(I)=A*CC*SIN((2.*FLOAT(I-1)+1.)*PI/(2.*RN))
 110    CONTINUE
        GO TO 300
C
 200    DO 210 I=1,M
        X(I)=A*BB*COS(FLOAT(I-1)*PI/RN)
        Y(I)=A*CC*SIN(FLOAT(I-1)*PI/RN)
 210    CONTINUE
C
 300    WRITE(6,320)
```

Figure 3.7. (*Partial listing, continued.*)

```
        DO 310 I=1,M
        U(I)=(1.-X(I)**2-Y(I)**2)/((1.-X(I))**2+Y(I)**2)
        V(I)=2.*Y(I)/((1.-X(I))**2+Y(I)**2)
        K=I-1
        WRITE(6,330) K,U(I),V(I)
310     CONTINUE
320     FORMAT(//,10X,'POLE LOCATIONS OF H(Z)*H(1/Z) IN 1/Z-PLANE',
   1    //,14X,'K;',5X,'REAL PART',11X,'IMAG PART',//)
330     FORMAT(12X,I3,3X,E17.8,3X,E17.8)
C
        CALL ZERO(X(1),X(N+1))
        CALL ZERO(Y(1),Y(N+1))
        L=0
        DO 400 I=1,M
        II=I
        J=1
410     IF((X(J).NE.U(I)).OR.(ABS(Y(J)).NE.ABS(V(I)))) GO TO 420
        IF(J.EQ.II) GO TO 400
        J=J+1
        GO TO 410
420     L=L+1
        X(L)=U(I)
        Y(L)=V(I)
400     CONTINUE
C
        L=0
        DO 430 I=1,N+1
        R=X(I)**2+Y(I)**2
        IF(R.LE.1.) GO TO 430
        L=L+1
        X(L)=X(I)/R
        Y(L)=-Y(I)/R
430     CONTINUE
C
        IF(L0.EQ.0) GO TO 500
        MAX=N/2+1
        DO 440 I=1,MAX
        IF(ABS(Y(I)).GT.10.E-6) GO TO 440
        INDEX=I
440     CONTINUE
        RX=X(INDEX)
        DO 450 I=INDEX,MAX-1
        X(I)=X(I+1)
        Y(I)=Y(I+1)
450     CONTINUE
C
500     CONTINUE
        NN=N/2
```

Figure 3.7. (*Partial listing, continued.*)

```
      DO 510 I=1,NN
      U(I)=-2.*X(I)
      V(I)=X(I)**2+Y(I)**2
510   CONTINUE
      CALL REALCO(NN,X,Y,U,V)
      IF(L0.EQ.0) GO TO 530
      IF(NN.NE.1) GO TO 529
      Y(2)=X(2)
      Y(3)=X(3)
529   CONTINUE
      Y(1)=1.
      DO 520 I=2,N+1
      X(I)=X(I)-Y(I-1)*RX
520   CONTINUE
530   CONTINUE
      WRITE(6,540)
540   FORMAT(//,10X,'COEFF OF STABLE DENOMINATOR IN Z, NOT IN 1/Z',
     1 /,14X,'STARTING FROM THE HIGHEST DEGREE COEFFICIENT',//)
      WRITE(6,550) ((I,X(I)),I=1,N+1)
550   FORMAT(12X,I3,';',3X,E17.8)
      STOP
      END
C
C
      SUBROUTINE REALCO(NN,X,Y,U,V)
      REAL X(200),Y(200),U(200),V(200)
      CALL ZERO(X(1),X(200))
      CALL ZERO(Y(1),Y(200))
      X(1)=1.
      X(2)=U(1)
      X(3)=V(1)
      IF(NN.EQ.1) RETURN
      M=2
100   MM=2*M+1
      DO 110 I=2,MM
      II=I
      IF(II.EQ.2) GO TO 111
      Y(I)=X(I)+X(I-1)*U(M)+X(I-2)*V(M)
      GO TO 110
111   Y(I)=X(I)+U(M)
110   CONTINUE
      DO 120 I=2,MM
      X(I)=Y(I)
120   CONTINUE
      IF(M.EQ.NN) RETURN
      M=M+1
      GO TO 100
      END
```

Figure 3.7. Listing of program CHEBYS.FOR for use in design of digital Chebyshev filter.

CHEBYS.FOR (see listing in Figure 3.7), the poles of $H(z)H(z^{-1})$ are at:

Real Part	Imaginary Part
0.18515219E+01	0.00000000E+00
0.10439507E+00	0.11733630E+01
0.75229985E−01	0.84555791E+00
0.54009622E+00	0.24558222E−08
0.75229985E−01	−0.84555791E+00
0.10439505E+00	−0.11733630E+01

Again CHEBYS.FOR yields the denominator polynomial $D(z)$ of $H(z) = K\,N(z)/D(z)$,

$$D(z) = z^3 - 0.69055619z^2 + 0.80189061z - 0.38920832.$$

For an nth-order Chebyshev filter $N(z)$ is $(z+1)^n$; therefore, here when $n = 3$,

$$N(z) = (1+z)^3,$$

$$K = H(1)\frac{D(1)}{N(1)} = \frac{D(1)}{N(1)} \quad (\text{since } H(1) = 1)$$

$$= \frac{0.72213039}{8}$$

$$= 0.09026630.$$

The magnitude and phase plots of the frequency response for the designed filter are shown in Figures 3.8A to H, along with frequency responses for larger n's.

Figure 3.8A. Plot of magnitude of frequency response of Chebyshev filter for $n = 3, 5, 7$.

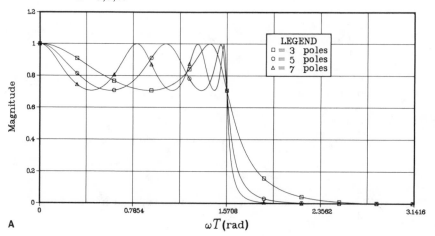

A ωT (rad)

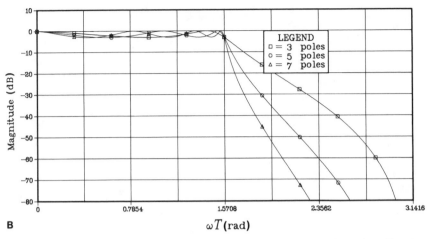

Figure 3.8B. Plot of magnitude of frequency response in db of Chebyshev filter for $n = 3, 5, 7$.

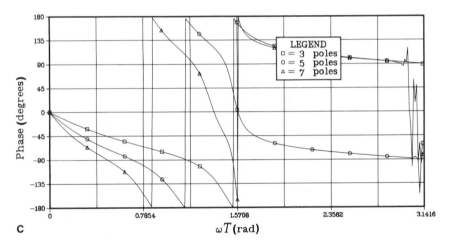

Figure 3.8C. Plot of phase response of Chebyshev filter for $n = 3, 5, 7$.

In the above problem, if greater minimum attenuation is desired at the transition frequency $f_1 = 2.5$ kHz, the order n of the filter has to be increased. Table 3.3 enables the designer to select n for a desired minimum attenuation at the transition frequency. Plots of the responses for various other choices of n, besides $n = 3$, are contained in Figures 3.8.

The next example illustrates in detail the principles underlying the analysis which led to the development of CHEBYS.FOR. The reader is advised to go over this example carefully, since it is expected to facilitate appreciably his comprehension of the design tools presented so far.

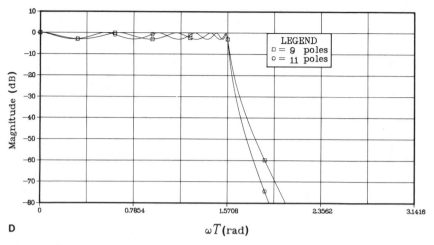

D ωT (rad)

Figure 3.8D. Plot of magnitude of frequency response in db of Chebyshev filter
for $n = 9,11$.

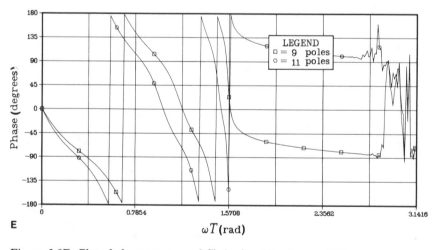

E ωT (rad)

Figure 3.8E. Plot of phase response of Chebyshev filter for $n = 9,11$.

Example 3.3. Design a lowpass digital filter satisfying the following specifications:

Passband: 0–2.5 kHz with ripple not exceeding 1 db.

Stopband: 3 kHz and *up* with a minimum attenuation of 10 db at 3 kHz.

The sampling frequency is 10 KHz.

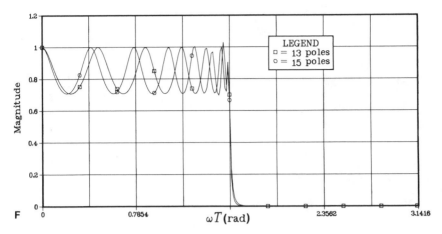

Figure 3.8F. Plot of magnitude of frequency response of Chebyshev filter for $n = 13,15$.

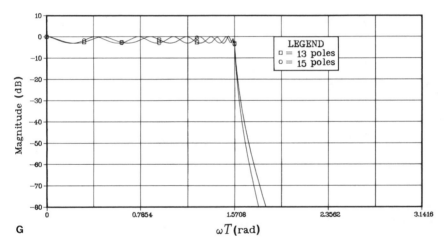

Figure 3.8G. Plot of magnitude of frequency response in db of Chebyshev filter for $n = 13,15$.

A lowpass digital filter of the Chebyshev type will be designed to satisfy the specifications. First, the ripple factor ε^2 is calculated from,

$$10\log(1 + \varepsilon^2) = 1.$$

Then,

$$\varepsilon^2 = 0.25892541 \quad \text{and} \quad \varepsilon = 0.50884714.$$

Second, the order of the filter will be calculated. Since $f_p = 2.5$ kHz and

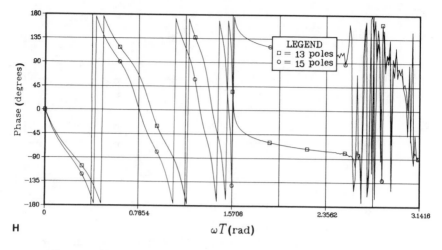

Figure 3.8H. Plot of phase response of Chebyshev filter for $n = 13,15$.

$T = 10^{-4}$ sec,

$$\tan \frac{\omega_p T}{2} = 1 = k.$$

At $\omega_s = (2\pi \times 3)$ krad/sec,

$$\tan \frac{\omega_s T}{2} = 1.37638192 \triangleq x_0.$$

Therefore,

$$10 \log \left[1 + 0.25892541 T_n^2(x_0) \right] = 10,$$
$$T_n(x_0) = 5.89568019.$$

From the Chebyshev polynomials $T_n(x)$, whose coefficients are given in

Table 3.3. Filter Order and Minimum Desired
Attenuation at Transition Frequency

Order n	$T_n(\tan 50°)$	Minimum desired attenuation in decibels at 2.50 kHz
3	3.195218306	-10.47705093
4	5.775272539	-15.33972174
5	10.57018529	-20.49990848
6	19.41884004	-25.75539768
7	35.71475948	-31.03974983
8	65.7075458	-36.3326911

Table 3.2, it is easy to calculate

$$T_1(x_0) = 1.376, \quad T_2(x_0) = 2.789, \quad T_3(x_0) = 6.301.$$

Therefore, the minimum order of the filter required is 3. Choose $n = 3$. From (3.13a), the poles of a normalized analog third-order Butterworth filter are at

$$s_{1b} = -1, \quad s_{2b}, s_{3b} = -\tfrac{1}{2} \pm j\sqrt{3}/2.$$

The locations of the poles for a normalized analog third-order Chebyshev filter are obtained from those of a Butterworth filter, as done next. From (3.26),

$$v = \frac{1}{n}\sinh^{-1}\frac{1}{\varepsilon} = \tfrac{1}{3}\sinh^{-1}\frac{1}{0.5088471399}$$
$$= 0.47599179,$$
$$\sinh v = 0.49417061,$$
$$\cosh v = 1.11543919.$$

The real part of a Chebyshev filter pole location is obtained by multiplying the real part of the corresponding Butterworth filter pole location by $\sinh v$. The imaginary part of a Chebyshev pole is obtained by multiplying the imaginary part of a corresponding Butterworth pole by $\cosh v$. In this problem, the poles of the third-order analog Chebyshev filter are at

$$s_{1c} = -0.4941706, \quad s_{2c} = -0.24708530 \pm j \cdot 96599868.$$

Therefore, the transfer function of the third order analog Chebyshev filter is

$$H_a(s) = \frac{K}{(s + 0.4941706)(s^2 + 0.49417061s + 0.99420459)},$$

where K is an arbitrary real constant.

The digital filter transfer function is obtained by applying the bilinear transformation

$$s \rightarrow k\frac{1 - z^{-1}}{1 + z^{-1}},$$

where $k = 1$ in this problem. It may then be easily verified that the required digital filter transfer function is

$$H(z^{-1}) = \frac{K_1(1 + z^{-1})^3}{1 - 0.34319322z^{-1} + 0.60439354z^{-2} - 0.20407467z^{-3}},$$

where K_1 is another real constant, related to K. Apply CHEBYS.FOR and confirm that the pole locations obtained above agree with those derived after application of the program.

A type of filter, referred to as inverse Chebyshev filters, displays monotonic behavior in the passband and equiripple behavior in the stopband of the magnitude response. The design of this type of filter is also based on the use of Chebyshev polynomials of the first kind. See Problem 25 at the end of this chapter. Filters which exhibit equiripple behavior both in the passband and stopband are termed *elliptic filters*. The analysis of this type of filter is quite complicated and requires an understanding of the Jacobi elliptic function. The interested reader is referred to reference [22] for information on the design of elliptic filters. A computer program for the design of digital elliptic filters has been developed in reference [24].

3.3. Impulse Invariant Transformation

This design procedure proceeds from a rational analog transfer function $H_a(s)$, obtained to satisfy design specifications. The objective is to derive from $H_a(s)$ a digital filter transfer function whose unit impulse response sequence is a regularly sampled sequence of the unit impulse response $h_a(t) \triangleq L^{-1}[H_a(s)]$ (L denotes Laplace transform and L^{-1} denotes inverse Laplace transform) of the analog filter. To be able to see how this objective is achieved, first obtain a partial fraction expansion of $H_a(s)$, which will be taken to be strictly proper [i.e., the degree of the denominator polynomial of $H_a(s)$ is greater than the degree of the numerator polynomial]; otherwise, aliasing problems will be severe. Let the poles of $H_a(s)$ be at $-s_1$, $-s_2, \ldots, -s_r$ and suppose that the multiplicity of the pole at $s = -s_i$ is m_i, $i = 1, 2, \ldots, r$. Then, if

$$H_a(s) = \frac{\sum_{k=0}^{n} a_k s^k}{\sum_{k=0}^{m} b_k s^k}, \quad m > n, \tag{3.32}$$

where the a_k's and b_k's are real constants, $\sum_{i=1}^{r} m_i = m$, then $H_a(s)$ can be expanded in the form

$$H_a(s) = \sum_{j=1}^{r} \sum_{i=0}^{m_j - 1} \frac{k_{m_j - i}^{(j)}}{(s + s_j)^{m_j - i}}. \tag{3.33}$$

On taking the inverse Laplace transform of (3.33),

$$h_a(t) = \sum_{j=1}^{r} \sum_{i=0}^{m_j - 1} \frac{k_{m_j - i}^{(j)}}{(m_j - i - 1)!} \left\{ t^{m_j - i - 1} e^{-s_j t} u(t) \right\}, \tag{3.34}$$

where $u(t)$ is the unit step function. Now sampling $h_a(t)$ every T sec the sampled sequence

$$\{ h_a(nT) \} = \left(h_a(0), h_a(T), h_a(2T), \ldots, h_a(nT), \ldots \right)$$

is formed and the transfer function $H(z)$ of the digital filter will be the

z-transform of $\{h_a(nT)\}$. Therefore,

$$H(z) = \sum_{j=1}^{r} \sum_{i=0}^{m_j-1} \frac{k_{m_j-i}^{(j)}}{(m_j-i-1)!} T^{m_j-i-1} \sum_{n=0}^{\infty} n^{m_j-i-1}\left(e^{-s_j T}z^{-1}\right)^n.$$
(3.35)

The above expression may be simplified by noting that

$$Z\{n^k f(n)\} = -z\frac{d}{dz}\left[Z\{n^{k-1}f(n)\}\right],$$
(3.36)

where $f(n)$ is a discrete-time function, $n \ge 0$, and $k > 0$ is any positive integer. In the special case, when the poles of $H_a(s)$ are simple, i.e., $m_i = 1$ for all i, so that $r = m$, $H(z)$ in (3.35) assumes the particularly simple form

$$H(z) = \sum_{j=1}^{m} k_1^{(j)} \sum_{n=0}^{\infty} \left(e^{-s_j T}z^{-1}\right)^n$$

$$= \sum_{j=1}^{m} \frac{k_1^{(j)}}{1-z^{-1}e^{-s_j T}}.$$
(3.37)

Therefore, in the case of simple poles, $H(z)$ is obtained from $H_a(s)$ by using the mapping function:

$$\frac{k_1^{(j)}}{s+s_j} \rightarrow \frac{k_1^{(j)}}{1-z^{-1}e^{-s_j T}}.$$
(3.38)

Since for rational functions with real coefficients, complex poles occur in conjugate pairs, it can be verified that if $s_r = \sigma_r + j\Omega_r$ and $k_1^{(r)} = k_{1r} + jk_{2r}$, then

$$\frac{k_1^{(r)}}{s+s_r} + \frac{\left[k_1^{(r)}\right]^*}{s+s_r^*} \rightarrow \frac{2k_{1r} - 2z^{-1}e^{-\sigma_r T}(k_{1r}\cos\Omega_r T - k_{2r}\sin\Omega_r T)}{1-2z^{-1}e^{-\sigma_r T}\cos\Omega_r T + z^{-2}e^{-2\sigma_r T}}.$$
(3.39)

To derive the above mapping function, make use of the fact that in the portion of the partial fraction expansion of $H_a(s)$ of (3.32), denoted by

$$\frac{k_1^{(r)}}{s+s_r} + \frac{k_1^{(t)}}{s+s_t},$$

$k_1^{(t)} = [k_1^{(r)}]^*$ when $s_t = s_r^*$; that is, the residue at the complex conjugate of a specified pole is the complex conjugate of the residue at the specified pole, when the rational function has real coefficients. Since the design procedure guarantees that the unit impulse response of the digital filter is a regularly sampled version of the impulse response of the analog filter, it follows that the frequency response of the digital filter is an aliased version of the frequency response of the analog filter. Quantitatively,

$$H(e^{j\omega T}) = \frac{1}{T} \sum_{k=-\infty}^{\infty} H_a\left(j\omega + jk\frac{2\pi}{T}\right).$$

Clearly, horizontal strips of width $2\pi/T$ in the s-plane are aliased or folded over each other in the z-plane. Therefore, for satisfactory matching of the frequency responses of the analog filter and the digital filter, derived by the impulse invariant transformation method, the analog filter must be band-limited to within π/T rad. Since the impulse invariant method is useful only when the effect of aliasing is small, this method is not suitable for the design of filters other than those of the lowpass type.

When the transfer function, $H_a(s)$ of the analog filter is in factored form and when the filter is digitized by replacing both the poles and zeros by the mapping relation,

$$(s + s_j) \rightarrow 1 - z^{-1}e^{-s_jT}$$

$$(s + \sigma_r + j\Omega_r)(s + \sigma_r - j\Omega_r) \rightarrow 1 - 2z^{-1}e^{-\sigma_rT}\cos\Omega_rT + e^{-2\sigma_rT}z^{-2},$$

$$(3.40)$$

the method of digitization is referred to as the matched z-transformation method. Comparing (3.40) and (3.38), (3.39), it is clear that the poles of the digital filter derived via the matched z-transformation method are identical to those obtained by the impulse invariant transformation method. However, the mapping relation of (3.40) applies to the zeros as well. This often causes problems in design, and therefore this method is less commonly used.

Another comment that will be made, before an illustrative example is presented, is the fact that it is possible to design by simple modifications of the impulse invariant method a digital filter whose response to a unit sampled step function is identically the same as the regularly sampled version of the unit step response of the analog filter prototype (in fact, other types of inputs besides the impulse or the step functions could also be considered). In digital control applications, designs based on the step-invariant method are quite popular.

Example 3.4. Obtain the digital transfer function via application of the impulse invariant transformation method, when the analog filter transfer function is specified to be

$$H_a(s) = \frac{1}{(s+1)(s^2+s+1)}.$$

Expanding $H_a(s)$ in partial fractions:

$$H_a(s) = \frac{1}{s+1} - \frac{\frac{1}{2} + j1/\sqrt{12}}{s + \left(\frac{1}{2} - j\sqrt{3}/2\right)} - \frac{\frac{1}{2} - j1/\sqrt{12}}{s + \left(\frac{1}{2} + j\sqrt{3}/2\right)}.$$

Then,

$$H(z) = \frac{1}{1 - z^{-1}e^{-T}} - \frac{\frac{1}{2} + j1/\sqrt{12}}{1 - z^{-1}e^{(-1/2+j\sqrt{3}/2)T}} - \frac{\frac{1}{2} - j1/\sqrt{12}}{1 - z^{-1}e^{(-1/2-j\sqrt{3}/2)T}}.$$

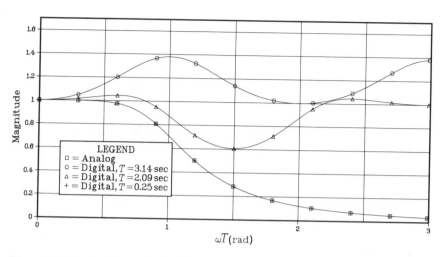

Figure 3.9. Plots of magnitude of frequency response of analog transfer function and the digitalized filter transfer function for different values of sampling time.

The plots of $|H(e^{j\omega T})|$ vs. ω for different values of the sampling period are given along with the plot of the magnitude response of the analog filter in Figure 3.9. It is seen that the digital filter magnitude response deviates from the analog filter magnitude response when the sampling frequency is not high enough.

3.4. Numerical Integration Method

This method provides a simple but not necessarily a satisfactory method for obtaining a difference equation describing the input–output relationship of a digital filter from the differential equation which characterizes the input–output behavior of a designed analog filter. Consider the analog transfer function $H_a(s)$ in (3.32) which is the ratio of the output transform $Y_a(s)$ and the input transform $X_a(s)$. Then,

$$\left(\sum_{k=0}^{m} b_k s^k \right) Y_a(s) = \left(\sum_{k=0}^{n} a_k s^k \right) X_a(s). \qquad (3.41)$$

On taking the Laplace inverse of the above equation under zero initial conditions,

$$\sum_{k=0}^{m} b_k \frac{d^k y_a(t)}{dt^k} = \sum_{k=0}^{n} a_k \frac{d^k x_a(t)}{dt^k}, \qquad (3.42)$$

where $y_a(t) \leftrightarrow Y_a(s)$ and $x_a(t) \leftrightarrow X_a(s)$ denote the input and output Laplace transform pairs. There are various ways for approximating the preceding differential equation by a difference equation. The Euler backward difference scheme replaces dy_a/dt by $[y(n) - y(n-1)]/T$, where

$$y(n) \triangleq y_a(t)|_{t=nT} \qquad (3.43)$$

and T is the sampling period. In the frequency domain this corresponds to the mapping function

$$s = \frac{1 - z^{-1}}{T} \qquad (3.44a)$$

or

$$z = \frac{1}{1 - sT}. \qquad (3.44b)$$

The transformation in (3.44b) maps the region, $\mathrm{Re}\, s < 0$, in the s-plane to the region $|z| < 1$. However, unlike the bilinear transformation the real frequency axis corresponding to $\mathrm{Re}\, s = 0$ in the s-plane is not mapped to $|z| = 1$, but instead to a circle

$$\left(\mathrm{Re}\, z - \tfrac{1}{2}\right)^2 + \left(\mathrm{Im}\, z\right)^2 = \tfrac{1}{4}, \qquad (3.45)$$

and, therefore, though the BIBO stability property remains invariant under the mapping, the frequency selective property of the analog filter changes after the transformation in (3.44a) is applied.

Example 3.5. Apply the transformation in (3.44a) with $T = \pi$ sec to a third-order normalized Butterworth analog filter, having a transfer function

$$H_a(s) = \frac{1}{(s+1)(s^2 + s + 1)}.$$

Then,

$$H(z) = \frac{1}{\left[(1 - z^{-1})/\pi + 1\right]\left\{\left[(1 - z^{-1})/\pi\right]^2 + (1 - z^{-1})/\pi + 1\right\}}$$

$$= \frac{\pi^3}{(\pi + 1 - z^{-1})\left[z^{-2} - (2 + \pi)z^{-1} + (\pi^2 + \pi + 1)\right]}.$$

The plots of $|H(e^{j\omega T})|$ vs. ωT and $\angle H(e^{j\omega T})$ vs. ωT are shown, respectively, in Figures 3.10A and 3.10B.

3.5. Frequency Transformations

The preceding sections described methods for finding a lowpass digital filter transfer function from a lowpass normalized analog filter transfer function. The objective here is to obtain transfer functions for other types of filters (highpass, bandpass, and bandstop), having the prescribed requirements, from the lowpass digital filter transfer function, already available. The

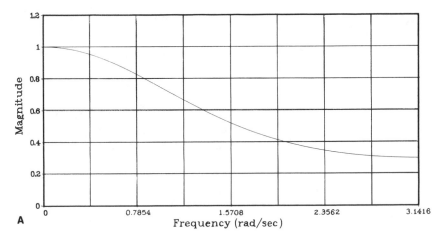

Figure 3.10A. Magnitude of frequency response of digital filter obtained by NIM method in Example 3.5.

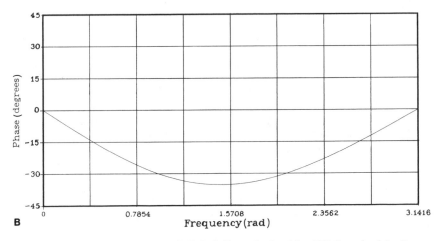

Figure 3.10B. Phase response of digital filter obtained by NIM method in Example 3.5.

problem is somewhat similar to, albeit more complicated than the problem occurring in the synthesis of analog filters [23, Chap. 11], where, in order to convert from a lowpass to any other type of filter, the complex frequency variable s is substituted by an appropriate reactance function. The complication is due to the fact that, for digital filters, the relevant range of frequencies is $-\pi/T \leq \omega \leq \pi/T$ (where T is the sampling period), instead of the whole frequency axis extending from $-\infty$ to $+\infty$. For the purpose here, instead of using reactance functions as in the analog case, one uses real rational functions in z^{-1} or z which have unit magnitude on $|z| = 1$. A real rational function $H(z)$ which has magnitude $|H(z)| = 1$ on $|z| = 1$ is referred to as a digital all-pass function and satisfies the relationship

$$H(z)H(z^{-1}) = 1.$$

Furthermore, for the BIBO stability property to remain invariant, it is necessary that these real rational functions be stable. The simplest such stable all-pass functions are of the following forms (see also Section 6.1):

$$g_1(z^{-1}) = \pm \frac{z^{-1} - \alpha}{1 - \alpha z^{-1}}, \quad \alpha \text{ real and } |\alpha| < 1, \tag{3.46a}$$

$$g_2(z^{-1}) = \pm \left(\frac{z^{-1} - \alpha}{1 - \alpha^* z^{-1}} \right) \left(\frac{z^{-1} - \alpha^*}{1 - \alpha z^{-1}} \right), \quad \alpha \text{ complex and } |\alpha| < 1. \tag{3.46b}$$

Certain applications are facilitated through the design and use of digital filters with complex coefficients [20]. In such designs, the restriction of realness on coefficients on the transformation function is eliminated. Here, as elsewhere in the text, coefficients are restricted to be real.

3.5.1. Digital – Digital Transformations

Let $H(z^{-1})$ be the transfer function of a lowpass digital filter having a cutoff frequency equaling ω_c rad/sec. For Butterworth filters, $|H(e^{j\omega T})|^2$ is given in (3.11), while for Chebyshev filters (3.20) applies, where ω_c is replaced by ω_p, the frequency at the end of the equiripple passband. Consider the transformed function

$$H_1(z^{-1}) \triangleq H[g_1(z^{-1})], \tag{3.47}$$

obtained after replacing z^{-1} by $g_1(z^{-1})$ of (3.46a) in $H(z^{-1})$. On $|z| = 1$, it is simple to verify that

$$\tan \frac{\omega T}{2} = -j \frac{1 - z^{-1}}{1 + z^{-1}} \bigg|_{z = e^{j\omega T}}. \tag{3.48}$$

In (3.11) and (3.20), $\tan(\omega T/2)$ is the lowpass frequency function which is zero when $\omega = 0$ and is ∞ when $\omega = \pi/T$. Replacing z^{-1} by $g_1(z^{-1})$ in (3.48), one gets the lowpass frequency function $f(\omega)$ of the transformed filter:

$$f(\omega) = -j \frac{1 - (z^{-1} - \alpha)/(1 - \alpha z^{-1})}{1 + (z^{-1} - \alpha)/(1 - \alpha z^{-1})} \bigg|_{z = e^{j\omega T}}$$

$$= \left(\frac{1 + \alpha}{1 - \alpha} \right) \tan \left(\frac{\omega T}{2} \right). \tag{3.49}$$

For the sake of brevity, the ensuing discussion refers to a lowpass Butter-worth prototype. From (3.48) and (3.49) it is clear that the cutoff frequency, ω_{ctl}, of the transformed lowpass filter, characterized by $H_1(z^{-1})$ of (3.47) is:

$$\omega_{ctl} = \frac{2}{T} \tan^{-1}\left(\frac{1-\alpha}{1+\alpha} \tan \frac{\omega_c T}{2}\right). \tag{3.50a}$$

Equation (3.50a) is obtained after replacing $\tan(\omega T/2)$ by $f(\omega)$ from (3.49) in (3.11). In (3.50a) ω_c is the cutoff frequency of the original lowpass filter. On solving (3.50a) for α, one gets

$$\alpha = \frac{\tan(\omega_c T/2) - \tan(\omega_{ctl} T/2)}{\tan(\omega_c T/2) + \tan(\omega_{ctl} T/2)}$$
$$= \frac{\sin[(\omega_c - \omega_{ctl})T/2]}{\sin[(\omega_c + \omega_{ctl})T/2]}. \tag{3.50b}$$

In the case of a normalized lowpass digital filter, $\tan(\omega_c T/2) = 1$, i.e., the cutoff frequency $\omega_c = \pi/2T$ is a quarter of the sampling frequency $2\pi/T$ in rad/sec, and in this case (3.50a) simplifies to

$$\omega_{ctl} = \frac{2}{T}\left(\tan^{-1}\frac{1-\alpha}{1+\alpha}\right). \tag{3.51}$$

To obtain the digital lowpass to digital highpass transformation, it is necessary to replace z^{-1} in (3.46a) by $-z^{-1}$ (since the replacement of $\tan(\omega T/2)$ in the lowpass magnitude response by $-\cot(\omega T/2)$ results in the highpass magnitude response) so that the transformation $g_1(z^{-1})$ of (3.46a) is replaced by

$$g_2(z^{-1}) = -\frac{(z^{-1}+\alpha)}{1+\alpha z^{-1}}. \tag{3.52a}$$

The value of α in (3.52a) is derivable from the specified cutoff frequency, ω_{cth} of the desired highpass filter following the application of the transfor-mation. To obtain the highpass filter having a cutoff frequency ω_{cth} rad/sec, it is necessary to use a lowpass filter of cutoff frequency $(\pi/T - \omega_{cth})$ rad/sec. Using this fact in (3.51), i.e., after replacing ω_{ctl} by $(\pi/T - \omega_{cth})$ and solving for α [proceeding, thereby, from the normalized lowpass case; otherwise (3.50) has to be used],

$$\alpha = \frac{1 - \tan(\pi/T - \omega_{cth})T/2}{1 + \tan(\pi/T + \omega_{cth})T/2}. \tag{3.52b}$$

On the other hand, if the transformation in (3.52a) is applied to a nonnor-malized lowpass filter having a cutoff frequency of ω_c rad/sec, replacement of ω_{ctl} by $\pi/T - \omega_{cth}$ in (3.50b) yields

$$\alpha = -\frac{\cos[(\omega_c + \omega_{cth})T/2]}{\cos[(\omega_c - \omega_{cth})T/2]}. \tag{3.52c}$$

[Note that if $(\omega_c + \omega_{cth})T$ and $(\omega_c - \omega_{cth})T$ in (3.52c) are interchanged, the magnitude response of the designed highpass filter would be invariant but the phase response would change by π.] In order to obtain the digital lowpass to digital bandpass transformation it should be noted that $\tan(\omega T/2)$ in the expression for the lowpass magnitude function should be replaced by

$$h(\omega) = \frac{\cos \omega_0 T - \cos \omega T}{\sin \omega T}, \qquad (3.53)$$

where ω_0 is the center frequency in rad/sec of the bandpass filter. Note that $h(\omega)$ of (3.53) satisfies

$$h(0) = -\infty,$$

$$h(\omega_0) = 0,$$

$$h(\pi/T) = +\infty.$$

From (3.48), for the lowpass case, on $|z| = 1$,

$$z^{-1} = \frac{1 - j \tan(\omega T/2)}{1 + j \tan(\omega T/2)} = e^{-j\omega T}.$$

Replacing $\tan(\omega T/2)$ by $h(\omega)$ in the center expression of the preceding equation, one gets the required transformation $g_3(z^{-1})$ so that $H[g_3(z^{-1})]$ becomes the transfer function of a digital bandpass filter:

$$
\begin{aligned}
g_3(z^{-1}) &= \left. \frac{1 - jh(\omega)}{1 + jh(\omega)} \right|_{e^{-j\omega T} \to z^{-1}} \\
&= -\frac{z^{-1}(z^{-1} - \cos \omega_0 T)}{1 - z^{-1}\cos \omega_0 T}.
\end{aligned}
\qquad (3.54)
$$

To get the digital bandpass filter transfer function from the normalized digital lowpass filter transfer function, derive the transformation $g_4(z^{-1})$ obtainable by the replacement of z^{-1} by $g_3(z^{-1})$ in (3.46a):

$$g_4(z^{-1}) = \frac{-z^{-1}(z^{-1} - \cos \omega_0 T)/(1 - z^{-1} \cos \omega_0 T) - \alpha}{1 + \alpha z^{-1}(z^{-1} - \cos \omega_0 T)/(1 - z^{-1} \cos \omega_0 T)} \qquad (3.55a)$$

with

$$\alpha = \frac{\cot[(\omega_2 - \omega_1)T/2] - 1}{\cot[(\omega_2 - \omega_1)T/2] + 1}, \qquad (3.55b)$$

where ω_1 and ω_2 are, respectively, the lower and upper cutoff frequencies of the digital bandpass filter. The value of α in (3.55b) is obtained by imposing the restriction

$$\omega_2 - \omega_1 = \omega_{ctl} \qquad (3.56)$$

in (3.51) and then solving for α. After simple algebraic manipulations

$g_4(z^{-1})$ in (3.55a) takes the generic form

$$g_4(z^{-1}) = -\frac{z^{-2} - [2\alpha_1 k_1/(k_1+1)]z^{-1} + (k_1-1)/(k_1+1)}{(k_1-1)/(k_1+1)z^{-2} - [2\alpha_1 k_1/(k_1+1)]z^{-1} + 1},$$

(3.57a)

where

$$\alpha_1 = \cos \omega_0 T,$$
$$k_1 = \cot[(\omega_2 - \omega_1)T/2].$$

(3.57b)

Constantinides [3] also considered a generalization of (3.57a) after replacing k_1 in (3.57b) by k in (3.58) and ignoring the restriction given in (3.56):

$$k = k_1 \tan \frac{\omega_{ctl} T}{2}$$

(3.58)

Only in the case $k = 1$, the restriction set in (3.56) has to be satisfied. Note that when $\alpha_1 = 0$ and $k = 1$, the magnitude characteristic of the bandpass filter is symmetric about the center frequency. It must be noted that the parameters ω_0, ω_1, and ω_2 in (3.57b) are not independent but constrained by

$$\alpha_1 = \cos \omega_0 T = \frac{\cos[(\omega_2 + \omega_1)T/2]}{\cos[(\omega_2 - \omega_1)T/2]},$$

(3.59)

which follows from the definition of ω_0 as the frequency in rad/sec at which (3.57a) following replacement of k_1 by k in (3.58) is zero. The frequency transformation required to design a digital bandstop filter from a digital lowpass prototype is derivable in an analogous manner and this case plus all the results described in this section are summarized in Table 3.4. The transformations are quite general and their applications preserve the magnitude characteristics (maximally flat, equiripple, etc.) of the lowpass prototype in the transformed filter.

Example 3.6. Consider the lowpass filter of Example 3.2 as prototype

$$H(z) = \frac{0.09026630(1 + z^{-1})^3}{1 - 0.69055619z^{-1} + 0.80189061z^{-2} - 0.38920832z^{-3}}.$$

Again, the sampling frequency is 9 kHz. Remember that for Chebyshev filters, ω_c is replaced by ω_p, the frequency at the end of the equiripple passband.

a. It is required to design a highpass filter from $H(z)$, with a cutoff frequency at 3.6 kHz. Noting that the cutoff frequency ω_p of $H(z)$ is $(2\pi \times 2.25)$ krad/sec and using (3.52c), where ω_p replaces ω_c (ω_p, for brevity, is sometimes referred to as the cutoff frequency, though it need

Table 3.4. Frequency Transformations for Designs of Different Types of Digital Filters from a Lowpass Digital Prototype Having a Cutoff Frequency ω_c rad/sec

Digital filter type	Replace z^{-1} in prototype by	Symbol description of transformed filter	Formulae
Lowpass	$\dfrac{z^{-1}-\alpha}{1-\alpha z^{-1}}$	Cutoff freq is ω_{ctl} rad/sec	$\alpha = \dfrac{\sin[(\omega_c - \omega_{ctl})T/2]}{\sin[(\omega_c + \omega_{ctl})T/2]}$
Highpass (See note below table)	$-\left(\dfrac{z^{-1}+\alpha}{1+\alpha z^{-1}}\right)$	Cutoff freq is ω_{cth} rad/sec	$\alpha = -\dfrac{\cos[(\omega_c \pm \omega_{cth})T/2]}{\cos[(\omega_c \mp \omega_{cth})T/2]}$
Bandpass	$-\left(\dfrac{z^{-2}-[2\alpha_1 k/(k+1)]z^{-1}+(k-1)/(k+1)}{[(k-1)/(k+1)]z^{-2}-[2\alpha_1 k/(k+1)]z^{-1}+1}\right)$	Center freq of passband is ω_0 rad/sec; upper and lower cutoff freqs are, respectively, ω_2 and ω_1 rad/sec	$\alpha_1 = \cos \omega_0 T$ $= \dfrac{\cos[(\omega_2 + \omega_1)T/2]}{\cos[(\omega_2 - \omega_1)T/2]}$ $k = \cot[(\omega_2 - \omega_1)T/2]\tan(\omega_c T/2)$
Bandstop	$\dfrac{z^{-2}-[2\alpha_1/(1+k)]z^{-1}+(1-k)/(1+k)}{(1-k)/(1+k)z^{-2}-[2\alpha_1/(1+k)]z^{-1}+1}$	Center freq of reject band is ω_0 rad/sec ω_2 and ω_1 are naturally defined	$\alpha_1 = \cos \omega_0 T$ $= \dfrac{\cos[(\omega_2 + \omega_1)T/2]}{\cos[(\omega_2 - \omega_1)T/2]}$ $k = \tan[(\omega_2 - \omega_1)T/2]\tan(\omega_c T/2)$

Note: α could also be replaced by $1/\alpha$ without changing the magnitude of the frequency response; it is stated that the transformation function chosen be stable.

not be the half-power frequency),

$$\alpha = -\frac{\cos(0.65\pi)}{\cos(-0.15\pi)} = 0.5095.$$

The required frequency transformation is [using (3.52c)]

$$z^{-1} \rightarrow -\frac{(z^{-1}+0.5095)}{1+0.5095z^{-1}}.$$

The transfer function of the highpass filter is:

$$H_1(z^{-1}) = \frac{0.0066(1-3z^{-1}+3z^{-2}-z^{-3})}{1+2.3605z^{-1}+2.1018z^{-2}+0.6884z^{-3}}.$$

b. It is required to obtain a bandpass filter with upper and lower cutoff frequencies at 3.8 and 3.4 kHz, respectively. From (3.59) and (3.58), where $\omega_{ptl} = (2\pi \times 2.25)$ krad/sec replaces ω_{ctl},

$$\alpha_1 = \frac{\cos 0.8\pi}{\cos 0.0444\pi} = -0.8170,$$

$$k = \cot(0.0444\pi)\tan(0.25\pi)$$
$$= 7.1154,$$

$$\frac{k-1}{k+1} = 0.7536, \quad \frac{2\alpha_1 k}{k+1} = -1.4326.$$

Therefore, the required frequency transformation is (using row 3 of Table 3.4)

$$z^{-1} \rightarrow \frac{-z^{-2}-1.4326z^{-1}-0.7536}{0.7536z^{-2}+1.4326z^{-1}+1}.$$

The transfer function of the bandpass filter is

$$H_2(z^{-1}) = \frac{.0006(1-3z^{-2}+3z^{-4}-z^{-6})}{1+4.7200z^{-1}+10.2422z^{-2}+12.7890z^{-3}+9.6874z^{-4}+4.2226z^{-5}+0.8465z^{-6}}$$

c. It is required to obtain a bandstop filter with upper and lower cutoff frequencies at 3.6 and 3.15 kHz, respectively. Here,

$$\alpha_1 = \frac{\cos 0.75\pi}{\cos 0.05\pi} = -0.7159,$$

$$k = \tan(0.05\pi)\tan(0.25\pi) = 0.1584,$$

$$\frac{1-k}{k+1} = 0.7265, \quad \frac{2\alpha_1}{k+1} = -1.2361.$$

The required frequency transformation is (using row 4 of Table 3.4)

$$z^{-1} \rightarrow \frac{z^{-2}+1.2361z^{-1}+0.7265}{0.7265z^{-2}+1.2361z^{-1}+1}.$$

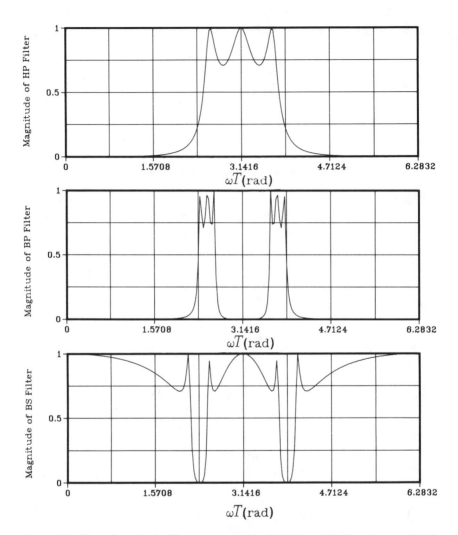

Figure 3.11. Plots of magnitude of frequency response of HP, BP, and BS filters in Example 3.6.

The transfer function of the bandstop filter is

$$H_3(z^{-1}) = \frac{\begin{aligned}0.6013(1 &+ 4.2955z^{-1} + 9.1505z^{-2} + 11.5266z^{-3} + 9.1505z^{-4} \\ &+ 4.2955z^{-5} + z^{-6})\end{aligned}}{\begin{aligned}(1 &+ 3.6462z^{-1} + 6.5164z^{-2} + 6.8307z^{-3} + 4.4205z^{-4} \\ &+ 1.6247z^{-5} + 0.2750z^{-6})\end{aligned}}$$

The plots of the magnitude of frequency responses of the filters obtained are plotted in Figure 3.11. Two comments are in order. First, after obtaining the relevant frequency transformation, the transfer function sought is obtainable with the aid of FTRANS.FOR, whose listing is provided in this chapter. Second, the stability of the filter transfer function must be checked using SCHUR1.FOR. (but see Problem 1 at the end of this chapter in order to save labor).

3.5.2. Analog–Digital Direct Transformations

It is possible to go directly from a designed analog lowpass filter to various types of digital filters using appropriate frequency transformations. Therefore, it is not necessary to go from the analog lowpass to the digital lowpass case prior to the use of the digital lowpass–digital highpass (bandpass, bandstop) transformations of the preceding subsection. If it is required, say, to design a digital lowpass filter with prescribed specifications, then an analog lowpass filter needs to be designed first and then the bilinear transformation needs to be used to get the required digital filter.

Example 3.7. Design a digital lowpass filter with cutoff frequency at 10 rad/sec from a second-order Butterworth type of analog lowpass filter. The sampling period T is given as $\pi/30$ sec.

The frequency function for a digital lowpass filter is $\tan(\omega T/2)$. The cutoff frequency of the associated analog filter corresponding to the prescribed cutoff frequency of the digital filter is

$$\Omega_c = \tan\left(\frac{10}{2} \times \frac{\pi}{30}\right) = \tan\frac{\pi}{6} = 0.5774.$$

The transfer function of a normalized second-order ($n = 2$) analog Butterworth filter is {refer to (3.9); the poles of $H_a(s)$ are the left-half plane zeros of $[1 + (-s^2)^n]$}

$$H_a(s) = \frac{K}{s^2 + \sqrt{2}\,s + 1},$$

where K is a suitable real constant. The transfer function of the second-order Butterworth filter having a cutoff frequency at 0.5774 rad/sec is

$$\hat{H}_a(s) = H_a\left(\frac{s}{0.5774}\right)$$

$$= \frac{K}{3s^2 + 2.4489s + 1}.$$

The transfer function of the required digital lowpass filter is

$$H(z) = \hat{H}_a\left(\frac{z-1}{z+1}\right) = \frac{0.6447(z^2 + 2z + 1)K}{4.1576z^2 - 2.5788z + 1}.$$

The same result is obtained after replacing s directly by $(1/0.5774)$ $(z-1)/(z+1)$ [see (3.6a)] in $H_a(s)$.

From the above example it is clear that a digital lowpass filter with a cutoff frequency at ω_c rad/sec can be obtained from a lowpass analog filter transfer function whose cutoff frequency is normalized to 1 rad/sec by replacing s as follows:

$$s \rightarrow \left(\cot \frac{\omega_c T}{2}\right) \frac{1 - z^{-1}}{1 + z^{-1}}. \tag{3.60}$$

The next example illustrates how one obtains a digital highpass filter from an analog lowpass filter.

Example 3.8. Design a digital highpass filter with a cutoff frequency at 20 rad/sec from a second-order Butterworth type of analog filter. The sampling period T is given as $\pi/30$ sec.

In the s-plane the lowpass analog filter cutoff frequency corresponding to the cutoff frequency of 20 rad/sec for the digital filter is

$$\Omega_c = \tan \frac{20 \times \pi}{2 \times 30} = \tan \frac{\pi}{3} = 1.732.$$

Again, the transfer function of the normalized lowpass second-order ($n = 2$) Butterworth filter is {refer to (3.9); the poles of $H_a(s)$ are formed from the left-half plane zeros of $[1 + (-s^2)^n]$}

$$H_a(s) = \frac{K}{s^2 + \sqrt{2}\,s + 1}.$$

Note that in (3.9), $|H_a(0)| = 1$. Here, the numerator of $H_a(s)$ is taken to be a suitable real constant in order to adjust the dc gain. The analog highpass filter transfer function is

$$\hat{H}_a(s) = H_a\left(\frac{\Omega_c}{s}\right) = \frac{Ks^2}{s^2 + 2.449s + 3}.$$

The digital highpass filter transfer function is

$$H(z) = \hat{H}_a\left(\frac{z-1}{z+1}\right) = \frac{0.15506K(z^2 - 2z + 1)}{z^2 + 0.6203z + 0.2405}.$$

The same result can be obtained directly after replacing s in $H_a(s)$ by $\Omega_c(z+1)/(z-1)$.

The above example illustrates that the frequency transformation required to design a highpass digital filter with a cutoff frequency at ω_c rad/sec from a lowpass analog filter whose cutoff frequency is normalized to 1 rad/sec is

$$s \rightarrow \left(\tan \frac{\omega_c T}{2}\right) \frac{z+1}{z-1}. \tag{3.61}$$

Next suppose that the lowpass analog to the bandpass digital frequency transformation is desired. Let the lowpass analog filter have its cutoff frequency normalized to 1 rad/sec and suppose that the specified upper and lower cutoff frequencies of the digital bandpass filter are, respectively, ω_2 and ω_1 radians per second. Combining the expression giving the frequency transformation from the digital lowpass case to the digital bandpass case in column 2 of Table 3.4 with the fact that $s \rightarrow (1 - z^{-1})/(1 + z^{-1})$ is the transformation from the analog lowpass to the digital lowpass case, it is simple to verify that the required analog lowpass to digital bandpass transformation is of the form

$$s \rightarrow k \frac{z^{-2} - 2\alpha_1 z^{-1} + 1}{1 - z^2}. \tag{3.62}$$

The above transformation indicates that a point $s = j\Omega$ on the analog frequency axis is related to a point $z = e^{j\omega T}$ on the unit circle in the z-plane via the following equation:

$$\Omega = k \frac{\alpha_1 - \cos \omega T}{\sin \omega T}. \tag{3.63}$$

Therefore,

$$\Omega = 0 \leftrightarrow \omega = \frac{1}{T} \cos^{-1} \alpha_1 \triangleq \omega_0,$$

where ω_0 is the center frequency of the digital filter. Furthermore,

$$-1 = k \left(\frac{\alpha_1 - \cos \omega_1 T}{\sin \omega_1 T} \right) \tag{3.64a}$$

$$1 = k \left(\frac{\alpha_1 - \cos \omega_2 T}{\sin \omega_2 T} \right) \tag{3.64b}$$

when the analog lowpass filter cutoff frequency is normalized to unity. Dividing (3.64a) by (3.64b), one gets

$$\frac{\alpha_1 - \cos \omega_1 T}{\alpha_1 - \cos \omega_2 T} = - \frac{\sin \omega_1 T}{\sin \omega_2 T}.$$

Therefore,

$$\begin{aligned} \alpha_1 &= \frac{\sin \omega_1 T \cos \omega_2 T + \cos \omega_1 T \sin \omega_2 T}{\sin \omega_1 T + \sin \omega_2 T} \\ &= \frac{\sin(\omega_1 + \omega_2) T}{2 \sin[(\omega_1 + \omega_2) T/2] \cos[(\omega_2 - \omega_1) T/2]} \\ &= \frac{\cos[(\omega_2 + \omega_1) T/2]}{\cos[(\omega_2 - \omega_1) T/2]} \end{aligned} \tag{3.65}$$

{since $\sin(\omega_1 + \omega_2) T = 2 \sin[(\omega_1 + \omega_2) T/2] \cos[(\omega_1 + \omega_2) T/2]$}.

Table 3.5. Frequency Transformations from Analog Lowpass with Cutoff at 1 rad/sec to Various Types of Digital Filters

Digital filter type	Digital filter cutoff frequency	Replace the analog filter variable s by	Description of symbols (sampling period is T)
Lowpass	ω_c	$k\dfrac{1-z^{-1}}{1+z^{-1}} = k\dfrac{z-1}{z+1}$	$k = \cot(\omega_c T/2)$
Highpass	ω_c	$k\dfrac{1+z^{-1}}{1-z^{-1}} = k\dfrac{z+1}{z-1}$	$k = \tan(\omega_c T/2)$
Bandpass	ω_1, ω_2 $(\omega_1 < \omega_2)$; ω_0 is center frequency	$k\dfrac{z^{-2}-2\alpha z^{-1}+1}{1-z^{-2}} = k\dfrac{z^2-2\alpha z+1}{z^2-1}$	$\alpha = \dfrac{\cos\big[(\omega_2+\omega_1)T/2\big]}{\sin\big[(\omega_2-\omega_1)T/2\big]}$ $= \cos\omega_0 T$ $k = \cot[(\omega_2-\omega_1)T/2]$
Bandstop	ω_1, ω_2 $(\omega_1 < \omega_2)$; ω_0 is center frequency	$k\dfrac{1-z^{-2}}{z^{-2}-2\alpha z^{-1}+1} = k\dfrac{z^2-1}{z^2-2\alpha z+1}$	$\alpha = \dfrac{\cos\big[(\omega_2+\omega_1)T/2\big]}{\cos\big[(\omega_2-\omega_1)T/2\big]}$ $k = \tan[(\omega_2-\omega_1)T/2]$
Multibandpass	$0 < \omega_{00} < \omega_{\infty 0}$ $< \omega_{01} < \omega_{\infty 1} \cdots$ $< \omega_{0(n-1)} < \pi/T$	$k\,\dfrac{1}{1-z^{-2}}\,\dfrac{\prod_{j=0}^{n-1}\left(z^{-2}-2\alpha_j z^{-1}+1\right)}{\prod_{j=0}^{n-2}\left(z^{-2}-2\beta_j z^{-1}+1\right)}$, $n \geq 2$	k is a real positive constant $\alpha_i = \cos\omega_{0i}T$ $\beta_i = \cos\omega_{\infty i}T$ # of passbands $= (n-1)$
Multibandstop	$0 < \omega_{\infty 0} < \omega_{00}$ $< \omega_{\infty 1} < \omega_{01} \cdots$ $< \omega_{\infty(n-1)} < \pi/T$	$k\,\dfrac{(1-z^2)\prod_{j=0}^{n-2}\left(z^{-2}-2\alpha_j z^{-1}+1\right)}{\prod_{j=0}^{n-1}\left(z^{-2}-2\beta_j z^{-1}+1\right)}$, $n \geq 2$	k is a real positive constant $\alpha_i = \cos\omega_{0i}T$ $\beta_i = \cos\omega_{\infty i}T$ # of stopbands $= (n-1)$

Substituting (3.65) in (3.64b),

$$k = \frac{\sin \omega_2 T \cos[(\omega_2 - \omega_1)T/2]}{\cos[(\omega_2 + \omega_1)T/2] - \cos \omega_2 T \cos[(\omega_2 - \omega_1)T/2]}$$

$$= \frac{2\sin(\omega_2 T/2)\cos(\omega_2 T/2)\cos[(\omega_2 - \omega_1)T/2]}{\cos(\omega_2 T/2)\cos(\omega_1 T/2)(1 - \cos \omega_2 T) - \sin(\omega_2 T/2)} .$$
$$\sin(\omega_1 T/2)(1 + \cos \omega_2 T)$$

Using the trigonometric identities

$$1 - \cos \omega_2 T = 2\sin^2 \frac{\omega_2 T}{2},$$

$$1 + \cos \omega_2 T = 2\cos^2 \frac{\omega_2 T}{2},$$

it follows that

$$k = \cot[(\omega_2 - \omega_1)T/2] \tag{3.66}$$

The expressions for α_1 and k in (3.65) and (3.66) are identical to the expressions for α_1 and k_1 in (3.59) and (3.57b), respectively. The results are summarized in Table 3.5 (for brevity, α_1 is replaced by α and, since this is defined in the table, no possibility of confusion occurs). The analog lowpass to digital bandstop frequency transformation can also be analogously obtained after noting that in this case $1/s$ instead of s should be replaced by the expression on the right of (3.62). The reader can verify that the counterparts of (3.63), (3.64), (3.65), and (3.66) for this case are (α is used instead of α_1):

$$\Omega = k \frac{\sin \omega T}{\cos \omega T - \alpha}, \tag{3.67}$$

$$-1 = k \frac{\sin \omega_2 T}{\cos \omega_2 T - \alpha}, \tag{3.68a}$$

$$1 = k \frac{\sin \omega_1 T}{\cos \omega_1 T - \alpha}, \tag{3.68b}$$

$$\alpha = \frac{\cos[(\omega_2 + \omega_1)T/2]}{\cos[(\omega_2 - \omega_1)T/2]}, \tag{3.69}$$

$$k = \tan \frac{(\omega_2 - \omega_1)T}{2}. \tag{3.70}$$

The preceding results along with some more general transformations required for the digital multipassband and multistopband cases are included in Table 3.5.

Example 3.9. It is required to design a bandpass filter with upper and lower cutoff frequencies at 1000 Hz and 800 Hz, respectively. The attenuation for frequencies smaller than 200 Hz and larger than 1500 Hz must be at least 9 db. The sampling period T is $1/5000$ sec.

Since $\omega_2/2\pi = 1000$ and $\omega_1/2\pi = 800$, using (3.65), (3.66), and (3.63), one gets

$$\alpha_1 = \frac{\cos 0.36\pi}{\cos 0.04\pi} = 0.4292,$$

$$k = \cot 0.04\pi = 7.9158,$$

$$\Omega_1 = k\left(\frac{\alpha_1 - \cos 0.32\pi}{\sin 0.32\pi}\right) = -0.9998,$$

$$\Omega_2 = k\left(\frac{\alpha_1 - \cos 0.4\pi}{\sin 0.4\pi}\right) = 0.9998 = -\Omega_1.$$

(You might wish to verify whether or not $\Omega_2 = -\Omega_1$ holds in general.) Also, it is given that $\omega_3/2\pi = 200$, and $\omega_4/2\pi = 1500$, where ω_3 and ω_4 are appropriately defined as implied in the problem. Again using (3.63),

$$\Omega_3 = k\left(\frac{\alpha_1 - \cos 0.08\pi}{\sin 0.08\pi}\right) = -17.1693,$$

$$\Omega_4 = k\left(\frac{\alpha_1 - \cos 0.6\pi}{\sin 0.6\pi}\right) = 6.1442.$$

It will be assumed that the type of magnitude response sought is maximally flat. First, a lowpass analog filter with a cutoff frequency normalized to 1 rad/sec (note that $\Omega_2 = 0.9998$) and having an attenuation of at least 9 db for frequencies larger than 6.1442 rad/sec has to be obtained. The order n of the Butterworth filter should be such that

$$10\log\left[1 + \Omega_4^{2n}\right] \geq 9$$

or

$$1 + \Omega_4 2^n \geq 10^{0.9} = 7.9433.$$

Therefore,

$$n \geq \frac{\log 6.9433}{2\log 6.1442} = 0.5337.$$

Therefore, $n = 1$ will do the job. The frequency normalized analog lowpass filter transfer function is

$$H_a(s) = \frac{K}{s+1},$$

where K is a real constant. The required digital bandpass filter transfer function is

$$H(z) = H_a\left(s \rightarrow k\frac{z^2 - 2\alpha_1 z + 1}{z^2 - 1}\right)$$

$$= \frac{(z^2 - 1)K}{1.1263z^2 - 0.8583z + 0.8737}. \tag{3.71}$$

$H(z)$ is obtainable from $H_a(s)$ by using the program FTRANS.FOR, whose listing is given in Figure 3.12. Use of SCHUR1.FOR verifies that the designed bandpass filter is BIBO stable. In (3.71), K may be conveniently chosen. The next example illustrates the use of FTRANS.FOR.

Example 3.10. Consider the sixth-order lowpass digital Butterworth filter designed in Example 3.1, whose transfer function $H(z)$ is given in (3.17a)–(3.17d). The lowpass cutoff frequency is $f_c = 2.25$ kHz, corresponding to an angle of $\pi/2$ rad. It is required to obtain a highpass filter having a cutoff frequency at $f_{cth} = 3$ kHz using the digital lowpass–digital highpass frequency transformation.

$$\omega_{cth}T = 2\pi \times 3 \times 10^3 \times \frac{1}{9 \times 10^3} = \frac{2\pi}{3}.$$

From the last column in the second row of Table 3.4,

$$\alpha = -\frac{\frac{1}{2}\cos(\pi/2 + 2\pi/3)}{\frac{1}{2}\cos(\pi/2 - 2\pi/3)} = \frac{1}{3.73205} \simeq 0.267949$$

Therefore, z^{-1} in prototype may be replaced by either (Table 3.4)

$$g_2(z^{-1}) = -\frac{(z^{-1} + \alpha)}{1 + \alpha z^{-1}}$$

or

$$g_2(z^{-1}) = -\frac{(z^{-1} + 1/\alpha)}{1 + (1/\alpha)z^{-1}}.$$

The magnitude of the frequency responses will be the same for both transformations; the first transformation is chosen in this example, since it is stable, so that

$$g_2(z^{-1}) = -\frac{(z^{-1} + 0.267949)}{(1 + 0.267949z^{-1})} = \frac{U(z^{-1})}{V(z^{-1})}.$$

$U(z^{-1})$ and $V(z^{-1})$ are, respectively, the numerator and denominator polynomials in z^{-1} of $g_2(z^{-1})$. Therefore, given $H(z)$ in (3.17a)–(3.17d), the transfer function $H[g_2(z)]$ has to be computed using FTRANS.FOR. The execution of FTRANS.FOR runs as follows (the variable x is used in place of z^{-1} in the actual printout; in FTRANS.FOR, $U(x)$ and $V(x)$ are, in general, polynomials of degree two and if the coefficient of the highest degree term is zero, then this fact is denoted by 0.0).

```
C       THIS PROGRAM OBTAINS THE RESULTING TRANSFER FUNCTION AFTER A
C       FREQUENCY TRANSFORMATION
C       GIVEN A TRANSFER FUNCTION H(Z)=A(Z)/B(Z)
C       AND A FREQUENCY TRANSFORMATION Y=U(X)/V(X) WITH
C       U(X)=U0+U1*X+U2*X**2   AND   V(X)=V0+V1*X+V2*X**2
C       THE OUTPUT WILL BE H1(X)=A(Y)/B(Y)
C
        DIMENSION R1(0:20),R2(0:20),P(0:20)
        DIMENSION A(0:20),B(0:20),S(0:20),S1(0:20)
        WRITE(6,20)
20      FORMAT(10X,'ENTER U(X) IN INCREASING POWERS OF X',/)
        READ(5,21) A1,B1,C1
        WRITE(6,22)
22      FORMAT(10X,'ENTER V(X) IN INCREASING POWERS OF X',/)
        READ(5,21) A2,B2,C2
21      FORMAT(3F)
        WRITE(6,23)
23      FORMAT(10X,'ENTER THE DEGREES OF THE NUM. AND DENOMIN.',/)
        READ(5,24) NNUM,NDEN
        IF((NNUM-NDEN).EQ.0) GO TO 400
        GO TO 2000
400     CONTINUE
24      FORMAT(2I)
        WRITE(6,25)
25      FORMAT(10X,'ENTER THE COEFF. OF NUMER. IN INCREASING POWERS',/)
        READ(5,26) (A(I),I=0,NNUM)
        WRITE(6,27)
27      FORMAT(10X,'ENTER THE COEFF. OF DENOM. IN INCREASING POWERS',/)
        READ(5,26) (B(I),I=0,NDEN)
26      FORMAT(5F)
        NDIF=NDEN-NNUM
        KCOUNT=1
        N=NNUM
500     CALL ZERO(S(0),S(20))
        DO 110 I=0,N
        CALL ZERO (R1(0),R1(20))
        CALL ZERO (R2(0),R2(20))
        I1=I
        I2=N-I
        CALL TRINOM(A1,B1,C1,I1,R1)
        CALL TRINOM(A2,B2,C2,I2,R2)
        IF((N-I).LT.I) GO TO 11
        N1=2*(N-I)
        N2=2*I
        CALL ZERO (P(0),P(20))
        CALL POLY(R1,R2,N1,N2,P)
        GO TO 220
11      N2=2*I
```

Figure 3.12. (*Partial listing, continued.*)

```
        N1=2*(N-I)
        CALL ZERO(P(0),P(20))
        CALL POLY(R2,R1,N2,N1,P)
220     DO 330 J=0,2*N
330     S(J)=S(J)+A(I)*P(J)
110     CONTINUE
        IF(KCOUNT.EQ.2) GO TO 1000
        NNUM=NNUM*2
        DO 510 I=0,NNUM
510     S1(I)=S(I)
        KCOUNT=KCOUNT+1
        N=NDEN
        DO 520 I=0,NDEN
520     A(I)=B(I)
        GO TO 500
1000    NDEN=NDEN*2
        WRITE(6,1500)
        DO 1250 L=0,NNUM
        IF(ABS(S1(L)).GT.0.) GO TO 1260
1250    CONTINUE
        GO TO 1265
1260    CNUM=S1(L)
1265    DO 1270 LL=0,NDEN
        IF(ABS(S(LL)).GT.0.) GO TO 1275
1270    CONTINUE
        GO TO 1280
1275    CDEN=S(LL)
        CONST=CNUM/CDEN
1280    CONTINUE
        WRITE(6,1281) CONST
1281    FORMAT(10X,'CONSTANT=',2X,F,//)
1500    FORMAT(/,10X,'THE COEFF. OF THE NUMERATOR ARE:',//)
        DO 1501 I=0,NNUM
        S1(I)=S1(I)/CNUM
1501    WRITE(6,1502) I,S1(I)
1502    FORMAT(10X,'A(',I2,')=',2X,F)
        WRITE(6,1600)
1600    FORMAT(//,10X,'THE COEFF. OF THE DENOMINATOR ARE:',//)
        DO 1601 I=0,NDEN
        S(I)=S(I)/CDEN
1601    WRITE(6,1602) I,S(I)
1602    FORMAT(10X,'B(',I2,')=',2X,F15.8)
        GO TO 2005
2000    WRITE(6,2001)
2001    FORMAT(5X,'THE DEGREES OF NUMERATOR AND DENOMINATOR ARE ')
        WRITE(6,2002)
2002    FORMAT(5X,'TAKEN TO BE EQUAL IF NOT THE REQUIRED ZERO COEFF.')
        WRITE(6,2003)
```

Figure 3.12. (*Partial listing, continued.*)

212

```
2003      FORMAT(5X,'WILL HAVE TO BE ADDED')
          GO TO 2005
2005      STOP
          END
C
C
C
          SUBROUTINE TRINOM(A,B,C,N,R)
          DIMENSION R(0:20),P(0:20)
          DO 1 I=0,N
          CALL ZERO(P(0),P(20))
          CALL POWER(A,B,N-I,P)
          DO 2 K=0,N-I
          I1=I
          CALL BINOM(N,I1,NBI)
          FCI=1.
          IF((I.NE.0).OR.(C.NE.0.)) FCI=C**I
2         R(K+2*I)=R(K+2*I)+P(K)*FCI*NBI
1         CONTINUE
          RETURN
          END
C
C
C
          SUBROUTINE POLY(PN,PM,N,M,Q)
          DIMENSION PN(0:20),PM(0:20),Q(0:20),P1(0:20)
          CALL ZERO(Q(0),Q(20))
          IF(M.GT.N) GO TO 15
          DO 100 K=0,M+N
          DO 10 J=0,K
10        Q(K)=Q(K)+PN(J)*PM(K-J)
          GO TO 100
100       CONTINUE
          RETURN
15        WRITE(6,1)
1         FORMAT(1X,'ERROR M CAN NOT BE GREATER THAN N',/)
          RETURN
          END
C
C
C
          SUBROUTINE POWER(A,B,K,P)
          DIMENSION P(0:20)
          DO 1 J=0,K
          J1=J
          K1=K
          CALL BINOM(K1,J1,NB)
1         P(J)=NB*B**J*A**(K-J)
```

Figure 3.12. (*Partial listing, continued.*)

```
            RETURN
            END
C
C
C
            SUBROUTINE FACTO(N,NF)
            NF=1.
            IF(N.EQ.0) GO TO 2
            DO 1 I=1,N
            NF=NF*I
1           CONTINUE
2           RETURN
            END
C
C
C
            SUBROUTINE BINOM(N,K,NB)
            CALL FACTO(N,NF1)
            M=N-K
            K1=K
            CALL FACTO(K1,KF2)
            CALL FACTO(M,MF3)
            NB=NF1/KF2/MF3
            RETURN
            END
```

Figure 3.12. Listing of program FTRANS.FOR to obtain a transfer function from a specified one following a frequency transformation.

ENTER $U(x)$ IN INCREASING POWERS OF x

$-.267949 \ -1. \ 0.0$

Enter $V(x)$ in increasing powers of

$1. \ .267949 \ 0.0$

Enter the degrees of the num. and denomin. of $H(x)$

6 6

Enter the Coeff. of Numer. of $H(x)$ in increasing powers

1. 6. 15. 20. 15. 6. 1.

Enter the Coeff. of Denom. of $H(x)$ in increasing powers

$1 \ -.00000545 \ .77769595 \ -.00000259 \ .11419942 \ -.00000017$
$.00175093$

THE COEFF. OF THE NUMERATOR ARE:

Constant =	$0.1456574 \times K \triangleq K_1$
$A(0) =$	1.0000000
$A(0) =$	-6.0000030
$A(2) =$	15.0000000
$A(3) =$	-20.0000100
$A(4) =$	15.0000000
$A(5) =$	-6.0000020
$A(6) =$	1.0000000

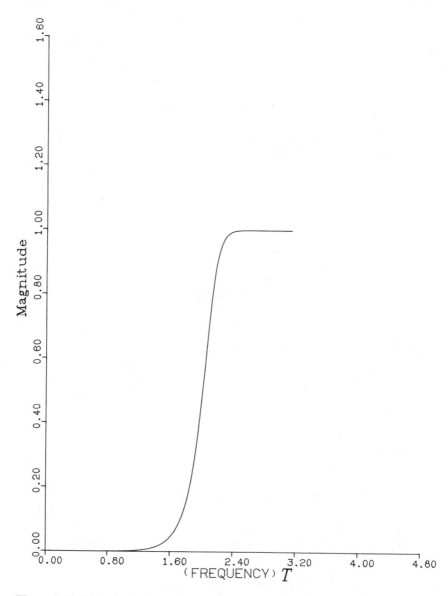

Figure 3.13. Magnitude of frequency response of derived highpass filter in Example 3.10.

THE COEFF. OF THE DENOMINATOR ARE:

$B(0) =$	1.00000000
$B(1) =$	1.98161413
$B(2) =$	2.25238472
$B(3) =$	1.46928734
$B(4) =$	0.59626459
$B(5) =$	0.13544086
$B(6) =$	0.01356374

The transformed filter transfer function is

$$H_2(z) = H\left[g_2(z^{-1})\right] = 0.145674K \frac{\sum_{k=0}^{6} A(k)z^{-k}}{\sum_{k=0}^{6} B(k)z^{-k}}$$

where the $A(k)$'s and $B(k)$'s are given above. The constant K may be chosen so that

$$H_2(-1) = 1,$$

which yields

$$K = 0.029588.$$

The plot of the magnitude of this frequency response for this highpass filter is given in Figure 3.13. It can be checked via use of SCHUR1.FOR that $H_2(z)$ characterizes a BIBO stable filter, as expected.

3.6. Time–Domain Design of IIR Filters

Given an impulse response sequence $\{h(k)\}$ whose z-transform is

$$H(z^{-1}) = \sum_{k=0}^{\infty} h(k)z^{-k} \qquad (3.72)$$

the objective is to approximate $H(z^{-1})$, in accordance with certain error criterion, by a rational function.

3.6.1. Padé Approximation Technique

One method for obtaining a rational approximant is via the Padé approximation technique. The $[r/m]$-order Padé approximant to $H(z^{-1})$ in (3.72) is of the form

$$\frac{P_r(z^{-1})}{Q_m(z^{-1})} = \frac{\sum_{k=0}^{r} p_k z^{-k}}{\sum_{k=0}^{m} q_k z^{-k}}, \qquad q_0 \neq 0. \qquad (3.73)$$

In (3.73), r and m are fixed *a priori*, and it may be assumed, without loss of generality, for exposition of the principles that

$$q_0 = 1. \tag{3.74}$$

The goal is to determine the remaining q_k's and the p_k's so that if the power series expansion of (3.73) about $z^{-1} = 0$ is

$$\frac{P_r(z^{-1})}{Q_m(z^{-1})} = \sum_{k=0}^{\infty} g(k) z^{-k}, \tag{3.75}$$

then

$$h(k) = g(k), \quad k = 0, 1, \ldots, m + r \tag{3.76a}$$

or, equivalently,

$$P_r(z^{-1}) - Q_m(z^{-1}) H(z^{-1}) \equiv O\left[(z^{-1})^{m+r+1}\right].^4 \tag{3.76b}$$

On equating coefficients in the preceding identity, starting from the zeroth power of z^{-1} and continuing up to the $(r + m)$th power of z^{-1}, one gets

$$
\begin{bmatrix}
h(-m) & h(-m+1) & \cdots & h(-1) & h(0) \\
h(-m+1) & h(-m+2) & \cdots & h(0) & h(1) \\
\cdots & \cdots & & \cdots & \cdots \\
h(r-m) & h(r-m+1) & \cdots & h(r-1) & h(r)
\end{bmatrix}
$$

$$
\begin{bmatrix}
q_m \\
q_{m-1} \\
\vdots \\
q_1 \\
1
\end{bmatrix}
=
\begin{bmatrix}
p_0 \\
p_1 \\
\vdots \\
p_{r-1} \\
p_r
\end{bmatrix}
\tag{3.77}
$$

and

$$
\underbrace{
\begin{bmatrix}
h(r-m+1) & h(r-m+2) & \cdots & h(r) & h(r+1) \\
h(r-m+2) & h(r-m+3) & \cdots & h(r+1) & h(r+2) \\
\cdots & \cdots & & \cdots & \cdots \\
h(r) & h(r+1) & & h(r+m-1) & h(r+m)
\end{bmatrix}
}_{H_1}
$$

$$
\begin{bmatrix}
q_m \\
q_{m-1} \\
\vdots \\
q_1 \\
1
\end{bmatrix}
=
\begin{bmatrix}
0 \\
0 \\
\vdots \\
0 \\
0
\end{bmatrix}.
\tag{3.78}
$$

In the above system of linear equations $h(k) = 0$, $k < 0$. The condition for existence of solution can be inferred from (3.78). For a solution to exist, it is necessary and sufficient that the rank of the coefficient matrix H_1 in (3.78) equal the rank of the matrix H [see (3.79a)] obtained after deletion of the

---

[4] $O[(z^{-1})^r]$ denotes that powers $[z^{-1}]^s$ of z^{-1} are absent for $s < r$.

last column from the coefficient matrix. The solution will be unique if and only if the $(m \times m)$ matrix obtained after the deletion of the last column in the coefficient matrix of (3.78) is nonsingular. When this is not the case, a factor common to the numerator and denominator of the rational approximant will be present.

Example 3.11. Consider

$$H(z^{-1}) = 1 + z^{-1} + z^{-2} + z^{-3} + 4z^{-4} + 5z^{-5} + \cdots.$$

Let $r = 1$ and $m = 2$. Then, from (3.78),

$$\begin{bmatrix} 1 & 1 & 1 \\ 1 & 1 & 1 \end{bmatrix} \begin{bmatrix} q_2 \\ q_1 \\ 1 \end{bmatrix} = \begin{bmatrix} 0 \\ 0 \end{bmatrix}. \tag{3.79a}$$

Clearly, $q_2 = -(1 + q_1)$. Also, from (3.77),

$$\begin{bmatrix} 0 & 0 & 1 \\ 0 & 1 & 1 \end{bmatrix} \begin{bmatrix} q_2 \\ q_1 \\ 1 \end{bmatrix} = \begin{bmatrix} p_0 \\ p_1 \end{bmatrix} \tag{3.79b}$$

Therefore,

$$\frac{P_r(z^{-1})}{Q_m(z^{-1})} = \frac{1 + (q_1 + 1)z^{-1}}{1 + q_1 z^{-1} - (1 + q_1)z^{-2}}$$

$$= \frac{1}{1 - z^{-1}} \quad \text{(after cancellation)}.$$

There are various ways to solve the system of equations in (3.77) and (3.78) in order to obtain $P_r(z^{-1})$ and $Q_m(z^{-1})$ in (3.73). The reader's attention is directed to Section 6.3 for this purpose.

One serious disadvantage of the Padé approximation technique in the time-domain design of recursive digital filters originates from the fact that an unstable filter might result. In general, there is no relation between the rational approximants obtained by the Padé technique and the important problem of stability. A criterion based on the impulse response sequence that provides sufficient conditions for establishing stability was given in reference [5]. Reduced order approximants may be obtained by removing the restriction implicit in (3.76a), and then the system of linear equations obtained may be solved using the method of least squares. This variant of the Padé approximation technique, which does not provide an exact fit as mentioned in (3.76a), has been considered in reference [6].

The general time-domain design problem of the IIR digital filter, then, is to find the $m + r + 1$ coefficients in (3.73) with $q_0 = 1$, such that the sequence $\{g(k)\}$ in (3.75) approximates in some sense a specified sequence $\{h(k)\}$ in (3.72) over a finite range of the independent variable. Naturally, the rational approximant is expected to satisfy the constraint of stability, besides, possibly, other desirable features, which could include a rapid decay of the filter impulse response outside a certain range. The actual design procedures, usually, involves some trial and error and are computer-aided.

3.7. Realization

Given a transfer function having real coefficients [see (3.2)]

$$H(z) = \frac{Y(z)}{X(z)} = \frac{\sum_{j=0}^{m} a_j z^{-j}}{1 + \sum_{j=1}^{n} b_j z^{-j}}, \tag{3.80}$$

the topology of structures realizing $H(z)$ is considered. Attention is primarily directed to the low-sensitivity lattice structure, which is derived using a discrete version of a well-known transformation, referred to in the literature as Richards' transformation. The elements used in a realization are adders, delays (denoted by square blocks enclosing z^{-1}), and multipliers (denoted by triangles associated with the values or coefficients of the respective multipliers).

3.7.1. Direct Realization

One mechanism to realize $H(z)$ in (3.80) is by writing it as

$$H(z) = \frac{W(z)}{X(z)} \frac{Y(z)}{W(z)} = \frac{1}{1 + \sum_{j=1}^{n} b_j z^{-j}} \cdot \sum_{j=0}^{m} a_j z^{-j}, \quad m \le n$$

(an auxiliary transform $W(z)$ has been introduced), realizing

$$\frac{W(z)}{X(z)} = \frac{1}{1 + \sum_{j=1}^{n} b_j z^{-j}} \tag{3.81a}$$

and

$$\frac{Y(z)}{W(z)} = \sum_{j=0}^{m} a_j z^{-j} \tag{3.81b}$$

separately, and, finally, cascading. The topology for this type of realization, referred to as a *direct form 2* realization, for the case when $n = m$ (this is no restriction) is shown in Figure 3.14A. The number of delays required in this type of realization is the minimum possible, and the coefficients of the multipliers can be read off directly from the specified rational function $H(z)$ in (3.80). Though the structure is canonic in the number of delays, it is known to possess poor sensitivity characteristics to coefficient quantization (see Chapter 5). There are other variants of this structure, which, because of poor performance under finite arithmetics constraints, will not be elaborated upon here. Note that the left-hand sides of (3.81a) and (3.81b) could have been interchanged to yield the realization in Figure 3.14B.

Example 3.12. Let the transfer function of a digital filter be specified as

$$H(z) = \frac{-\frac{15}{16} - \frac{23}{16} z^{-1} - \frac{9}{16} z^{-2} - \frac{1}{16} z^{-3}}{1 + \frac{1}{4} z^{-1} - \frac{1}{4} z^{-2} - \frac{1}{16} z^{-3}}.$$

The *direct form 2* realization of $H(z)$ is shown in Figure 3.15A. Figure

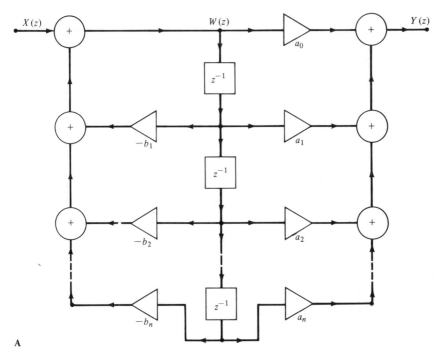

A

Figure 3.14A. Direct form 2 realization of $H(z)$ in (3.80).

3.15B gives another realization of $H(z)$, referred to as the direct form 1 realization, which requires twice the number of delays present in the direct form 2 realization.

3.7.2. Cascade Realization

In order to realize a digital filter transfer function $H(z)$ in cascade form, it is factorized in the form

$$H(z) = \prod_{i=1}^{k} \left(\frac{\alpha_{0i} + \alpha_{1i}z^{-1} + \alpha_{2i}z^{-2}}{1 + \beta_{1i}z^{-1} + \beta_{2i}z^{-2}} \right), \tag{3.82}$$

where for each i the biquadratic rational function [enclosed within parentheses in (3.82)]

$$H_i(z) = \frac{\alpha_{0i} + \alpha_{1i}z^{-1} + \alpha_{2i}z^{-2}}{1 + \beta_{1i}z^{-1} + \beta_{2i}z^{-2}} \tag{3.83}$$

has real coefficients. $H_i(z)$ can be realized in direct form as in Figure 3.14 with $m = n = 2$, and then the k second-order sections may be cascaded to form the composite filter. Of course, $H_i(z)$ may also be realized in alternate forms before cascading. For high-order filters, a significant saving (in

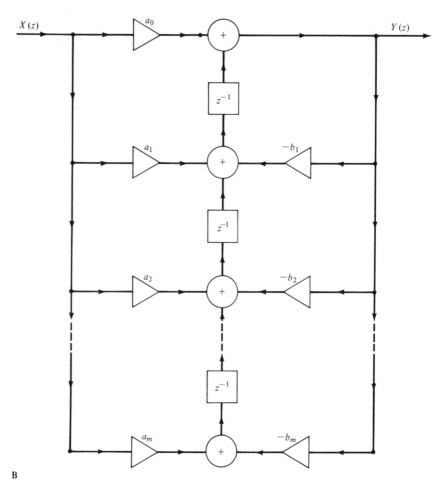

B

Figure 3.14B. Alternate direct form 2 realization of $H(z)$ in (3.80); again $m = n$.

comparison with the direct form of realization) in the number of coefficients bits can be achieved to ensure stability and satisfactory filter frequency characteristics, via use of the cascaded topology or the parallel form of realization, discussed next.

3.7.3. Parallel Form of Realization

To realize $H(z)$ in parallel form, it is written in the form

$$H(z) = \gamma_{00} + \sum_{i=1}^{k} \left(\frac{\gamma_{0i} + \gamma_{1i} z^{-1}}{1 + \delta_{1i} z^{-1} + \delta_{2i} z^{-2}} \right), \qquad (3.84)$$

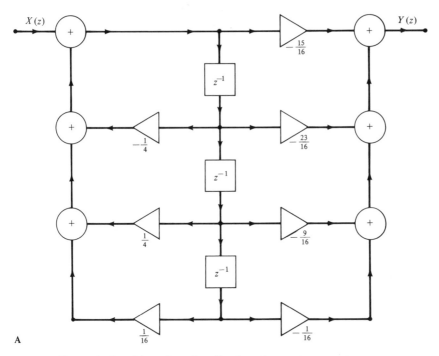

Figure 3.15A. Direct form 2 realization of $H(z)$ in Example 3.12.

where the biquadratic function within parentheses has real coefficients. Each such function can be realized in a suitable form, and then these second-order sections can be connected in parallel to yield the composite filter.

3.7.4. Ladder Form of Realization

These realizations are based on various continued fraction expansions of the transfer function. Each type of continued fraction expansion, when it exists, yields an associated realization. This form of realization is restrictive in the sense that only certain classes of stable digital filters can be synthesized. Furthermore, even though the resulting filters are canonic with respect to both delays and multipliers, the sensitivity properties of such structures might not be satisfactory. Therefore, the philosophy underlying this approach will be briefly explained through a discussion of a particular type of synthesized ladder structure. When certain conditions are satisfied by the coefficients of the specified transfer function (though these conditions may be tested for *a priori*, the algebraic tests involve computations of the determinants of a sequence of matrices which could be expensive timewise), the specified transfer function $H(z)$ in (3.80) may be expandable in one of

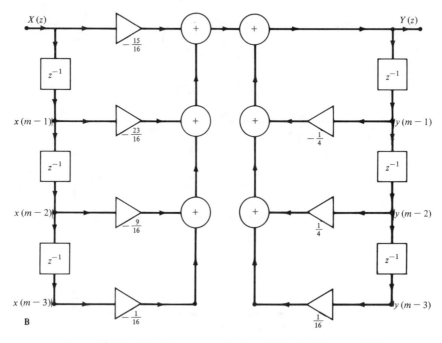

Figure 3.15B. Direct form 1 realization of $H(z)$ in Example 3.12.

the following two forms:

$$H(z) = A_0 + \cfrac{1}{B_1 z^{-1} + \cfrac{1}{A_1 + \cfrac{1}{\ddots + \cfrac{1}{B_n z^{-1} + \cfrac{1}{A_n}}}}} \qquad (3.85)$$

or

$$H(z) = \hat{A}_0 + \cfrac{1}{\hat{B}_1 z + \cfrac{1}{\hat{A}_1 + \cfrac{1}{\ddots + \cfrac{1}{\hat{B}_n z + \cfrac{1}{\hat{A}_n}}}}} \qquad (3.86)$$

The forms in (3.85) and (3.86) are types of continued fraction expansions.

The expansion in (3.85) is obtained by applying the divide-invert-divide sequence of operations repetitively on the rational function $H(z)$, around $z^{-1} = \infty$, while (3.86) is the outcome of applying a similar sequence of operations around $z^{-1} = 0$. Note that $H(z)$ in (3.85) can be rewritten as

$$H(z) = A_0 + \frac{1}{B_1 z^{-1} + 1/H_1(z)}, \qquad (3.87)$$

where $H_1(z)$ has an expansion similar in form to the expansion of $H(z)$ in (3.85), but it is of order $(n-1)$ instead of n. The realization of $H(z)$ in terms of $H_1(z)$ is shown in Figure 3.16A. The cycle of operations is repeated on $H_1(z)$ and continued until the complete realization of $H(z)$ is obtained in terms of adders, scalar multipliers, and delay elements. Delay-free loops might appear in the realization and for techniques to eliminate such delay-free loops, the reader may consult reference [25].

Example 3.13. Consider the transfer function specified in the previous example. The divide-invert-divide steps required to obtain the representation in (3.85) are described below:

$$-\tfrac{1}{16}z^{-3} - \tfrac{1}{4}z^{-2} + \tfrac{1}{4}z^{-1} + 1 \overline{\big)\; -\tfrac{1}{16}z^{-3} - \tfrac{9}{16}z^{-2} - \tfrac{23}{16}z^{-1} - \tfrac{15}{16}}$$
$$\underline{-\tfrac{1}{16}z^{-3} - \tfrac{1}{4}z^{-2} + \tfrac{1}{4}z^{-1} + 1}$$
$$-\tfrac{5}{16}z^{-2} - \tfrac{27}{16}z^{-1} - \tfrac{21}{16}$$

$$\tfrac{1}{5}z^{-1} \quad \leftarrow \text{CONTINUED}$$
$$-\tfrac{5}{16}z^{-2} - \tfrac{27}{16}z^{-1} - \tfrac{21}{16} \overline{\big)\; -\tfrac{1}{16}z^{-3} - \tfrac{1}{4}z^{-2} + \tfrac{1}{4}z^{-1} + 1}$$
$$\underline{-\tfrac{1}{16}z^{-3} - \tfrac{27}{80}z^{-2} - \tfrac{21}{80}z^{-1}}$$
$$\tfrac{7}{80}z^{-2} + \tfrac{41}{80}z^{-1} + 1$$

$$-\tfrac{25}{7} \qquad \leftarrow \text{CONTINUED}$$
$$\tfrac{7}{80}z^{-2} + \tfrac{41}{80}z^{-1} + 1 \overline{\big)\; -\tfrac{5}{16}z^{-2} - \tfrac{27}{16}z^{-1} - \tfrac{21}{16}}$$
$$\underline{-\tfrac{5}{16}z^{-2} - \tfrac{205}{112}z^{-1} - \tfrac{25}{7}} \qquad \tfrac{49}{80}z^{-1}$$
$$\tfrac{1}{7}z^{-1} + \tfrac{253}{112} \overline{\big)\; \tfrac{7}{80}z^{-2} + \tfrac{41}{80}z^{-1} + 1}$$
$$\underline{\tfrac{7}{80}z^{-2} + \tfrac{1771}{1280}z^{-1}} \qquad -\tfrac{256}{1561}$$
$$-\tfrac{223}{256}z^{-1} + 1 \overline{\big)\; \tfrac{1}{7}z^{-1} + \tfrac{253}{112}}$$
$$\tfrac{1}{7}z^{-1} - \tfrac{256}{1561}$$
$$\leftarrow \text{CONTINUED} \qquad \underline{\tfrac{60,515}{24,976}}$$

$$-\tfrac{49,729}{138,320}z^{-1}$$
$$\tfrac{60,515}{24,976} \overline{\big)\; -\tfrac{223}{256}z^{-1} + 1}$$
$$\underline{-\tfrac{223}{256}z^{-1}} \qquad \tfrac{60,515}{24,976}$$
$$1 \overline{\big)\; \tfrac{60,515}{24,976}}$$
$$\underline{\tfrac{60,515}{24,976}}$$
$$0$$

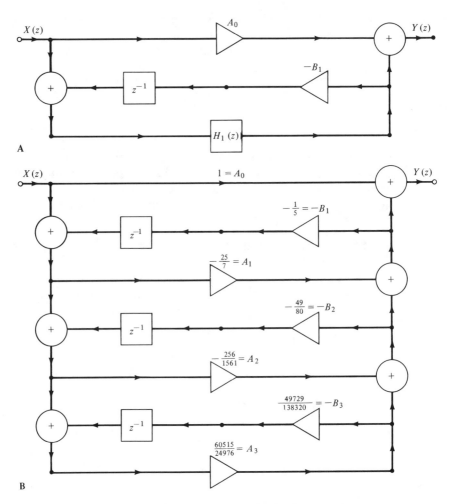

Figure 3.16A. Realization of $H(z)$ in (3.87). B. Realization of $H(z)$ in Example 3.13.

The resulting continued fraction expansion (CFE) is:

$$H(z) = 1 + \cfrac{1}{\frac{1}{5}z^{-1} + \cfrac{1}{-\frac{25}{7} + \cfrac{1}{\frac{49}{80}z^{-1} + \cfrac{1}{-\frac{256}{1561} + \cfrac{1}{+\frac{49,729}{138,320}z^{-1} + \cfrac{1}{\frac{60,515}{24,976}}}}}}}.$$

The complete realization of $H(z)$ is shown in Figure 3.16B. The multiplier values are also indicated in terms of the literal coefficients of the expansion in (3.85).

Next, note that $H(z)$ in (3.86) can be rewritten as

$$H(z) = \hat{A}_0 + \cfrac{1}{\hat{B}_1 + \cfrac{1}{H_2(z)}}, \tag{3.88}$$

where $H_2(z)$ has an expansion similar in form to the expansion of $H(z)$ in (3.86). However, like $H_1(z)$ in (3.87), $H_2(z)$ is of order $(n-1)$, i.e., its order has reduced by 1 from the order of $H(z)$. The realization of $H(z)$ in terms of $H_2(z)$ is shown in Figure 3.17A. The cycle of operations is repeated on $H_2(z)$ and continued till the desired final configuration is obtained.

Example 3.14. The transfer function dealt with in the previous two examples is again considered. The steps required to obtain the representation in (3.86) are described below:

$$-16z^3 - 4z^2 + 4z + 1 \overline{) \begin{array}{l} -\frac{15}{16} \\ 15z^3 + 23z^2 + 9z + 1 \end{array}}$$

$$15z^3 + \tfrac{15}{4}z^2 - \tfrac{15}{4}z - \tfrac{15}{16} \qquad -\tfrac{64}{77}z$$

$$\tfrac{77}{4}z^2 + \tfrac{51}{4}z + \tfrac{31}{16} \quad \overline{)\, -16z^3 - 4z^2 + 4z + 1}$$

$$-16z^3 - \tfrac{816}{77}z^2 - \tfrac{124}{77}z$$

$$\tfrac{508}{77}z^2 + \tfrac{432}{77}z + 1$$

CONTINUED

$$\tfrac{508}{77}z^2 + \tfrac{432}{77}z + 1 \overline{) \begin{array}{l} \frac{5929}{2032} \\ \tfrac{77}{4}z^2 + \tfrac{51}{4}z + \tfrac{31}{16} \end{array}}$$

$$\tfrac{77}{4}z^2 + \tfrac{2079}{127}z + \tfrac{5929}{2032} \qquad -\tfrac{258{,}064}{141{,}603}z$$

$$-\tfrac{1839}{508}z - \tfrac{249}{254} \quad \overline{)\, \tfrac{508}{77}z^2 + \tfrac{432}{77}z + 1}$$

$$\tfrac{508}{77}z^2 + \tfrac{84{,}328}{47{,}201}z \qquad -\tfrac{1{,}127{,}307}{1{,}190{,}752}$$

$$\tfrac{2344}{613}z + 1 \quad \overline{)\, -\tfrac{1839}{508}z - \tfrac{249}{254}}$$

$$-\tfrac{1839}{508}z - \tfrac{1{,}127{,}307}{1{,}190{,}752}$$

$$-\tfrac{315}{9376}$$

$$-\tfrac{315}{9376} \overline{) \begin{array}{l} -\frac{21{,}977{,}344}{193{,}095}z \\ \tfrac{2344}{613}z + 1 \end{array}} \qquad \leftarrow \text{CONTINUED}$$

$$\tfrac{2344}{613}z \qquad -\tfrac{315}{9376}$$

$$1 \quad \overline{)\, -\tfrac{315}{9376}}$$

$$-\tfrac{315}{9376}$$

$$0$$

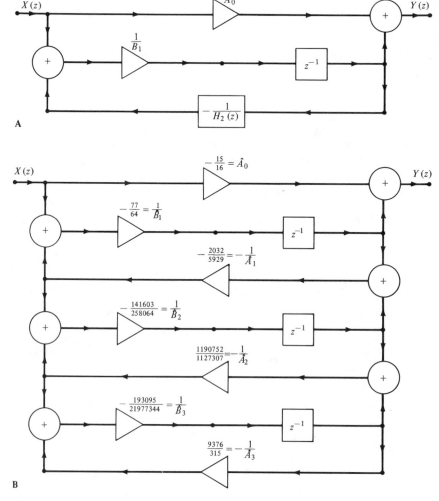

Figure 3.17A. Realization of $H(z)$ in (3.88). B. Realization of $H(z)$ in Example 3.14.

The desired expansion of $H(z)$ is

$$H(z) = -\tfrac{15}{16} + \cfrac{1}{-\tfrac{64}{77}z + \cfrac{1}{\tfrac{5929}{2032} + \cfrac{1}{-\tfrac{258,064}{141,603}z + \cfrac{1}{-\tfrac{1,127,307}{1,190,752} + \cfrac{1}{-\tfrac{21,977,344}{193,095}z + \cfrac{1}{-\tfrac{315}{9376}}}}}}}.$$

The complete realization of $H(z)$ is shown in Figure 3.17B. The multiplier values are also indicated in terms of the literal coefficients of the expansion in (3.86).

3.7.5. Lattice Form of Realization

The lattice form of realization is the outcome of a special case of an approach suggested by Vaidyanathan and Mitra [10] to design low-sensitivity digital filters. Apart from the generality of their approach, their synthesis is implemented in the z-domain, eliminating, thereby, the need for bilinear transformation which is required to arrive at a digital filter structure from an analog prototype. In this section, some of the fundamental ideas in the approach of Vaidyanathan and Mitra are discussed, especially in the context of arriving at the lattice form of realization, originally proposed by Gray and Markel [11] in a more complicated setting. First, the important concept of discrete bounded real function is defined.

Definition 3.1. A real rational function $H(z)$ will be called a discrete bounded real function (BRF) if it satisfies the following properties:

a. $H(z)$ has no poles in $|z| \geq 1$.
b. $|H(z)| \leq 1$ when $|z| = 1$.

When condition b in the above definition is replaced by a more stringent condition, a discrete lossless bounded real function (LBRF), sometimes also referred to as a discrete stable allpass function results.

Definition 3.2. A real rational function $H(z)$ will be called a discrete lossless bounded real function (LBRF) if it satisfies the following properties:

a. $H(z)$ has no poles in $|z| \geq 1$.
b. $|H(z)| = 1$ when $|z| = 1$.

Consider next the chain matrix characterization of a digital 2-port (or 2-pair), shown in Figure 3.18. The input and output transform variables $X_1(z)$, $X_2(z)$ and $Y_1(z), Y_2(z)$, of the digital two-pair are interrelated by

Figure 3.18. A digital two-pair.

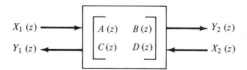

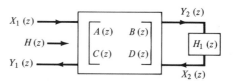

Figure 3.19. A constrained 2-pair.

parameters $A(z)$, $B(z)$, $C(z)$, and $D(z)$, forming the chain matrix in the manner shown next:

$$\begin{bmatrix} X_1(z) \\ Y_1(z) \end{bmatrix} = \begin{bmatrix} A(z) & B(z) \\ C(z) & D(z) \end{bmatrix} \begin{bmatrix} Y_2(z) \\ X_2(z) \end{bmatrix}. \tag{3.89}$$

Consider the realization of a transfer function $H(z)$ in the form of a constrained 2-pair as shown in Figure 3.19, where $H_1(z)$ is the constraining transfer function. Straightforward analysis yields (see Problem 16)

$$H_1(z) = \frac{C(z) - A(z)H(z)}{B(z)H(z) - D(z)}. \tag{3.90}$$

For the synthesis to terminate, the extracted 2-pair should be chosen so that remainder transfer function $H_1(z)$ has a lower order than $H(z)$. Also, for the cycle to be repeated on $H_1(z)$, it must have the same general properties as $H(z)$. The case when $H(z)$ is an LBRF is considered next. The theorem, given next, is crucial to the development of a cascaded lattice type of structure for discrete lossless bounded real functions.

Theorem 3.1. *Let $H(z)$ be an LBRF. Then*

$$H_1(z) = z\frac{H(\infty) - H(z)}{H(\infty)H(z) - 1} \tag{3.91}$$

is also an LBRF. Furthermore, $H_1(z)$ is of lower order than $H(z)$.

PROOF: Note that $H(z)$ is of the form

$$H(z) = \frac{a_0 z^n + a_1 z^{n-1} + \cdots + a_n}{a_n z^n + a_{n-1}z^{n-1} + \cdots + a_0}. \tag{3.92}$$

The form in (3.92) occurs because, in an LBRF, the poles must be reciprocals of the zeros. Since all poles of $H(z)$ are in $|z| < 1$ and $H(\infty) = a_0/a_n$, it follows that

$$|H(\infty)| < 1. \tag{3.93}$$

Poles of $H_1(z)$ can occur when

$$|H(z)| = \left|\frac{1}{H(\infty)}\right| > 1. \tag{3.94}$$

Since $H(z)$ is an LBRF, $|H(z)| = 1$, when $|z| = 1$ and $H(z)$ has no poles in $|z| \geq 1$. Then,

$$|H(z)| > 1 \quad \text{for } |z| < 1, \qquad |H(z)| < 1 \quad \text{for } |z| > 1. \tag{3.95}$$

Equation (3.95) follows from the application of the maximum modulus theorem. Therefore, (3.94) holds when $|z| < 1$, which implies that all poles of $H_1(z)$ occur in $|z| < 1$. Also, it is trivial to verify that since

$$|H(z)| = 1 \quad \text{if and only if} \quad |z| = 1,$$

from (3.91)

$$|H_1(z)| = 1 \quad \text{when } |z| = 1.$$

Therefore, $H_1(z)$ is an LBRF. The fact that $H_1(z)$ is of lower order than $H(z)$ follows from direct substitution of $H(\infty) = a_0/a_n$ along with the form of $H(z)$ from (3.92) in (3.91). This completes the proof of the theorem.

On comparing terms in (3.90) and (3.91), it follows that the digital 2-pair in Figure 3.19 may be characterizable (not uniquely) by the chain matrix

$$\mathbf{H_1}(z) = \begin{bmatrix} 1 & z^{-1}H(\infty) \\ H(\infty) & z^{-1} \end{bmatrix}. \tag{3.96}$$

The digital 2-pair, characterized by $\mathbf{H_1}(z)$ in (3.96), is realizable as the lattice shown in Figure 3.20. This lattice is the two multiplier form of the Gray–Markel lattice. After extraction of the 2-pair, the remainder $H_1(z)$ is also lossless bounded real, and it is of lower degree than $H(z)$. A 2-pair, characterized by $\mathbf{H_2}(z)$ is then extracted from $H_1(z)$, leaving a remainder

Figure 3.20. The Gray and Markel lattice (2-multiplier form).

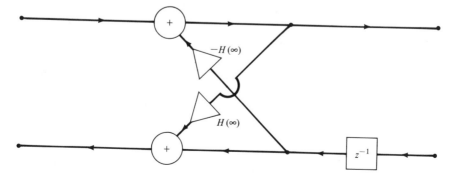

function $H_2(z)$, also lossless bounded real and of lower degree than $H_1(z)$. The process is continued until the remainder function $H_n(z)$ is $+1$ or -1. The complete procedure is illustrated next via a nontrivial example.

Example 3.15. It is required to realize as a cascade of lattices the LBRF of degree $n = 4$, given below:

$$H(z) = -\frac{3z^4 - z^3 + 5z^2 + z + 8}{8z^4 + z^3 + 5z^2 - z + 3},$$

$$H(\infty) = -\frac{3}{8}$$

Applying successively the previous theorem, one gets the sequence $\{H_k(z)\}$ of LBRF's:

$$H_1(z) = \frac{11z^3 - 25z^2 - 11z - 55}{55z^3 + 11z^2 + 25z - 11},$$

$$H_1(\infty) = \frac{11}{55}.$$

Then, defining

$$H_k(z) = z\frac{H_{k-1}(\infty) - H_{k-1}(z)}{H_{k-1}(\infty)H_{k-1}(z) - 1}, \quad k = 2,3,4,$$

one gets

$$H_2(z) = -\frac{17z^2 + 10z + 33}{33z^2 + 10z + 17},$$

$$H_2(\infty) = -\frac{17}{33},$$

$$H_3(z) = -\frac{z + 5}{5z + 1},$$

$$H_3(\infty) = -\frac{1}{5},$$

$$H_4(z) = -1.$$

The complete realization is shown in Figure 3.21

Note that the magnitude of the multiplier coefficients, $H(\infty)$ and $H_k(\infty)$, $k = 1, 2, \ldots, n - 1$, are each less than 1; these conditions can be shown to be necessary and sufficient for stability, i.e., the zeros of the denominator polynomial in z of $H(z)$ all lie in $|z| < 1$. Therefore, no separate test for stability is necessary; if the magnitude of the multiplier coefficients in each cascade is less than unity, the filter is stable, while if a multiplier coefficient having a magnitude greater than or equal to unity is encountered, the realization need not be completed as the filter is unstable.

231

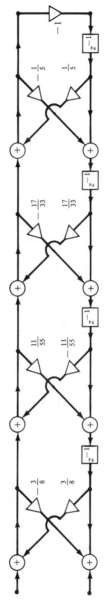

Figure 3.21. Complete realization of transfer function in Example 3.15.

3.8. Conclusions and Suggestions

The main items discussed in this chapter are the various types of frequency domain methods for the design of IIR filters, frequency transformations which allow the design of digital filters having specified types of response characteristics from a digital lowpass prototype, frequency transformations that enable the designer to get a desired type of digital filter from a lowpass analog prototype and different structures used in the realization of digital filters. For an overview of frequency domain design methods of IIR filters, the reader is referred to reference [1]. Here, the direct design of IIR digital filters in the frequency domain is considered following the choice of suitable magnitude-squared functions. These functions have known forms associated with the maximally flat (Butterworth) and equiripple (Chebyshev) types of responses and are derivable from the corresponding functions characterizing, respectively, Butterworth and Chebyshev types of analog filters via use of the bilinear transformation. In the lowpass case the pole locations in the z-plane for these filter types are explicitly computed in terms of the desired filter specifications like filter order, cutoff frequency, and sampling period in the case of Butterworth filters and the additional specification of ripple factor in the case of Chebyshev filters. In both cases, filters of order n have zeros of order n at $z = -1$. The cutoff frequency plays the same role in both analog and digital cases while the order n is usually chosen to achieve the desired minimum attenuation at a stopband frequency. The complete designs of the digital Butterworth and Chebyshev filters via the methods described are adequately illustrated by complete examples and the plots of the magnitude as well as the phase responses of the designed filters have been included. It should be noted that, though the bilinear transformation in (3.6) maps the entire imaginary axis in the s-plane to the unit circle in the z-plane (thus eliminating aliasing errors), the frequency scales are distorted relative to one another due to the phenomenon of warping. The nonlinear nature of the map that describes warping can be observed by graphing $\Omega = \tan(\omega T/2)$ in the (Ω, ω) plane. This warping limits the use of bilinear transformation, unless measures are taken to compensate [2, p. 223].

Other methods used to design IIR digital filters in the frequency domain include the impulse-invariant method (and step-invariant method), numerical integration method, and the matched z-transform method. The advantages and disadvantages of these methods and also of that based on the use of bilinear transformation are summarized in Table 3.6. It is fair to say that when the preservation of the impulse response is crucial, the impulse-invariant method is recommended while the bilinear transformation method is usable in practically all remaining situations. The use of those two techniques is usually preferred over the numerical integration and matched z-transform methods.

Table 3.6. Advantages and Disadvantages of Different Methods for Designing IIR Filters in the Frequency Domain

Method of digital filter design	Advantages	Disadvantages
1. Use of bilinear transformation	1a. Can use to advantage the vast resources of analog design techniques 1b. No aliasing 1c. Stability preserved	1a. Warping of frequency scales 1b. Neither the impulse response nor the phase response is preserved
2. Impulse-invariant method	2a. The impulse response of digital filter is a sampled version of the analog filter 2b. Convenient to use and implement	2a. The frequency response may differ widely from that of the analog filter due to aliasing 2b. Gain of the filter varies with sampling rate and possibility of overflow in the computer program exists
3. Numerical integration method	3a. The digital filter is derivable easily from the analog filter by an algebraic substitution	3a. The frequency axis of the analog filter does not, generally, map to the unit circle in the z-plane 3b. Stability not preserved, in general
4. Matched z-transformation	4a. Easy to use	4a. Possibility of aliasing 4b. Unsuitable when analog filter is all-pole 4c. The analog filter transfer function must be in factored form

The presentation of the digital to digital frequency transformations is largely influenced by the work of Constantinides [3]. In the design of any one of different types (highpass, bandpass, bandstop) of digital filters from a lowpass digital filter prototype characterized by the transfer function $H(z^{-1})$, it is necessary to use an appropriate transformation of the delay variable z^{-1},

$$z^{-1} \rightarrow g(z^{-1})$$

in order to calculate the composite transfer function, $H[g(z^{-1})]$, which then characterizes the filter being designed. It should be noted that when $g(z^{-1})$

is an allpass stable transfer function, then the stability of $H(z^{-1})$ implies stability of $H[g(z^{-1})]$; therefore, no separate stability check is necessary on the filter obtained via the frequency transformation (see Problem 1). An easy-to-read exposition of frequency transformations (digital to digital and analog to digital) can also be found in reference [4].

Techniques for directly designing IIR filters in the time domain are based on the approximation of a desired impulse response sequence $\{h(k)\}$, by the coefficient sequence $\{g(k)\}$ of the power series expansion of the filter transfer function ($\{h(k)\}$ and $g(k)\}$ are understood to be real sequences),

$$H(z^{-1}) = \frac{P_r(z^{-1})}{Q_m(z^{-1})} = \sum_{k=0}^{\infty} g(k) z^{-k}$$

so that a cost function (let $\{w(k)\}$ be a positive weighting sequence associated with the error sequence $\{g(k) - h(k)\}$)

$$E = \sum_{k=0}^{N-1} [g(k) - h(k)]^2 w(k)$$

is minimized over $0 \le k \le N-1$, subject, of course, to constraints like stability on $H(z^{-1})$. When r and m are, respectively, the degrees in z^{-1} of $P_r(z^{-1})/Q_m(z^{-1})$, N is chosen to equal $m + r + 1$, and the constraint $g(k) = h(k)$, $k = 0, 1, \ldots, m + r$, is imposed, then the rational approximant in z^{-1} becomes the Padé approximant. A large volume of literature exists on this type of approximant, and for details regarding properties and scopes for recursive computation of Padé approximants, the reader is referred to Chapter 6. Since the design of filters based on the Padé approximation theory is based solely on time domain specifications, the frequency responses of such filters might be unacceptable. Therefore, such filter performances need to be improved via use of computer-aided optimization techniques. Since Padé approximants need not be stable, attention has to be directed to the stabilization problem as well.

Unlike the case of Padé approximants, computer-aided techniques for filter design are based on the application of certain optimization procedures, iterative in nature, which minimize a cost function in contrast to the exact computation of filter coefficients by solving a set of linear equations. In reference [7], an unconstrained optimization algorithm is used to minimize a squared-error criterion involving the magnitude of the frequency response following the choice of a cascaded form of realization of second-order transfer function blocks and the incorporation of a strategy which ensures not only stability but the property of minimum phase (i.e., the zeros are also within the unit circle) in the designed filter. Deczky [8] essentially generalized the results in reference [7] by considering higher than second-order error criteria and incorporating group delay considerations in the error criterion in addition to the magnitude of the frequency response. For references and information on other computer-aided design approaches, the reader is referred to reference [2, pp. 267–282].

Finally, the different types of implementation techniques of digital filters have been presented. Direct form 2, cascade, and parallel forms of realizations can realize arbitrary digital filter transfer functions, and these forms are canonic in the sense that a minimum number of adders, multipliers, and delays are required and the topology of structure for each form of realization is fixed. Since the choice of a form of realization is considerably influenced by the sensitivity to finite arithmetic constraints, it will be seen in Chapter 5 that the direct form is usually avoided because the accuracy requirements on the filter coefficients are very severe. The cascade form is useful for filters whose transmission zeros are on the unit circle, since then the number of multipliers can be significantly reduced [9]. When implementing a digital filter in the cascaded form under finite arithmetic conditions, it is necessary to direct attention to the problems of *pairing* poles and zeros in the second-order blocks and the *ordering* of these basic building blocks. Error analysis (see Chapter 6) is helpful for the purpose. In the cascaded form of realization, it is also necessary to have scaling multipliers between individual second-order blocks in the cascade to prevent the filter variables from taking extreme values. Like the cascade form, the parallel form realizes each pair of complex conjugate poles separately and is generally preferred over the direct form from parameter quantization standpoints.

The ladder form of realization has been covered more for the sake of completeness rather than for any potential utilitarian value. As in passive synthesis, where special classes of driving-point functions (for example RC, LC, and RL) are realizable in ladder form via the Cauer synthesis procedure by repeated application of division (using Euclid's algorithm for polynomials), inversion, and division steps; only a special class of digital filter transfer functions is realizable in ladder form. For additional insight, see references [12] and [13]. Low sensitivity digital filters can be realized in lattice form. Furthermore, the notion of passivity has been shown to be closely linked to the realization of the low sensitivity property in digital filters [10], [14]. In this book a brief introduction to the realization of low sensitivity digital has been given through the concept of a lossless bounded real function, which is shown to be realizable in a finite number of steps by repeated application of Theorem 3.1. This theorem could be viewed as a discrete domain version of the celebrated theorem due to Richards, which has been used in a variety of applications including the transformerless synthesis (Bott–Duffin method [15, pp. 85–90]) of passive driving-point immittance functions and the cascade synthesis of distributed networks [16, p. 339]. The advanced reader is invited to consult references [10] and [17] for additional insight into the cascade lattice realization of any stable digital filter transfer function. A more advanced and generalized treatment of the topic of realization of any stable, passive digital rational transfer function in a cascaded topology appears in the recently published paper of Rao and Kailath [18].

Problems

1. Let $A(z^{-1})/B(z^{-1})$ be a stable real rational function where $A(z^{-1})$ and $B(z^{-1})$ are polynomials in the delay variable z^{-1}. Consider the mapping

$$z^{-1} \to \frac{a(z^{-1})}{b(z^{-1})},$$

where $a(z^{-1})/b(z^{-1})$ is a stable allpass real rational function. Show that $A[a(z^{-1})/b(z^{-1})]/B[a(z^{-1})/b(z^{-1})]$ is also a stable real rational function. What happens in the case of an unstable allpass function?

2. The function

$$H_i(z) = \frac{\left(z^{-1} - \alpha_i\right)}{1 - \alpha_i^* z^{-1}}$$

has a property that $|H_i(z)| = 1$ on $|z| = 1$. Does this property hold for the following functions, each generated from a finite number k of functions of the type $H_i(z)$:

a. $\sum_{i=1}^{k} H_i(z)$.
b. $\prod_{i=1}^{k} H_i(z)$.
c. $\sum_{i=1}^{k-1} \dfrac{H_i(z)}{H_{i+1}(z)}$.

3. A lowpass digital filter is required to have a maximally flat magnitude characteristic and a cutoff frequency of 3.9 kHz. The attenuation at the frequency 4.50 kHz is to be more than 10 db. The sampling frequency is 20 kHz. Design a filter satisfying the given specifications.

4. a. Let $y(k) = \cos k\theta$, for an integer k. Show that

$$y(k+1) + y(k-1) = 2\cos\theta \, y(k).$$

Let $x(k) = \sin k\theta$, for an integer k. Show that

$$x(k+1) - x(k-1) = 2\sin\theta \, y(k)$$

b. Verify that $\cos n\theta \neq 0$, when $\theta = \theta_1 + j\theta_2$, $\theta_2 \neq 0$ with θ_1, θ_2 real.
c. Let

$$H(z) = \frac{\sum_{k=0}^{m} a(k) z^{-k}}{\sum_{k=0}^{n} b(k) z^{-k}}$$

be the transfer function of a linear time-invariant digital filter. Using the results proved in a and c, establish that the magnitude-squared response function

$$\left| G\left(e^{j\omega T} \right) \right|^2 \triangleq H\left(e^{j\omega T} \right) H\left(e^{-j\omega T} \right)$$

can be expressed in the form

$$\left| G\left(e^{j\omega T} \right) \right|^2 = \frac{\Sigma_k c(k) \tan^{2k}(\omega T/2)}{\Sigma_k d(k) \tan^{2k}(\omega T/2)},$$

where $c(k)$'s and $d(k)$'s are real constants like $a(k)$'s and $b(k)$'s. You might like to use the trigonometric identity

$$\cos \theta = \frac{1 - \tan^2(\theta/2)}{1 + \tan^2(\theta/2)}.$$

5. A lowpass filter is required to satisfy the following specifications.
   a. The passband equiripples are 0.15 db.
   b. The cutoff frequency is 100 Hz.
   c. The stopband response is monotonic and the stopband attenuation is greater than 15 db between 600 and 900 Hz.
   d. The sampling frequency is 5 kHz.
   Design a suitable filter.

6. Consider the transfer function of a normalized fourth order analog Butterworth filter:

$$H_a(s) = \frac{1}{(s^2 + 0.765s + 1)(s^2 + 1.848s + 1)}.$$

Let the sampling frequency be 100 Hz. Find the transfer function of the digital filter via
   a. the bilinear transformation method,
   b. the impulse invariant transformation method,
   c. the matched z-transformation method.
   For each of the above cases obtain the plot of the magnitude characteristic of the frequency response for the obtained digital filter transfer function.

7. Design an eighth-order Chebyshev bandpass filter with the upper and lower cutoff frequencies at 6 kHz and 4 kHz, respectively. The percentage passband ripple is 10% and the sampling frequency is 18 kHz.

8. Fettweis [19] proposed a design method for maximally flat digital filters by making use of the familiar bilinear transformation. Consider the transcendental function:

$$F(s) = \tanh\left(\mu \tanh^{-1} s \right)$$

$F(s)$ can be expanded as a continued fraction:

$$F(s) = \frac{\mu|}{\left|\dfrac{1}{s}\right.} + \frac{(\mu^2 - 1)|}{\left|\dfrac{3}{s}\right.} + \frac{(\mu^2 - 4)|}{\left|\dfrac{5}{s}\right.} + \cdots + \frac{(\mu^2 - m^2)|}{|(2m+1)/s} + \cdots .$$

Define the nth convergent $F_n(s)$ as

$$F_n(s) = \frac{\mu|}{\left|\dfrac{1}{s}\right.} + \frac{(\mu^2 - 1)|}{\left|\dfrac{3}{s}\right.} + \cdots + \frac{\left[\mu^2 - (n-1)^2\right]|}{\left|\dfrac{(2n-1)}{s}\right.}$$

Form

$$H_{an}(s) = \frac{(1+s)^n}{[1 + F_n(s)] d_e(s)},$$

where $F_n(s) = d_o(s)/d_e(s)$. Then form $H_n(z^{-1})$ after replacing s by $(1 - z^{-1})/1 + z^{-1}$ in $H_a(s)$. Calculate $H_3(z^{-1})$, $H_5(z^{-1})$, or $H_7(z^{-1})$.

9. Consider the second-order recursive digital filter, characterized by the transfer function

$$H(z^{-1}) = \frac{a_0}{1 + b_1 z^{-1} + b_2 z^{-2}}.$$

Show that the filter is BIBO stable for all values of b_1 and b_2 in the region inside the triangle shown below.

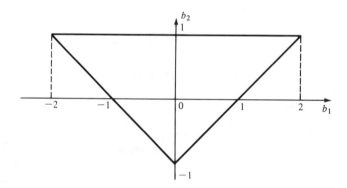

10. The magnitude-squared transfer function of a lowpass sine Butterworth filter is written in the form

$$H(z) H(z^{-1})\big|_{z = e^{j\omega T}} = \left| H(e^{j\omega T}) \right|^2 = \frac{1}{1 + [\sin(\omega T/2)/\sin(\omega_c T/2)]^{2n}},$$

where ω_c is the cutoff frequency in radians per second. For $n = 3$, obtain the stable rational function $H(z^{-1})$.

11. Consider the transfer function of an IIR filter (a, b, c, are suitable real constants):

$$H(z^{-1}) = \frac{1-c}{1-a^{-1}z^{-1}} + \frac{c}{1-b^{-1}z^{-1}}.$$

For each of the cases $c = 0$ and $c = 1$, do the following: Obtain the power series expansion of $H(z^{-1})$ in the form

$$H(z^{-1}) = \sum_{k=0}^{\infty} h_k z^{-k}.$$

Consider the truncated sum or partial sum:

$$H_m(z^{-1}) = \sum_{k=0}^{m-1} h_k z^{-k}.$$

Rewrite $H_m(z^{-1})$ as a rational function by summing in each of the two cases the geometric progression. Note the distribution of zeros of $H_m(z^{-1})$ for the cases $c = 0$ and $c = 1$. Comment on what happens to the loci of zeros as c varies continuously from 0 to 1 and try to be as quantitative as you can.

12. Let $\{h(k)\}$ be a discrete-time sequence whose z-transform is denoted by $H(z)$. Define a new sampled sequence $\{s(k)\}$ obtained by using one of every N samples of $\{h(k)\}$, i.e., $s(k) = h(Nk)$, where N is a positive integer. Show that the z-transform of $\{s(k)\}$ is

$$S(z) = \frac{1}{N} \sum_{k=0}^{N-1} H(z^{1/N} e^{-j2\pi k/N}).$$

13. One channel of a telephone communication system can be modeled as follows, where $p(t)$ is the signal, $n(t)$ is the noise, and $y(t)$ is the signal recovered after filtering and reconstruction.

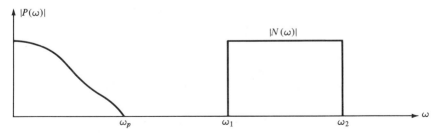

The frequency content of $p(t)$ and $n(t)$ is given as

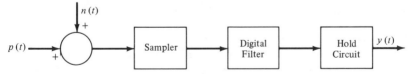

where $\omega_p = 15,000$ rad/sec, $\omega_1 = 30,000$ rad/sec, and $\omega_2 = 60,000$ rad/sec. The objective of the filter is to reduce the noise amplitude by a factor of at least 10 while not affecting the signal amplitude very much. Determine a suitable sampling frequency ω_s, a suitable cutoff frequency ω_c, and suitable higher frequency specification for the digital filter and give reasons for these selections. Give a sketch of the overall frequency response of the filter in the range $0 \le \omega \le 2\omega_s$.

14. Design a lowpass digital filter of the Butterworth type to satisfy the following specifications and give the corresponding difference equation which must be implemented:

> Cutoff frequency, $f_c = 10$ Hz.
> Filter gain at $f_2 = 25$ Hz is to be less than 0.38.
> Filter gain at 0 Hz $= 1$.
> Sampling period, $T = 0.01$ sec.

15. Let $H(z^{-1})$ be the transfer function of a first-order lowpass filter, which is realized with one delay element. Replace the delay element by the following function in the transformation:

$$z^{-1} \to \pm \frac{z^{-2} + b(1-a)z^{-1} + a}{az^{-2} + b(1-a)z^{-1} + 1},$$

where $|a| < 1$ and $|b| < 1$. State when the transformation is a lowpass–bandpass and when it is lowpass–bandstop. Investigate the claim that the midband frequency of the bandpass filter is controlled by b only whereas the parameter a affects the bandwidth of the bandpass filter.

16. Using the equations describing the constrained 2-pair in Figure 3.19,

$$X_1(z) = A(z)Y_2(z) + B(z)X_2(z),$$
$$Y_1(z) = C(z)Y_2(z) + D(z)X_2(z),$$
$$X_2(z) = H_1(z)Y_2(z),$$

prove equation (3.89).

17. Take any fourth-order lossless bounded real rational transfer function.
a. Realize the transfer function completely in cascaded form using Theorem 3.1.
b. Are you able to infer the stability property of your filter from the multiplier values in your realization?

18. Let

$$Z(s) = \frac{s(s^2 + 2)}{(s^2 + 1)}$$

be a specified reactance function, i.e., a function which is realizable as the driving-point function of a network containing passive inductors and capacitors.
a. Form

$$\rho(s) = \frac{Z(s) - 1}{Z(s) + 1}.$$

b. Form $\hat{\rho}(z)$ by replacing s by $(1 - z^{-1})/(1 + z^{-1})$ in $\rho(s)$.
c. If $\hat{\rho}(z)$ is a lossless bounded real function (LBRF), realize it as a cascade of lattices.

19. If $G(z)$ is an LBRF, what condition must be satisfied by α and β, so that

$$G_a(z) = \frac{\alpha G(z) + \beta}{\beta G(z) + \alpha}, \quad \alpha, \beta \text{ real,}$$

is also lossless bounded real.

20. Verify whether or not the following 1-multiplier lattice is equivalent as a digital 2-pair to the lattice used in each extraction step of cascade synthesis of an LBRF in Section 3.7. To do this, relate $X_1(z), Y_1(z)$ to $Y_2(z), X_2(z)$ via the chain matrix:

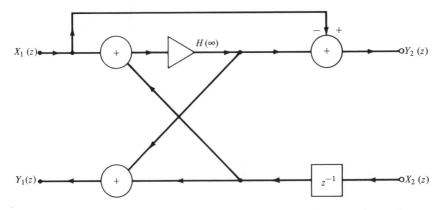

21. **a.** Show that the following structure realizes a second order allpass digital filter transfer function. Calculate the transfer function.

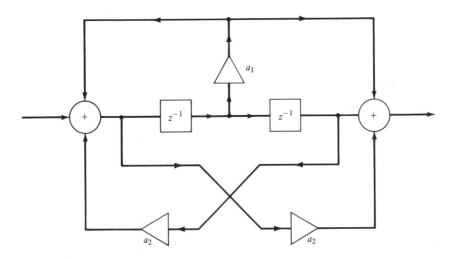

b. Can the number of multipliers be reduced from three to two, with possible increase in the number of delays, so that the transfer function in a is again realized, but by a different structure? Justify your answer.

Chapter 3. Infinite Impulse Response Digital Filter Design

22. A lowpass digital IIR filter is required to satisfy the following specifications, where α db denotes the attenuation of the magnitude response in decibels. The sketch below is a typical "brick-wall" type of filter characteristics specifications.

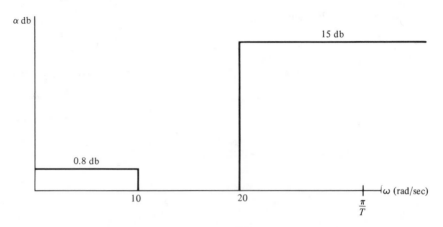

The maximum attenuation in the passband is 0.8 db and the minimum attenuation in the stopband is 15 db. The sampling period is 0.01 sec. Calculate the minimum order of the Chebyshev filter.

23. In the following figures are shown three pole-zero patterns and three possible magnitude of frequency responses of LSI systems. Match the pole-zero patterns to the frequency responses by writing the same letter for corresponding pairs. Note that $z = e^{j\omega T}$ on $|z| = 1$.

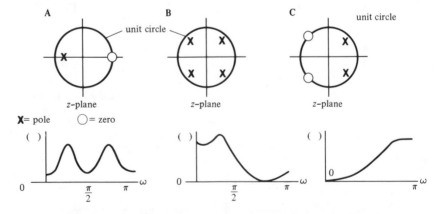

24. Consider the frequency transformations in Table 3.4 for the bandpass and bandstop types of digital filters. Prove that the magnitude characteristics of the bandpass and bandstop filters are symmetric about the center frequencies in the passband and stopband, respectively, when $\alpha_1 = 0$ and $k = 1$. If the cutoff frequency of the prototype lowpass filter is ω_c, calculate the center frequency, and the lower and upper cutoff frequencies for the bandpass filter obtained using the transformation in Table 3.4 with $\alpha_1 = 0$ and $k = 1$.

25. The magnitude-squared frequency response function of a digital Chebyshev lowpass filter having a monotonic characteristic in the passband and an equiripple behavior in the stopband is given by

$$\left| H(e^{j\omega T}) \right|^2 = \cfrac{1}{1 + \varepsilon^2 \cfrac{T_n^2\left[\tan(\omega_s T/2)\big/\tan(\omega_p T/2)\right]}{T_n^2\left[\tan(\omega_s T/2)\big/\tan(\omega T/2)\right]}} ,$$

where T is the sampling period and ω_p and ω_s are the frequencies in rad/sec shown in the sketch below. Filters having such frequency response functions are called inverse Chebyshev filters.

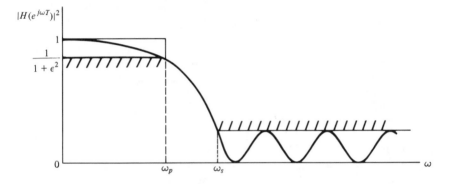

If the minimum attenuation desired in the stopband is 40 db, $\omega_p = 1000$ rad/sec, $\omega_s = 1100$ rad/sec, $T = 0.0001$ sec, and $\varepsilon = 0.25$, calculate the order of the filter required to satisfy these specifications.

References

1. Rader, C. M., and Gold, B. 1967. Digital filter design techniques in the frequency domain. Proc. IEEE, 55:149–171.
2. Rabiner, L. R., and Gold, B. 1975. *Theory and Application of Digital Signal Processing.* Prentice-Hall, Englewood Cliffs, NJ.
3. Constantinides, A. G. 1970. Spectral transformations for digital filters. Proc. IEEE, 117:110–115.
4. Chen, C. T. 1979. *One-Dimensional Digital Signal Processing.* Marcel Dekker, New York.
5. Brophy, F., and Salazar, A. C. 1973. Considerations of the Padé approximant technique on the synthesis of recursive digital filters. IEEE Trans. Acoust. Speech Signal Proc., 21:500–505.
6. James, R. Hastings, and Mehra, S. K. 1977. Extensions of the Padé approximation technique for the design of recursive digital filters. IEEE Trans. Acoust. Speech Signal Proc., 25:501–509.

7. Stieglitz, K. 1970. Computer-aided design of recursive digital filters. IEEE Trans. Audio Electroacoust., 18:123–129.

8. Deczky, A. G. 1972. Synthesis of recursive digital filters using the minimum p-error criterion. IEEE Trans. Audio Electroacoust., 20:257–263.

9. Jackson, L. B., Kaiser, J. F., and McDonald, H. S. 1968. An approach to the implementation of digital filters. IEEE Trans. Audio Electroacoust., 16:413–421.

10. Vaidyanathan, P. P., and Mitra, S. K. 1984. A general theory and synthesis procedure for low sensitivity digital filter structures, ECE Report No. 82-09, Dept. of Electrical and Computer Eng., University of California, Santa Barbara, Sep. 1982. Some of the results in this report are now available in Proc. IEEE, 72:404–423.

11. Gray, A. H., and Markel, J. D., 1973. Digital lattice and ladder filter synthesis. IEEE Trans. Audio Electroacoust., 21:491–500.

12. Mitra, S. K., and Sherwood, R. J. 1973. Digital ladder networks. IEEE Trans. Audio Electroacoust., 21:30–36.

13. Mitra, S. K., and Sherwood, R. J. 1972. Canonic realizations of digital filters using the continued fraction expansion. IEEE Trans. Audio Electroacoust., 20:185–194.

14. Gray, A. H. 1980. Passive cascaded lattice digital filters. IEEE Trans. Circuits Systems, 27:337–344.

15. Hazony, D. 1963. *Elements of Network Synthesis*. Reinhold, Chapman and Hall, London.

16. Temes, G. C., and Lapatra, J. W. 1977. *Introduction to Circuit Synthesis and Design*, McGraw Hill Book Co., New York, 1977.

17. Mitra, S. K., Kamat, P. S., and Huey, D. C. 1977. Cascaded lattice realization of digital filters. Int. J. Circuit Theory Appl., 5:3–11.

18. Rao, S. K., and Kailath, T. 1984. Orthogonal digital filters for VLSI implementation. IEEE Trans. Circuits Systems, 31:933–945.

19. Fettweis, A. 1972. A simple design of maximally flat delay digital filters. IEEE Trans. Audio Electroacoust., 20:112–114.

20. Crystal, T. H., and Ehrman, L. 1968. The design and application of digital filters with complex coefficients. IEEE Trans. Audio Electroacoust., 16:315–320.

21. Nguyen, T. V. 1984. A triangle of coefficients for Chebyshev polynomials. Proc. IEEE, 72:982–983.

22. Antoniou, A. 1979. *Digital Filters: Analysis and Design*. McGraw-Hill, New York.

23. Van Valkenburg, M. E. 1982. *Analog Filter Design*. Holt, Rinehart and Winston, New York.

24. Gray, A. H., and Markel, J. D. 1976. A computer program for designing digital elliptic filters. IEEE Trans. Acoust. Speech Signal Proc., 24:529–538.

25. Szczupak, J., and Mitra, S. K. 1975. Detection, location and removal of delay-free loops in digital filter configurations. IEEE Trans. Acoust. Speech Signal Proc., 23:558–562.

Chapter 4
Design of Finite Impulse Response Filters

4.1. Introduction

Finite impulse response (FIR) filters have been referred to in the literature as moving average filters, transversal filters, and nonrecursive filters. The word "nonrecursive," has more to do with the nature of implementation, but since filters that have a finite impulse response sequence are usually implemented nonrecursively, it is probably justifiable, at least for the sake of brevity, to treat FIR as synonymous with nonrecursive. This type of filter possesses certain desirable characteristics. A nonrecursive filter is inherently stable—if not causal, it can be made so by inserting an appropriate time delay—and it can be designed so as to have an exactly linear phase characteristic in the passband. The possible disadvantages associated with a nonrecursive FIR filter are that high order is usually required to meet sharp cutoff specifications and that the choice of this order is usually determined by trial and error. The disadvantage of high order, which would increase the computational complexity in implementation, has to some extent been overcome by the development of fast algorithms.

For the convenience of the reader, in Section 4.2 precise definitions and some of the important issues will be highlighted before proceeding to the main objective of design. In Section 4.3, a natural and conceptually straightforward analytical technique for designing nonrecursive digital filters based on truncation or windowing is presented. In Section 4.4, a design technique based on filter specifications in terms of samples of the desired frequency response via use of polynomial interpolation on the unit circle is discussed. Section 4.5 is devoted to the exposition of an optimal minimax design technique based on the Remez exchange algorithm. In Section 4.6, attention is given to several structures for realizing nonrecursive filters. The pros and cons of the various design techniques are briefly summarized within Section 4.7. The reader should find the problems at the end of the chapter to be of added value.

4.2. Preliminaries

Definition 4.1. A nonrecursive filter is a discrete-time filter for which any sample $y(k)$ of the output sequence $\{y(k)\}$ is explicitly determined as a weighted sum of a finite number of past and present input samples of the input sequence $\{x(k)\}$. For linear time-invariant nonrecursive filters, the topic of this chapter, these weighting factors are constants.

Example 4.1. The filter whose input–output behavior is described by the following difference equation is a linear time-invariant nonrecursive filter.

$$y(k) = 3x(k) + 0.5x(k-1) - 2x(k-2)$$

The transfer function, $H(z)$, of this filter is obtained by taking the z-transform of the above difference equation under zero initial conditions. If $\{y(k)\} \leftrightarrow Y(z)$ and $\{x(k)\} \leftrightarrow X(z)$ are z-transform pairs, then

$$Y(z) = (3 + 0.5z^{-1} - 2z^{-2})X(z)$$

$$H(z) = \frac{Y(z)}{X(z)} = 3 + 0.5z^{-1} - 2z^{-2}$$

The impulse response sequence $\{h(k)\}$ of this nonrecursive filter is finite.

$$\{h(k)\} = (3, 0.5, -2)$$

Definition 4.2. A finite impulse response (FIR) filter is a filter whose impulse response sequence $\{h(k)\}$ has a finite support, that is, $h(k) = 0$ for $k > N_1$ and $k < N_2$ with $N_1 \geq N_2$, for some finite integers N_1, N_2. If the filter is causal, N_1 and N_2 are further restricted to be nonnegative.

FIR filters are generally realized nonrecursively, although it is also possible to realize a FIR filter recursively, that is, the output samples may be computable as a weighted sum of not only the past and present input samples but also the past output samples. The reader has already encountered such a possibility in the beginning of Chapter 3. Since FIR filters are usually realized nonrecursively, for the sake of brevity nonrecursive filters and FIR filters will be treated as synonymous.

Fact 4.1. *The transfer function $H(z)$ of any nonrecursive filter is a polynomial in the delay variable z^{-1} and has no poles in the z-plane, except possibly at $z = 0$. Consequently, a nonrecursive filter is always bounded-input bounded-output (BIBO) stable.*

One of the main characteristics of FIR digital filters is that they can be designed to have linear phase. FIR filters, it may be inferred from Definition 4.2, are causal when $h(k) = 0$, $k < 0$; otherwise, they will be noncausal.

This is consistent with the meaning of causality described earlier. For a real finite impulse response sequence $\{h(k)\}$ to produce linear phase filters, certain symmetry conditions have to be imposed on $\{h(k)\}$. Consider a causal sequence of length N, whose z-transform is

$$H(z) = \sum_{k=0}^{N-1} h(k) z^{-k}. \tag{4.1}$$

Then the frequency response (sampling period is T) is

$$H(e^{j\omega T}) = \sum_{k=0}^{N-1} h(k) e^{-jk\omega T}. \tag{4.2}$$

Consider the case when N is an odd integer and suppose that $\{h(k)\}$ has even symmetry about $k = (N-1)/2$, that is,

$$h(k) = h(N-1-k), \quad k = 0, 1, \ldots, (N-3)/2. \tag{4.3}$$

Then, substituting (4.3) in (4.2) and simplifying,

$$H(e^{j\omega T}) = e^{-j[(N-1)/2]\omega T}\left[h\left(\frac{N-1}{2}\right) + 2\sum_{k=1}^{(N-1)/2} h\left(\frac{N-1}{2} - k\right)\cos k\omega T\right]. \tag{4.4}$$

On the other hand, if $\{h(k)\}$ is a noncausal sequence of odd length N having even symmetry about $k = 0$, that is,

$$h(-k) = h(k), \quad k = 1, \ldots, \frac{N-1}{2},$$

then

$$H(e^{j\omega T}) = h(0) + \sum_{k=1}^{(N-1)/2} 2h(k)\cos k\omega T. \tag{4.5}$$

Thus, a zero-phase filter frequency response is produced. Various other symmetry conditions on $\{h(k)\}$ yielding linear phase frequency responses are summarized in Table 4.1. (Note from Problem 18 that linear phase frequency response means that $H(e^{j\omega T})$ is the product of a pure delay term and another term that is either pure real or pure imaginary.) From (4.4) and (4.5) it becomes clear that when N is odd, then, after the design of a noncausal FIR filter is completed, a causal filter with linear phase or constant group delay can be obtained by delaying the finite sequence associated with the noncausal filter by $(N-1)/2$ units; in other words, by multiplying the transfer function

$$H(z) = h(0) + \sum_{k=1}^{(N-1)/2} h(k)[z^k + z^{-k}] \tag{4.6}$$

of the noncausal filter by $z^{-(N-1)/2}$, a causal filter transfer function is obtained.

Table 4.1. Various Symmetry Conditions on Elements of a Sequence $\{h(k)\}$ of Finite Length N Required to Attain Linear Phase Frequency Responses

Odd N Causal Even Symmetry

$$H(e^{j\omega T}) = e^{-j[(N-1)/2]\omega T}\left[h\left(\frac{N-1}{2}\right) + 2 \sum_{k=1}^{(N-1)/2} h\left(\frac{N-1}{2} - k\right)\cos k\omega T \right]$$

$h(k) = h(N-1-k)$

$k = 0,1,\ldots,\dfrac{N-3}{2}$

$h\left(\dfrac{N-1}{2}\right)$ is arbitrary

Odd N Noncausal Even Symmetry

$$H(e^{j\omega T}) = h(0) + 2\sum_{k=1}^{(N-1)/2} h(k)\cos k\omega T$$

$h(-k) = h(k)$

$k = 1,2,\ldots,\dfrac{N-1}{2}$

$h(0)$ is arbitrary

Odd N Causal Odd Symmetry

$h(k) = -h(N-1-k)$

$$H(e^{j\omega T}) = -je^{-j[(N-1)/2]\omega T}\sum_{k=1}^{(N-1)/2} 2h\left(\frac{N-1}{2} + k\right)\sin k\omega T$$

$k = 0,1,\ldots,\dfrac{N-3}{2}$

$h\left(\dfrac{N-1}{2}\right) = 0$

Odd N Noncausal Odd Symmetry

$h(-k) = -h(k)$

$$H(e^{j\omega T}) = -j2\sum_{k=1}^{(N-1)/2} h(k)\sin k\omega T$$

$k = 1,2,\ldots,\dfrac{N-1}{2}$

$h(0) = 0$

Table 4.1. (*Continued*)

Even N $h(k) = h(N-1-k)$ $k = 0,1,\ldots,\dfrac{N}{2} - 1$	Causal Even Symmetry	$H(e^{j\omega T}) = e^{-j[(N-1)/2]\omega T} \displaystyle\sum_{k=0}^{(N/2)-1} 2h(k)\cos\left[\frac{N-1}{2} - k\right]\omega T$	
Even N $h(k-\tfrac{1}{2})$ $= h(-k+\tfrac{1}{2})$ $k = 1,2,\ldots,\dfrac{N}{2}$	Noncausal Even Symmetry	$H(e^{j\omega T}) = 2 \displaystyle\sum_{k=1}^{N/2} h(k-\tfrac{1}{2})\cos[k-\tfrac{1}{2}]\omega T$	
Even N $h(k) = -h(N-1-k)$ $k = 0,1,\ldots,\dfrac{N}{2} - 1$	Causal Odd Symmetry	$H(e^{j\omega T}) = je^{-j[(N-1)/2]\omega T}\left[\displaystyle\sum_{k=0}^{(N/2)-1} 2h(k)\sin\left\{\frac{N-1}{2} - k\right\}\omega T\right]$	
Even N $h(k-\tfrac{1}{2})$ $= -h(-k+\tfrac{1}{2})$ $k = 1,2,\ldots,\dfrac{N}{2}$	Noncausal Odd Symmetry	$H(e^{j\omega T}) = -j2 \displaystyle\sum_{k=1}^{N/2} h(k-\tfrac{1}{2})\sin[k-\tfrac{1}{2}]\omega T$	

4.3. Finite Impulse Response Filter Design by Windowing and Truncation

The windowing and truncation method of FIR filter design exploits the fact that the frequency response function $H(e^{j\omega T})$ satisfies the Dirichlet conditions, so that the Fourier series expansion may be obtained. The coefficients of this Fourier series expansion are taken to form the impulse response sequence of a digital filter. Since this sequence is, in general, infinite, it is necessary to truncate the sequence so that the designed filter has a finite impulse response and is implementable nonrecursively. Truncation is achieved by a technique known as windowing, and the design of windows is critical for satisfactorily approximating the desired filter characteristics. To wit, consider the frequency response function $H(e^{j\omega T})$ for an ideal lowpass filter having a cutoff frequency ω_c rad/sec and a linear phase characteristic as shown in Figure 4.1A, B.

Let the Fourier series expansion for $H(e^{j\omega T}) = |H(e^{j\omega T})|\underline{/H(e^{j\omega T})} = e^{-jk_0\omega T}$ be (note that k_0 is a real constant),

$$e^{-jk_0\omega T} = \sum_{k=-\infty}^{\infty} h(k)e^{-jk\omega T}.$$

Therefore,

$$h(k) = \frac{T}{2\pi}\int_{-\omega_c}^{\omega_c} e^{-jk_0\omega T}e^{jk\omega T}\,d\omega$$

$$= \begin{cases} \dfrac{\sin[(k-k_0)\omega_c T]}{(k-k_0)\pi}, & k \neq k_0 \\[3mm] \dfrac{\omega_c T}{\pi}, & k = k_0 \end{cases} \qquad (4.7)$$

For the sake of brevity, take $k_0 = 0$, normalize $\omega_c T$ to 1 rad, and then consider the task of truncating

$$H(z) = \sum_{k=-\infty}^{\infty} h(k)z^{-k} \qquad (4.8a)$$

Figure 4.1A and B.

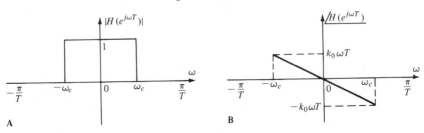

A

B

or, in this case,

$$H(z) = \frac{1}{\pi} + \sum_{k=1}^{\infty} \frac{\sin k}{k\pi}(z^k + z^{-k}), \qquad (4.8b)$$

where $\{h(k)\} \leftrightarrow H(z)$ is a z-transform pair. Equation (4.8b) is obtained upon the substitution of (4.7) with $k_0 = 0$, $\omega_c T = 1$ in (4.8a), noting that $h(k) = h(-k)$, $k = 1,2,\ldots,\infty$. The most obvious window function is the rectangular window function,

$$w(k) = \begin{cases} 1, & |k| \le \dfrac{N-1}{2}, \quad N \text{ odd} \\ 0, & |k| > \dfrac{N-1}{2} \end{cases}, \qquad (4.9)$$

so that the windowed impulse response and the transfer function after windowing are, respectively,

$$\hat{h}(k) = h(k)w(k), \qquad (4.10)$$

$$\hat{H}(z) = \sum_{k=-\infty}^{\infty} h(k)w(k)z^{-k}$$

$$= \frac{1}{\pi} + \sum_{k=1}^{(N-1)/2} \frac{\sin k}{k\pi}(z^k + z^{-k}). \qquad (4.11)$$

Figure 4.2 (A to E) shows the shape of plots of $|\hat{H}(e^{j\omega T})|$ vs ωT for $N = 25$, 51, 101, 151, and 201, respectively. Figure 4.2 (F to J) shows the shape of plot of $20\log_{10}|\hat{H}(e^{j\omega T})|$ vs ωT for $N = 25$, 51, 101, 151, and 201, respectively. (Note that $\hat{H}(e^{j\omega T})$ is a real valued function.) It is clear that truncation introduces ripples in the frequency response and although the ripple frequency increases with N, the maximum height of the ripple is almost constant. This ripple is referred to as Gibbs' phenomenon. In general, a truncated Fourier series, $\hat{H}(e^{j\omega T})$ [see (4.11)], tends to have excessive ripples near the points where the function $H(e^{j\omega T})$ [see (4.8)], being approximated, has sharp discontinuities. The approximant, however, is best in the integral least-square error sense. From the orthogonality property of the sequence of basis functions in a Fourier series expansion, it is simple to show that (remember that $\omega_c T = 1$)

$$\frac{T}{2\pi} \int_{-\omega_c}^{\omega_c} |H(e^{j\omega T}) - \hat{H}(e^{j\omega T})|^2 \, d\omega$$

$$= \left[\sum_{k=-\infty}^{\infty} |h(k)|^2 - \sum_{k=[-(N-1)]/2}^{(N-1)/2} |h(k)|^2 \cdot \frac{1}{\pi} \right]$$

The ripples in $|\hat{H}(e^{j\omega T})|$ are essentially caused by the ripples in the Fourier transform $W(e^{j\omega T})$ (i.e., the z-transform $W(z)$ evaluated at $z = e^{j\omega T}$) of the

252

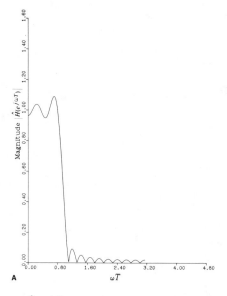

A

Figure 4.2A. Plot of $|\hat{H}(e^{j\omega T})|$ versus ωT with rectangular window and $N = 25$.

Figure 4.2B. Plot of $|\hat{H}(e^{j\omega T})|$ versus ωT with rectangular window and $N = 51$.

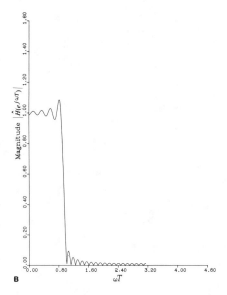

B

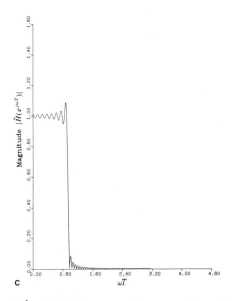

Figure 4.2C. Plot of $|\hat{H}(e^{j\omega T})|$ versus ωT with rectangular window and $N = 101$.

Figure 4.2D. Plot of $|\hat{H}(e^{j\omega T})|$ versus ωT with rectangular window and $N = 151$.

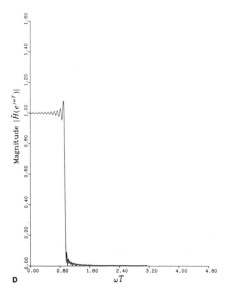

254

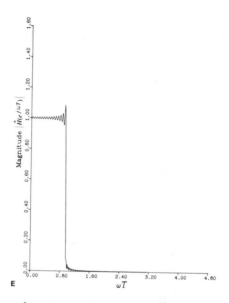

Figure 4.2E. Plot of $|\hat{H}(e^{j\omega T})|$ versus ωT with rectangular window and $N = 201$.

Figure 4.2F. Plot of $20\log_{10}|\hat{H}(e^{j\omega T})|$ versus ωT with rectangular window and $N = 25$.

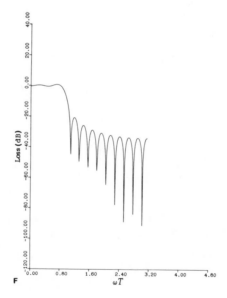

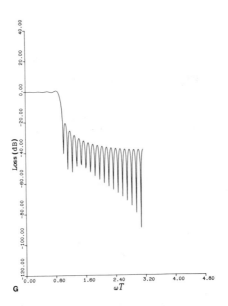

G

Figure 4.2G. Plot of $20\log_{10}|\hat{H}(e^{j\omega T})|$ versus ωT with rectangular window and $N = 51$.

Figure 4.2H. Plot of $20\log_{10}|\hat{H}(e^{j\omega T})|$ versus ωT with rectangular window and $N = 151$.

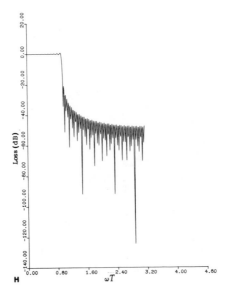

H

256

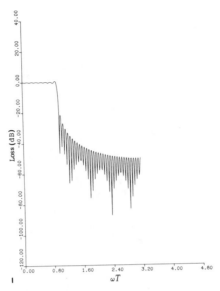

Figure 4.2I. Plot of $20\log_{10}|\hat{H}(e^{j\omega T})|$ versus ωT with rectangular window and $N = 101$.

Figure 4.2J. Plot of $20\log_{10}|\hat{H}(e^{j\omega T})|$ versus ωT with rectangular window and $N = 201$.

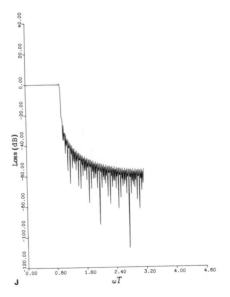

window function $w(k)$ in (4.9; see Problem 3):

$$W(e^{j\omega T}) = \sum_{k=[-(N-1)]/2}^{(N-1)/2} e^{-jk\omega T}$$

$$= \frac{\sin(\omega NT/2)}{\sin(\omega T/2)}.$$

Note that, since (4.10) holds, $\hat{H}(e^{j\omega T})$ can be calculated by the frequency domain circular convolution of $W(e^{j\omega T})$ and $H(e^{j\omega T})$:

$$\hat{H}(e^{j\omega T}) = \frac{1}{2\pi} \int_{-\pi}^{\pi} H(e^{j\theta}) W(e^{j(\omega T-\theta)}) \, d\theta.$$

The width of the main lobe, centered around $\omega = 0$, of $W(e^{j\omega T})$ when plotted as a function of ω, is equal to $4\pi/TN$. It may be verified that the distance between the positive and negative peaks, in the plot of $\hat{H}(e^{j\omega T})$ vs ω, adjacent to a discontinuity of $H(e^{j\omega T})$ is the width $4\pi/TN$ of the main lobe of $W(e^{j\omega T})$. The function $W(e^{j\omega T})$ also has side lobes. The widths of these lobes decrease with increase of N, though the heights increase in a manner so as to keep the area under each lobe practically invariant. From the last equation, it is then verifiable that, in the vicinity of a discontinuity, the ripple oscillations occur more rapidly as N increases, but the ripple overshoot remains almost fixed (at about 8.95%). This large overshoot causes problems even when sharp cutoffs (low transition bandwidths) are attained. The transition bandwidth is approximately equal to the width, $4\pi/TN$, of the main lobe, and in order to attain sharp cutoffs N is often required to be very high. The height of the first side lobe on each side of the main lobe is about one fifth the height of the main lobe.

In order to tackle the problem of ripples caused by the side lobes of $W(e^{j\omega T})$ for the rectangular window, various other windows were proposed. Some of these windows are listed in Table 4.2. The Fourier transform of the Hanning and Hamming windows have negligible side-lobe levels; however, the width of the central main lobe, for a fixed N and T, is greater than that for the rectangular window. The highest side lobe for the Hamming window is about one third the highest side lobe for the Hanning window. However, the heights of the side lobes for the Hanning window fall off more rapidly than do those of the Hamming window. Also, for both these windows the main lobes are four times as wide as the side lobes, excepting the split side lobes nearest the main lobes. The Blackman window also has decreased side-lobe levels; however, the width of its main lobe for chosen N and T is a factor of 3.5 greater than that for the rectangular window. The Blackman window listed in Table 4.2 is actually a special case of the window

$$w(k) = a_0 + 2\sum_{r=1}^{\infty} a_k \cos\frac{2\pi kr}{N-1}, \quad |k| \le \frac{N-1}{2}$$

$$= 0, \quad |k| > \frac{N-1}{2},$$

Table 4.2. Various Types of Window Functions

Window type	Window function, $w(k)$, $-\left(\dfrac{N-1}{2}\right) \le k \le \left(\dfrac{N-1}{2}\right)$, N odd Also $w(k) = 0$, $\|k\| > \dfrac{N-1}{2}$
Rectangular	1
Hanning	$\dfrac{1}{2}\left[1 + \cos\dfrac{2\pi k}{N-1}\right]$
Hamming	$0.54 + 0.46\cos\dfrac{2\pi k}{N-1}$
Blackman	$0.42 + 0.5\cos\dfrac{2\pi k}{N-1} + 0.08\cos\dfrac{4\pi k}{N-1}$
Fejer-Cesaro	$1 - \dfrac{2\|k\|}{N-1}$
Lanczos	$\left(\dfrac{\sin[2k\pi/(N-1)]}{2k\pi/(N-1)}\right)^{L}$, $L > 0$
Dolph-Chebyshev (frequency function)	$W(e^{j\omega T}) = \dfrac{\cos\left[(N-1)\cos^{-1}(\alpha\cos(\omega T/2))\right]}{\cosh\left[(N-1)\cosh^{-1}(\alpha)\right]}$, $\alpha > 0$ $[W(e^{j\omega T})]$ is the Fourier transform of $\{w(k)\}$ and T is the sampling period
Papoulis	$\dfrac{1}{\pi}\left\|\sin\left(\dfrac{2\pi k}{N-1}\right)\right\| + \left(\dfrac{2\|k\|}{N-1}\right)\cos\dfrac{2\pi k}{N-1}$
Kaiser	$\dfrac{I_0\left[\beta[(N-1)/2]T\sqrt{1-[2k/(N-1)]^2}\right]}{I_0[\beta[(N-1)/2]T]}$ ($\beta > 0$, I_0 is modified Bessel function of the first kind and order zero; T is the sampling period)
Tukey	1, $\|k\| \le \alpha N_1$, $0 < \alpha < 1$, $N_1 \triangleq \dfrac{N-1}{2}$ $\dfrac{1}{2}\left[1 + \cos\left\{\dfrac{(k-\alpha N_1)\pi}{(1-\alpha)N_1}\right\}\right]$, $\alpha N_1 \le \|k\| \le N_1$
Parzen	$1 - 24\left(\dfrac{k}{N-1}\right)^2 + 48\left(\left\|\dfrac{k}{N-1}\right\|\right)_3$, $\|k\| \le \dfrac{N-1}{4}$ $\dfrac{1}{2}\left(1 - \dfrac{2\|k\|}{N-1}\right)$, $\dfrac{N-1}{4} \le \|k\| \le \dfrac{N-1}{2}$

subject to the constraint

$$a_0 + 2\sum_{k=1}^{\infty} a_k = 1.$$

The properties of the windows considered so far hint at an important fact. The dual objectives of realizing minimum main lobe width (for low transition bandwidth of filter) and minimum area under side lobes (to control undesirable ripple effects) are incompatible. The Fejer window (also referred to as the Cesaro-Bartlett window) has the disadvantage that the associated rise time is very long, that is, when the window is used for smoothing, the filter passband or stopband frequency response magnitudes are approached very slowly. The Lanczos window is an improvement over the Fejer window in this respect, and it also succeeds in smoothing out the ripples caused by truncation of the Fourier series representation for a periodic function in the vicinity of the discontinuity. Figure 4.3(A to R) shows the shape of the plots of the frequency response when some of these windows are used in place of the rectangular window, as in Figures 4.1 and 4.2.

The Dolph-Chebyshev window (also listed in Table 4.2) has the property of smallest main-lobe width for a given side-lobe level in its Fourier transform. Also, the side-lobe levels are of equal height. The Fourier transform of the Dolph-Chebyshev window is also a periodic function of frequency, which is one reason why the side-lobe levels are small. Nonrecursive filter design via the use of this type of window has been treated by Helms [12].

Kaiser's window is very close to optimum. By adjusting a parameter, α, of the window, the side lobes can be diminished at the cost of increased transition bandwidth ($\alpha \triangleq \beta[(N-1)/2]T$ in Table 4.2) for a fixed value of N. With α defined in this manner, the window function for the Kaiser window is

$$w(k) = \frac{I_0\left[\alpha\sqrt{1 - [2k/(N-1)^2]}\right]}{I_0(\alpha)}, \qquad |k| \leq \frac{N-1}{2}$$

$$= 0, \qquad |k| > \frac{N-1}{2}.$$

The parameter α dictates a frequency response tradeoff between the peak height of the side lobe ripples and the width of the main lobe. Here, the window function can be quite conveniently computed using a computer program and the power series expansion of the Bessel function, $I_0(\alpha)$, of zeroth order:

$$I_0(\alpha) = 1 + \sum_{k=1}^{\infty} \left[\frac{(\alpha/2)^k}{k!}\right]^2.$$

A computer program, KAISER.FOR, is listed in Figure 4.4. This program can be used to compute $w(k)$ for a Kaiser window with specified α and N. (Note that, for convenience, the width, N, of the window in the program has been replaced in the program by $2N+1$.) In the continuous case the prolate spheroidal functions are known to approximate well the desired properties of an ideal window function, that is, in a certain sense both the window width and the width of its transform are narrow, to extent possible. Kaiser's

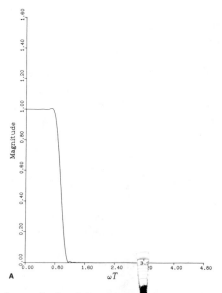

Figure 4.3A. Plot of magnitude of frequency response with Hanning window and $N = 51$.

Figure 4.3B. Plot of loss in db versus ωT with Hanning window and $N = 51$.

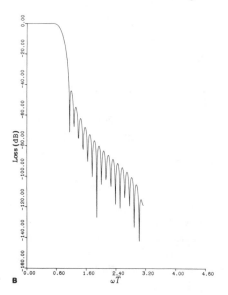

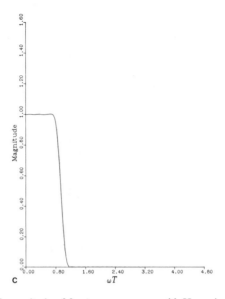

Figure 4.3C. Plot of magnitude of frequency response with Hamming window and $N = 51$.

Figure 4.3D. Plot of loss in db versus wT with Hamming window and $N = 51$.

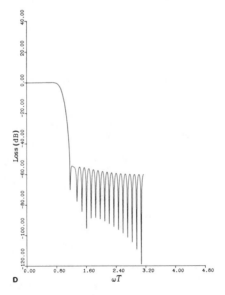

262

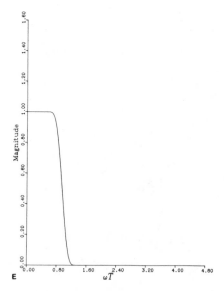

E

Figure 4.3E. Plot of magnitude of frequency response with Blackman window and $N = 51$.

Figure 4.3F. Plot of loss in db versus ωT with Blackman window and $N = 51$.

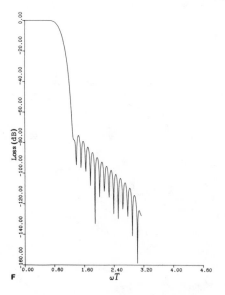

F

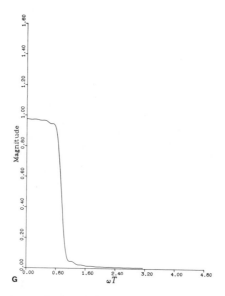

Figure 4.3G. Plot of magnitude of frequency response with Fejer window and $N = 51$.

Figure 4.3H. Plot of loss in db versus ωT with Fejer window and $N = 51$.

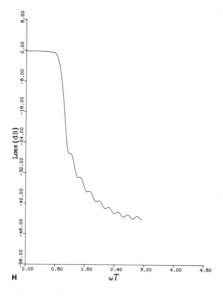

264

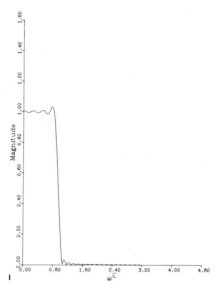

I

Figure 4.3I. Plot of magnitude of frequency response with Lanczos window and $N = 51$, $L = 0.5$ (see Table 4.2).

Figure 4.3J. Plot of loss in db versus ωT with Lanczos window and $N = 51$, $L = 0.5$.

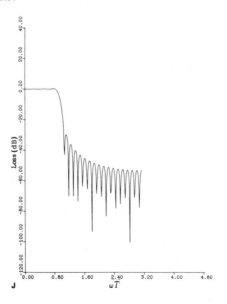

J

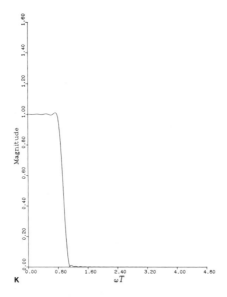

Figure 4.3K. Plot of magnitude of frequency response with Lanczos window and $N = 51$, $L = 1.0$ (see Table 4.2).

Figure 4.3L. Plot of loss in db versus ωT with Lanczos window and $N = 51$, $L = 1.0$.

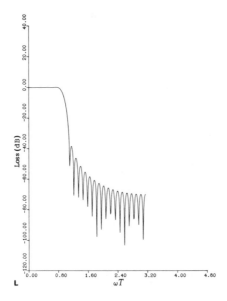

266

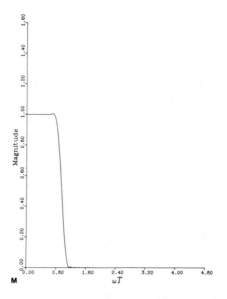

M

Figure 4.3M. Plot of magnitude of frequency response with Lanczos window and $N = 51$, $L = 1.5$ (see Table 4.2).

Figure 4.3N. Plot of loss in db versus ωT with Lanczos window and $N = 51$, $L = 1.5$.

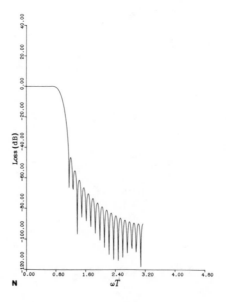

N

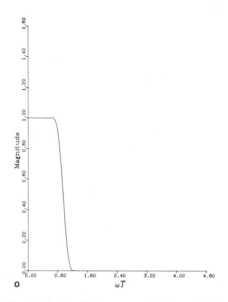

Figure 4.3O. Plot of magnitude of frequency response with Lanczos window and $N = 51$, $L = 2.0$ (see Table 4.2).

Figure 4.3P. Plot of loss in db versus ωT Lanczos window and $N = 51$, $L = 2.0$.

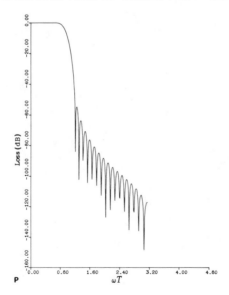

268

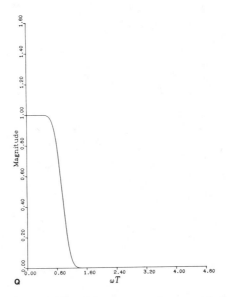

Figure 4.3Q. Plot of magnitude of frequency response with Lanczos window and $N = 51$, $L = 5.0$ (see Table 4.2).

Figure 4.3R. Plot of loss in db versus ωT with Lanczos window and $N = 51$, $L = 5.0$.

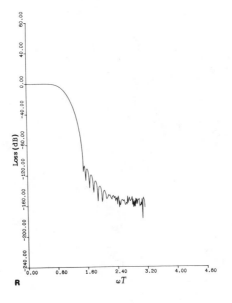

```
C
C       A FORTRAN PROGRAM (KAISER.FOR) TO COMPUTE THE COEFFICIENT,
C       W(K), K=0,1,...,N, FOR THE KAISER WINDOW OF LENGTH (2*N+1).
C       THE VALUES FOR 'N' & 'ALPHA' SHOULD BE GIVEN AND
C       W(-K)=W(K), K=0,1,...,N.
C
        DIMENSION W(0:120)
        WRITE(6,110)
110     FORMAT(2X,'ENTER VALUES FOR N AND ALPHA IN ONE LINE')
        READ(5,120) N,ALPHA
120     FORMAT(I,F)
        WRITE(6,130)
130     FORMAT(2X,'COEFFICIENTS FOR THE KAISER WINDOW ARE ;')
        CALL BESS(ALPHA,ABESS)
        DO 200 I=0,N
        B=ALPHA*SQRT(1.-(FLOAT(I)/FLOAT(N))**2)
        CALL BESS(B,BBESS)
        W(I)=BBESS/ABESS
        WRITE(6,210) I,W(I)
200     CONTINUE
210     FORMAT(4X,'W(',I3,')=',2X,E17.8)
        STOP
        END
C
C
C
        SUBROUTINE BESS(X,XBESS)
        S=1.
        K=1
        E1=10000.
        E2=E1
100     KF=IFACT(K)
        E1=E2
        E2=((X/2.)**K/KF)**2
        S=S+E2
        ERR1=E2
        ERR2=E2/E1
        IF(ERR1.LT.0.00001) GO TO 200
        IF(ERR2.LT.0.00001) GO TO 200
        IF(K.GT.100) GO TO 300
        K=K+1
        GO TO 100
200     XBESS=S
        RETURN
300     WRITE(6,400)
400     FORMAT(5X,'AFTER 100 ITERETATIONS NO CONVERGENCE REACHED',//)
        RETURN
```

Figure 4.4. (*Partial listing, continued.*)

```
        END
C
C
C
        FUNCTION IFACT(K)
        IF(K.EQ.0) GO TO 200
        KPROD=1
        DO 100 I=1,K
100     KPROD=KPROD*I
        IFACT=KPROD
        RETURN
200     IFACT=1
        RETURN
        END
```

Figure 4.4. Listing of program KAISER.FOR to compute the Kaiser window
          function.

window provides a good approximation in the discrete case to this continuous window function. Because of its near optimal property, the Kaiser
window is used to design a nonrecursive filter in the following example. See
also Problems 11 and 12 at the end of this chapter.

Example 4.2. Consider the transfer function over one period of an ideal
digital differentiator (the sampling period, T, is for convenience normalized
to 1), where $-\pi \le \omega \le \pi$

$$H(e^{j\omega}) = j\omega, \quad |\omega| < \omega_c$$

$$= 0, \quad |\omega| > \omega_c.$$

The Fourier series expansion of $H(e^{j\omega})$ is easily verified to be

$$H(e^{j\omega}) = \sum_{k=-\infty}^{\infty} h(k)e^{-jk\omega},$$

where

$$h(k) = \frac{1}{2\pi} \int_{-\pi}^{\pi} H(e^{j\omega})e^{jk\omega} \, d\omega$$

$$= -\frac{1}{\pi} \left[\frac{\sin k\omega_c}{k^2} - \frac{\omega_c \cos k\omega_c}{k} \right].$$

Take the cutoff frequency to be $\omega_c = \pi/4$ radians (remember that $T = 1$) and
suppose that the width of the Kaiser window is $N = 51$. Figure 4.5A shows
the plots of the frequency responses when $\alpha = 1$ and $\alpha = 4$ and these
approximants are superimposed on the plot for $|H(e^{j\omega})|$ specified. Figure
4.5B shows similar plots for different values of N when $\alpha = 4$.

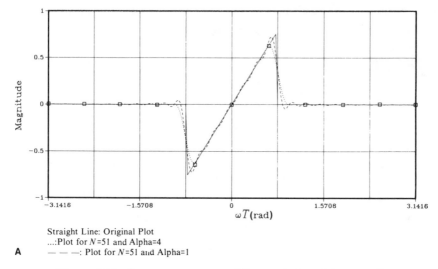

Figure 4.5A. Frequency response of Kaiser windowed differentiator.

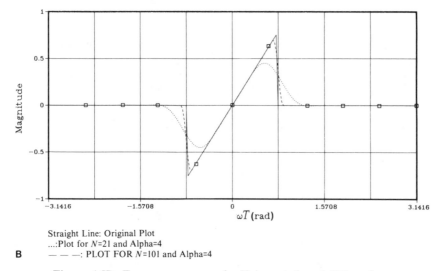

Figure 4.5B. Frequency response for Kaiser windowed differentiator.

4.4. Finite Impulse Response Filter Design by the Frequency Sampling Method

It has been noted that a causal FIR filter design can be completed by suitably delaying the finite sequence associated with a noncausal filter. For example, the transfer function in (4.6), when multiplied by $z^{-(N-1)/2}$, is transformed into a transfer function of a FIR causal filter. The starting point in the design by the frequency sampling method is the specification of N frequency samples $(\hat{H}(0), \hat{H}(1),\ldots, \hat{H}(N-1))$ which can be related to the impulse response sequence $(h(0), h(1),\ldots, h(N-1))$ via the discrete Fourier transform (DFT).

$$\hat{H}(k) = \sum_{r=0}^{N-1} h(r)e^{-j2\pi kr/N}. \tag{4.12a}$$

The inverse discrete Fourier transform (IDFT) yields

$$h(r) = \frac{1}{N}\sum_{k=0}^{N-1} \hat{H}(k)e^{j2\pi kr/N}. \tag{4.12b}$$

In order to obtain the transfer function $H(z)$ from the specified sequence $\hat{H}(k)$, carry out the following manipulations:

$$H(z) = \sum_{r=0}^{N-1} h(r)z^{-r}$$

$$= \frac{1}{N}\sum_{r=0}^{N-1}\sum_{k=0}^{N-1} \hat{H}(k)e^{j2\pi kr/N}z^{-r}.$$

After interchanging the order of summations, the preceding equation can be rewritten as

$$H(z) = \frac{1}{N}\sum_{k=0}^{N-1} \hat{H}(k)\sum_{r=0}^{N-1} [e^{j2\pi k/N}z^{-1}]^r$$

$$= \frac{1}{N}\sum_{k=0}^{N-1} \hat{H}(k)\frac{1-e^{j2\pi k}z^{-N}}{1-e^{j2\pi k/N}z^{-1}}.$$

Since $e^{j2\pi k} = 1$, the above equation simplifies to

$$H(z) = \frac{1-z^{-N}}{N}\sum_{k=0}^{N-1} \frac{\hat{H}(k)}{1-e^{j2\pi k/N}z^{-1}}. \tag{4.13}$$

Interestingly, (4.13) decomposes the transfer function $H(z)$ of a FIR filter as the product of the transfer function $(1-z^{-N})/N$ of a FIR filter and the parallel combination of N IIR filters. Each of these IIR filters is a first-order section, and the impulse response sequence associated with the kth section is easily verifiable to be $\{\hat{H}(k)e^{j2\pi kr/N}\}$, for $k = 0,1,2,3,\ldots, N-1$. This impulse response sequence is periodic with a period N. The frequency response of the filter is obtainable as

$$H(e^{j\omega T}) = \frac{1-e^{-jN\omega T}}{N}\sum_{k=0}^{N-1} \frac{\hat{H}(k)}{1-e^{j2\pi k/N}e^{-j\omega T}}.$$

Routine algebraic manipulation simplifies the preceding equation to

$$H(e^{j\omega T}) = \frac{e^{-j\omega T[(N-1)/2]}}{N} \sum_{k=0}^{N-1} \hat{H}(k) \frac{e^{-j(\pi/N)k}\sin[N\omega T/2]}{\sin[(\omega T/2)-(k\pi/N)]}.$$

$$(4.14)$$

From (4.14), it is readily checked that

$$H(e^{j(2\pi r/N)}) = \hat{H}(r), \quad r = 0,1,2,\ldots,N-1, \qquad (4.15)$$

which is consistent with the hypothesis that filter design by the frequency sampling method is based on specification of samples, finite in number, of the desired frequency response at N uniformly spaced points around the unit circle. To offer certain flexibilities in the design method, which is often handicapped by the designer's inability to meet the passband and stopband cutoff frequency requirements, the uniformly spaced samples are sometimes taken to be at

$$\omega T = \frac{2\pi(r+\frac{1}{2})}{N}, \quad r = 0,1,2,\ldots,N-1 \qquad (4.16)$$

instead of at

$$\omega T = \frac{2\pi r}{N}, \quad r = 0,1,2,\ldots,N-1,$$

as done in (4.15). Although arbitrary cutoff frequencies may be approximated by making N large enough, this approach is impractical from the computational complexity standpoint. Another disadvantage is due to the fact that although the desired response is interpolated at uniformly spaced sampling points, the error of interpolation between sampling points could be significant, especially in the design of those filters whose frequency responses change widely across bands. In such cases, a popular strategy is to allow the frequency samples occurring in transition bands to be unspecified variables whose values are determined by an optimization algorithm that minimizes a suitable error function like, for example, the maximum absolute error in the passband or stopband.

Example 4.3. 1. The samples of the frequency response are specified to be

$$\hat{H}(r) = e^{[-j(N-1)r\pi]/N}, \quad r = 0,1,\ldots,6 \quad \text{and} \quad r = 15,16,\ldots,19,20$$

$$\hat{H}(r) = 0, \quad r = 7,8,\ldots,13,14.$$

The sequence $\{h(k)\}$ is calculated by applying the IDFT:

$$h(k) = \frac{1}{N} \sum_{r=0}^{N-1} \hat{H}(r) e^{j2\pi rk/N}.$$

Here $N = 21$, $\hat{H}(0)$ is real, and since

$$\hat{H}(r) = \hat{H}^*(N - r), \quad r = 1, \ldots, \left(\frac{N-1}{2}\right),$$

the sequence $\{h(k)\}$ is a real sequence. It is readily verifiable that, for this problem,

$$h(k) = h(N - 1 - k), \quad k = 0, 1, \ldots, \left(\frac{N-3}{2}\right).$$

Define

$$\hat{H}(z) = \sum_{k=0}^{N-1} h(k) z^{-k}.$$

Then, on applying the constraint on $h(k)$ to $H(z)$, one gets

$$\hat{H}(z) = z^{[-(N-1)]/2} \left[h\left(\frac{N-1}{2}\right) + \sum_{k=0}^{(N-3)/2} h\left(\frac{N-3}{2} - k\right) \left\{ z^{k+1} + z^{-(k+1)} \right\} \right].$$

The magnitude of the frequency response is

$$|\hat{H}(e^{j\omega T})| = \left\| \left[h\left(\frac{N-1}{2}\right) + \sum_{k=0}^{(N-3)/2} 2h\left(\frac{N-3}{2} - k\right) \cos(k+1)\omega T \right] \right\|.$$

The plot of the amplitude of $|\hat{H}(e^{j\omega T})|$ vs. ωT is given in Figure 4.6A.

Figure 4.6A. Frequency response for 21 tap digital FIR filter.

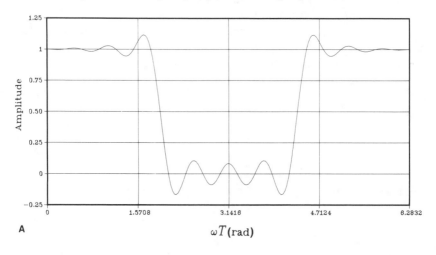

A ωT (rad)

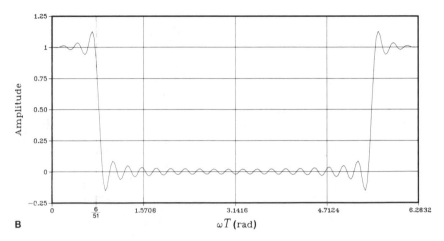

Figure 4.6B. Frequency response for 51 tap digital filter.

2. The samples of the frequency response are specified as in Example 1 above; N is changed to 51. The plot of the amplitude of $|\hat{H}(e^{j\omega T})|$ vs. ωT is given in Figure 4.6B.

4.5. Finite Impulse Response Equiripple Filters

The windowing and the frequency sampling techniques discussed so far suffer from the disadvantage that the desired cutoff frequencies in the pass- or stopbands may not be accurately realized. The method discussed here overcomes this drawback. Let the transfer function of the filter to be designed be

$$H(z) = \sum_{k=-N}^{k=N} h(k)z^{-k} \quad \text{(noncausal)} \tag{4.17a}$$

or

$$H(z) = \sum_{k=0}^{2N} h(k)z^{-k} \quad \text{(causal)}. \tag{4.17b}$$

The frequency response function in each of the above cases, when $h(k) = h(-k)$ in (4.17a) and $h(k) = h(2N - k)$ in (4.17b), for $k = 0, 1, \ldots, N$, is

$$H(e^{j\omega T}) = h(0) + \sum_{k=1}^{N} 2h(k)\cos k\omega T \quad \text{(noncausal)} \tag{4.18a}$$

or

$$H(e^{j\omega T}) = e^{-j\omega NT}\left[\sum_{k=0}^{N-1} 2h(k)\cos[(N-k)\omega T] + h(N)\right] \quad \text{(causal)},$$

(4.18b)

where T is the sampling period and $\omega = 2\pi f$, f being the frequency variable in hertz. The problem of designing a lowpass linear phase FIR filter of length $2N+1$ in an equiripple manner reduces to finding the $h(k)$'s so that, typically,

$$|H(e^{j\omega T})| = h(0) + \sum_{k=1}^{N} 2h(k)\cos k\omega T \qquad (4.19)$$

minimizes the cost value (or error value in the present context),

$$\|\phi(f_i)\| = \max_{0 \le f \le 0.5} |\phi(f)| = \max_{0 \le f \le 0.5} \left|W(f)\left[D(e^{j2\pi fT}) - |H(e^{j2\pi fT})|\right]\right|,$$

(4.20)

(Note that the frequency scale has been normalized with respect to the sampling frequency and therefore the frequency range of interest is $0 \le f \le 0.5$; also, f_i in (4.20) denotes any one of a set of frequencies where the maxima defined in (4.20) is attained.)

where $W(f)$ is a weight function and $D(e^{j2\pi fT})$ is the magnitude of the frequency response of the ideal filter over the digital frequency range, $0 \le f \le 0.5$. The theoretical basis for using (4.19) as an approximating function is the fact that the sequence of functions $\{\cos k\omega T\}_{k=0}^{N}$ or $\{\cos 2\pi fkT\}_{k=0}^{N}$ is said to satisfy the Haar conditions, that is, the basis $\{1, \cos 2\pi fT, \cos 4\pi fT, \cos 6\pi fT, \ldots, \cos 2\pi NfT\}$ forms a Haar subspace, defined next.

Definition 4.3. Let $C[a, b]$ denote the space of continuous real-valued functions $g_k(f)$ defined on the closed interval $[a, b]$ or $a \le f \le b$ with norm,

$$\|g_k\| = \max_{a \le f \le b} |g_k(f)|.$$

An $(N + 1)$-dimensional subspace of $C[a, b]$, having a basis $\{g_0, g_1, g_2, \ldots, g_N\}$, is said to be a Haar subspace on $[a, b]$ if for every $(N+1)$ distinct points $f_0, f_1, f_2, \ldots, f_N$ in $[a, b]$ the determinant of the $(N+1) \times (N+1)$ matrix,

$$\begin{bmatrix} g_0(f_0) & g_1(f_0) & \cdots & g_N(f_0) \\ g_0(f_1) & g_1(f_1) & \cdots & g_N(f_1) \\ \vdots & \vdots & & \vdots \\ g_0(f_N) & g_1(f_N) & \cdots & g_N(f_N) \end{bmatrix} \qquad (4.21)$$

is nonzero. In that event, the set of functions $\{g_0, g_1, \ldots, g_N\}$ is also said to form a Chebyshev set.

Example 4.4. 1. The set of functions $\{1, f, f^2, \ldots, f^N\}$, where f is the frequency variable, forms a Chebyshev set on any interval, since

$$
\det \begin{bmatrix} 1 & f_0 & f_0^2 & \cdots & f_0^N \\ 1 & f_1 & f_1^2 & \cdots & f_1^N \\ \vdots & \vdots & \vdots & & \\ 1 & f_N & f_N^2 & \cdots & f_N^N \end{bmatrix} = \prod_{\substack{j=0,1,\ldots,i-1 \\ i=1,2,\ldots,N}} (f_i - f_j)
$$

is always nonsingular for every set of distinct $(N+1)$-tuple, $\{f_0, f_1, \ldots, f_N\}$. The preceding determinant is also referred to as the Vandermonde determinant.

2. The set of functions $\{f, e^f\}$ does not form a Chebyshev set on $0 \le f \le 3$.

An important property for Chebyshev sets, which is essential for the development of the theory in this section, is summarized next.

Fact 4.2. *The set* $\{g_k(f)\}_{k=0}^N$ *is said to form a Chebyshev set in* $0 \le f \le 0.5$ *if*

$$
\sum_{k=0}^N a_k g_k(f) - \sum_{k=0}^N b_k g_k(f)
$$

has at most N zeros in $0 \le f \le 0.5$, where the collection of parameters $\{a_k\}_{k=0}^N$ of the approximating function, $\sum_{k=0}^N a_k g_k(f)$, is different from the collection of parameters $\{b_k\}_{k=0}^N$ of the approximating function $\sum_{k=0}^N b_k g_k(f)$. The space to which the parameters belong is the ordinary $(N+1)$-dimensional euclidean space.

When the basis functions in an approximant form a Haar subspace, then the Chebyshev approximation problem described in (4.20) has a unique solution. This unique solution is characterized by the "alternation" theorem, stated next.

Alternation Theorem 4.1. *Let S be any compact subset of $[0, \frac{1}{2}]$ or $0 \le f \le \frac{1}{2}$. In order for $|H(e^{j2\pi fT})| = h(0) + \sum_{k=1}^N 2h(k)\cos 2\pi fkT$ to be the unique best approximation on S to $D(e^{j2\pi fT})$ or $D(f)$, for brevity, it is necessary and sufficient that the cost function or error function $\phi(f)$ in (4.20) display on S at least $N+2$ alternations. Thus, if $|\phi(f_i)| \triangleq \max_{f \in S} |\phi(f)|$, then $\phi(f_{i+1}) = -\phi(f_i) = \pm \|\phi\|$ with $f_i < f_{i+1}$, $f_i \in S$, $i = 0, 1, \ldots, N$ and also $f_{N+1} \in S$. Note that $\|\phi\|$ denotes that L_∞-norm of $\phi(f)$, i.e., the minimum value of the cost in (4.20) obtained as a result to the solution of the Chebyshev approximation problem.*

Definition 4.4. In the preceding theorem, the points f_i are called extremal points and the set of points $\{f_0, f_1, \ldots, f_{N+1}\}$ is often called an alternant of $\phi(f)$.

Fact 4.3. *The alternant of the Chebyshev polynomial of the first kind, $T_n(x)$, which possesses $(n+1)$ extremal points in $-1 \le x \le 1$, consist of the points $\cos r\pi/n$, $r = 0, 1, \ldots, n$, which are the roots of the polynomial $(1 - x^2)U_{n-1}(x)$ $(n > 0)$, where*

$$U_n(x) = \frac{\sin\left[(n+1)\cos^{-1}x\right]}{\sin\left[\cos^{-1}x\right]}$$

is the Chebyshev polynomial of the second kind, degree n.

Exchange Algorithm. *The unique solution referred to above may be arrived at in an iterative manner using the Remez exchange algorithm. Consider the kth step in this algorithm, where for a given set of $(N+2)$ frequency points, $(f_0^{(k)}, f_1^{(k)}, \ldots, f_{N+1}^{(k)})$, and an error function magnitude $|\rho^{(k)}|$, say (the error function is forced to have this magnitude with alternating sign), one requires the solution of*

$$W\left(f_r^{(k)}\right)\left[h^{(k)}{}_{(0)} + \sum_{j=1}^{N} 2h^{(k)}(j)\cos 2\pi j f_r^{(k)}T - D\left(f_r^{(k)}\right)\right] = -(-1)^r \rho^{(k)}$$

$$(4.22a)$$

for $r = 0, 1, \ldots, N+1$.

After writing the set of equations in (4.22a) in matrix form,

$$\begin{bmatrix} 1 & \cos 2\pi f_0^{(k)}T & \cos 4\pi f_0^{(k)}T & \cdots & \cos 2\pi N f_0^{(k)}T & \dfrac{1}{W\left(f_0^{(k)}\right)} \\[2ex] 1 & \cos 2\pi f_1^{(k)}T & \cos 4\pi f_1^{(k)}T & \cdots & \cos 2\pi N f_1^{(k)}T & \dfrac{-1}{W\left(f_1^{(k)}\right)} \\[2ex] \vdots & \vdots & \vdots & & \vdots & \vdots \\[2ex] 1 & \cos 2\pi f_{N+1}^{(k)}T & \cos 4\pi f_{N+1}^{(k)}T & \cdots & \cos 2\pi N f_{N+1}^{(k)}T & \dfrac{(-1)^{N+1}}{W\left(f_{N+1}^{(k)}\right)} \end{bmatrix}$$

$$\times \begin{bmatrix} h^{(k)}(0) \\ 2h^{(k)}(1) \\ \vdots \\ 2h^{(k)}(N) \\ \rho^{(k)} \end{bmatrix} = \begin{bmatrix} D\left(f_0^{(k)}\right) \\ D\left(f_1^{(k)}\right) \\ \vdots \\ D\left(f_N^{(k)}\right) \\ D\left(f_{N+1}^{(k)}\right) \end{bmatrix}, \qquad (4.22b)$$

one can solve explicitly for $\rho^{(k)}$. *Note the coefficient matrix in the preceding set of linear equations is nonsingular, since* $\{\cos 2\pi r f_r^{(k)}\}_{r=0}^{N}$ *forms a Chebyshev set* (*see Problem* 16).

$$\rho^{(k)} = \frac{\displaystyle\sum_{i=0}^{N+1} (-1)^i \left[\prod_{\substack{r=0 \\ r \ne i}}^{N+1} 1/\left(\cos 2\pi f_r^{(k)}T - \cos 2\pi f_i^{(k)}T\right) \right] D\left(f_i^{(k)}\right)}{\displaystyle\sum_{i=0}^{N+1} 1/W\left(f_i^{(k)}\right)\left[\prod_{\substack{r=0 \\ r \ne i}}^{N+1} 1/\left(\cos 2\pi f_r^{(k)}T - \cos 2\pi f_i^{(k)}T\right) \right]}.$$

(4.23a)

From (4.22), *for* $r = 0, 1, \ldots, N+1$,

$$H_r^{(k)} \triangleq h^{(k)}(0) + \sum_{j=1}^{N} 2h^{(k)}(j)\cos 2\pi j f_r^{(k)}T$$

$$= D\left(f_r^{(k)}\right) - \frac{(-1)^r \rho^{(k)}}{W\left(f_r^{(k)}\right)}.$$

(4.23b)

The function $H^{(k)}(e^{j2\pi fT})$, *which interpolates to* $H_r^{(k)}$ *over* $f_r^{(k)}$, $r = 0, 1, \ldots, N$, *is* (*this function, described below in* (4.24), *also interpolates to* $H_{N+1}^{(k)}$, *since it satisfied* (4.23b) *when* $r = N+1$),

$$H^{(k)}\left(e^{j2\pi fT}\right) = \frac{\displaystyle\sum_{r=0}^{N} \left[\beta_r / \left(\cos 2\pi fT - \cos 2\pi f_r^{(k)}T\right) \right] H_r^{(k)}}{\displaystyle\sum_{r=0}^{N} \left[\beta_r / \left(\cos 2\pi fT - \cos 2\pi f_r^{(k)}T\right) \right]}, \quad (4.24)$$

where

$$\beta_r = (-1)^r \prod_{\substack{i=0 \\ i \ne r}}^{N} \frac{1}{\cos 2\pi f_i^{(k)}T - \cos 2\pi f_r^{(k)}T}.$$

Define the error function $\phi^{(k)}(f)$ *at the kth step as in* (4.25 *below*). *Evaluate* $\phi^{(k)}(f)$ *at a finite number of points, sufficiently dense, in* $0 \le f \le 0.5$. *If the error magnitude is less than or equal to* $|\rho^{(k)}|$, *then the best approximation has been found. If the error magnitude is greater than* $|\rho^{(k)}|$ *for some f in* [0, 0.5], *then a new set of* $N+2$ *frequencies,* $\{f_r^{(k+1)}\}$ *must be chosen. The choice of this new set of frequencies is based on the following consideration. The error function at the kth step is defined by,*

$$\phi^{(k)}(f) = -\left[h^{(k)}(0) + \sum_{j=1}^{N} 2h^{(k)}(j)\cos 2\pi j fT - D(f) \right] W(f).$$

(4.25)

The function $\phi^{(k)}(f)$ possesses at least N zeros in the interval, $0 \le f \le 0.5$ and has at most $(N+1)$ extremal points in $0 \le f \le 0.5$ (including the endpoints). In the case of a lowpass filter, these extremal points can be coupled with the bandedge frequencies f_p and f_s in the pass- and stopbands, respectively, (see Figure 4.7) to yield at most $N+3$ frequencies at which the error function attains its maximum or minimum. Then points $\{f_r^{(k+1)}\}$ are chosen to include the sequence of points at which $\phi^{(k)}(f)$ has alternately a maximum or minimum in $0 < f < 0.5$ together, possibly, with $f = 0$, or $f = 0.5$ plus f_p and f_s.

For the convenience of the reader, the exchange algorithm, which enables one to arrive at an optimal solution to the minimax (equiripple) approximation problem in FIR filter design, is summarized next.

Summary of Algorithm. *The various steps in the algorithm for the design of an optimal lowpass filter that approximates the desired frequency response $D(f)$ in the pass- and stopbands ($D(f) = 1$ in the passband, $D(f) = 0$ in the stopband, for an ideal lowpass filter) with a specified weighting function $W(f)$, in an equiripple manner so that the maximum error magnitude $|\rho^{(k)}|$ is minimized (note that $\phi^{(k)}(f)$ in (4.25) is a function of f, while $|\rho^{(k)}|$ in (4.22) is the maximum of the evaluated values, $|\phi^{(k)}(f_r^{(k)})|$, for any $f_r^{(k)}$ in $0 \le f_r^{(k)} \le 0.5$) are given next.*

Step 1.   Select a suitable value of N in (4.19); set $k = 0$.

Step 2.   Choose $N + 2$ distinct values of frequency, $\{f\} = (f_0^{(k)}, f_1^{(k)}, \ldots, f_{(N+1)}^{(k)})$, so that $0 \le f \le 0.5$, including 0 or 0.5 (depending upon where the error is larger) and the bandedge frequencies f_p and f_s. In the $k = 0$ case, the other values may be selected arbitrarily, while when $k > 0$ these are selected to be the extremal points of $\phi^k(f)$.

Step 3.   Calculate $\rho^{(k)}$ and $h^{(k)}(0), h^{(k)}(1), \ldots, h^{(k)}(N)$ in (4.22b). For this purpose, you might use, instead, (4.23a) and (4.24) for computational efficiency.

Step 4.   Compute $\phi^{(k)}(f)$ in (4.25) for at least $20N$ equally spaced points in $0 \le f \le 0.5$. If $|\phi^{(k)}(f)| > |\rho^{(k)}|$ for some f, go to step 5; otherwise go to step 6.

Step 5.   Set $k = k + 1$ and go to step 2.

Step 6.   Take $h^{(k)}(0), h^{(k)}(1), \ldots, h^{(k)}(N)$ to define the optimal filter after replacing $h(j)$ by $h^{(k)}(j)$, $j = 0, 1, \ldots, N$ in (4.19).

Step 7.   Stop.

The above algorithm has been implemented in a computer program, EQFIR.FOR, which is readily available in the software package compiled by the IEEE Acoustics, Speech and Signal Processing Group. The complete

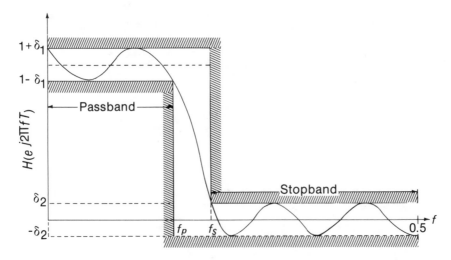

Figure 4.7. Typical frequency response of FIR equiripple lowpass filter.

package is available on magnetic tape, purchaseable at a very modest cost. Complete descriptions of all programs, including EQFIR.FOR are documented in a book that accompanies the tape. EQFIR.FOR can be used to design a wide variety of optimal FIR filters, including

Option 1: lowpass, highpass, bandpass, and bandstop filters

Option 2: differentiators

Option 3: Hilbert transformers

The input to the program is contained in four cards. The first card specifies sequentially the order of the filter, filter type option, number of pass- and stopbands, and punch option; the second card specifies the bandedges for each band, in a sequential manner; the third card specifies the desired value of the frequency response in each band; and the fourth card specifies the ripple weights[1] in each band. The examples below illustrate the convenience with which the program may be used. The runs were made on a PDP-10 general purpose computer. In the lowpass case, the amplitude response in the passband ($0 \le f \le f_p$) generally oscillates between $(1 + \delta_1)$ and $(1 - \delta_1)$ and in the stopband ($f_s \le f \le 0.5$) it oscillates between $+ \delta_2$ and $- \delta_2$. See Figure 4.7.

---

[1] The weight associated with ripple factor δ_i in band i.

Example 4.5. Design an optimal lowpass 3-point FIR filter with passband and stopband edges at $f_p = 1/2\pi$ Hz and $f_s = 1.2/2\pi$ Hz, respectively, which approximates an ideal lowpass filter characteristic. The ripple weights are equal to unity in both bands. These inputs are read as follows:

$$3 \quad 1 \quad 2 \quad 0$$

$$0. \quad .15915494 \quad .19098594 \quad .5$$

$$1 \quad 0$$

$$1. \quad 1.$$

The output data are given below (after minor changes in format for the sake of consistency).

Finite Impulse Response (FIR)

Linear Phase Digital Filter Design

Remez Exchange Algorithm

Lowpass Filter

Filter Length = 3

Impulse Response

$$h(0) = 0.32461160\text{E}+00 = h(2)$$
$$h(1) = 0.20698609\text{E}+00 = h(1)$$

	Band 1	Band 2
Lower bandedge	0.0000000	0.1909859
Upper bandedge	0.1591549	0.5000000
Desired value	1.0000000	0.0000000
Weighting	1.0000000	1.0000000
Deviation	0.4422371	0.4422371
Deviation in dB	3.1807333	− 7.0868963

Extremal Frequencies

0.1591549	0.1909859	0.5000000

The magnitude of the frequency response of the resulting filter is plotted in Figure 4.8A. This response when the length is changed to 5 is plotted in Figure 4.8B.

Example 4.6. Design an optimal bandpass 5-point FIR filter with stopband cutoff frequencies of 0.19098594 and 0.36605638 Hz and passband cutoff

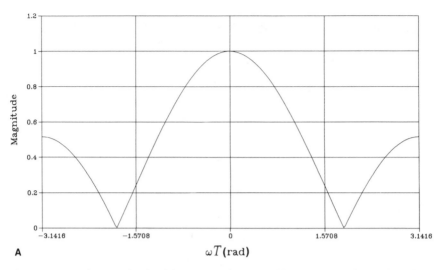

Figure 4.8A. The magnitude of frequency response of lowpass FIR filter of length 3 in example 4.5.

Figure 4.8B. The magnitude of frequency response of lowpass FIR filter of length 5 in Example 4.5.

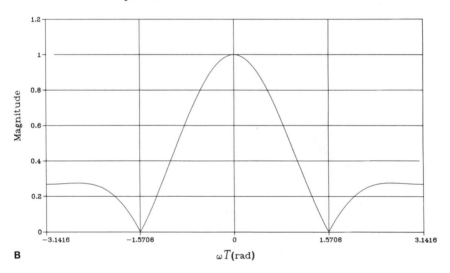

frequencies of 0.23873242 and 0.3183099 Hz, with ripple weights equal to unity in each band, which will approximate an ideal bandpass filter characteristic.

The input data are

$$5 \quad 1 \quad 3 \quad 0$$

$$0. \quad .19098594 \quad .23873242 \quad .3183099$$

$$.36605638 \quad .5$$

$$0 \quad 1 \quad 0$$

$$1. \quad 1. \quad 1.$$

The output data are

Finite Impulse Response (FIR)

Linear Phase Digital Filter Design

Remez Exchange Algorithm

Bandpass Filter

Filter Length = 5

Impulse Response

$$h(0) = -0.19885225\text{E}+00 = h(4)$$
$$h(1) = -0.12086971\text{E}+00 = h(3)$$
$$h(2) = \quad 0.21688761\text{E}+00 = h(2)$$

	Band 1	Band 2	Band 3
Lower bandedge	0.0000000	0.2387324	0.3660564
Upper bandedge	0.1909859	0.3183099	0.5000000
Desired value	0.0000000	1.0000000	0.0000000
Weighting	1.0000000	1.0000000	1.0000000
Deviation	0.4225563	0.4225563	0.4225563
Deviation in dB	-7.4823082	3.0613893	-7.4823082

Extremal Frequencies—Maxima of the Error Curve

| 0.0000000 | 0.1909859 | 0.3183099 | 0.3660564 |

The magnitude of frequency response of the resulting filter is plotted in Figure 4.9A, and the plot in decibels is given in Figure 4.9B. Similar plots when the length of the filter is increased to 19 are given in Figure 4.9C, D. Finally, Figure 4.9E, F shows the corresponding plots for the 22-point filter. The input and output data for the 22-point filter are recorded below.

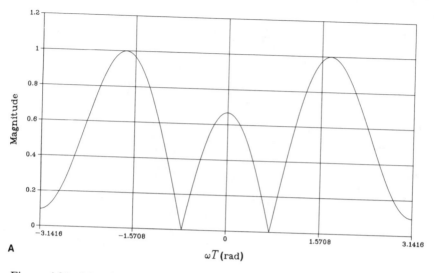

A

Figure 4.9A. Magnitude of frequency response of bandpass FIR filter of length 5 in Example 4.6.

Figure 4.9B. The loss in db for the frequency response of bandpass FIR filter of length 5 in Example 4.6.

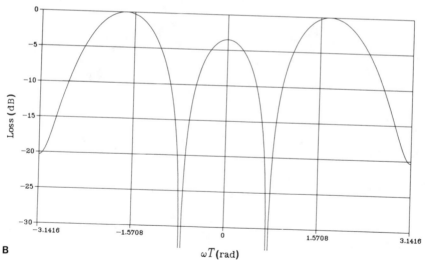

B

286

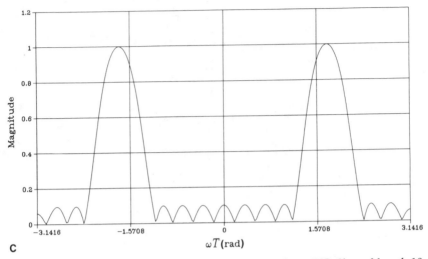

C

ωT(rad)

Figure 4.9C. Magnitude of frequency response of bandpass FIR filter of length 19 in Example 4.6.

Figure 4.9D. The loss in db for the frequency response of bandpass FIR filter of length 19 in Example 4.6.

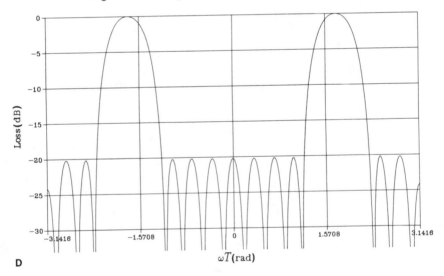

D

ωT(rad)

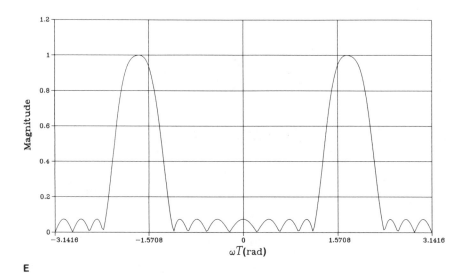

E

Figure 4.9E. Magnitude of frequency response of bandpass FIR filter of length 22 in Example 4.6.

Figure 4.9F. The loss in db for the frequency response of bandpass FIR filter of length 22 in Example 4.6.

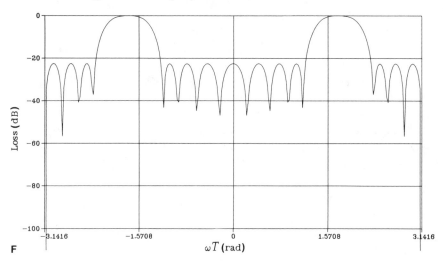

F

The input data for 22-point filter are as follows:

$$22 \quad 1 \quad 3 \quad 0$$

$$0. \quad .19098594 \quad .23873242 \quad .3183099$$

$$.36605638 \quad .5$$

$$0 \quad 1 \quad 0$$

$$1. \quad 1. \quad 1.$$

The output data for 22-point filter are as follows:

Impulse Response

$$
\begin{aligned}
h(0) &= -0.30037307\mathrm{E}-01 = h(21) \\
h(1) &= 0.31554722\mathrm{E}-01 = h(20) \\
h(2) &= 0.16492475\mathrm{E}-01 = h(19) \\
h(3) &= 0.22989931\mathrm{E}-01 = h(18) \\
h(4) &= 0.23399305\mathrm{E}-01 = h(17) \\
h(5) &= -0.83173965\mathrm{E}-01 = h(16) \\
h(6) &= -0.11649043\mathrm{E}-02 = h(15) \\
h(7) &= 0.17535555\mathrm{E}+00 = h(14) \\
h(8) &= -0.70064098\mathrm{E}-01 = h(13) \\
h(9) &= -0.21180583\mathrm{E}+00 = h(12) \\
h(10) &= 0.16676427\mathrm{E}+00 = h(11)
\end{aligned}
$$

	Band 1	Band 2	Band 3
Lower bandedge	0.0000000	0.2387324	0.3660564
Upper bandedge	0.1909859	0.3183099	0.5000000
Desired value	0.0000000	1.0000000	0.0000000
Weighting	1.0000000	1.0000000	1.0000000
Deviation	0.0806203	0.0806203	0.0806203
Deviation in dB	−21.8711110	0.6734625	−21.8711110

Extremal Frequencies

0.0000000	0.0681818	0.1221591	0.1676136	0.1909859
0.2387324	0.2756642	0.3183099	0.3660564	0.3887837
0.4313973	0.4768519			

The Remez optimization procedure discussed and also illustrated through examples above is efficient for the design of linear phase FIR digital differentiators and FIR digital Hilbert transformers. Among other results it has been concluded in references [17] and [18] that for efficient direct form realization, the values of N, which represent the number of samples in the filter unit impulse response sequence, should be, whenever possible, odd for FIR digital Hilbert transformers and even for FIR digital differentiators.

4.6. Realization of Finite Impulse Response Filters

In this section, the various possibilities for realization of a FIR filter are briefly discussed. The transfer function to be realized is of the form given in (4.1),

$$H(z) = \sum_{k=0}^{N-1} h(k)z^{-k}.$$

4.6.1. The Direct Form Realization

Let $\{y(k)\} \leftrightarrow Y(z)$ and $\{x(k)\} \leftrightarrow X(z)$ be, respectively, the output and input transform pairs. Then

$$y(k) = \sum_{r=0}^{N-1} h(r)x(k-r). \tag{4.26}$$

The direct form realization of the preceding difference equation is given in Figure 4.10A, and it is seen to contain $(N-1)$ delay elements, N multipliers, and summers. The transposed direct form structure is shown in Figure 4.10B and it is also seen to contain $(N-1)$ delay elements, N multipliers, and summers. The computations can be implemented in various other ways, each leading to a theoretically equivalent network structure. Computation of each output point via the direct form realization can be done with N multiplications and $(N-1)$ additions, so that if M output points are to be computed NM multiplications and $(N-1)M$ additions are sufficient. However, the multiplicative complexity in the computation of M output points can be considerably reduced. Using algebraic computational complexity theory it can be shown that the number of multiplications necessary and sufficient to compute

$y(0) = h(0)x(0)$

$y(1) = h(0)x(1) \; + h(1)x(0)$

$y(2) = h(0)x(2) \; + h(1)x(1) + \; h(2)x(0)$

$$\vdots \qquad \vdots \qquad \quad \vdots \qquad \qquad \vdots$$

$y(M) = h(0)x(M) + h(1)x(M-1) + \cdots + h(N-1)x(M-N+1)$

equals M (where $h(r)$ or $x(r)$ is 0 when r falls outside the region of support of sequence $\{h(r)\}$ or $\{x(r)\}$) when $\{h(k)\}$ is a sequence of length N and $\{x(k)\}$ is a sequence of length $M-N-1$; the reduction in the number of multiplications, of course, may be accompanied with an increase in the number of additions, as discussed in Chapter 2.

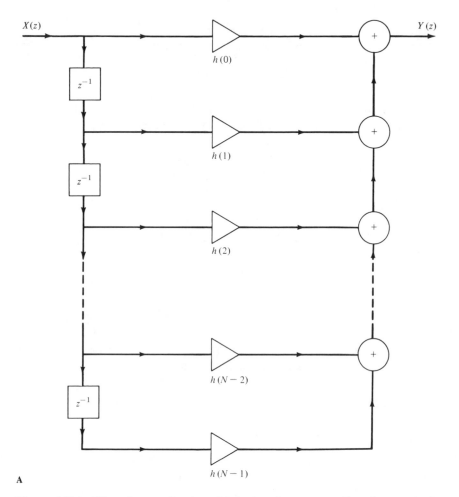

A

Figure 4.10A. Direct form realization of finite impulse response filter characterized
        by (4.26).

When the impulse response sequence is constrained in order to satisfy
certain desired filter characteristics, reduction in the complexity of the
realization may be forthcoming. For example, when the causal FIR digital
filter is a linear phase filter, it has been seen that

$$h(k) = h(N-1-k), \quad k = 0, 1, \dots, \frac{N}{2} - 1 \text{ for even } N,$$

$$k = 0, 1, \dots, \frac{N-3}{2} \text{ for odd } N,$$

(4.27)

and Figures 4.11 and 4.12, respectively, show that the number of multipliers
required is $N/2$ when N is even and $(N+1)/2$ when N is odd.

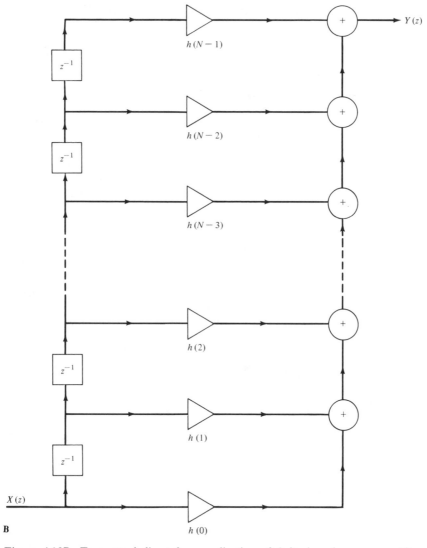

Figure 4.10B. Transposed direct form realization of finite impulse response filter characterized by (4.26).

4.6.2. Cascade Form Realization

The cascade form realization is based on the factorization of $H(z)$ in (4.1) as a product of linear and/or quadratic factors, where each factor has real coefficients:

$$H(z) = \prod_{k=1}^{\lfloor N/2 \rfloor} \left(a_{0k} + a_{1k}z^{-1} + a_{2k}z^{-2} \right), \tag{4.28}$$

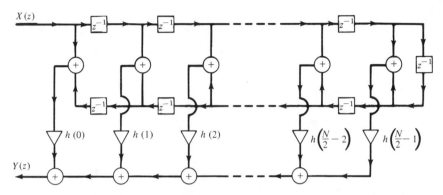

Figure 4.11. Realization of FIR filter of length N (N is even) with symmetry conditions in (4.27) imposed on the impulse response sequence.

$\lfloor x \rfloor$ denotes the largest integer less than or equal to x. When N is even, one of the coefficients in the set $(a_{21}, a_{22}, \ldots, a_{2(N/2)})$ in the preceding representation for $H(z)$ is zero, since a polynomial of odd degree $N-1$ with real coefficients must have at least one real zero. Each factor, $a_{0k} + a_{1k}z^{-1} + a_{2k}z^{-2}$ is realizable in direct form, and each of these component structures may be connected in cascade to yield a cascade form realization of $H(z)$. The order in which the components are connected in cascade might influence the filter response under finite arithmetic conditions.

4.6.3. Realization of Frequency Sampling Filters

In Section 4.4 it was seen that the transfer function of a frequency sampling filter is of the form

$$H(z) = (1 - z^{-N}) \sum_{k=0}^{N-1} \frac{\hat{H}(k)/N}{1 - z^{-1}e^{j2\pi k/N}} .$$

This can be realized as the cascade of a comb filter, having a transfer function, $1 - z^{-N}$, and a parallel combination of first-order IIR filters, each having a transfer function, relating input and output transforms $X_k(z)$ and $Y_k(z)$, respectively, by

$$H_k(z) = \frac{\hat{H}(k)/N}{1 - z^{-1}e^{j2\pi k/N}} \triangleq \frac{Y_k(z)}{X_k(z)} . \tag{4.29}$$

The poles of $H_k(z)$ occur at the zeros of the comb filter. $H_k(z)$ may be realized as in Figure 4.13A, but this realization involves complex multipliers $e^{j2\pi k/N}$ and $\hat{H}(k)/N$. In most applications, the filter impulse response

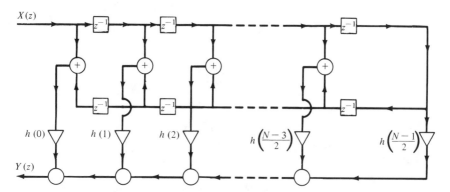

Figure 4.12. Realization of FIR filter of length N (N is odd) with symmetry conditions in (4.27) imposed on the impulse response sequence.

Figure 4.13A. Realization of the complex pole in (4.29). B. Realization of a pair of complex conjugate poles contained in (4.31).

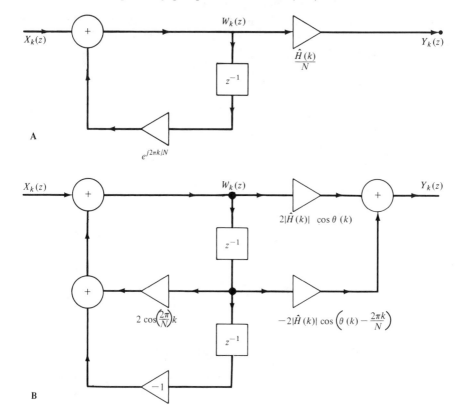

sequence is real, implying that (see also Example 4.3)

$$\hat{H}(k) = \hat{H}^*(N-k), \tag{4.30}$$

and $\hat{H}(0)$ is real. With the preceding restriction on the frequency samples, let

$$\hat{H}(k) = |\hat{H}(k)|e^{j\theta(k)}.$$

Then it is simple to verify that for filters with a real impulse response sequence and even N (when N is odd, the upper index in the summation is changed from $(N/2)-1$ to $(N-1)/2$ in (4.31)),

$$H(z) = \frac{(1-z^{-N})}{N}\left[\sum_{k=1}^{(N/2)-1}\frac{2|\hat{H}(k)|[\cos\theta(k)-z^{-1}\cos(\theta(k)-2\pi k/N)]}{1-2z^{-1}\cos[(2\pi/N)k]+z^{-2}}\right.$$

$$\left.+\frac{\hat{H}(0)}{1-z^{-1}}+\frac{\hat{H}(N/2)}{1+z^{-1}}\right] \tag{4.31}$$

($\hat{H}(0)$ is real and, from (4.30), $\hat{H}(N/2)$ is also real). The above transfer function is realizable as a cascade of a comb filter and a parallel combination of $((N/2)-1)$ second-order IIR filters and two first-order IIR filters. The kth second-order IIR filter or resonator is realizable, as in Figure 4.13B, where all the multiplier coefficients are now real. If, furthermore, the filter has a linear phase characteristic,

$$\theta(k) = -k\pi, \quad k = 0,1,\dots,\frac{N}{2}-1, \tag{4.32}$$

then it can be shown that

$$\hat{H}\left(\frac{N}{2}\right) = 0, \tag{4.33}$$

so that $H(z)$ in (4.31) simplifies to

$$H(z) = \frac{(1-z^{-N})}{N}\left[\frac{\hat{H}(0)}{1-z^{-1}}\right.$$

$$\left.+\sum_{k=1}^{(N/2)-1}\frac{(-1)^k 2|\hat{H}(k)|\{1-(\cos(2\pi k/N))z^{-1}\}}{1-2(\cos(2\pi k/N))z^{-1}+z^{-2}}\right]. \tag{4.34}$$

In case the linear phase characteristic is of the form

$$\theta(k) = -\pi k\frac{N-1}{N}, \quad k = 0,1,\dots,\frac{N}{2}-1, \tag{4.35}$$

then instead of (4.34), $H(z)$ in (4.31) simplifies to

$$H(z) = \frac{1 - z^{-N}}{N}$$

$$\times \left[\frac{H(0)}{1 - z^{-1}} + \sum_{k=1}^{(N/2)-1} \frac{(-1)^k 2|H(k)|\cos(k\pi/N)(1 - z^{-1})}{1 - 2\cos(2\pi k/N)z^{-1} + z^{-2}} \right].$$

$$(4.36)$$

The realization of $H(z)$ in (4.36), involves fewer multipliers than the realization of $H(z)$ in (4.34), because the second-order IIR filters in (4.36) have a common factor, $(1 - z^{-1})$, which can be cascaded with the parallel bank of second-order IIR filters, each having a transfer function with no finite zeros in the z^{-1} plane (i.e., the numerator of the transfer function, viewed as a rational function in z^{-1}, is a constant).

4.6.4. Hardware Realization of Finite Impulse Response Filter

So far, the discussion has centered around the various approaches toward the design of digital filters. Although software simulation of a designed digital filter is useful in the analysis of its performance, hardware implementation offers real-time capabilities that are difficult to attain in software. The need for hardware implementation of digital filters has been spurred by technological advances in the fabrication of low-cost, high-speed logic components, due to the progress achieved in very large-scale integrated (VLSI) circuit technology. Windsor and Toldalagi [14] consider a hardware implementation, described in the block diagram of Figure 4.14. An analog signal $x(t)$ is first prefiltered in order to remove high-frequency noise components that could otherwise introduce serious aliasing error. This filtered signal is, then, passed through an analog/digital (A/D) converter, whose sampling frequency is about 2 to 5 times the highest frequency component of the signal, and the discrete samples, following quantization, are stored in digital form in random access memory (RAM). Either a RAM or a programmable read only memory (PROM) stores the filter coefficients. The order and the symmetry conditions imposed on the filter impulse response determine the number of distinct filter coefficients, and therefore the number of memory locations required. A clock and counter circuit presents the contents of the RAM and the PROM to a multiplier, which itself is part of a multiplier accumulator (MAC), and the output sequence is obtained following the digital convolution of the filter coefficients and the input. As mentioned in reference [14], various practical considerations enter into the hardware design. Since the analog prefilter must provide high attenuation to the high-frequency noise components, it must have satisfactory rolloff characteristics (24 dB/octave can be achieved by a fourth-order

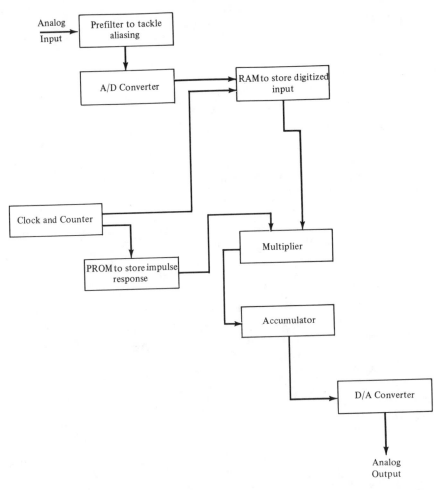

Figure 4.14. Block diagram illustrating implementation of nonrecursive digital filter
in hardware.

analog Butterworth filter or a lower order analog Chebyshev filter). For
storage of signal and filter coefficients in RAM and PROM, respectively,
usually 16 bits per word are sufficient. The MAC typically has a 16×16 bit
multiplier and a 35-bit accumulator, which provides 3 bits of extended
precision to handle overflows from the addition of multiple 32-bit products.
Sometimes an external multiplier chip may be required for faster filter
throughput, depending upon the sampling frequency and the filter order.
Other related factors that enter into the hardware implementations are the
use of data pointers for correct retrieval of words, representing the input
and filter coefficients, from memory and scaling of coefficients before

storage to tackle possible accumulator overflow problems. High-order non-recursive filters are sometimes implemented in hardware via use of the residue number system and the Chinese remainder theorem (CRT; see Chapter 2). The digitized inputs and filter coefficients may be coded as residues with respect to a suitable number of primes prior to the operations of multiplication and additions, which can subsequently be implemented in parallel in the residue number system. The residue coded outputs can then be converted to natural numbers by the CRT, whose hardware implementation is also available [14].

4.7. Summary

Given that the desired frequency characteristic is a periodic function satisfying the Dirichlet conditions, it can be expanded in a Fourier series. This Fourier series will, in general, have an infinite number of coefficients. On truncating it to a finite number of terms so that a finite impulse response filter results, one encounters the Gibbs phenomenon in the frequency domain. The truncation is implemented by means of a window. Ideally, the window as well as its Fourier transform must be narrow, so that the speed in implementation may be coupled with the fidelity of reproduction of the desired frequency response. Since this ideal is unattainable, the design philosophy for windows is based on some clever guidelines. The Fourier transform of the most natural and straightforward window—the rectangular window—has a main lobe and side lobes. The main lobe gives rise to the transition band, and the side lobes are responsible for the occurrence of Gibbs phenomenon, whenever a function has a jump or discontinuity. To smooth out the undesirable effects due to Gibbs phenomenon, various windows have been introduced, which have the effect of introducing weighted multipliers to the corresponding coefficients of the original Fourier series. A variety of windows are listed in Table 4.2 and some of their special features are discussed in the text. Some of the problems at the end of this chapter provide added information on the topic of windowing. A very clear exposition of this topic has been provided by Hamming [1]. Insights into the philosophy of design via the windowing technique are also provided in the earlier book by Blackman and Tukey [2]. See also reference [13].

The frequency sampling technique provides a realization of a FIR filter of N sample duration as a cascade of a comb filter (having a transfer function, $1 - z^{-N}$) and a parallel bank of N complex pole resonators, so that the filter output is obtainable as a weighted sum of the output of each parallel branch. The weighting factor in the kth branch is the kth specified sample of the frequency response. The frequency response samples are specified in the pass- and stopbands, while frequency response samples in the transition bands may be varied in order to improve aspects of the filter frequency response. It has been found that in the design of lowpass filters by using the frequency sampling method the order of the filter required to achieve a peak ripple of 0.01% is about 50% less than those required by the use of good windows like Kaiser's.

Since the FIR realization based on the frequency sampling method is based on the exact cancellation of a pole by a zero, which is difficult, if not impossible, to ensure under finite arithmetic conditions, it has been recommended that the poles and zeros of the recursively realized transfer functions be moved to within the unit circle, by replacing z^{-1} with rz^{-1}, where $0 < r < 1$. The chosen value of r is determined from the number of bits used

Table 4.3. Summary of Advantages and Disadvantages of Three Types of Finite Impulse Response Filter Design Methods

Type of Design	Advantages	Disadvantages
Windowing or truncation method	Very straightforward; tends to minimize the least-squares error between the desired response and the filter response	Fourier coefficients have to be evaluated Difficult to predict the type of window and the filter order required for approximating satisfactorily a prescribed frequency response Occurrence of ripples due to Gibbs phenomenon in the neighborhood of a discontinuity
Frequency sampling method	Suitable for narrowband filters Unspecified frequency samples in transition band may be used to advantage in design	Lack of flexibility in the specification of passband and stopband cutoff frequencies Inefficient when the number N_0 of specified nonzero frequency samples is large. Ideally, $N_0 < < N/2$ Possible stability problems due to IIR filters required in implementation
Equiripple method	Convenient for meeting cutoff frequency specifications Length of the filter need not be very large Iterative as opposed to analytic technique; has minimum transition width among types that meet equivalent design specifications	No analytical formula exists for determining a priori the length of the filter to meet design specifications of cutoff frequencies and ripples in pass as well as stopbands Optimization of filter response under finite wordlength coefficients becomes difficult to handle

to represent the coefficients of the filter. The frequency sampling method was originally proposed by Gold and Rader [3] and thoroughly analyzed by Rabiner and Schafer [4]. The fact that the frequency sampling technique is closely related to the windowing of the Fourier series expansion technique was pointed out in reference [5]. For extensions and applications of the frequency sampling method, see references [6] and [7].

The equiripple filter design approach for FIR filters was adopted by various authors, and Parks and McClellan [8] used the Remez exchange algorithm in order to design FIR filters via the Chebyshev approximation theory. They also developed a computer program based on the use of the exchange algorithm, exploiting the special features of the problem of lowpass filter design [9], where approximants to a function that is piecewise constant on two disjoint intervals (the passband and the stopband) are required to be constructed. As a result of this, an optimal filter of length

Table 4.4. FIR and IIR Filters: Advantages and Disadvantages

No.	Property	FIR Filters	IIR Filter
1	Stability	Inherently stable	Design requires incorporation of stability constraint
2	Analog equivalent	No meaningful analog equivalent	Linked to analog filters via the bilinear transformation
3	Phase linearity	Linear phase property useful in applications like speech processing	The phase response is non-linear
4	Sensitivity to coefficient inaccuracy	Low sensitivity permits FIR filter implementation with small word size, typically, 12 to 16 bits	Higher sensitivity requires implementation with larger word size, typically, 16 to 24 bits
5	Speed of implementation (time complexity)	Slow; usually, a large number of multiplications and additions are required	Very efficient with few coefficients and low computational complexity in implementation
6	Storage (space complexity)	Requires high amount of ROM storage, because the number of coefficients are usually large	Low storage requirements; only a small fraction of FIR spatial complexity
7	Adaptivity	Well suited for adjusting to changes in real time	Not so well suited to changing of response characteristics in response to external requirements

$2N + 1$ must have an error curve that exhibits either $N + 2$ or $N + 3$ extrema over the digital frequency interval $0 \leq f \leq 0.5$. Since the Remez exchange algorithm is iterative, the search, in this situation, for a new set of extremal points is considerably expedited. Since in addition to attaining an extremal value at each of the ripple frequencies the error curves attain an extremal value at the two edges of the transition band, the maximum number of error "maxima" obtainable is $N + 3$ instead of the minimum number, $N + 2$, required for optimality by Chebyshev theory. Therefore, such filters were called "extraripple filters" by Parks and McClellan. A unified approach to the design of linear phase FIR filters, optimal in the minimax sense, using the powerful Remez exchange algorithm was presented in reference [10]. For an useful review of the FIR digital filter design technique based on the equiripple method, see reference [11].

Table 4.3 provides a brief summary of the advantages and disadvantages of the three basic FIR filter design techniques discussed in this chapter.

Finally, Table 4.4 summarizes the advantages and disadvantages of IIR and FIR types of filters, both of which have now been covered. For a comparison between equivalent FIR and IIR digital filters based on the number of multiplications per sample required to realize these filters, see reference [16].

Problems

1. Consider the number of equi-spaced samples N to be even, and suppose that the impulse response sequence $\{h(k)\}$ of length N of a FIR filter is noncausal, having even symmetry. Note that the origin, in this case, has to be located between the $N/2$ and $(N/2)+1$th samples. Show that the frequency response

    $$H(e^{j\omega T}) = \sum_{k = -(N/2)+1}^{N/2} h\left(k - \tfrac{1}{2}\right) e^{-j\omega T(k - \tfrac{1}{2})}$$

    can be written as

    $$H(e^{j\omega T}) = 2 \sum_{k=1}^{N/2} h\left(k - \tfrac{1}{2}\right) \cos\left[k - \tfrac{1}{2}\right] \omega T.$$

    T is, as usual, the sampling period.

2. Suppose that the impulse response sequence $\{h(k)\}$ of length N (with N even) of a FIR filter is causal, having odd symmetry. Show that the frequency response

    $$H(e^{j\omega T}) = \sum_{k=0}^{N-1} h(k) e^{-j\omega Tk}$$

    can be written as

    $$H(e^{j\omega T}) = je^{-j[(N-1)/2]\omega T} \sum_{k=0}^{N/2-1} 2h(k)\sin\left(\frac{N-1}{2} - k\right)\omega T.$$

    Again, T is the sampling period.

3. Calculate the Fourier transform or spectrum $W(e^{j\omega T})$ of $\{w(k)\}$ ($T > 0$ is the sampling period),

$$W(e^{j\omega T}) = \sum_{k=-\infty}^{\infty} w(k) e^{-j\omega Tk}$$

when $\{w(k)\}$ is (see Table 4.2)

a. the rectangular window function sequence;
b. the Fejer window function sequence.

Note that the spectrum of the rectangular window function has negative lobes, while the spectrum of the Fejer window function is a nonnegative function of ω.

4. From the result of Problem 3, can you conclude that the Fejer window sequence of appropriate length is obtainable by convolving the rectangular window function of Table 4.2 with itself and then dividing the result by N? Justify your answer.

5. Do any of the other window functions of Table 4.2 besides the triangular window function have a spectrum with no negative lobes? How do the areas within the negative lobes of the Blackman window function spectrum compare with the areas within the negative lobes of the rectangular window function spectrum? Can you use your last observation to justify the presence of relatively less prominent ripples when the Blackman window function is used in FIR filter design instead of the rectangular window function?

6. Show that the width of the main lobe in the spectrum of the Hamming window function (see Table 4.2) is twice the width of the main lobe in the spectrum of the rectangular window function for any fixed N.

7. Show that the following sets of functions in the independent variable x form Chebyshev sets on $[0,1]$.
a. $\{1 \quad x^2\}$ in a linear space of dimension 2.
b. $\{1 \quad x^3\}$ in a linear space of dimension 2.
c. $\{1 \quad x \quad e^{(x^4-1)/x^2}\}$ in a linear space of dimension 3.
d. $\{1 \quad \sin x \quad \cos x \quad \sin 2x \quad \cos 2x \cdots \sin nx \quad \cos nx\}$ in a linear space of dimension $(2n+1)$.
Which of the above does not form a Chebyshev set on $[-1,+1]$? Justify your answer.

8. In the design of FIR filters by the windowing technique, study the possibility of using each of the following functions as a window function $w(k)$, $-\dfrac{(N-1)}{2} \leq k \leq \left(\dfrac{N-1}{2}\right)$, where N is an odd integer.

a. $w(k) = \left[1 - \left\{\dfrac{2k}{N-1}\right\}^2\right]^{\alpha}$, $\quad \alpha = 20$
(Landau window)

b. $w(k) = \dfrac{1}{1 + \{2\beta k/(N-1)\}^2}$, $\quad \beta = 10$

(Abel-Poisson window)

c. $w(k) = \exp\left[-\beta\left\{\dfrac{2k}{N-1}\right\}^2\right]$, $\quad \beta = 10$

(Weierstrass window)

d. $w(k) = \left[\dfrac{\sin\{\beta k\pi/[10(N-1)]\}}{\beta\sin\{k\pi/[10(N-1)]\}}\right]^4$, $\quad \beta = 20$

(Jackson window)

You might consider the special cases when $N = 51, 101$, and 201 in your study.

9. A digital filter that is a Hilbert transformer has a frequency response over one period given by

$$H(e^{j\omega T}) = \begin{cases} -j, & 0 < \omega < \dfrac{\pi}{T} \\ 0, & \omega = 0 \\ j, & -\dfrac{\pi}{T} < \omega < 0 \end{cases}$$

Let $T = 1$.

   a. Obtain the Fourier coefficients of $H(e^{j\omega T})$.

   b. Calculate the coefficients for the 21 tap (length, N, of the filter is 21) nonrecursive filter obtained using the Hamming window.

   c. Plot the magnitude of the frequency response of the filter in (b).

   d. Approximately how many taps are required to obtain the same usable bandwidth using the rectangular window?

10. Repeat Problem 9 when the digital filter is a fullband differentiator, characterized by

$$H(e^{j\omega T}) = j\omega, \quad -\dfrac{\pi}{T} \le \omega \le \dfrac{\pi}{T}.$$

11. Study the Kaiser window, defined in Table 4.2. The power series expansion of the modified Bessel function of the first kind, $I_0(\alpha)$, is

$$I_0(\alpha) = 1 + \sum_{k=1}^{\infty}\left[\dfrac{(\alpha/2)^k}{k!}\right]^2, \quad \alpha \triangleq \dfrac{\beta(N-1)T}{2}.$$

Interpret what happens in each of the following cases:

   a. $\alpha = 0$

   b. $\alpha = 5.4414$

   c. $\alpha = 8.885$

(Observe that $w(0) = 1$, $w\left(\dfrac{N-1}{2}\right) = w\left(-\left(\dfrac{N-1}{2}\right)\right) = \dfrac{1}{I_0(\alpha)}$, where $w(k)$ is the Kaiser window function.)

12. Consider a periodic function, with period L:

$$g(x) = 1, \quad 0 < x < \dfrac{L}{2}$$

$$\quad\quad = -1, \quad \dfrac{L}{2} < x < L$$

It may be verified that the Fourier series expansion for $g(x)$ is

$$g(x) = \frac{-2j}{\pi} \sum_{k=-\infty}^{\infty} b_k e^{j2\pi kx/L},$$

where

$$b_k = \frac{1}{k}, \quad k = \pm 1, \pm 3, \pm 5, \ldots,$$

$$b_k = 0, \quad k = 0, \pm 2, \pm 4, \ldots.$$

Use each of the following window functions to truncate $g(x)$ so that the b_k's are ignored for $|k| > 50$.

(i) Rectangular window
(ii) Hamming window
(iii) Kaiser window

Plot the magnitude of the truncated function in each case when $L = \frac{1}{2}$.

13. **a.** Find the impulse response sequence characterizing a filter with transfer function

$$H(z) = \frac{(1 - z^{-16})^2}{(1 + z^{-1})^2}.$$

b. Plot the magnitude of the frequency response, $|H(e^{j\omega T})|$, in the interval $-\frac{\pi}{T} \leq \omega \leq \frac{\pi}{T}$. You may take $T = 1$.

c. Obtain the pole-zero plot for the filter.

14. The impulse response sequence, $\{h(k)\}$, of a FIR filter is specified by

$$h(k) = k + 1, \quad k = 0, 1, \ldots, N;$$

$$h(2N - k) = h(k), \quad k = N + 1, \ldots, 2N.$$

a. Show that the transfer function $H(z)$ of the FIR filter may be written in the form

$$H(z) = \frac{(1 - z^{-(N+1)})^2}{(1 - z^{-1})^2}.$$

b. From $H(z)$ in (a), prove that the difference equation relating the input sequence $\{x(k)\}$ to the output sequence $\{y(k)\}$ in a recursive implementation of the FIR filter is

$$y(k) = -y(k-2) + 2y(k-1) + x(k - 2N - 2)$$
$$-2x(k - N - 1) + x(k).$$

c. Plot the magnitude of the frequency response, $|H(e^{j\omega T})|$ in the interval, $-\frac{\pi}{T} \leq \omega \leq \frac{\pi}{T}$.

15. Similar to what was illustrated in Problem 14 for the lowpass case, pole-zero cancellation in a transfer function is sometimes used to obtain recursive implementations of FIR highpass or bandpass filters.

a. Show that the transfer function

$$H_1(z) = \frac{(1 + z^{-11})^2}{(1 + z^{-1})^2}$$

characterizes a highpass filter; obtain the finite impulse response sequence associated with $H_1(z)$, plot the magnitude of its frequency response, and derive the difference equation relating the input and output sequences in a recursive implementation of the FIR filter.

b. Show that the transfer functions

$$H_2(z) = \frac{(1 - z^{-12})}{(1 + z^{-2})}$$

and

$$H_3(z) = \frac{(1 - z^{-12})(1 + z^{-1})}{(1 + z^{-3})}$$

characterize bandpass filters of the FIR type.

16. Let $\{g_0(f), g_1(f), \ldots, g_N(f)\}$ be a Chebyshev set on $0 \le f \le 0.5$. Show that the following matrix is nonsingular when $f_0 < f_1 < f_2 < \cdots < f_{N+1}$ are $N + 2$ distinct points ordered on $[0, 0.5]$.

$$\begin{bmatrix} g_0(f_0) & g_1(f_0) & \cdots & g_N(f_0) & 1 \\ g_0(f_1) & g_1(f_1) & \cdots & g_N(f_1) & -1 \\ \vdots & \vdots & & \vdots & \vdots \\ g_0(f_{N+1}) & g_1(f_{N+1}) & \cdots & g_N(f_{N+1}) & (-1)^{N+1} \end{bmatrix}$$

Hint: Use the fact that the approximating function $\sum_{k=0}^{N} a_k g_k(f)$, where the parameters $\{a_k\}$ belong to the Euclidean space, can change signs at most N times in $[0, 0.5]$.

17. In a frequency sampling filter the specified frequency samples are $H(k)$, $k = 0, 1, \ldots, N - 1$, where N is even. It is specified that $H(k) = H^*(N - k)$, $k = 1, 2, \ldots, N - 1$ and $H(0)$ is real, so that the impulse response sequence $\{h(k)\}$ is real. Furthermore, the phase $\theta(k)$ is linear, where $\theta(k)$ is the argument of $H(k)$.

$$\theta(k) = -k\pi, \quad k = 0, 1, \ldots, \frac{N}{2} - 1.$$

Show that

$$h(k) \triangleq \frac{1}{N} \sum_{r=0}^{N-1} H(r) e^{j(2\pi k r/N)},$$

simplifies to

$$h(k) = \frac{H(0)}{N} + \frac{2}{N} \sum_{r=1}^{(N/2)-1} (-1)^r |H(r)| \cos \frac{2\pi k r}{N}.$$

18. The frequency response

$$H(e^{j\omega T}) = \sum_{k=0}^{N-1} h(k) e^{-j\omega k T}$$

of a causal FIR filter of length N and sampling period T is of the linear phase type, provided $H(e^{j\omega T})$ can be expressed in the form

$$H(e^{j\omega T}) = G(e^{j\omega T})e^{j(a - b\omega T)},$$

where the function $G(e^{j\omega T})$ is real-valued and a, b are constants. The impulse response sequence $\{h(k)\}$ is, of course, real. Show that the only solutions for a and b are (N is odd):

$$a = 0 \quad \text{or} \quad \frac{\pi}{2}, \quad b = \left(\frac{N-1}{2}\right).$$

Show also that when $a = 0$, then $h(k) = h(N-1-k)$, and when $a = \pi/2$, then $h(k) = -h(N-1-k)$ for $k = 0,1,\ldots, \frac{N-1}{2}$, N odd.

19. Let the frequency response of a linear phase FIR filter of length N (N odd) and sampling period T be

$$H(e^{j\omega T}) = \sum_{k=0}^{(N-1)/2} h(k)\cos \omega kT.$$

Note that the frequency response, in this problem, is a real-valued function. Show that the number of extrema of $H(e^{j\omega T})$ cannot exceed $(N+1)/2$. What is the number of extrema in the case when N is even and

$$H(e^{j\omega T}) = \sum_{k=1}^{N/2} h(k)\sin\left[\omega T\left(k - \tfrac{1}{2}\right)\right].$$

In this problem, the sequence $\{h(k)\}$ is related, but not identically equal, to the impulse response sequence.

20. [17] Consider a real sequence $\{x(k)\}$ obtained by uniformly sampling a continuous-time signal with a sampling period of 1 sec. Define the Fourier transform of $\{x(k)\}$ to be $X(e^{j\omega})$. Construct the complex sequence $\{x_c(k)\}$, where

$$x_c(k) = x(k) + j\hat{x}(k),$$

whose Fourier transform, $X_c(e^{j\omega})$ is

$$X_c(e^{j\omega}) = \begin{cases} 2X(e^{j\omega}), & 0 < \omega < \pi \\ 0, & \pi < \omega < 2\pi \end{cases}.$$

Show that the real sequence $\{\hat{x}(k)\}$ can be obtained by linear filtering the sequence $\{x(k)\}$ with a system having a frequency response

$$H_d(e^{j\omega}) = \begin{cases} -j & 0 < \omega < \pi \\ +j & \pi < \omega < 2\pi \end{cases}.$$

As seen in Section 2.4, $H_d(e^{j\omega})$ is the spectrum of an ideal Hilbert transformer. FIR filters are suitable for approximating an ideal Hilbert transformer. Check if the unit impulse response of the system described above is

$$h_d(k) = \begin{cases} \dfrac{2}{\pi}\sin^2\left(\dfrac{k\pi}{2}\right), & k \neq 0 \\ 0, & k = 0 \end{cases}.$$

21. [18] A digital differentiator is an integral part of many practical systems. The frequency response of the ideal digital differentiator with a delay of τ samples is

$$H_d(e^{j\omega}) = \begin{cases} j\omega e^{-j\omega\tau}, & 0 \le \omega \le \pi \\ j(\omega - 2\pi)e^{-j\tau(\omega - 2\pi)}, & \pi < \omega \le 2\pi \end{cases}.$$

FIR filters are suitable for approximating the frequency response of an ideal digital differentiator.

a. Calculate the unit impulse response sequence for the ideal digital differentiator with delay τ.

b. Check if when $\tau = 0$, the unit impulse response simplifies to

$$h_d(k) = \begin{cases} 0, & k = 0 \\ \cos \pi k / k, & k \ne 0 \end{cases}.$$

22. Consider the input/output relationship of a FIR digital filter,

$$y(k) = a_0 x(k) + a_1 x(k-1),$$

where

$$a_0 = 129, \quad a_1 = -47.$$

Given,

$$x(10) = 20, \quad x(9) = 91,$$

it is required to compute $y(10)$ via the CRT using the following steps.

Step 1. Choose {19   23   29   31} to be a modulii set.

Step 2. Represent a_0, a_1, $x(10)$ and $x(9)$ in the residue number system (RNS) with respect to the modulii set.

Step 3. Compute the representation for $y(10)$ in the RNS by implementing the specified difference equation in the RNS.

Step 4. Reconstruct $y(10)$ from its representation in RNS.

Check to see that the result agrees with that attained from standard operations performed in the ring of integers.

Refer to reference [15] in this chapter for design details of FIR digital filters using the RNS and CRT.

23. (Contributed by L. F. Chaparro and M. Kanefsky) Consider a 9-point discrete-time signal, $x(k)$, $1 \le k \le 9$, whose values are the integers of your social security number.

a. Generate a signal $s(k)$, $0 \le k \le 40$ as follows. First, obtain a signal $y(k)$, $-3 \le k \le 43$ as follows:

$$y(k) = \begin{cases} x\left(\dfrac{k}{4}\right), & k = 4, 8, 12, \ldots, 36 \\ 0, & \text{otherwise} \end{cases}$$

Then generate $s(k)$ for $0 \le k \le 40$ (define $s(k)$ to be zero otherwise) as follows:

$$s(k) = \sum_{i=-3}^{3} h(i) y(k-i),$$

where

$$h(0) = 1, \quad h(1) = h(-1) = \frac{2+\sqrt{2}}{4},$$

$$h(2) = h(-2) = 1/2, \quad h(3) = h(-3) = \frac{2-\sqrt{2}}{4}.$$

Plot the signal $s(k)$ vs. k. Also take a 64-point DFT of $\{s(k)\}$ (in doing this, take $s(k) = 0$, $k > 40$ or $k < 0$) and plot the magnitude of this DFT.

b. Pass the generated signal $s(k)$ through a causal FIR filter, whose unit impulse response sequence $g(k)$, is defined by

$$g(k) = \begin{cases} \dfrac{1}{\pi} \dfrac{\sin(28-k)\pi/2}{(28-k)}, & 1 \le k \le 55 \text{ and } k \ne 28 \\ \dfrac{1}{2}, & k = 28 \end{cases}$$

and calculate the output after filtering using linear convolution.

c. Generate a signal $\hat{s}(k)$ as follows:

$$\hat{s}(k) = \begin{cases} s(k) + 5\cos\pi k \sin\dfrac{\pi}{16}k, & 0 \le k \le 48 \\ 0, & \text{otherwise} \end{cases}$$

Plot $\hat{s}(k)$ vs. k. Also take a 64-point DFT of $\{\hat{s}(k)\}$ and plot the magnitude of this DFT.

d. Pass $\hat{s}(k)$ through the 55-point FIR filter of part (b), as a reasonable first step to decipher your social security number.

e. Compute the output of the filter in part (d) via
   (1) linear convolution techniques;
   (2) use of 64-point DFT and IDFT;
   (3) use of 128-point DFT and IDFT.
Explain the difference between the outputs. Is your social security number evident? Explain.

f. Deconvolve using the output in part (e) and the impulse response in part (a). Then complete the final step which you think will lead to your social security number.

References

1. Hamming, R. W. 1977. Digital Filters. Prentice Hall, Englewood Cliffs, NJ.
2. Blackman, R. B., and Tukey, J. W. 1959. The Measurement of Power Spectra. Dover Publications, New York.
3. Gold, B., and Rader, C. M. 1969. Digital Processing of Signals. McGraw-Hill, New York, pp. 78–85.
4. Rabiner, L. R., and Schafer, R. W. 1971. Recursive and nonrecursive realizations of digital filters designed by frequency sampling techniques. IEEE Trans. Audio Electroacoustics, 19:200–207.
5. McCreary, T. J. 1972. On frequency sampling digital filters. IEEE Trans. Audio Electroacoustics, 20:222–223.

6. Echard, J. D., and Boorstyn, R. R. 1972. Digital filtering for radar signal processing applications. IEEE Trans. Audio Electroacoustics, 20:42–52.

7. Burlage, D. W., Houts, R. C., and Vaughn, G. L. 1974. Time-domain design of frequency-sampling digital filters for pulse shaping using linear programming techniques. IEEE Trans. Acoustics, Speech, Signal Processing, 22:180–185.

8. Parks, T. W., and McClellan, J. H. 1972. Chebyshev approximation for nonrecursive digital filters with linear phase. IEEE Trans. Circuit Theory, 19:189–194.

9. Parks, T. W., and McClellan, J. H. 1972. A program for the design of linear phase finite impulse response digital filters. IEEE Trans. Audio Electroacoustics, 20:195–199.

10. McClellan, J. H., and Parks, T. W. 1973. A unified approach to the design of optimum FIR linear phase digital filters. IEEE Trans. Circuit Theory, 6:697–701.

11. Rabiner, L. R., McClellan, J. H., and Parks, T. W. 1975. FIR digital filter design techniques using weighted Chebyshev approximation. Proc. IEEE, 63:595–610.

12. Helms, H. D. 1968. Nonrecursive digital filters; design methods for achieving specifications on frequency response. IEEE Trans. Audio Electroacoustics, 16:336–342.

13. Harris, F. J. 1978. On the use of windows for harmonic analysis with the discrete Fourier transform. Proc. IEEE, 66:51–83.

14. Windsor, B., and Toldalagi, P. March 3, 1983. Simplify FIR-filter design with a cookbook approach. EDN, pp. 119–128.

15. Jenkins, W. K., and Leon, B. J. 1977. The use of residue number systems in the design of finite impulse response digital filters. IEEE Trans. Circuits Systems, 24:191–201.

16. Rabiner, L. R., Kaiser, J. F., Herrmann, O., and Dolan, M. T. 1974. Some comparisons between FIR and IIR digital filters. Bell Syst. Tech. J., 53:305–330.

17. Rabiner, L. R., and Schaefer, R. W. 1974. On the behavior of minimax FIR digital Hilbert transformers. Bell Syst. Tech. J., 53:363–390.

18. Rabiner, L. R., and Schaefer, R. W. 1974. On the behavior of minimax relative error for digital differentiators. Bell Syst. Tech. J., 53:333–362.

Chapter 5
Error Analysis

5.1. Introduction

A digital filter can be simulated on a general-purpose computer, minicomputer, microprocessor, or it can be constructed with a special-purpose dedicated hardware. The representation of input data, filter coefficients, and the results of arithmetic operations required to be performed on these are constrained by finite machine word length, that is, a word is built from a specified finite number (say 16 or 8) of binary digits (0 or 1) or bits. The choice of word length is influenced by the conflicting factors of cost and accuracy desired. The filter performance is, in turn, influenced by the topology of realization, the type of arithmetic used to implement the operations, and the machine word length.

In Section 3.7 various types of structures realizing the transfer function in (3.80) were introduced. The transfer function in (3.80) characterizes an IIR filter, describable in the time-domain by the difference equation

$$y(n) = \sum_{j=0}^{m} a_j x(n-j) - \sum_{j=1}^{n} b_j y(n-j),$$

where $\{y(k)\}, \{x(k)\}$ are, respectively, the output and input sequences. In implementing the preceding difference equation to compute $\{y(n)\}$, the most common sources of error under the constraint of finite word length are attributable to the need for quantization of $\{x(k)\}$ into a set of discrete levels, the representation of the sequences of filter coefficients, $\{a_k\}$ and $\{b_k\}$, by only a finite number of bits and the occurrence of roundoff errors due to rounding or truncation after performing the arithmetic operations of addition and multiplication. The rounding and truncation (or chopping) are nonlinear operations and could lead to limit cycle oscillations. Also, limited accuracy and adder overflows could lead to overflow oscillations.

Section 5.2 describes the technical devices for representing real numbers by a finite number of bits. The two modes of representation discussed are the fixed-point arithmetic mode (also called positional number system) and the floating-point arithmetic mode. The fixed-point binary system is emphasized because it is widely used for real-time applications. However, though the fixed-point binary arithmetic offers advantages in the contexts of economy and speed, the floating-point binary arithmetic leads to increased dynamic range for a fixed word length, resulting in greater accuracy. The analysis of floating-point errors is generally more complicated than fixed-point errors and is somewhat outside the intended scope of this book.

In Section 5.3, an expression for the output noise power due to fixed-point input quantization error is derived. The derivation is based on an additive noise model, where the noise is assumed to be white. Since minimal usage of the theory of random signals is required, the inclusion of the derivation may be justifiable, especially in view of the fact that the form of the end result is important and occurs in several other contexts. At the heart of the expression for steady-state output noise power is a contour integral. The feasibility for computing this contour integral via an algebraic procedure, efficiently and accurately implementable, is discussed. It is emphasized that computation of residues following factorization of polynomials is not necessary for the purpose.

Section 5.4 gives a brief exposition of the effects of coefficient inaccuracy in the performance of designed filters. To preserve properties of interest, like stability, it is necessary to have sufficient accuracy in the representation of the filter coefficients. This accuracy is related also to the sampling rate and the dynamic order of complexity of the filter. Coefficient inaccuracy caused by quantization could also adversely affect the filter frequency response.

Section 5.5 discusses the effect of roundoff error due to multiplication. The discussion is centered around fixed-point filters. The influence of the topology of filter structures on the output noise power due to product roundoff error is brought out. The calculation of this noise power requires computation of contour integrals, similar in form to the one discussed in Section 5.3.1.

Section 5.6 contains a brief discussion of the possibility of occurrence of limit cycle oscillations, adder overflow oscillations and schemes that are commonly adopted to eliminate such problems. The concluding remarks of Section 5.7 are followed by several exercises, whose solution should reinforce comprehension of the material covered in this chapter.

5.2. Representation of Numbers

The two most common methods for representing numbers to implement a digital filter are those based on fixed-point and floating-point arithmetic.

5.2.1. Fixed-Point Representation

Any finite nonnegative real number N can be represented in the form,

$$N = \sum_{i=-\infty}^{m} a_i r^i \quad (m \text{ is an integer}) \tag{5.1}$$

where the positive integer r is the *radix* and the coefficient a_i belongs to the set $(0, 1, \ldots, r-2, r-1)$. The decimal (radix 10) representation of numbers, is, probably, most prevalent in day-to-day ordinary usage of arithmetic. When implementing arithmetic operations on the computer, it is very convenient to adopt the binary representation (radix 2). Unless specified otherwise, this binary representation of numbers will be implied. Therefore, the coefficient a_i (referred to as a *bit*) in (5.1) will be either 0 or 1. A string of 1's and 0's (corresponding to the a_i's for $i \geq 0$), then, denotes a nonnegative integer, while a second string of 1's and 0's (corresponding to the a_i's in (5.1) for $i < 0$), separated from the first by a binary point (similar to a decimal point), can represent any fractional part. An extra coefficient or bit is used to distinguish between positive and negative numbers. This sign bit is assigned the leading position; when it is 1 the sign associated with the number is negative, and when the sign bit is 0, the complementary situation is interpreted. The reader might easily check that

$$0101 = +\left(1 \times 2^2 + 0 \times 2^1 + 1 \times 2^0\right)$$

$$11011 = -\left(1 \times 2^3 + 0 \times 2^2 + 1 \times 2^1 + 1 \times 2^0\right)$$

$$111.101 = -\left(1 \times 2^1 + 1 \times 2^0 + 1 \times 2^{-1} + 0 \times 2^{-2} + 1 \times 2^{-3}\right)$$

are, the binary representations of numbers $+5$, -11, and -3.625 in the decimal representation. Since the sign bit allows the coverage of both nonnegative and negative real numbers, fixed-point numbers, just discussed, are said to have the *sign* and *magnitude* representation.

Negative numbers may be represented in two other ways besides the sign and magnitude method. The *one's* (1's) *complement* representation of a negative number is obtained after replacing 0's by 1's and 1's by 0's the bits in the sign and magnitude representation of the absolute value of the specified negative number. For example, the 1's complement representation of -10.625 is obtained, first, by writing the sign and magnitude representation of 10.625, which is 01010.101, and then by changing 1's to 0's and vice-versa. Therefore, the 1's complement representation of -10.625 is 10101.010.

Another way to represent negative numbers is by the *two's* (2's) *complement* method. The 2's complement is obtainable from the 1's complement by adding 1 to the least significant bit of the 1's complement representation. For example, from the previous paragraph, the 2's complement representation of -10.625 may be derived to be 10101.011. The number of bits (including the sign bit) required to represent a number constitutes the word length, and the digital filtering operations are implemented with a chosen word length. Multiplication of 2 integers each of b bits (excluding the sign bit) results in an integer of $2b$ bits, and if this is truncated to b bits, the resulting representation of the product becomes meaningless. To overcome this problem, in the fixed-point implementation of digital filtering operations, the numbers are always scaled to be less than unity in magnitude. The 2's complement coding of a negative fractional number y is, then, obtained as the sign and magnitude representation of the decimal number, $2 - |y|$, where $|y|$ is the absolute value of y. In this system, 1 binary bit is assigned to the left of the binary point, when $|y| < 1$. Positive fractions are represented as in the other two systems. A negative fraction is, typically, represented as a $(b+1)$-bit binary number,

$$1 \cdot a_1 a_2 \cdots a_b, \tag{5.2a}$$

which is interpreted as a positive number, $1 + \sum_{i=1}^{b} a_i 2^{-i}$, and which represents a decimal number having a value

$$-\left[2 - \left(1 + \sum_{i=1}^{b} a_i 2^{-i}\right)\right] = \sum_{i=1}^{b} a_i 2^{-i} - 1. \tag{5.2b}$$

Note that an explicit sign bit is not used in this system. Now it is clear that the dynamic range of a decimal number representable by a $(b+1)$-bit binary number in 2's complement form extends over -1 to $(1 - 2^{-b})$.

The 2's complement representation is widely used in digital signal processing. Addition and subtraction are particularly convenient to implement using this type of representation. The signs of the numbers to be added are inherently included in the process of adding (or subtracting). In case of a carryover, the carryover bit is discarded. The following examples illustrate the processes of addition and subtraction in 2's complement mode. The dot in (5.2a) will be discarded for notational convenience.

Example 5.1

Decimal	2's Complement
-0.84375	100101
$+0.12500$	000100
-0.71875	101001 $\leftrightarrow -0.71875$

                                              ↑

Indicates that sum must be negative

Example 5.2

    (a)

	Decimal	2's Complement
	+0.84375	011011
	+0.12500	000100
	+0.96875	011111 ↔ +0.96875

                              ↑
                     Indicates that sum must be positive

    (b)

	Decimal	2's Complement
	−0.84375	100101
	+0.96875	011111
	+0.12500	1000100 ↔ +0.12500

                              ↑
                     Overflow
                     (discarded)

The next example illustrates how in the 2's complement mode, overflow in the intermediate stage or stages of calculation caused due to the sum of two fractions exceeding unity does not affect the final answer when the final solution falls within the scaled range (absolute value less than unity).

Example 5.3

Decimal		2's Complement
+0.84375		011011
+0.50000		010000
+1.34375	(Overflow)	101011
−0.56250		101110
+0.78125		011001 ↔ 0.78125

The problem of overflow occurs only in addition but not in multiplication, since each number is scaled, a priori, to have a magnitude less than 1. However, a truncation or rounding error may be introduced in multiplication. (See Section 5.2.3 for information on these sources of error.) Also, a multiplication is more complicated to implement than addition even when using 2's complement arithmetic, although hardware to carry out multiplications is now well developed. The fixed-point arithmetic using the 2's complement representation is well suited for real-time applications, primarily because it leads to economical and fast hardware. The main disadvantage of the fixed-point arithmetic is the limitation on the range of numbers that can be represented. For example, overflow in addition can lead to incorrect answers. The limitation of dynamic range is overcome by the use of the floating-point arithmetic.

5.2.2. Floating-Point Representation

The binary (radix 2) floating-point representation of a decimal number N is of the form

$$N = 2^c m$$

where the mantissa m is either positive or signed (sometimes for representing negative numbers) and the integer (positive or negative) c is the characteristic or exponent. The mantissa, m, is chosen so that, $0.5 \leq |m| < 1$ and then the floating-point binary number is said to be normalized. This normalization also provides uniqueness to the representation. In floating-point arithmetic, multiplication and division are comparatively simple, but unlike in fixed-point arithmetic, addition and subtraction are somewhat more difficult to implement.

The characteristic c and mantissa m are coded in binary and the following example illustrates how the product of two decimal numbers is computable by floating-point arithmetic. Whether a number is represented in decimal or floating-point binary should be clear from context, and for the sake of brevity, no notational distinction is made between the binary point and the decimal point. Also for clarity, in the ensuing discussions the mantissa is restricted to be positive.

Example 5.4

$$N_1 = 3.00 = 2^{10.0}0.110$$

$$N_2 = 2.50 = 2^{10.0}0.101$$

$$N_1 N_2 = 2^{10.0 + 10.0}[(0.110)(0.101)]$$

$$= 2^{100.0}(0.011110).$$

Note that the mantissa in the above product is positive and is not in the range $\frac{1}{2} \leq m < 1$; in order to bring it to this range, the characteristic is decremented and the mantissa is shifted one place to the left, and this yields

$$N_1 N_2 = 2^{011.0}(0.111100)$$

$$= 7.5.$$

To add or subtract two floating-point binary numbers, it is necessary to adjust the mantissa, m_1, of the smaller number until the characteristics c_1

and c_2 are equal and then the sum is formed by adding m_1 to the mantissa m_2 of the larger number in order to obtain $m = m_1 + m_2$ as the mantissa of the sum; the characteristic of the sum is, $c = c_1 = c_2$, provided $0.5 \le m < 1$. If $m = (m_1 + m_2)$ is not in the range $0 \le m < 1$, then the characteristic has to be adjusted to bring the mantissa to the proper range, as illustrated below.

Example 5.5

$$N_1 = 3.00 = 2^{10.0}0.110$$

$$N_2 = 1.25 = 2^{01.0}0.101.$$

In this case, the smaller number, N_2, is rewritten as

$$N_2 = 2^{10.0}0.0101.$$

Therefore,

$$N_1 + N_2 = 2^{10.0}0.110 + 2^{10.0}0.0101.$$

Therefore $c = 10.0$, and $m = 1.0001$. Since $m \ge 1$, the characteristic has to be increased to $\hat{c} = 11.0$ and the mantissa is decreased to $\hat{m} = 0.10001$ by shifting each bit in mantissa m one place to the right. Then,

$$N_1 + N_2 = 2^{11.0}0.10001$$
$$= (8)(\tfrac{17}{32})$$
$$= 4.25$$

In floating-point arithmetic, truncation or rounding errors exist in multiplication as well as in addition unlike in fixed-point arithmetic where such errors are possible only in multiplication. Floating-point arithmetic, however, provides much greater dynamic range and accuracy than fixed-point arithmetic.

5.2.3. Truncation, Rounding Errors, and Models

Fixed-Point Mode. First, consider a binary number of $(b + 1)$ bits, where the first bit is the sign bit. The numbers representable by these bits are called quantization levels and the gap between two successive quantization levels is called the quantization step, Q. Clearly, the number 2^{-b}, realized by assigning 0 value to all but the least significant bit, will equal the quantization step

$$Q = 2^{-b}. \tag{5.3}$$

Let a number, N, be represented in the fixed-point binary mode as (here N will be assumed to have a scaled magnitude less than unity)

$$N = a_0 \cdot a_1 a_2 \cdots a_b \cdots a_k, \quad b < k \tag{5.4}$$

so that there are $(k+1)$ bits including the sign bit, a_0. *Truncation* (or *chopping*) of a fixed-point number to $(b+1)$ bits or b data bits, where $b < k$, involves the dropping of all bits beyond a_b in (5.4). For example, if

$$N = 1.110011001,$$

and we wish to truncate N to 5 data bits (or 6 bits, including the sign bit), then the truncated number is

$$N_t = 1.11001.$$

Let

$$e_t \triangleq N_t - N \qquad (5.5)$$

denote the error after truncation. It is clear that

$$|e_t| < Q, \qquad (5.6)$$

where $Q = 2^{-b}$ and a_b is the least significant bit beyond which all other bits are dropped in the truncation process. The sign of the error depends on the scheme used to code the number. For the sign and magnitude type of representation,

$$\begin{aligned} -Q < e_t \leq 0 \quad \text{for } N \geq 0 \\ 0 \leq e_t < Q \quad \text{for } N < 0, \end{aligned} \qquad (5.7)$$

while it can be shown that for the 2's complement type of representation, in (5.2b),

$$-Q < e_t \leq 0 \quad \text{for all } N. \qquad (5.8)$$

Rounding of a fixed-point number N to b data bits is the process of adding a 1 to the $(b+1)$-th data bit of N in (5.4) and then truncating to b data bits. For example, if

$$N = 0.110011001,$$

and we wish to round N to 5 data bits, then it is necessary to obtain

$$\begin{aligned} \hat{N} &= 0.110011 + 0.000001 \\ &= 0.110100, \end{aligned}$$

and then $\hat{N}$ is truncated to 5 data bits in order to obtain the 5 data bit rounded version, N_r, of the number specified.

$$N_r = 0.11010.$$

Again, let

$$e_r \triangleq N_r - N \qquad (5.9)$$

denote the error after rounding. It is clear that

$$|e_r| \leq \frac{Q}{2}. \qquad (5.10)$$

Since rounding is based on the magnitude of the number, (5.10) applies to all representations. The nonlinear characteristics of the truncation (or

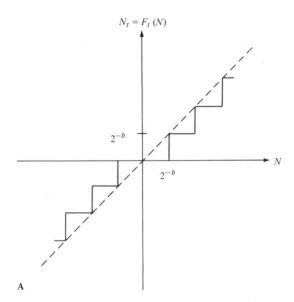

Figure 5.1A. Plot of nonlinear characteristic function, $F_t(N)$, representing trunca-tion error (sign and magnitude; see 5.7).

chopping) and rounding processes are shown in Figure 5.1. Note that e_r (and, also, for that matter e_t) is a random variable. For convenience in analysis, e_r is assumed to be an uniformly distributed random variable in the interval $-\dfrac{Q}{2} < e_r < \dfrac{Q}{2}$. Its probability distribution is shown in Figure 5.2A. It has zero mean and its variance is

$$\sigma_{e_r}^2 = \frac{1}{Q} \int_{-Q/2}^{Q/2} e_r^2 \, de_r = \frac{Q^2}{12}. \tag{5.11}$$

If b bits (excluding the sign bit) are used, then $Q = 2^{-b}$ and

$$\sigma_{e_r}^2 = \frac{1}{12} 2^{-2b}. \tag{5.12}$$

Floating-Point Mode. In floating-point arithmetic, the relative error (*error due to rounding is again considered*) E_r is more important than e_r.

$$E_r \triangleq \frac{N_r - N}{N} = \frac{e_r}{N}. \tag{5.13}$$

Since N_r and N are representable as

$$N_r = 2^c m_r, \quad N = 2^c m,$$

it follows that

$$E_r = \frac{m_r - m}{m}. \tag{5.14}$$

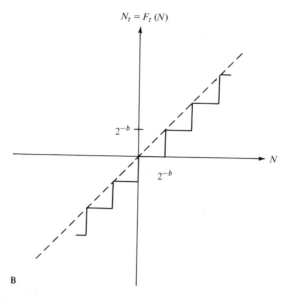

Figure 5.1B. Plot of nonlinear characteristic function, $F_t(N)$, representing truncation error (2's complement; see 5.8).

Figure 5.1C. Plot of nonlinear characteristic function, $F_r(N)$, representing rounding error (see 5.10).

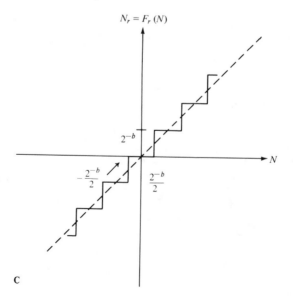

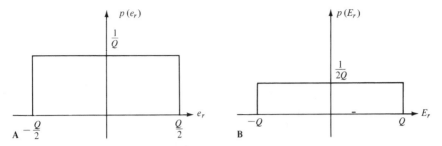

Figure 5.2A. Probability density function, $p(e_r)$ of fixed-point rounding error e_r for both sign and magnitude and 2's complement codings. B. Probability density function $p(E_r)$ of floating-point rounding error E_r for both sign and magnitude and 2's complement codings.

Defining $\hat{e}_r \triangleq m_r - m$ and substituting in (5.14),

$$E_r = \frac{\hat{e}_r}{m}. \qquad (5.15)$$

Since $|\hat{e}_r| < \frac{Q}{2}$, where Q is, again, the quantization step size, and $0.5 \le |m| < 1$, it follows that

$$|E_r| < Q. \qquad (5.16)$$

Note that the range of errors in the floating-point case is twice the range of errors in the fixed-point mode. Subject to the assumption that E_r is an uniformly distributed random variable over the interval $-Q < E_r < Q$ (its probability density function is shown in Figure 5.2B), it has zero mean and a variance

$$\sigma_{E_r}^2 = \frac{1}{2Q} \int_{-Q}^{Q} E_r^2 \, dE_r = \frac{Q^2}{3}. \qquad (5.17)$$

The probability density functions for roundoff errors due to fixed-point and floating-point arithmetics are sketched in Figure 5.2. Note that both addition and multiplication might introduce roundoff errors in floating-point arithmetic. Let N_r denote the $(b+1)$-bit (one of which is the sign bit of the mantissa), mantissa floating-point representation of the number N. Then, the actual sum and product of two numbers, N_1 and N_2, are

$$(N_1 + N_2)_r = (N_1 + N_2)(1 + \varepsilon)$$

$$(N_1 N_2)_r = (N_1 N_2)(1 + \delta),$$

where the relative errors ε, δ fall in the ranges $-Q \le \varepsilon < Q$ and $-Q \le \delta < Q$, where $Q = 2^{-b}$.

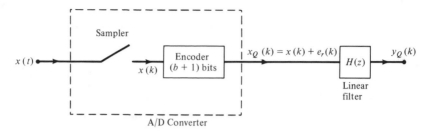

Figure 5.3. Output of linear discrete filter when subjected to quantized input from A/D converter.

5.3. Output Noise Power Due to Input Quantization Error

The first source of error because of finite arithmetic is due to the process of quantization in the analog-digital (A/D) converter. The derivation of an expression for the output noise power as a result of quantization of the input signal is essential for estimating the necessary word length which will guarantee a satisfactory ratio of signal to noise power. This derivation will be carried out here for the fixed-point mode of representation. The quantized output sequence $\{x_Q(k)\}$ from the A/D converter may be viewed as the superposition of the true input sequence $\{x(k)\}$ and a noise $\{e_r(k)\}$ modeling the quantization process:

$$x_Q(k) = x(k) + e_r(k). \tag{5.18}$$

Introduction of a scaling factor, f, where $0 < f \leq 1$, which multiplies $x(k)$, may be necessary to prevent either adder overflow or the exceeding of the permissible dynamic range by the input sequence. Next, suppose that $\{x_Q(k)\}$ becomes the input to a discrete stable linear time-invariant system, characterized by the transfer function $H(z)$ or impulse response sequence $\{h(k)\}$, as shown in Figure 5.3. The output sequence, $\{y_Q(k)\}$, is then the superposition of two sequences,

$$\{y_Q(k)\} = \{h(k)\} * \{x(k)\} + \{h(k)\} * \{e_r(k)\},$$

obtained by convolving $\{x_Q(k)\}$ in (5.18) with the impulse response sequence $\{h(k)\}$. Any element of the output noise sequence $\{\hat{y}(k)\}$, obtained by only convolving $\{h(k)\}$ with $\{e_r(k)\}$, is a random variable. The autocorrelation function, $\phi_{\hat{y}}$ of this output sequence is defined by the statistical expectation:

$$
\begin{aligned}
\phi_{\hat{y}}(k) &= E\left\{ \left[\sum_j h(j)e_r(n-j) \right] \left[\sum_i h(i)e_r(n-k-i) \right] \right\} \\
&= \sum_i h(i) \sum_j h(j) E\{ e_r(n-j)e_r(n-k-i) \} \\
&= \sum_i h(i) \sum_j h(j) \phi_e(k+i-j),
\end{aligned}
\tag{5.19}
$$

where the autocorrelation function $\phi_e(k)$ of the input noise, assumed to be white, is zero for all k except $k = 0$. Furthermore,

$$\phi_e(0) = \sigma_e^2,$$

where σ_e^2 is the variance of the input noise. Therefore,

$$\phi_{\hat{y}}(k) = \sigma_e^2 \sum_i h(i)h(k+i) \tag{5.20}$$

and the variance of the output noise is

$$\sigma_{\hat{y}}^2 = \phi_{\hat{y}}(0) = \sigma_e^2 \sum_i h^2(i). \tag{5.21}$$

From Parseval's relation, it follows that

$$\frac{\sigma_{\hat{y}}^2}{\sigma_e^2} = \sum_i h^2(i) = \frac{1}{2\pi j} \oint_C z^{-1} H(z)H(z^{-1})\, dz, \tag{5.22}$$

where the contour of integration C is the unit circle, traversed in the counterclockwise direction. There are numerical and algebraic techniques for evaluating the integral in (5.22), and an algebraic technique that can be efficiently implemented is discussed next.

5.3.1. Evaluation of $\dfrac{1}{2\pi j}\oint_C H(z)H(z^{-1})\dfrac{dz}{z}$ [1]

It has been seen that the computation of output noise variance due to input signal quantization requires the evaluation of an integral

$$I_0 \triangleq \frac{1}{2\pi j} \oint_C H(z)H(z^{-1})\, \frac{dz}{z}, \tag{5.23}$$

where the filter transfer function, $H(z)$ is of the form

$$H(z) = \frac{\displaystyle\sum_{k=0}^{m} a_k z^{-k}}{1 + \displaystyle\sum_{k=1}^{n} b_k z^{-k}},$$

and it may be assumed to be stable (after checking for stability), that is,

$$1 + \sum_{k=1}^{n} b_k z^{-k} \neq 0, \quad |z| \geq 1.$$

Since the filter is stable, the contour of integration, in the evaluation of I_0, may be taken to be $|z| = 1$. Suppose that the power series expansion of the causal filter transfer function about the point $z^{-1} = 0$ or $z = \infty$ (note that $H(z)$ is analytic at $z^{-1} = 0$) is

$$H(z) = \sum_{k=0}^{\infty} h_k z^{-k}, \tag{5.24}$$

where the impulse-response sequence $\{h_k\}$ is recursively computable from

$$h_k = a_k - \sum_{i=1}^{n} b_i h_{k-i}, \quad k \geq 0, \tag{5.25}$$

where $a_i = 0$, $i \geq m+1$. Define

$$I_k \triangleq \sum_{i=0}^{\infty} h_i h_{i+k}. \tag{5.26}$$

Note that

$$I_k = I_{-k}. \tag{5.27}$$

$\{I_k\}$ is referred to as a covariance sequence. Define $b_0 \triangleq 1$. Multiply both sides of (5.25) by h_{k-j} sum each side of the resulting equation from $k = 0$ to $k = \infty$, substitute (5.26) and rearrange terms to get (remember that $a_i = 0$, $i \geq m+1$),

$$\sum_{i=0}^{n} b_i I_{k-i} = \sum_{i=0}^{m-k} h_i a_{i+k}, \quad k \geq 0. \tag{5.28}$$

After defining

$$d_k \triangleq \sum_{i=0}^{m-k} h_i a_{i+k}, \quad k \geq 0 \tag{5.29}$$

and using (5.27), the preceding set of linear equations may be put in matrix form,

$$A\mathbf{I} = \mathbf{d}, \tag{5.30}$$

where

$$A = \begin{bmatrix} b_0 & b_1 & b_2 & b_3 & \cdots & & & & b_n \\ b_1 & b_2 + b_0 & b_3 & b_4 & \cdots & & & b_n & 0 \\ b_2 & b_3 + b_1 & b_4 + b_0 & b_5 & \cdots & & b_n & 0 & 0 \\ b_3 & b_4 + b_2 & b_5 + b_1 & b_6 + b_0 & \cdots & b_n & 0 & 0 & 0 \\ \vdots & \vdots & \vdots & & & & & & \\ b_n & b_{n-1} & & & \cdots & & & b_1 & b_0 \end{bmatrix}$$

with $b_0 \triangleq 1$ and

$$\mathbf{I} \triangleq [I_0 \quad I_1 \quad \cdots \quad I_n]^t$$

$$\mathbf{d} \triangleq [d_0 \quad d_1 \quad \cdots \quad d_n]^t.$$

Note that the element a_{ij}, in the ith row and jth column of matrix A, $i = 1, 2, \ldots, n+1$, $j = 1, 2, \ldots, n+1$ is given by

$$a_{ij} = \begin{cases} b_{i-1} & \text{for } j = 1 \\ b_{i+j-2} + b_{i-j} & \text{for } j > 1 \text{ and } b_k = 0, \ k < 0 \text{ or } k > n. \end{cases}$$

I_0 (and $I_1, I_2, \ldots, I_n$, if necessary) may then be computed by solving the set of linear equations in (5.30) using any efficient method. More generally, it may be noted that the computation of

$$I_k = \frac{1}{2\pi j} \oint_{|z|=1} H(z) H(z^{-1}) z^k \frac{dz}{z}$$

is required in applications concerned with model reduction procedures for approximating high-order filters with low-order ones.

Dugre and Jury [2] proposed a modification of the above procedure to obtain an algorithm that does not require a separate test for stability.

5.4. Effect of Inaccuracy in Representation of Coefficient

When the transfer function in (3.80) is to be digitally implemented, errors due to the constraints of finite arithmetic occur in the representation of coefficient sequences $\{a_k\}$ and $\{b_k\}$. Let a_{kQ} and b_{kQ} be the quantized values of a_k and b_k, respectively. Then

$$a_{kQ} = a_k + \alpha_k \tag{5.31a}$$

$$b_{kQ} = b_k + \beta_k, \tag{5.31b}$$

where α_k, β_k are quantization error terms. The errors due to coefficient quantization could be serious but the effects could be analyzed conveniently. It has been proved that, in general, the effect of coefficient inaccuracy due to the finite arithmetic constraints is more pronounced for a high-order filter when it is realized in the direct form than when it is realized in the parallel, cascade, or lattice forms. A simple example, given next, illustrates an adverse effect of coefficient quantization with a specified number of bits allowed in the coefficient representation. This adverse effect manifests itself in the loss of stability property; in other cases, appreciable deviation of the frequency response from that expected under infinite precision conditions could be interpreted as an adverse effect.

Example 5.6. Consider the realization of the transfer function

$$H(z) = \frac{1}{z^{-1} + \frac{11}{8}} + \frac{1}{z^{-1} + \frac{9}{8}} \tag{5.32a}$$

$$= \frac{2z^{-1} + \frac{5}{2}}{\left(z^{-1} + \frac{11}{8}\right)} \cdot \frac{1}{\left(z^{-1} + \frac{9}{8}\right)} \tag{5.32b}$$

$$= \frac{2z^{-1} + \frac{5}{2}}{z^{-2} + \frac{5}{2}z^{-1} + \frac{99}{64}}. \tag{5.32c}$$

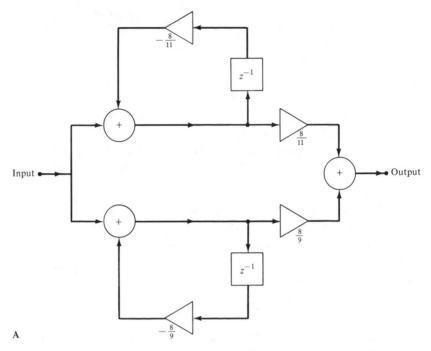

A

Figure 5.4A. Realization of transfer function in Example 5.6 in parallel form.

The realization of $H(z)$ in (5.32a) is possible as a parallel connection of two first-order filters as shown in Figure 5.4A; the realization of $H(z)$ in (5.32b) is possible as a cascade of two first-order filters as shown in Figure 5.4B. Both these realizations are possible with five bits (including the sign bit) needed to represent each coefficient in fixed-point arithmetic. However, the realization of $H(z)$ in (5.32c) in any form requires as many as 8 bits

Figure 5.4B. Realization of transfer function in Example 5.6 in cascade form.

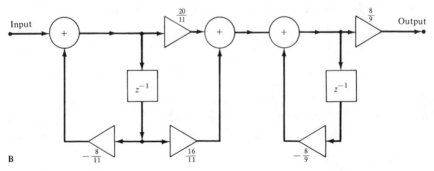

B

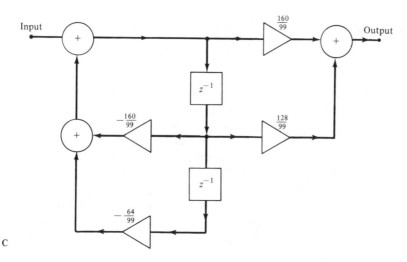

Figure 5.4C. Realization of transfer function in Example 5.6 in direct form 2.

(including sign bit) just to represent the coefficient $\frac{99}{64}$ as

$$
\overset{\text{sign bit}}{\underset{\downarrow}{}}
$$

$$
\tfrac{99}{64} = 01.100011.
$$

The coefficients of the rational function

$$
\hat{H}(z) = \frac{2z^{-1} + \tfrac{5}{2}}{z^{-2} + \tfrac{5}{2}z^{-1} + \tfrac{3}{2}},
$$

obtained by replacing the problem-causing coefficient $\frac{99}{64}$ of $H(z)$ by its approximant $\frac{3}{2}$ may be representable by 3-bit binary numbers. However, $\hat{H}(z)$ is unstable and therefore its realization is useless.

Particularly when the filter has poles close to the unit circle, instability could easily occur as a consequence of coefficient inaccuracy. In fact, the problem could be critical when either the sampling rate or the order of the filter or both are high. Doubling the order of the filter will require approximately twice as many digits accuracy for the representation of the denominator coefficients (b_k's) of the filter transfer function so that stability is preserved. Similarly, doubling the sample rate for an nth order filter requires about $0.3n$ additional bits in the representation of the b_k's. Kaiser [3] demonstrated that for an nth order lowpass filter operating with a sampling period T and having distinct poles at $z = e^{p_k T}$, Re $p_k < 0$, stability is guaranteed if the number N of bits used satisfy

$$N > \left\lceil -\log_2\left(\frac{5\sqrt{n}}{2^{n+2}} \prod_{k=1}^{n} p_k T \right) \right\rceil,\tag{5.33}$$

where $\lceil x \rceil$ denotes the smallest integer greater than x. Extensions of this and related results are possible to filters of highpass, bandpass, or bandstop types having multiple order poles. Kaiser's error bound is overly pessimistic because his approach is based on the deterministic theory. Knowles and Olcayto [4] formulated their analysis using statistical methods and presented an input/output model for coefficient quantization. They also showed how the variance of error due to coefficient quantization might be calculated. Although good agreement between theoretical calculations and experimental measurements have been observed for high-order filters, the validity of the assumption of random coefficient error when the order of the filter is low has been questioned.

Coefficient sensitivity studies can be done quite easily. For example, because of the importance of the stability problem, it is desirable to compute pole-sensitivity due to changes in the coefficients b_j in (3.2). To do this, write the denominator polynomial in z^{-1} of the filter transfer function in the factored form:

$$1 + \sum_{k=1}^{n} b_k z^{-k} = \prod_{k=1}^{n} \left(1 - \frac{z^{-1}}{z_k} \right).$$

Then,

$$\log\left[1 + \sum_{k=1}^{n} b_k z^{-k} \right] = \sum_{k=1}^{n} \log\left(1 - \frac{z^{-1}}{z_k} \right),$$

Straightforward manipulation yields,

$$\frac{\partial z_i}{\partial b_k} = \frac{z_i^{k+1}}{\displaystyle\prod_{\substack{k=1 \\ k \neq i}}^{n} \left(1 - \frac{z_i}{z_k} \right)}.$$

Let Δz_i denote a small change in the location of the pole at $z^{-1} = z_i$ due to small change Δb_k in the coefficient b_k. Define

$$\alpha_k \triangleq \frac{\Delta b_k}{b_k}, \qquad \beta_k \triangleq \frac{\Delta z_i}{z_i},$$

then, subject to the assumption that the poles are clustered near $|z| = 1$, it follows that

$$\beta_k \simeq \frac{\Delta b_k}{\displaystyle\prod_{\substack{r=1 \\ r \neq i}}^{n} (z_r - z_i)}.$$

Therefore, the closer the poles are clustered towards the unit circle, the smaller will be the change required in the coefficient b_k for potential instability problems.

5.5. Multiplication Roundoff Error

The products resulting from multiplications of data by coefficients within a digital filter must be rounded or truncated to a small number of digits. The data to be multiplied is presumed to be already in digital form and the rounding or truncation of the products takes place at various points within the digital filter. The resulting error is usually referred to as roundoff noise, irrespective of whether the products are rounded or truncated. The level and character of the roundoff noise depends on the number of digits available for representing data and filter coefficients, the employment of fixed-point or floating-point (or block floating-point [5]) modes of arithmetic and the topology of the structure chosen to realize the filter. Various structures, which realize a specified transfer function, under infinite precision conditions behave very differently when subjected to finite arithmetic constraints. In a digital filter, the effect of rounding due to multiplication is modeled after assuming that the errors due to the multiplications are noise sources satisfying certain conditions with regard to their statistical independence. The ensuing analysis will be based on the fixed-point arithmetic mode of operation and the quantization will be via rounding.

Since, in roundoff noise analysis the data $x(k)$ to be multiplied by a coefficient a_k is assumed to be in digital form and rounded, it will be denoted by $x_r(k)$ (as before, the subscript r stands for "rounded"). Then, the rounding process

$$y_r(k) = \left(x_r(k) \cdot a_k\right)_r, \qquad (5.34)$$

can be modeled as in Figure 5.5, where roundoff noise, $e(k)$ is modeled as a random variable with uniform probability density function, having zero mean and a variance of $2^{-2b}/12$, when $(b+1)$ is the total number of bits (including the sign bit) used in the fixed-point mode of representation. Errors might occur whenever there are multiplications. This sequence,

Figure 5.5. Fixed-point product (multiplication) roundoff noise model.

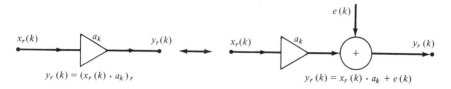

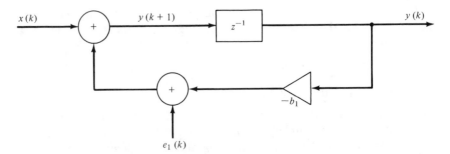

Figure 5.6. Realization of (5.35) using the fixed-point product roundoff noise model for multiplication.

$\{e(k)\}$, of errors is assumed to satisfy the following conditions:

1. Any two different samples from the same noise source are uncorrelated, that is, each noise source is modeled as a discrete stationary white random process having zero mean and a variance of $Q^2/12$, where $Q = 2^{-b}$,
2. any two different noise sources are uncorrelated, and
3. each noise source is uncorrelated with the input sequence.

The above set of assumptions makes it possible to calculate the total noise variance at the output of the filter as a superposition of the variances due to each noise source with the other noise sources and input deactivated. The procedure will be illustrated via discussions proceeding from a simple first-order realization to structures of increasing complexity and variety.

Consider a first-order digital filter characterized by the difference equation

$$y(k+1) = -b_1 y(k) + x(k), \quad |b_1| < 1. \tag{5.35}$$

There is one multiplication involved and a realization of the filter involving the fixed-point product roundoff noise model of Figure 5.5 is shown in Figure 5.6. Under zero input conditions, the difference equation relating the noise source $e_1(k)$ to the output $y(k)|_{x(k)=0} \triangleq \hat{y}_1(k)$ due to this noise source is

$$\hat{y}_1(k+1) = -b_1 \hat{y}_1(k) + e_1(k).$$

The transfer function of the linear system which generates the preceding input/output relation between the input noise source $e_1(k)$ (since $e_1(k)$ is an infinite-energy signal, note that its z-transform does not exist) and output $\hat{y}_1(k)$ is

$$H_1(z) = \frac{1}{(z+b_1)}.$$

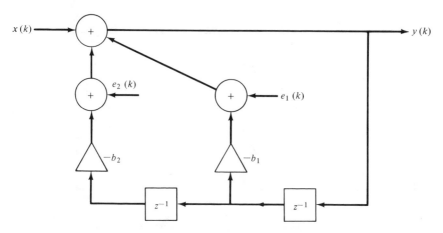

Figure 5.7. Realization of a second-order digital filter using the fixed-point product roundoff noise model.

If $\{h_1(k)\} = Z^{-1}[H_1(z)]$, then by Parseval's theorem,

$$\sum_k h_1^2(k) = \frac{1}{2\pi j} \oint_{|z|=1} H_1(z) H_1(z^{-1}) \frac{dz}{z}.$$

Therefore, the output variance due to $e_1(k)$, having a variance $2^{-2b}/12$, is (see (5.22))

$$\sigma_{\hat{y}_1}^2 = \frac{2^{-2b}}{12} \sum_k h_1^2(k)$$

$$= \frac{2^{-2b}}{12} \cdot \frac{1}{2\pi j} \oint_{|z|=1} H_1(z) H_1(z^{-1}) \frac{dz}{z}.$$

Next, consider the second order filter structure, shown in Figure 5.7 where each of the two multiplications have been modeled by the fixed-point product roundoff noise model of Figure 5.5. The two noise sources are denoted by $e_1(k)$ and $e_2(k)$. Deactivate $x(k)$ and $e_2(k)$ and then relate $\hat{y}_1(k)$ to $e_1(k)$ in Figure 5.8A:

$$\hat{y}_1(k) = -b_1\hat{y}_1(k-1) - b_2\hat{y}_1(k-2) + e_1(k). \qquad (5.36a)$$

Then, deactivating $x(k)$ and $e_1(k)$, relate $\hat{y}_2(k)$ to $e_2(k)$ in Figure 5.8B:

$$\hat{y}_2(k) = -b_1\hat{y}_2(k-1) - b_2\hat{y}_2(k-2) + e_2(k). \qquad (5.36b)$$

The transfer functions $H_1(z)$ and $H_2(z)$ of the linear systems that generate the input/output relations in (5.36a) and (5.36b), respectively, are

$$H_1(z) = \frac{1}{1 + b_1 z^{-1} + b_2 z^{-2}}$$

$$H_2(z) = \frac{1}{1 + b_1 z^{-1} + b_2 z^{-2}} = H_1(z).$$

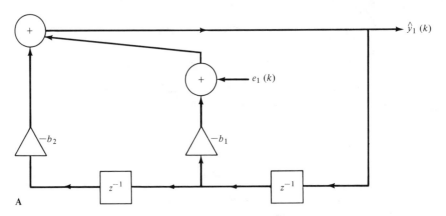

Figure 5.8A. Structure used to calculate output variance due to $e_1(k)$ only.

The coefficients b_1 and b_2 are assumed to satisfy the stability conditions, that is,

$$\left| \frac{-b_1 \pm \sqrt{b_1^2 - 4b_2}}{2} \right| < 1.$$

Then, since the output error variance due to $e_1(k)$ is equal to the output error variance due to $e_2(k)$, the total output error variance (input $x(k) = 0$)

$$\sigma_{\hat{y}}^2(k) = \frac{2^{-2b}}{6} \frac{1}{2\pi j} \oint_{|z|=1} H(z) H(z^{-1}) \frac{dz}{z},$$

where

$$H(z) \triangleq H_1(z) = H_2(z).$$

Figure 5.8B. Structure used to calculate output variance due to $e_2(k)$ only.

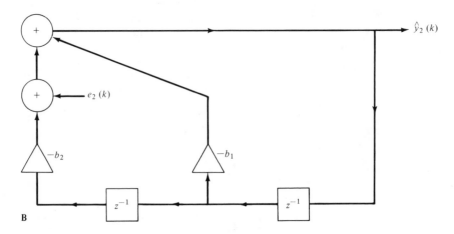

331

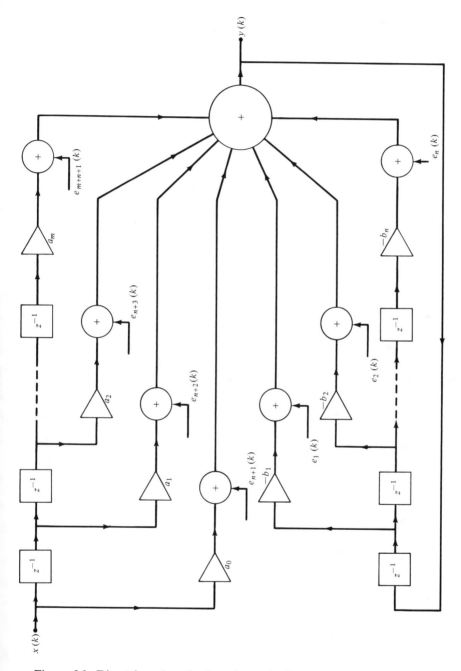

Figure 5.9. Direct form 1 realization of transfer function in (3.2) using the fixed-point product roundoff noise model for each multiplication.

Consider next the direct form 1 realization of the transfer function $H(z)$ in (3.2) as shown in Figure 5.9, where each multiplication of data by filter coefficient has been substituted by the fixed-point product roundoff noise model of Figure 5.5. Note that all noise sources are combined at the output adder. It is simple to see that the transfer function of the linear system, whose input is a noise source, $e_i(k)$ (the external input $x(k)$ and the other noise sources, $e_j(k)$, $j \neq i$, are deactivated) and its output is

$$y(k)\Big|_{\substack{x(k)=0 \\ e_j(k)=0, \quad j \neq i \\ j=1,2,\ldots,m+n+1}} = \hat{y}_i(k),$$

is given by

$$G(z) = \cfrac{1}{1 + \sum_{k=1}^{n} b_k z^{-k}}.$$

Note that $G(z)$ is related to $H(z)$ in (3.80) by

$$G(z) = \cfrac{H(z)}{\sum_{k=0}^{m} a_k z^{-k}}.$$

Since the total number of noise sources is $(m + n + 1)$, the total output noise variance is (it is understood that $H(z)$ is stable and $x(k) = 0$)

$$\sigma_{\hat{y}}^2 = \frac{2^{-2b}}{12} (m + n + 1) \frac{1}{2\pi j} \oint_{|z|=1} G(z)G(z^{-1}) \frac{dz}{z}. \tag{5.37}$$

In (5.37), it is understood that none of the $(m + n + 1)$ coefficients in (3.80) are absent, that is, each is capable of contributing to the output roundoff noise.

Next, consider the realization of $H(z)$, written in the form of (3.82), as a cascade of second-order blocks. Suppose that each second-order component in the cascade is realized as a direct form 1 structure. From (5.37), it is clear that the direct form 1 realization of (3.80), taking into account product roundoff noise error, can be modeled as in Figure 5.10, where the equivalent noise source $e(k)$ has a variance

$$\sigma_e^2 = (m + n + 1) \frac{2^{-2b}}{12}.$$

Therefore, for a second-order block ($m = n = 2$), the preceding expression specializes to (e is replaced by e_i, $i = 1, 2, \ldots, K$ and e_i is shown in Figure 5.11),

$$\sigma_{e_i}^2 = \frac{5}{12} 2^{-2b}. \tag{5.38}$$

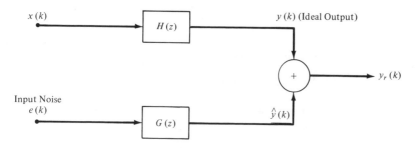

Figure 5.10. Roundoff error in direct form 1 realization.

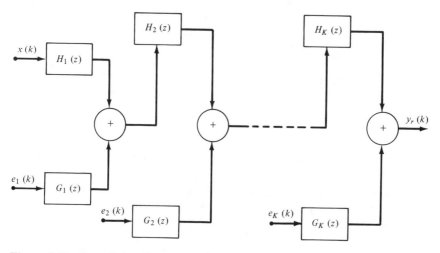

Figure 5.11. Cascaded realization of $H(z)$ in (3.82) and (3.83) taking into account product roundoff error noise.

Therefore, the cascade realization of $H(z)$ in (3.82), taking into account product roundoff noise error is modeled as in Figure 5.11, where

$$G_i(z) = \frac{H_i(z)}{\alpha_{0i} + \alpha_{1i}z^{-1} + \alpha_{2i}z^{-2}} \qquad (5.39)$$

($\alpha_{0i} = 1$ in (3.82)) and $e_i(k)$ has a variance equaling $\sigma_{e_i}^2$ for $i = 1, 2, \ldots, K$ (to avoid notational ambiguity the symbol k in (3.82) has been replaced by K in the present context). $H_i(z)$ has been defined in (3.83). It has five potential noise sources, so that (5.38) applies. From Figure 5.11, it can be readily verified that the transfer function, $H_{e_i}(z)$ of the linear system whose input is $e_i(k)$ (with the external input along with remaining noise sources deactivated) and the corresponding output is $\hat{y}_i(k)$, takes the form

$$H_{e_i}(z) = G_i(z) \prod_{k=i+1}^{K} H_k(z) \qquad (5.40a)$$

for $i = 1, 2, \ldots, K - 1$ and

$$H_{e_K}(z) = G_K(z). \tag{5.40b}$$

Therefore, the variance of the output due to noise source $e_i(k)$ is

$$\sigma_{\hat{y}_i}^2 = \frac{5}{12} 2^{-2b} \frac{1}{2\pi j} \oint_{|z|=1} H_{e_i}(z) H_{e_i}(z^{-1}) \frac{dz}{z}. \tag{5.41a}$$

The total output error variance is

$$\sigma_{\hat{y}}^2 = \sum_{i=1}^{K} \sigma_{\hat{y}_i}^2. \tag{5.41b}$$

Finally consider the parallel realization of the transfer function, $H(z)$, expressed in the form of (3.84) (set $\gamma_{00} = 0$ and replace k in (3.84) by K here). Define (δ_{0i} in (5.42a) and (5.42b) is not necessarily 1),

$$\hat{H}_i(z) = \frac{\gamma_{0i} + \gamma_{1i} z^{-1}}{\delta_{0i} + \delta_{1i} z^{-1} + \delta_{2i} z^{-2}} \tag{5.42a}$$

(the five multipliers in the direct form 1 realization of $\hat{H}_i(z)$ are potential sources of multiplication roundoff noise and therefore, (5.38) applies) and

$$\hat{G}_i(z) = \frac{1}{\delta_{0i} + \delta_{1i} z^{-1} + \delta_{2i} z^{-2}} \tag{5.42b}$$

for $i = 1, 2, \ldots, K$. Realize each second-order block in accordance with the direct form 1 structure. Then, if the output error variance caused by the input noise $e_i(k)$ to the ith second-order block is represented by $\sigma_{\hat{y}_i}^2(k)$ the total output noise variance of the structure in Figure 5.12 is

$$\sigma_{\hat{y}}^2 = \sum_{i=1}^{K} \sigma_{\hat{y}_i}^2,$$

where

$$\sigma_{\hat{y}_i}^2 = \frac{5}{12} 2^{-2b} \frac{1}{2\pi j} \oint_{|z|=1} \hat{G}_i(z) \hat{G}_i(z^{-1}) \frac{dz}{z}. \tag{5.43}$$

The *principle* involved in computing the output error variance due to roundoff noise for structures other than those dealt with here is identical; each multiplication of data and coefficient is modeled in the generic form shown in Figure 5.5, the output error variance due to each noise source with the input and remaining noise sources deactivated is calculated and because of the underlying assumptions on statistical independence mentioned earlier the total output error variance is computed as a superposition of the individual output error variances. The treatment of roundoff error analysis for filters using floating-point arithmetic is somewhat different and more complicated than that for filters using fixed-point arithmetic. The interested reader is referred to the papers by Sandberg [6] and Liu and Kaneko [7] for information on this topic. This section will be concluded with the presentation of a comprehensive example which will illustrate how the output error variance changes with the structure adopted for realization.

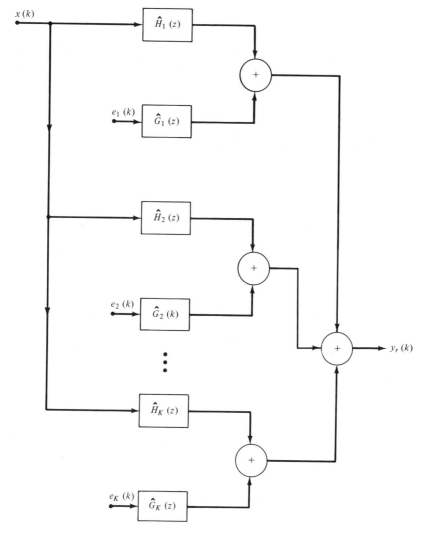

Figure 5.12. Parallel realization of $H(z)$ in (3.84) taking into account product roundoff error noise.

Example 5.7. For the specified transfer function (note that it is stable)

$$H(z) = \frac{z^{-1} + 0.2z^{-2}}{1 - 0.2z^{-1} - 0.24z^{-2}}, \qquad (5.44)$$

the output error variance $\sigma_{\hat{y}}^2$ will be calculated for various realization structures. Integrals of the form

$$\frac{1}{2\pi j} \oint_{|z|=1} H(z) H(z^{-1}) \frac{dz}{z}$$

336

```
C
C       *************************************************************************
C
C       A FORTRAN PROGRAM (FWERR.FOR) TO COMPUTE THE ERROR RESULTING
C       FROM THE FINITE WORDLENGTH DIGITAL FILTER.
C
C       TO RUN THIS PROGRAM, ISSUE
C       EX FWERR.FOR,PRG:IMSL.REL/SEARCH
C
C       *************************************************************************
C
        REAL A(0:49),H(0:49),AA(50,50),B(0:49),D(0:49)
        REAL WKAREA(50)
C
C       PROVIDE DATA FOR "M", "N", "A(I)", & "B(I)"
C
        WRITE(6,100)
100     FORMAT(4X,'ENTER THE DEGREE "M" OF THE NUMERATOR POLYNOMIAL')
        READ(5,110) M
110     FORMAT(I)
        WRITE(6,120)
120     FORMAT(/,4X,'ENTER THE DEGREE "N" OF THE DENOMINATOR POLY.')
        READ(5,110) N
        WRITE(6,130)
130     FORMAT(/,4X,'ENTER NUMERATOR COEFF. "A(I)" OF A(I)*Z**(-I)',
     *       /,4X,'FROM I=0 TO I=M')
        READ(5,140) (A(I),I=0,M)
140     FORMAT(F)
        WRITE(6,150)
150     FORMAT(/,4X,'ENTER DENOMINATOR COEFF. "B(I)" OF B(I)*Z**(-I)',
     *       /,4X,'FROM I=0 TO I=N WITH B(0)=1.')
        READ(5,140) (B(I),I=0,N)
C
C       FORM THE SEQUENCE "H(K)", K=0,1,2,...,M
C
        DO 200 K=0,M
        BH=0.
        DO 210 I=1,N
        INDEX=K-I
        IF(INDEX.LT.0) GO TO 210
        BH=BH+B(I)*H(INDEX)
210     CONTINUE
        H(K)=A(K)-BH
200     CONTINUE
C
C       FORM THE SEQUENCE "D(K)", K=0,1,2,...,N
C
        MIN=MIN0(M,N)
```

Figure 5.13. (*Partial listing, continued.*)

```
        DO 300 K=0,N
        IF(K.GT.MIN) GO TO 320
        HA=0.
        DO 310 I=0,M-K
310     HA=HA+H(I)*A(I+K)
        D(K)=HA
        GO TO 300
320     D(K)=0.
300     CONTINUE
C
C       FORM THE MATRIX "AA(I,J)"
C
        DO 400 I=1,N+1
400     AA(I,1)=B(I-1)
        DO 410 J=2,N+1
        DO 410 I=1,N+1
        INDEX1=I+J-2
        INDEX2=I-J
        IF(INDEX2.LT.0) GO TO 420
        AA(I,J)=B(INDEX1)+B(INDEX2)
        GO TO 410
420     AA(I,J)=B(INDEX1)
410     CONTINUE
        WRITE(6,430)
430     FORMAT(4X,'THE VALES FOR "H(K)" ARE')
        WRITE(6,440) (H(K),K=0,M)
440     FORMAT(8X,5F9.5)
        WRITE(6,450)
450     FORMAT(/,4X,'THE VALES FOR "D(K)" ARE')
        WRITE(6,440) (D(K),K=0,N)
        WRITE(6,460)
460     FORMAT(/,4X,'THE VALES FOR MATRIX "AA(I,J)"  ARE')
        WRITE(6,440) ((AA(I,J),J=1,N+1),I=1,N+1)
C
C       SOLVE THE LINEAR EQ. AA*X=D
C
        CALL LEQT1F(AA,1,N+1,50,D,0,WKAREA,IER)
        WRITE(6,470)
470     FORMAT(/,4X,'THE VALUE FOR THE ERROR IS')
        WRITE(6,440) D(0)
        STOP
        END
```

Figure 5.13. Computer listing of program FWERR.FOR to evaluate the generic integral of Section 5.3.1.

will be calculated via the computer program FWERR.FOR, which implements the procedure given in Section 5.3.1 for evaluation of the type of integral characterized above. The listing for FWERR.FOR is given in Figure 5.13. Let the quantization step size be $Q = 2^{-2b}$.

Direct Form 1. The realization of $H(z)$ in (5.44) is given in Figure 5.14A, and it contains three scalar multipliers, each capable of contributing to the

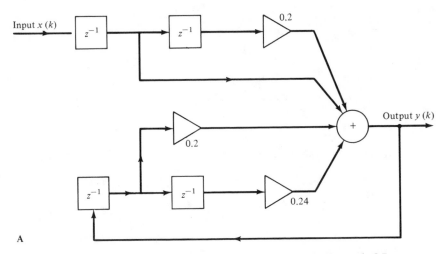

Figure 5.14A. Direct Form 1 realization of $H(z)$ in Example 5.7.

output noise. Let

$$G(z) = \frac{1}{1 - 0.2z^{-1} - 0.24z^{-2}}.$$

Then, using FWERR.FOR,

$$\frac{1}{2\pi j} \oint_{|z|=1} G(z)G(z^{-1}) \frac{dz}{z} = 1.14007.$$

Therefore, applying (5.37), taking into account the fact that because of the three multipliers there are only three possible noise sources instead of $(2 + 2 + 1) = 5$, it follows that

$$\sigma_y^2 = 3.42021 \frac{2^{-2b}}{12}.$$

Direct Form 2. The realization for $H(z)$ of (5.44) in this form is given in Figure 5.14B. Applying FWERR.FOR,

$$\frac{1}{2\pi j} \oint_{|z|=1} H(z)H(z^{-1}) \frac{dz}{z} = 1.30568. \qquad (5.45)$$

From Figure 5.14B, the transfer function between each of the roundoff noise sources associated with the multipliers, 0.2 and 0.24, (on the left) and the output is $H(z)$, while the transfer function between the roundoff noise source associated with the multiplier 0.2 (on the right) and the output is 1. Realizing that

$$\frac{1}{2\pi j} \oint_{|z|=1} \frac{1}{z} dz = 1, \qquad (5.46)$$

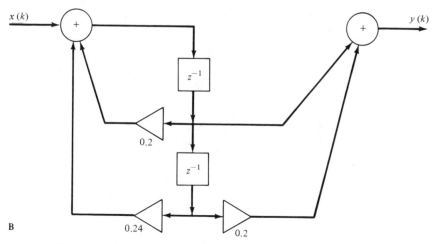

Figure 5.14B. Direct Form 2 realization of $H(z)$ in Example 5.7.

it is possible to write

$$\sigma_{\hat{y}}^2 = (2 \times 1.30568 + 1) \frac{Q^2}{12}$$

$$= 3.61136 \frac{Q^2}{12}.$$

where $Q = 2^{-b}$.

Cascade 1. Rewriting $H(z)$ in (5.44) as

$$H(z) = \frac{z^{-1}}{1+0.4z^{-1}} \cdot \frac{1+0.2z^{-1}}{1-0.6z^{-1}}, \tag{5.47}$$

a cascade realization of the type shown in Figure 5.14C is possible. The transfer functions between roundoff noise error sources associated with

Figure 5.14C. Cascade 1 realization of $H(z)$ in Example 5.7.

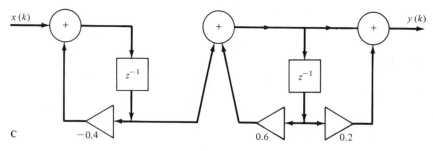

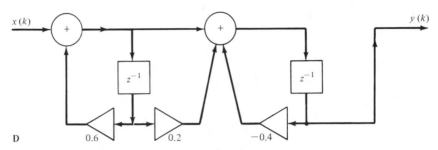

Figure 5.14D. Cascade 2 realization of $H(z)$ in Example 5.7.

scalar multipliers -0.4, 0.6, and 0.2 and the output are, respectively,

$$H(z) = \frac{z^{-1}}{1+0.4z^{-1}} \frac{1+0.2z^{-1}}{1-0.6z^{-1}} \triangleq H_1(z)H_2(z)$$

$$H_2(z) \triangleq \frac{1+0.2z^{-1}}{1-0.6z^{-1}},$$

and

$$H_3(z) \triangleq 1.$$

Applying FWERR.FOR,

$$\frac{1}{2\pi j}\oint_{|z|=1} H_2(z)H_2(z^{-1})\frac{dz}{z} = 2. \tag{5.48}$$

Therefore, from (5.45), (5.46), and (5.48),

$$\sigma_{\hat{y}}^2 = (1.30568+2+1)\frac{Q^2}{12}$$

$$= 4.30568\frac{Q^2}{12}.$$

Cascade 2. Rewriting $H(z)$ in (5.44) as

$$H(z) = \frac{1+0.2z^{-1}}{1-0.6z^{-1}} \cdot \frac{z^{-1}}{1+0.4z^{-1}},$$

a cascade realization of the type shown in Figure 5.14D is possible. The transfer functions between roundoff noise error sources associated with the scalar multipliers 0.6, 0.2, and -0.4 and the output are, respectively, $H(z)$, $H_1(z)$, and $H_1(z)$, where

$$H_1(z) \triangleq \frac{z^{-1}}{1+0.4z^{-1}}.$$

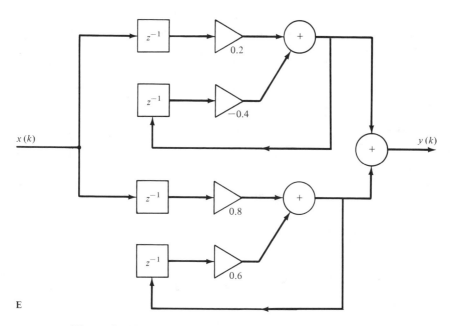

Figure 5.14E. Parallel 1 realization of $H(z)$ in Example 5.7.

Applying FWERR.FOR,

$$\frac{1}{2\pi j}\oint_{|z|=1} H_1(z)H_1(z^{-1})\frac{dz}{z} = 1.19048. \tag{5.49}$$

Therefore, using (5.45) and (5.49),

$$\sigma_{\hat{y}}^2 = (1.30568 + 2 \times 1.19048)\frac{Q^2}{12}$$

$$= 3.68664\frac{Q^2}{12}$$

Parallel 1. Rewriting $H(z)$ in (5.44) as

$$H(z) = \frac{0.2z^{-1}}{1+0.4z^{-1}} + \frac{0.8z^{-1}}{1-0.6z^{-1}}, \tag{5.50}$$

the realization in Figure 5.14E is obtained. Define

$$G_1(z) \triangleq \frac{1}{1+0.4z^{-1}}, \qquad G_2(z) \triangleq \frac{1}{1-0.6z^{-1}}.$$

Applying FWERR.FOR,

$$\frac{1}{2\pi j}\oint_{|z|=1} G_1(z)G_1(z^{-1})\frac{dz}{z} = 1.19048,$$

$$\frac{1}{2\pi j}\oint_{|z|=1} G_2(z)G_2(z^{-1})\frac{dz}{z} = 1.56250.$$

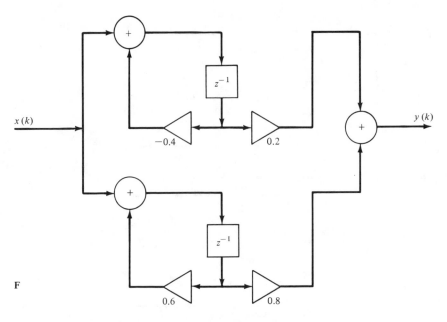

Figure 5.14F. Parallel 2 realization of $H(z)$ in Example 5.7.

Since each of the blocks connected in parallel (see Figure 5.14E) has 2 and not 5 multipliers,

$$\sigma_{\hat{y}_1}^2 = (2 \times 1.19048) \frac{Q^2}{12}$$

$$\sigma_{\hat{y}_2}^2 = (2 \times 1.56250) \frac{Q^2}{12}$$

$$\sigma_{\hat{y}}^2 = \sigma_{\hat{y}_1}^2 + \sigma_{\hat{y}_2}^2 = 5.50596 \frac{Q^2}{12} .$$

Parallel 2. The transfer function $H(z)$ in (5.50) can also be realized as in Figure 5.14F. Define

$$H_a(z) = \frac{0.2z^{-1}}{1 + 0.4z^{-1}}, \qquad H_b(z) = \frac{0.8z^{-1}}{1 - 0.6z^{-1}}$$

Applying FWERR.FOR,

$$\frac{1}{2\pi j} \oint_{|z|=1} H_a(z) H_a(z^{-1}) \frac{dz}{z} = 0.04762 \qquad (5.51)$$

$$\frac{1}{2\pi j} \oint_{|z|=1} H_b(z) H_b(z^{-1}) \frac{dz}{z} = 1. \qquad (5.52)$$

The transfer function between roundoff error noise sources associated with multipliers, -0.4, 0.2, 0.6, and 0.8 and the output are, respectively, $H_a(z)$, 1, $H_b(z)$, and 1. Therefore, using (5.46), (5.51, and (5.52),

$$\sigma_{\hat{y}}^2 = (0.04762 + 1 + 1 + 1)\frac{Q^2}{12}$$

$$= 3.04762 \frac{Q^2}{12}.$$

5.6. Limit Cycle Oscillations

Limit cycles could occur due to the nonlinearities inherent in the finite arithmetic operations in the *recursive* implementation of digital filters. A limit cycle or a periodic output sequence of length (or period) N is said to exist if the elements of the output sequence $\{y(k)\}$ satisfy

$$y(k + N) = y(k), \quad k > k_0,$$

where N is the smallest integer for which the above condition holds. This periodic output sequence might be a self-sustained zero-input response to an initial state in which the digital filter might find itself. It could also be the zero-state response to a constant input or even the response to a periodic input sequence. In the latter case, the limit cycle length is an integer multiple of the length of the periodic input sequence. The effects of the nonlinearities under the constraints of finite arithmetics are determined not only by the type of input sequence (unforced or forced) but also by the form and structure of the computational algorithm chosen to implement the digital filter. The two basically different types of limit cycles that occur in the implementation of digital filters with a feedback path under finite arithmetic conditions are:

1. the multiplication roundoff limit cycle, and
2. adder overflow limit cycle.

Limit cycles are not necessarily disadvantageous. For example, the limit cycle phenomenon could be used to advantage in the design of digital oscillators. However, in a great variety of applications like, for example, the processing of speech, the occurrence of limit cycle oscillations when the input samples to the filter are temporarily deactivated are highly undesirable. Limit cycles, by their very nature, are generated by quantization error sequences that are highly correlated. Therefore, studies on occurrence and bounds on the limit cycle amplitude are based on deterministic as opposed to stochastic approaches. Here, roundoff limit cycle oscillations only in fixed-point implementation of recursive digital filters are considered, since they may generally be neglected in realizations implemented via floating-point arithmetic. Subsequently, adder overflow limit cycles are described.

Table 5.1. Zero-Input Response for $y(k) = \alpha y(k-1)$
Due to Rounding with $y(-1) = 10$
and Various α

k	$y_r(k)$ ($\alpha = 0.8$)	$y_r(k)$ ($\alpha = -0.8$)	$y_r(k)$ ($\alpha = 0.9$)	$y_r(k)$ ($\alpha = -0.9$)
0	8	-8	9	-9
1	6	6	8	8
2	5	-5	7	-7
3	4	4	6	6
4	3	-3	5	-5
5	2	2	5	5
6	2	-2	5	-5
7	2	2	5	5
8	2	-2	5	-5

5.6.1. Multiplication Roundoff Limit Cycle Oscillations

This type of limit cycle describes the fact that a filter with all its poles within the unit circle might not have its output going to zero when the input is removed. To wit, consider the first-order filter, characterized by the difference equation

$$y(k) = \alpha y(k-1) + x(k), \qquad (5.53)$$

where $\{ y(k) \}$ and $\{ x(k) \}$ are, respectively, the output and input sequences. The zero-input response of the filter under infinite precision condition of implementation will go toward 0 as k increases, provided $|\alpha| < 1$. Under finite arithmetic conditions, for example, when each data value is rounded to the nearest integer, things may be quite different. The output sequence $\{ y_r(k) \}$ for a certain initial condition, $y(-1) = 10$, and zero input, when products are rounded, is shown in Table 5.1. The amplitude intervals within which the periodic outputs are confined are called *deadbands*. The limit cycles occurring within the deadbands are of two forms: constant magnitude and sign for positive α or constant magnitude with alternating sign for negative α. Within the deadband, the mutliplier α, then, has an effective value of ± 1 giving rise to an effective pole of the first-order filter at $z = \pm 1$.

Table 5.2 gives the output sequence, $\{ y_r(k) \}$, $k \geq 0$, due to rounding, when $y(-1) = 0$ and the causal input sequence $\{ x(k) \} = (10 \quad 0 \quad 0 \quad 0 \quad \cdots)$, for various values of α. Other nonzero input sequences, constant or nonconstant, can be similarly considered. Nonzero inputs occur in the study of limit cycles of higher-order filters realized as cascades of first-order (or second-order) digital filter sections. In the cascade type of realization, only the first section in the cascade is constrained to have zero input, when the input to the overall filter is set to zero.

Table 5.2. Zero-State Response for $y(k) = \alpha y(k-1) + x(k)$ Due to Rounding when $\{x(k)\} = (10 \quad 0 \quad 0 \quad 0 \quad \cdots)$ and Various α

k	$y_r(k)$ ($\alpha = 0.9$)	$y_r(k)$ ($\alpha = 0.7$)	$y_r(k)$ ($\alpha = 0.6$)	$y_r(k)$ ($\alpha = 0.5$)	$y_r(k)$ ($\alpha = -0.9$)
0	10	10	10	10	10
1	9	7	6	5	-9
2	8	5	4	3	8
3	7	4	2	2	-7
4	6	3	1	1	6
5	5	2	1	1	-5
6	5	1	1	1	5
7	5	1	1	1	-5
8	5	1	1	1	5

It is clear that the output in all the cases dealt with in Tables 5.1 and 5.2 is periodic beyond certain k's. In fact it is known that for a first-order filter described in (5.53), the "deadband" within which limit cycle oscillations can exist with zero input to the filter is the interval, $[-v, v]$ where v is the largest integer satisfying

$$v \le \frac{0.5}{1 - |\alpha|}.$$

The limit cycle oscillation is controllable via increase of the word length of the processor. When this proves to be more expensive than desired, a small "dither signal,"

$$d(k) = (-1)^k \beta,$$

where β is approximately half of the quantization interval is sometimes added with the objective of assisting the filter output to jump across the threshold of quantization. Then a small steady-state oscillation about the final desired output is attained.

Consider, for example, the addition of a dither signal

$$d(k) = (-1)^k 0.5$$

to the right-hand side of (5.53) with $x(k) = 0$, for all k. Consider, then, the solution of the following difference equation

$$y(k) = 0.9y(k-1) + (-1)^k 0.5 \tag{5.54}$$

after rounding to the nearest integer, given that $y(-1) = 10$.

Table 5.3 substantiates that the "deadband effect" is appreciably reduced with a dither signal.

Table 5.3. Solution to (5.54)
After Rounding to
Nearest Integer

k	$y_r(k)$	k	$y_r(k)$
0	10	15	2
1	9	16	2
2	9	17	1
3	8	18	1
4	8	19	0
5	7	20	1
6	7	21	0
7	6	22	1
8	6	23	0
9	5	24	1
10	5	25	0
11	4	26	1
12	4	27	0
13	3	28	1
14	3		

5.6.2. Adder Overflow Oscillations

A known cause of this phenomenon is due to the fact that the digital filter realization of the addition operation in the fixed-point arithmetic can cause overflow, thereby creating a severe nonlinearity. With adder overflow, the two binary numbers to a digital adder might total to another number too large for representation by the number of bits available. In 2's complement arithmetic, the carry bit which results to the left of the binary point would be incorrectly interpreted as a negative number, possibly causing severe distortion. These overflow oscillations could be of very large amplitude. The problem is overcome or reduced by increasing the word length or by using saturation arithmetic [8]. It may also be recalled that partial sums may be permitted to overflow as long as the final sum is less than unity in nonsaturating adders based on the 2's complement arithmetic.

To understand the phenomenon of overflow oscillations, consider the second-order digital filter in Figure 5.15, where the function $f(\cdot)$ accounts for adder or accumulator overflow. The input/output relationship of the digital filter is governed by the following non-autonomous nonlinear difference equation,

$$y(k) = f[-b_1 y(k-1) - b_2 y(k-2) + x(k)] + e(k), \qquad (5.55)$$

where $e(k)$ accounts for any multiplication roundoff error that may be present. The input/output relationship of the corresponding ideal filter having no adder overflow and no quantization error (due to either trunca-

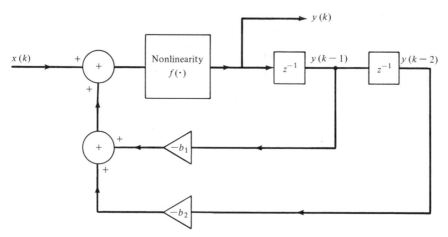

Figure 5.15. Second-order digital filter with adder overflow nonlinearity.

tion or rounding of the multiplier products) is:

$$y(k) = -b_1 y(k-1) - b_2 y(k-2) + x(k). \qquad (5.56)$$

Simple stability analysis reveals that for the ideal filter, stability is guaranteed if and only if the multiplier coefficients, $-b_1$ and $-b_2$ lie within an open triangular region of the parameter space (see Problem 9, Chapter 3). However, in a practical implementation of the digital filter, the nonlinearity modeling the adder overflow may be responsible for certain self-sustaining oscillations that may be observed even when the input is zero. The nature of the nonlinear function, $f(\cdot)$, depends upon the type of arithmetic used to actually implement the filter on a digital computer. Typical nonlinear characteristics of the adder are sketched in Figures 5.16A to C. Note that the magnitude of $f(\cdot)$ is always constrained to be less than or equal to unity. Ebert, Mazo and Taylor [8] showed that in the undriven case $[x(k) = 0$ in (5.55)] and with quantization error ignored $[e(k) = 0$ in (5.55)], a necessary and sufficient condition for absence of self-sustained oscillation, when the adder overflow nonlinearity is of the type shown in Figure 5.16A, is:

$$|b_1| + |b_2| < 1. \qquad (5.57)$$

The inequality in (5.57) may be justified by considering the two possibilities for oscillation in the undriven system governed by

$$y(k) = f\left[-b_1 y(k-1) - b_2 y(k-2)\right].$$

The period 1 oscillation occurs when,

$$y(k) \equiv y \text{ for all } k, \qquad (5.58a)$$

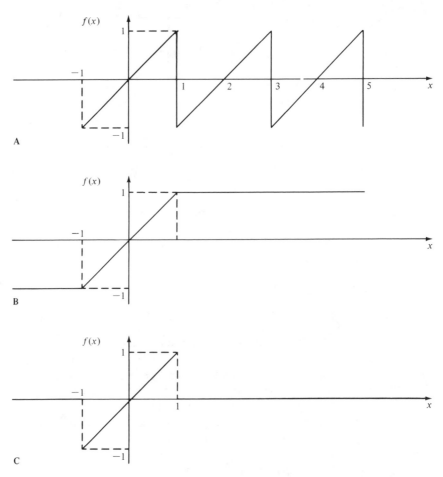

Figure 5.16. Various types of adder overflow nonlinearities: (A) 2's complement arithmetic, (B) saturation arithmetic and (C) zeroing arithmetic.

and the period 2 oscillation occurs when,

$$y(k) \equiv (-1)^k y \text{ for all } k, \tag{5.58b}$$

where y is a certain constant.

The conditions in (5.58a) and (5.58b) are avoidable using the nonlinearity in Figure 5.16A provided the inequality in (5.57) is satisfied. Since, this may be quite restrictive in design, it was further shown in [6] as well as in [8] that use of the saturation nonlinearity (Figure 5.16B) leads to stable operation whenever linear theory would predict it to be so. For a more general treatment of limit cycles due to adder overflow for various types of nonlinearities controlling the input/output description of (5.55), see [19].

The treatment of forced (nonzero input) adder overflow limit cycles is beyond the scope of this book. The interested reader may refer to [20] for information on that topic.

5.7. Conclusions and Additional Comments

The most important feature that distinguishes the theory of digital filters from analog filters is associated with the fact that digital quantities are only represented up to a finite and, generally, fixed precision. The effects of the imposition of this constraint of finite arithmetic are that the representation of all physical signals must be viewed as corrupted by noise, the coefficients of a designed filter can be represented only up to a certain accuracy, the results of arithmetic operations performed in the filter have usually to be rounded, giving rise to additional noise, and the operations on signals which are outside the normal dynamic range of representation could cause overflow problems, which introduce nonlinear characteristics in a digital filter that would operate linearly without the constraints of finite arithmetic. The importance of the topic and the impressive development of its underlying theory over the past several years motivate its inclusion in this book.

Arithmetic operations are performed after the representation of operands. In ordinary arithmetic stemming from human endeavor, the numbers are given the so-called fixed-point or positional decimal (radix 10) representation. The first American high-speed computers also used decimal arithmetic, but the radix 2 or binary representation is now most prevalent for performing computational chores with a processor. The radix point is placed between the integer part and the fractional part of a number and as the name implies, this point in the fixed-point or positional number notation remains fixed so that the programmer knows where the radix point is located in the numbers he manipulates. For greater dynamic range, there are advantages to letting the position of the radix point float or be a dynamical variable and to carry with each number an indication of the corresponding radix point position. Reference [9, Chapter 4] would be useful to the reader as a supplement to knowledge of the number systems, and this reference also provides the historical development of number representations. Oppenheim [5] proposed the block floating-point arithmetic, which has some of the advantages of both fixed-point and floating-point arithmetics.

Quantization of the input signal, necessitated by the analog to digital conversion, is equivalent to adding a noiselike term. Bennett [10] was the first researcher to document the effects of input quantization in a statistical manner. In order to carry out the analysis, it is necessary to assume that the additive noise is characterizable by very simple (i.e., white) statistics. The validity of the statistical approach has been justified by simulated experiments on the computer.

The problem of coefficient quantization can be analyzed using deterministic techniques as was done by Kaiser [3]. Though the adverse effects of
coefficient quantization could be instability or significant deterioration in
the frequency response, these are usually well understood and, therefore,
controllable a priori, with adequate exercise of caution. Generally, a rule of
thumb is used whereby three or four bits are added to the minimum number
calculated to ensure stability and normally this word length provides
sufficient accuracy to meet the frequency response specifications [11, pp.
134–135]. Also, the changes in locations of poles and zeros due to coefficient
rounding might be calculated and after assigning suitable sensitivity measures the frequency response variation can be assessed for different realizations. It has been concluded that either a cascade or a parallel combination
of first and second-order digital filters is almost always preferable in the
realization of higher-order filters than direct forms of realizations in order to
assure that small errors in the representation of coefficients do not result in
significant movements of pole (and/or zero) positions.

The realization of a given digital filter also affects the quantization noise
caused by multiplication roundoff errors. These errors, are nonlinear in
nature because roundoff is a function of the amplitude of the signal being
processed. To facilitate analysis, a noise source is associated with each
multiplier and certain assumptions of statistical independence are made.
The development closely parallels the original work of Knowles and Edwards
[12]. A simplified treatment is available in the text book by Ahmed and
Natarajan [13] and also in the paper by Jackson [14]. The computation of
the "square integral" figures conspicuously in both input quantization and
multiplication roundoff error analysis. A convenient method, which does
not require polynomial factorization and residue computation, for evaluating this generic integral is given. An example is included to illustrate the
procedure for calculating output roundoff error variance for different realizations. Though, not deducible from this worked-out example, the cascade
and parallel configurations usually yield better roundoff noise performance
than direct realizations.

Even when the input to a digital filter is zero, there may be limit cycle
oscillations caused by roundoff in the multipliers. The roundoff is a nonlinear operation, and even though the transfer function of the linear filter
(under infinite precision, without roundoff errors) might have all its poles
within the unit circle, this filter, after rounding effects are considered, may
not be asymptotically stable because a periodic output may be maintained
with zero input to the system. The occurrence of limit cycles is demonstrated by a first-order filter example. Jackson [15] investigated both first
and second-order filters for this type of limit cycle oscillations. The work of
Parker and Hess [16] is also relevant in this context. Various bounds exist
[17, 18] on limit cycles in fixed-point implementations of digital filters
(roundoff limit cycle is generally neglected for floating-point).

Oscillations in a digital filter also could occur due to adder overflow. Due to limited dynamic range in the fixed-point, 2's complement arithmetic, the addition of two numbers could result in a very undesirable overflow. The adder overflow limit cycles may be eliminated by replacing a 2's complement adder by what is referred to as the "saturating adder" [8].

Problems

1. It is stated that in the 1's complement binary fixed-point mode of representation, the negation of a 1's complement number can be accomplished by conjugating (replacing 1's by 0's and vice versa) all of the bits.
   a. Represent the decimal numbers 143 and -143, using the 1's complement scheme and a 16-bit (including a sign bit) fixed-point representation.
   b. What is the range of decimal numbers that can be represented using 16 bits (including a sign bit) in a 1's complement fixed-point mode?

2. It is stated that in the 2's complement binary fixed-point mode of representation, the negation of a 2's complement number can be accomplished by conjugating (see Problem 1 above) all of the bits, adding the binary representation of 2^{-b}, where $b+1$ is the number of bits including the sign bit and ignoring any overflow from the addition problem.
   a. Represent the decimal numbers 143 and -143, using the 2's complement scheme and a 16-bit (including a sign bit) fixed-point representation.
   b. What is the range of decimal numbers that can be represented using 16 bits (including a sign bit) in a 2's complement fixed-point mode?

3. Obtain the binary 2's complement fixed-point representation for the following decimal numbers using 7 data bits and a sign bit.
   a. 0.4375
   b. -0.4375
   c. 0.8515625
   d. -0.8515625

4. a. Truncate and round the fixed-point numbers in Problem 3 to 4 data bits.
   b. Compute the corresponding truncation and roundoff errors.

5. In a 32-bit binary floating-point representation scheme for numbers, 6 bits are used to store the characteristic or exponent, 24 bits are used to store the mantissa and 1 bit each is used for the sign bit of number and the sign bit for exponent. What is the range of decimal numbers that can be represented via this scheme? The exponent and mantissa may be coded in the sign-and-magnitude, 1's complement or 2's complement scheme, as in the fixed-point mode.

6. Given

$$H(z) = \frac{1 - 0.4z^{-1}}{1 - 0.9z^{-1} + 0.18z^{-2}},$$

compute

$$I_0 = \frac{1}{2\pi j}\oint_{|z|=1} H(z)H(z^{-1})\frac{dz}{z},$$

using the procedure outlined in Section 5.3.1, and check your result with that obtained by calculating residues.

7. An input sequence $\{x(k)\}$, $|x(k)| < 1$ for all k, is encoded in 2's complement binary fixed-point arithmetic using 4 bits, one of which is the sign bit. The quantized sequence is passed through a digital filter, characterized by the transfer function

$$H(z) = \frac{1-0.4z^{-1}}{1-0.9z^{-1}+0.18z^{-2}}.$$

   a. Find the steady-state output noise power due to the input quantization.
   b. It is required to reduce the output noise power computed in (a) by at least 50%. Find the number of bits necessary for the purpose.
   c. Calculate the output noise power if the number of bits used is 16.

8. Consider the transfer function

$$H(z) = \frac{1-0.4z^{-1}}{1-0.9z^{-1}+0.18z^{-2}}.$$

   a. Plot the magnitude of the frequency response for $H(z)$.
   b. Find the fixed-point representation of the coefficients by rounding to 4 bits. Calculate the pole locations of the resulting transfer function, and obtain its magnitude of frequency response.

9. The digital filter transfer function

$$H(z) = \frac{0.85z^{-2}}{(1-0.4z^{-1})(1-0.6z^{-1})}$$

   is realized in
   a. direct form 2
   b. parallel form
   c. cascaded forms as shown below.

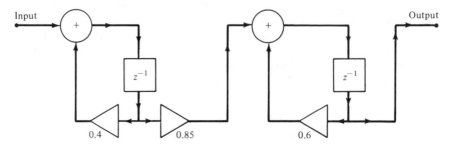

Cascade realization of transfer function.

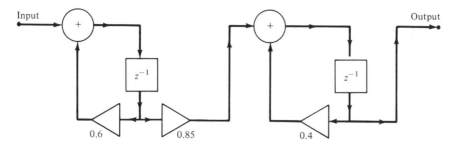

Alternate cascade realization of transfer function.

Calculate the output noise power due to product *roundoff* error in each case.

10. A recursive first-order digital filter is represented by the following linear difference equation:

$$y(k) = x(k) - 0.95y(k-1).$$

Assume that the filter output $y(k)$ is rounded to the nearest integer. Let $y(-1) = 0$. Calculate the output sequence $\{y(k)\}_{k=0}^{100}$ assuming that the input sequence, $\{x(k)\}_{k=0}^{\infty}$ is $\{10,0,0,\ldots,0\}$. For what range of the initial condition, $y(-1)$, do you expect the deadband effect?

11. Consider the direct form 2 realization of the first-order transfer function

$$H(z) = \frac{1 + 0.625z^{-1}}{1 - 0.825z^{-1}}.$$

Given that a reasonable value of the scalar multiplier s_0 (see below) to avoid

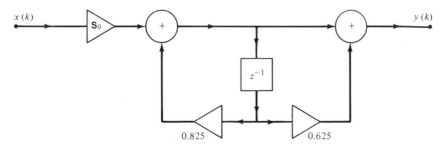

Realization of transfer function with input scaling.

overflow in the input adder is given by

$$s_0^2 = \frac{1}{J}, \quad J = \frac{1}{2\pi j} \oint_{|z|=1} \frac{z^{-1}\,dz}{B(z)B(z^{-1})},$$

where $B(z)$ is the denominator expression of $H(z)$, calculate the value of s_0 in this problem.

References

1. Dugre, J. P., Beex, A. A. L., and Scharf, L. L. 1980. Generating covariance sequences and the calculation of quantization and roundoff error variances in digital filters. IEEE Trans. Acoustics, Speech Signal Process., 28:102–104.

2. Dugre, J. P., and Jury, E. I. 1982. A note on the evaluation of complex integrals using filtering interpretations. IEEE Trans. Acoustics, Speech Signal Process., 30:804–807.

3. Kaiser, J. F. 1965. Some practical considerations in the realization of linear digital filters. Proc. 3rd Annual Allerton Conf. Circuit System Theory, pp. 621–633.

4. Knowles, J. B., and Olcayto, E. M. 1968. Coefficient accuracy and digital filter response. IEEE Trans. Circuit Theory, 15:1197–1207.

5. Oppenheim, A. V. 1970. Realization of digital filters using block floating-point arithmetic. IEEE Trans. Audio Electroacoustics, 18:130–136.

6. Sandberg, I. W. 1967. Floating-point roundoff accumulation in digital filter realization. Bell Sys. Tech. J., 46:1775–1791.

7. Liu, B., and Kaneko, T. 1969. Error analysis of digital filters realized with floating-point arithmetic. Proc. IEEE, 57:1735–1747.

8. Ebert, P. M., Mazo, J. E., and Taylor, M. G. 1967. Overflow oscillations in digital filters. Bell Sys. Tech. J., 46:1775–1791.

9. Knuth, D. E. 1969. The Art of Computer Programming: Seminumerical Algorithms. Vol. 2. Addison Wesley, Reading, MA.

10. Bennett, W. R. 1948. Spectra of quantized signals. Bell Sys. Tech. J., 27:446–472.

11. Terrell, T. J. 1980. Introduction to Digital Filters. Macmillan Press, London.

12. Knowles, J. B., and Edwards, R. 1965. Effects of finite-word-length computer in a sampled data feedback system. Proc. Inst. Elec. Eng., 112:1197–1207.

13. Ahmed, N., and Natarajan, T. 1983. Discrete-Time Signals and Systems. Reston Publishing Co., Reston, VA.

14. Jackson, L. B. 1970. On the interaction of roundoff noise and dynamic range in digital filters. Bell Sys. Tech. J., 49:159–184.

15. Jackson, L. B. 1969. An analysis of limit cycles due to multiplication rounding in recursive digital (sub) filters. Proc. 7th Annual Allerton Conf. Circuit System Theory, pp. 69–78.

16. Parker, S. R., and Hess, S. F. 1971. Limit cycle oscillations in digital filters. IEEE Trans. Circuit Theory, 18:687–697.

17. Sandberg, I. W., and Kaiser, J. F. 1972. A bound on limit cycles in fixed-point implementation of digital filters. IEEE Trans. Audio Electroacoustics, 20:110–112.

18. Long, L. J., and Trick, T. N. 1973. An absolute bound on limit cycles due to roundoff error in digital filters. IEEE Trans. Audio Electroacoustics, 21:27–30.

19. Willson Jr., A. N. 1972. Limit cycles due to adder overflow in digital filters. IEEE Trans. Circuit Theory, 19:342–346.

20. Samueli, H., and Willson Jr., A. N. 1982. Almost period p sequences and the analysis of forced overflow oscillations in digital filters. 29:510–515.

Chapter 6
Structure and Properties of Matrices in Various Digital Filtering Problems

6.1. Introduction

This chapter is concerned with selected applications, where the structural properties of characterizing matrices are exploited to develop and implement fast algorithms for solving the problems of interest. Some of these problems require the solution of a system of linear equations where the coefficient matrix is a Toeplitz matrix or a Hankel (also referred to as orthosymmetric) matrix. In these matrices the elements exhibit certain interdependence, which in turn provide certain structures to the matrices, thereby enabling matrix operations of interest like inversion and triangular decomposition, (which lead to solutions of the given systems of linear equations) to be implemented faster than would be the case with general matrices of the same order. Often it is required to compute a matrix vector product, where the matrix is endowed with a certain structure. It was seen in Section 2.5.2, for example, that the cyclic convolution of two finite sequences may be obtained by computing a matrix vector product, where the matrix is a circulant.

In Section 6.2 the important problem of stabilization of unstable digital filters is introduced. It is important that this be attained without appreciable change in frequency response of the original filter. In principle, this is possible to achieve through the use of allpass functions, as discussed in Section 6.2.1. This approach leads to a simple stabilization technique which only requires an accurate root-finder. A conceptually richer approach is based on the planar least-squares inverse (PLSI) technique, introduced in Section 6.2.2. The application of this technique to the stabilization problem is shown to require the solution of a system of linear equations, whose characterizing matrix is symmetric Toeplitz.

In Section 6.3 the problem of approximating a formal power series expansion by a rational function having numerator and denominator polynomials of specified degrees r and m, respectively, via the Padé technique is considered. This is shown to reduce to the problem of solving a system of linear equations, where the characterizing matrix possesses a Hankel structure. A three-term recurrence relation, useful for computing Padé approximants, is derived after exploiting the Hankel structure of the coefficient matrix. Links with the theory of polynomials, orthogonal over the real line, are briefly pointed out.

The technique for stabilization of unstable digital filters via the discrete Hilbert transform (DHT) is considered in Section 6.4. The DHT was introduced in Section 2.4, and in the computation of the DHT via the matrix approach it was seen that a matrix vector product need be computed where the matrix, a special case of Toeplitz matrix, is a circulant.

In Section 6.5 concluding remarks are made and the attention of the interested reader is directed to sources that provide additional information on the topics covered in this chapter. The problems at the end of the chapter should serve to reinforce in the reader's mind some of the material covered.

6.2. Stabilization of Unstable Digital Filters

An important property to be attained before the completion of the design of an IIR filter is stability. Suppose that after a preliminary design, the IIR filter is found to be unstable. In many applications an engineer might be called upon to stabilize the unstable filter without appreciable change in the magnitude of the frequency response.

6.2.1. Stabilization via Allpass Functions

Consider the rational function

$$H_{1k}(z) = \frac{z^{-1} + a_k}{1 + z^{-1}a_k^*}, \tag{6.1}$$

where a_k is a complex constant of magnitude, $|a_k| < 1$, so that the rational function has a pole at $z = (-a_k)^*$, the complex conjugate of $(-a_k)$, and this pole is within the unit circle, defined by $|z| < 1$. $H_{1k}(z)$ has one zero at $z = -(1/a_k)$, and the location of this zero is often referred to as being "*reciprocal*" to the location of the pole with respect to the unit circle, $|z| = 1$. Evaluate the magnitude of $H_{1k}(z)$ when z takes an arbitrary value, $z = e^{j\omega}$, ω real, on the unit circle, $|z| = 1$.

$$|H_{1k}(e^{j\omega})| = \left| \frac{a_k + e^{-j\omega}}{1 + e^{-j\omega}a_k^*} \right|$$

$$= \left| \frac{(1 + a_k e^{j\omega})e^{-j\omega}}{[1 + (a_k e^{j\omega})^*]} \right| \tag{6.2}$$

$$= 1.$$

Similarly, for the rational function,

$$\hat{H}_{1k}(z) = \frac{z^{-1} + a_k^*}{1 + z^{-1}a_k},\qquad(6.3)$$

$$|\hat{H}_{1k}(e^{j\omega})| = 1.\qquad(6.4)$$

Both $H_{1k}(z)$ and $\hat{H}_{1k}(z)$ are first-order functions with complex coefficients. Next, consider the second-order rational function with real coefficients.

$$\begin{aligned}H_{2k}(z) &= H_{1k}(z)\hat{H}_{1k}(z)\\ &= \frac{z^{-2} + (2\,\mathrm{Re}\,a_k)z^{-1} + |a_k|^2}{|a_k|^2 z^{-2} + (2\,\mathrm{Re}\,a_k)z^{-1} + 1}.\end{aligned}\qquad(6.5)$$

Since $H_{1k}(z)$ and $\hat{H}_{1k}(z)$, respectively, satisfy (6.2) and (6.4), it follows that

$$\left|H_{2k}(e^{j\omega})\right| = 1.\qquad(6.6)$$

Note that the zeros of $H_{2k}(z)$ are at $z = -(1/a_k)$ and at $z = -(1/a_k^*)$, while its poles are at $z = -a_k$ and $z = -a_k^*$. Again, the zeros of $H_{2k}(z)$ are conjugate reciprocals to the poles of $H_{2k}(z)$ with respect to the unit circle $|z| = 1$. $H_{1k}(z)$, $\hat{H}_{1k}(z)$, and $H_{2k}(z)$ are special cases of stable digital allpass functions, formally defined next. (Although an allpass function was encountered in Section 3.5.1 and in Problem 1 at the end of Chapter 3, this is the right place to define it formally.)

Definition 6.1. A rational function $H(z)$ in the complex variable z will be called a stable digital allpass function, provided $|H(e^{j\omega})| = 1$ for $-\pi \le \omega \le \pi$ (and therefore for all real values of ω because of periodicity) and all its poles lie within the unit circle, $|z| < 1$. (In the present context, no confusion will occur if the qualifying words "stable digital" are omitted and this will henceforth be done for the sake of brevity.)

The preceding discussion may be generalized to the following result, which is trivial to prove.

Fact 6.1. *The product of several allpass functions is also an allpass function. Consider next the following $2n$th order functions.*

$$H_1(z) = \prod_{k=1}^{2n} \frac{z^{-1} + a_k}{1 + z^{-1}a_k^*},\qquad |a_k| < 1,\quad k = 1,2,\ldots,2n;\qquad(6.7)$$

$$H_2(z) = \prod_{k=1}^{n} \frac{(z^{-1} + a_k)(z^{-1} + a_k^*)}{(1 + z^{-1}a_k^*)(z^{-1}a_k + 1)},\qquad |a_k| < 1,\quad k = 1,2,\ldots,n.$$

$$\qquad(6.8)$$

$H_1(z)$ has complex coefficients and all coefficients of $H_2(z)$ are real. On applying Fact 6.1, and noting that a typical term in each of the above

products is an allpass function, it follows that $H_1(z)$ *and* $H_2(z)$ *are allpass functions.*

Consider next an unstable digital filter characterized by the transfer function

$$\hat{H}(z) = H_s(z) \prod_{k=1}^{m} \frac{1}{z^{-1} + b_k}, \qquad (6.9)$$

where $H_s(z)$ has all its poles in $|z| < 1$ and $|b_k| < 1$, $k = 1, \ldots, m$. Thus, the poles at $z = -(1/b_k)$, $k = 1, \ldots, m$ are unstable. The function $H(z)$ is formed after multiplying $\hat{H}(z)$ by an allpass function:

$$H(z) = \hat{H}(z) \prod_{k=1}^{m} \frac{z^{-1} + b_k}{1 + z^{-1}b_k^*}, \qquad |b_k| < 1, \quad k = 1, \ldots, m. \quad (6.10)$$

Note that $H(z)$, in reduced form (i.e., with relatively prime numerator and denominator after exact cancellation of the unstable poles by the numerator $\prod_{k=1}^{m}(z^{-1} + b_k)$, of the allpass function), is stable, is of the same order as $\hat{H}(z)$ and, importantly,

$$|H(e^{j\omega})| = |\hat{H}(e^{j\omega})|$$

for all values of the real variable, ω. Therefore, we conclude that an unstable digital filter can be stabilized without altering its magnitude response by replacing the poles outside the unit circle by their respective conjugate reciprocals. Note, however, that if $\hat{H}(z)$ has a pole on $|z| = 1$, that is, $|b_k| = 1$ for some k in (6.9), then its conjugate reciprocal will also be on $|z| = 1$ and therefore $\hat{H}(z)$ cannot be stabilized by this approach. Otherwise, the technique is very simple and is convenient to apply provided an accurate "root-finder" is available for locating the poles of the filter transfer function. The procedure for stabilization in Section 6.2.2 is conceptually interesting; the underlying mathematical technique also finds other diverse applications, some of which will be cited later.

6.2.2. Stabilization via the Planar Least-Squares Inverse (PLSI) Technique

Consider a specified polynomial in z^{-1},

$$B(z^{-1}) = \sum_{k=0}^{n} b_k z^{-k}, \qquad (6.11)$$

which may be viewed as the z-transform of a causal sequence, $(b_0 \ b_1 \ \cdots \ b_{n-1} \ b_n)$ of finite length $(n+1)$. Attend to the problem of finding another polynomial $A(z^{-1})$ of finite specified degree m,

$$A(z^{-1}) = \sum_{k=0}^{m} a_k z^{-k}, \qquad (6.12)$$

such that if

$$B(z^{-1})A(z^{-1}) = \sum_{k=0}^{m+n} c_k z^{-k}, \qquad (6.13)$$

then the coefficients a_k, $k = 0, 1, \ldots, m$ are chosen to minimize the cost function

$$J \triangleq \left(\sum_{k=1}^{m+n} c_k^2 \right) + (1 - c_0)^2$$

$$= 1 - 2c_0 + \sum_{k=0}^{m+n} c_k^2. \qquad (6.14)$$

When the coefficients a_k, $k = 0, 1, \ldots, m$ are calculated so that J is minimized, $A(z^{-1})$ is referred to as the planar least-squares inverse of $B(z^{-1})$, since the cost function is the sum of the squares of the difference between each element in the sequence $(c_0 \quad c_1 \quad \cdots \quad c_{m+n})$ and the corresponding element in the desired ideal sequence $(1 \quad 0 \quad 0 \quad \cdots \quad 0 \quad 0)$ associated with the exact inverse, $\hat{A}(z^{-1})$, defined by

$$B(z^{-1})\hat{A}(z^{-1}) \triangleq 1.$$

In order that J is minimized, the following set of equalities must hold:

$$\frac{\partial J}{\partial a_k} = 0, \quad k = 0, 1, \ldots, m. \qquad (6.15)$$

Calculate the partial derivative in (6.15) after substituting

$$c_0 = a_0 b_0 \qquad (6.16a)$$

$$c_k = \sum_{r=0}^{m} a_r b_{(k-r)}, \quad k = 1, 2, \ldots, m+n \qquad (6.16b)$$

in (6.14); $b_r = 0$, $r < 0$, or $r > n$. (Equations (6.16a) and (6.16b) are obtained after the substitution of (6.11) and (6.12) in (6.13) and then equating the coefficients of respective powers of z^{-1} on both sides of the equation.) Then

$$\frac{\partial J}{\partial a_0} = -2b_0 + \sum_{k=0}^{m+n} 2 \left[\sum_{r=0}^{m} a_r b_{(k-r)} \right] b_k = 0$$

$$\frac{\partial J}{\partial a_i} = \sum_{k=0}^{m+n} 2 \left[\sum_{r=0}^{m} a_r b_{(k-r)} \right] b_{(k-i)} = 0, \quad i = 1, \ldots, m.$$

After defining

$$q_{-r+s} \triangleq \sum_{k=0}^{m+n} b_{(k-r)} b_{(k-s)}, \qquad (6.17)$$

for $s = 0, 1, \ldots, m$ with $b_i = 0$, $i < 0$ or $i > n$, the preceding set of equations, following the interchange in the order of summations, can be written as

$$
\begin{bmatrix}
q_0 & q_{-1} & q_{-2} & \cdots & q_{-m} \\
q_1 & q_0 & q_{-1} & \cdots & q_{-m+1} \\
q_2 & q_1 & q_0 & \cdots & q_{-m+2} \\
\vdots & \vdots & \vdots & \vdots & \vdots \\
q_m & q_{m-1} & q_{m-2} & \cdots & q_0
\end{bmatrix}
\begin{bmatrix}
a_0 \\ a_1 \\ a_2 \\ \vdots \\ a_m
\end{bmatrix}
=
\begin{bmatrix}
b_0 \\ 0 \\ 0 \\ \vdots \\ 0
\end{bmatrix}. \qquad (6.18)
$$

The square coefficient matrix of order $(m+1)$ in (6.18) is a Toeplitz matrix, which is formally defined next.

Definition 6.2. A matrix $Q = [q_{ij}]$, where q_{ij} is an element in the ith row and jth column, $1 \le i,\ j \le (m+1)$, is a Toeplitz matrix, provided it has only $(2m+1)$ independent elements, defined by

$$
q_{ij} = q_{(i-j)}, \qquad 1 \le i,\ j \le (m+1)
$$

(or $q_{ij} = q_{-j+i+1}$ is equally satisfactory). The matrix Q is a symmetric Toeplitz matrix, provided

$$
q_{ij} = q_{|i-j|}, \qquad 1 \le i,\ j \le m+1.
$$

Note that a Toeplitz matrix need not be symmetric; however, it is symmetric about its secondary diagonal, that is, the diagonal from the top right-hand corner to the bottom left-hand corner. Also, in the present context, from (6.17)

$$
q_{-r} = q_r, \qquad r = 0,1,\ldots, m, \qquad (6.19)
$$

and therefore the square coefficient matrix in (6.18) is a symmetric Toeplitz matrix.

In the design of digital filters and in the construction of rational approximants via the Padé technique, for example, the coefficient b_0 in (6.11) is often normalized to 1. In the PLSI $A(z^{-1})$ sought, if the coefficient a_0 is also normalized to 1, the counterparts of (6.14) and (6.15) become, respectively,

$$
J = \sum_{k=1}^{m+n} c_k^2, \qquad c_0 = 1, \qquad (6.20a)
$$

$$
\frac{\partial J}{\partial a_k} = 0, \qquad k = 1,2,\ldots, m, \qquad (6.20b)
$$

and (6.18) is replaced by (6.21):

$$
\begin{bmatrix}
q_0 & q_{-1} & q_{-2} & \cdots & q_{-m+1} \\
q_1 & q_0 & q_{-1} & \cdots & q_{-m+2} \\
\vdots & \vdots & \vdots & \vdots & \vdots \\
q_{m-1} & q_{m-2} & q_{m-3} & \cdots & q_0
\end{bmatrix}
\begin{bmatrix}
a_1 \\ a_2 \\ \vdots \\ a_m
\end{bmatrix}
= -
\begin{bmatrix}
q_1 \\ q_2 \\ \vdots \\ q_m
\end{bmatrix}. \qquad (6.21)
$$

Therefore, to obtain the PLSI of $B(z^{-1})$ (with either a_0 unspecified or a_0 set to unity), remembering that from (6.19) $\{q_r\}$ must have an even symmetry about $r = 0$, it is necessary to solve a system of linear equations of the form

$$Qa = g, \qquad (6.22)$$

where Q is a symmetric Toeplitz matrix, a is the unknown vector to be computed, while g is a specified vector. Any nonsingular square matrix of order m can be inverted using $0(m^3)$ operations via Gaussian elimination. However, the Levinson algorithm [1] exploits the Toeplitz structure of the matrix to solve for the unknown vector, using only $0(m^2)$ operations. The improved computational time complexity is also accompanied by improved spatial complexity interpreted in terms of reduced storage. These benefits become especially significant when the matrix Q is of high order and when particular applications require repeated use of the algorithm for solution.

Figure 6.1 gives the flowchart for using the Levinson recursion to solve a system of linear equations, characterized in (6.22), when Q is a symmetric Toeplitz matrix of order $(m+1)$ as in (6.18), $a = (a_0 \quad a_1 \quad \cdots \quad a_m)^t$ is the unknown vector to be computed and g is any specified column matrix having $(m+1)$ entries. The algorithm is simple to program and a listing is provided in Figure 6.2 under a program name LEVINS.FOR. For calculating a PLSI of a specified $B(z^{-1})$, it is necessary to compute the q_i's defined in (6.17), and LEVINS.FOR is particularly geared to solve the PLSI problem represented in (6.18). Variants of this particular characterization can be easily accommodated by trivial changes in the program.

Before discussing the important properties of the solution obtained, the mechanics of the algorithm given in the flowchart of Figure 6.1 will be briefly illustrated by a nontrivial example. You are also invited to consult Problems 12 and 13 at the end of this chapter for a summary of some of the principal results in references [1] and [19].

Example 6.1. It is required to find the PLSI $A(z^{-1})$, with $m = 3$ in (6.12), of

$$B(z^{-1}) = 6 + 7z^{-1} + 11z^{-2} + 2z^{-3}.$$

Here,

$$b_0 = 6, \ b_1 = 7, \ b_2 = 11, \ b_3 = 2, \ b_i = 0, \ i > 3, \ i < 0$$

$$q_k = \sum_{i=0}^{3} b_{i+k}b_i = q_{-k}, \ k = 0,1,2,3$$

$$q_0 = b_0^2 + b_1^2 + b_2^2 + b_3^2 = 210$$

$$q_1 = b_1 b_0 + b_2 b_1 + b_3 b_2 = 141$$

$$q_2 = b_2 b_0 + b_3 b_1 = 80$$

$$q_3 = b_3 b_0 = 12.$$

362

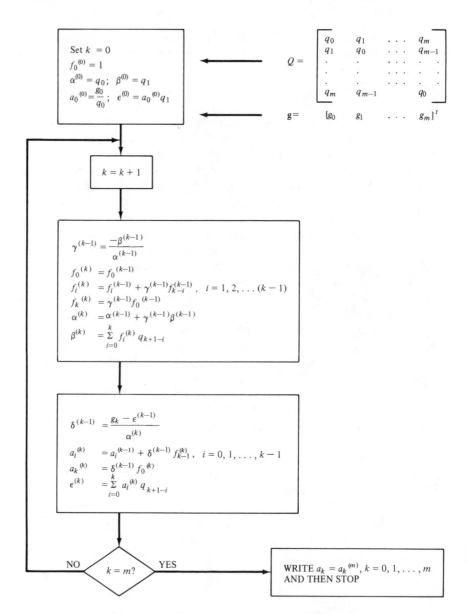

Figure 6.1. Flowchart for Levinson's algorithm.

```
C
C
C      ****************************************************************
C
C      A FORTRAN PROGRAM (LEVINS.FOR) TO IMPLEMENT THE LEVINSON'S
C      ALGORITHM TO SOLVE THE TOEPLITZ SYSTEM OF LINEAR EQUATIONS
C      IN 1-D SYSTEM.
C
C      MAXIMUM DIMENSION FOR THE TOEPLITZ MATRIX IS 100*100,
C      THAT IS, MAXIMUM OF M=100.
C      IF LARGER THAN 100*100 DIMENSION IS REQUIRED, ALL THE REAL
C      ARRAYS SHOULD HAVE DIMENSION EXTENDED.
C
C      ****************************************************************
C
       REAL F(0:99),FF(0:99),A(0:99),G(0:99),Q(0:99)
C
C      GIVE THE DIMENSION OF THE TOEPLITZ MATRIX
C
       WRITE(6,10)
 10    FORMAT(4X,'ENTER THE DIMENSION OF TOEPLITZ MATRIX, M')
       READ(5,20) M
 20    FORMAT(I)
       M1=M-1
C
C      GIVE THE DATA FOR Q(I) & G(I)
C
       WRITE(6,30)
 30    FORMAT(//,4X,'ENTER VALUES FOR Q(I), I=0,1,...,M-1')
       READ(5,40) (Q(I),I=0,M1)
 40    FORMAT(F)
       WRITE(6,50)
 50    FORMAT(//,4X,'ENTER VALUES FOR G(I), I=0,1,...,M-1')
       READ(5,40) (G(I),I=0,M1)
C
C      SET THE INITIAL CONDITIONS
C
       K=0
       F(0)=1.
       ALPHA=Q(0)
       BETA=Q(1)
       A(0)=G(0)/Q(0)
       EPSILN=A(0)*Q(1)
C
 100   CONTINUE
C
       DO 110 I=0,K
       FF(I)=F(I)
 110   CONTINUE
```

Figure 6.2. (*Partial listing, continued.*)

364

```
C
C        COMPUTE THE "K"TH VALUE FOR "GAMMA"
C
         GAMMA=-BETA/ALPHA
C
C        COMPUTE THE "K+1"TH VALUES FOR "F(I)"
C
         F(0)=FF(0)
         F(K+1)=GAMMA*FF(0)
C
         IF(K.EQ.0) GO TO 300
         DO 200 I=1,K
200      F(I)=FF(I)+GAMMA*FF(K+1-I)
C
C        COMPUTE THE "K+1"TH VALUES FOR "ALPHA" & "BETA"
C
300      ALPHA=ALPHA+GAMMA*BETA
         BETA=0.
         DO 400 I=0,K+1
400      BETA=BETA+F(I)*Q(K+2-I)
C
C        COMPUTE THE "K"TH VALUE FOR "DELTA"
C
         DELTA=(G(K+1)-EPSILN)/ALPHA
C
C        COMPUTE THE "K+1"TH VALUES FOR "A(I)"
C
         DO 500 I=0,K
500      A(I)=A(I)+DELTA*F(K+1-I)
         A(K+1)=DELTA*F(0)
C
C        COMPUTE THE "K+1"TH VALUE OF "EPSILN"
C
         EPSILN=0.
         DO 600 I=0,K+1
600      EPSILN=EPSILN+A(I)*Q(K+2-I)
C
         IF(K.EQ.M1-1) GO TO 700
         K=K+1
         GO TO 100
C
700      WRITE(6,800)
800      FORMAT(///,4X,'SOLUTIONS ARE')
         WRITE(6,810) ((I,A(I)),I=0,M1)
810      FORMAT(4X,'A(',I2,')=',E17.8)
         STOP
         END
```

Figure 6.2. Listing of computer program LEVINS.FOR to solve a system of linear equations, characterized by a symmetric Toeplitz matrix.

Therefore, for this problem, (6.18) becomes

$$
\begin{bmatrix}
210 & 141 & 80 & 12 \\
141 & 210 & 141 & 80 \\
80 & 141 & 210 & 141 \\
12 & 80 & 141 & 210
\end{bmatrix}
\begin{bmatrix}
a_0 \\ a_1 \\ a_2 \\ a_3
\end{bmatrix}
=
\begin{bmatrix}
6 \\ 0 \\ 0 \\ 0
\end{bmatrix}.
$$

The next objective is to solve the above system of linear equations by applying Levinson's algorithm.

$k = 0$ (Initial Conditions)

$$f_0^{(0)} = 1, \quad \alpha_0 = q_0 = 210, \quad \beta_0 = q_1 = 141$$

$$a_0^{(0)} = \frac{g_0}{q_0} = \frac{6}{210} \quad \text{as } g_0 = b_0 = 6$$

$$\varepsilon^{(0)} = a_0^{(0)} q_1 = \frac{141}{35} \simeq 4.0286.$$

(For brevity, the numerical calculations will be implemented with sufficient, but not exact, accuracy.)

$k = 1$

$$\gamma^{(0)} = \frac{-\beta^{(0)}}{\alpha^{(0)}} = -0.6714$$

$$f_0^{(1)} = f_0^{(0)} = 1$$

$$f_1^{(1)} = +\gamma^{(0)} f_0^{(0)} = -0.6714$$

$$\alpha^{(1)} = \alpha^{(0)} - \frac{\beta^{(0)}}{\alpha^{(0)}} \beta^{(0)} = 115.33$$

$$\beta^{(1)} = f_0^{(1)} q_2 + f_1^{(1)} q_1 = -14.667$$

$$a_0^{(1)} = a_0^{(0)} + \frac{g_1 - \varepsilon^{(0)}}{\alpha^{(1)}} f_1^{(1)} = 0.052$$

$$a_1^{(1)} = \frac{g_1 - \varepsilon^{(0)}}{\alpha^{(1)}} f_0^{(1)} = -0.03493$$

$$\varepsilon^{(1)} = a_0^{(1)} q_2 + a_1^{(1)} q_1 = -0.7644.$$

The reader is invited to verify the subsequent calculations.

$k = 2$

$$f_0^{(2)} = 1.0, \qquad f_1^{(2)} = -0.7568, \qquad f_2^{(2)} = 0.1272,$$

$$\alpha^{(2)} = 113.465, \qquad \beta^{(2)} = -30.611,$$

$$a_0^{(2)} = 0.05286, \qquad a_1^{(2)} = -0.04, \qquad a_2^{(2)} = 0.00674,$$

$$\varepsilon^{(2)} = -1.615.$$

$k = 3$

$$f_0^{(3)} = 1.0, \qquad\qquad f_1^{(3)} = -0.7225,$$

$$f_2^{(3)} = -0.0769, \qquad\quad f_3^{(3)} = 0.2698,$$

$$\alpha^{(3)} = 105.2, \qquad\qquad \beta^{(3)} = 23.22,$$

$$a_0^{(3)} = 0.0570, \qquad\qquad a_1^{(3)} = -0.0412,$$

$$a_2^{(3)} = -0.0044, \qquad\quad a_3^{(3)} = 0.0154,$$

$$\varepsilon^{(3)} = 1.322.$$

Required PLSI (up to the numerical accuracy used) is $A(z^{-1}) = 0.0570 - 0.0412z^{-1} - 0.0044z^{-2} + 0.0154z^{-3}$.

Since the above example introduces the reader to the manner in which Levinson's recursion is implemented, in the next example only the problem formulation and the final answer arrived at is reported. In order to test his comprehension, the reader is advised to apply the algorithm in order to arrive at this answer.

Example 6.2. Given that

$$B(z^{-1}) = 6 + 17z^{-1} + 11z^{-2} + 2z^{-3},$$

it is required, first, to find a PLSI,

$$A(z^{-1}) = a_0 + a_1 z^{-1} + a_2 z^{-2}.$$

The counterpart of (6.18) for this problem is

$$\begin{bmatrix} 450 & 311 & 100 \\ 311 & 450 & 311 \\ 100 & 311 & 450 \end{bmatrix} \begin{bmatrix} a_0 \\ a_1 \\ a_2 \end{bmatrix} = \begin{bmatrix} 6 \\ 0 \\ 0 \end{bmatrix}.$$

Then, using LEVINS.FOR, the required PLSI is

$$A(z^{-1}) = 0.0335 - 0.0345z^{-1} + 0.0164z^{-2}.$$

For the sake of brevity, the coefficients have been truncated. Finally, obtain the PLSI

$$\hat{A}(z^{-1}) = \hat{a}_0 + \hat{a}_1 z^{-1} + \hat{a}_2 z^{-2}$$

of $A(z^{-1})$ arrived at above.

Again after routine manipulation, (6.18) yields

$$\begin{bmatrix} 0.00260 & -0.0017 & 0.00055 \\ -0.00170 & 0.0026 & -0.0017 \\ 0.00055 & -0.0017 & 0.0026 \end{bmatrix} \begin{bmatrix} \hat{a}_0 \\ \hat{a}_1 \\ \hat{a}_2 \end{bmatrix} = \begin{bmatrix} 0.0335 \\ 0 \\ 0 \end{bmatrix}.$$

On applying again, LEVINS.FOR, the *double PLSI* (DPLSI), $\hat{A}(z^{-1})$, of $B(z^{-1})$ is

$$\hat{A}(z^{-1}) = 26.24 + 23.63z^{-1} + 9.90z^{-2}.$$

The Levinson recursion shown in Figure 6.1 for the case when Q is symmetric is known to be the same as certain well-known recursions for the so-called Szegö orthogonal polynomials. In fact, if one associates a polynomial

$$C^{(i)}(z) \triangleq z^i \sum_{k=0}^{i} a_k^{(i)} z^{-k}, \quad i = 0,1,2,\ldots \tag{6.23}$$

with the sequence of coefficients $\{a_k^{(i)}\}_{k=0}^i$ at the ith step, then it can be shown that the set of polynomials $\{C^{(i)}(z)\}$ form an orthogonal set over the unit circle, $|z| = 1$. The recurrence relation for these polynomials is

$$C^{(i)}(z) = C^{(i-1)}(z) + \delta^{(i-1)} z^i D^{(i-1)}(z^{-1}), \tag{6.24}$$

where

$$D^{(i)}(z) = z^i \sum_{k=0}^{i} f_k^{(i)} z^{-k}, \tag{6.25}$$

$\{f_k^{(i)}\}_{k=0}^i$ being the sequence of auxiliary coefficients (see Figure 6.1) in the kth step. The sequence of polynomials $\{D^{(i)}(z)\}$ also satisfies the following three-term recurrence relation:

$$D^{(i)}(z) = D^{(i-1)}(z) + \gamma^{(i-1)} z^i D^{(i-1)}(z^{-1}). \tag{6.26}$$

If the matrix Q is symmetric and positive definite (which is the case, since Q can be regarded as the covariance matrix of a segment $\{b_0 \quad b_1 \quad \cdots \quad b_n\}$ of a discrete-time stationary process), then *all the roots of $C^{(i)}(z)$ lie inside the unit circle.* This fact is verifiable for the PLSI polynomials in the examples given so far. The mathematical developments leading up to this important theoretical result are outside the scope of this book. The interested reader is referred to the works of Szegö [11] and Akhiezer [18].

Other Applications. There are various other problems in digital filter design and signal processing that require the solution of a system of linear equations of the type considered in this subsection. In geophysical exploration it is often necessary to alter the shape of a known signal. The design of shaping filters is commonly done in the time domain. For a known input sequence and a desired output sequence a finite impulse response filter is designed based on the least-mean-square error criterion, which minimizes the error energy between the desired output and the actual output. A system of linear equations of the type encountered in (6.21) is obtained, whose solution yields the filter coefficients. The reader interested in learning about how the system of linear equations is formulated in the design of digital filters for the analysis of time-series recordings in seismic signal processing should refer to reference [10, pp. 141–149]. Another important area of application, where equations of the type in (6.21) occur, is in speech processing.

Owing to technical innovations permitting the implementation of complex digital operations in real time economically, linear prediction (LP) has been established as the most useful tool for the processing of speech. LP requires the estimate of the best linear approximant, in the least-squares sense of the present sample $x(k)$, of a discrete-time stationary process. Mathematically, this demands the determination of coefficients $a_1, a_2, \ldots, a_N$ so that the mean-square value of the error $e(k)$ is minimized in

$$x(k) = - \sum_{i=1}^{N} a_i x(k-i) + e(k).$$

It can be shown that the problem is transformable to one that requires the solution of a system of linear equations, in which the coefficient matrix is symmetric Toeplitz. More specifically, if the autocorrelation sequence for the process $x(k)$ is $\{R_k\}$, then the system of linear equations that has to be solved for the a_i's is

$$\sum_{i=1}^{N} a_i R_{(k-i)} = - R_k, \quad k = 1, 2, \ldots, N.$$

After the a_i's are obtained, the preceding recurrence relation may be rewritten as

$$R_k = - \sum_{i=k-1}^{k-N} a_{k-i} R_i$$

in order that the autocorrelation values R_k, $k > N$ may be calculated. The power spectrum as a function of frequency is obtainable (to within a constant multiplier) by evaluating

$$S(z) = \left| 1 + \sum_{k=1}^{N} a_k z^{-k} \right|^{-2}$$

on $|z| = 1$. The interested reader is referred to reference [16, pp. 398–407] for additional details. The preceding discussion should convince the reader of the importance of exploiting the symmetric Toeplitz structure of the coefficient matrix to obtain solutions more efficiently, especially since the diverse areas of application benefit from the use of the fast recursive technique (Levinson's algorithm and its variants [16, pp. 411–413]). Problems 11 to 15 at the end of this chapter alert the reader to some of the applications requiring the use of Levinson's algorithm. The survey article by Schroeder [20] is particularly geared toward applications in speech processing.

6.3. Padé Approximants (and Hankel Structure)

Certain rational approximants to a specified power series expansion have found numerous applications in diverse areas of continuous and discrete linear systems. In this section these approximants, referred to as Padé approximants, are characterized as solutions to a system of linear equations with a coefficient matrix possessing a Hankel structure. This structural property can be exploited to develop fast recursive algorithms for solving the system of linear equations characterizing the problem. The basic mathematical problem was introduced in Section 3.5.1. Here some minor notational changes are introduced to facilitate derivation of the desired recursion. Let $h(z)$ be analytic around $z = 0$ with a specified power series expansion.

$$h(z) = h_0 + h_1 z + h_2 z^2 + \cdots = \sum_{k=0}^{\infty} h_k z^k. \tag{6.27}$$

Given nonnegative integers, r and m, a rational approximant

$$\frac{A_r(z)}{B_m(z)} = \frac{a_0 + a_1 z + \cdots + a_r z^r}{b_0 + b_1 z + \cdots + b_m z^m} \tag{6.28}$$

$(A_r(z)/B_m(z)$ is the $[r/m]$ Padé approximant to $h(z)$, provided the condition in (6.29) is satisfied) is sought so that

$$A_r(z) - h(z) B_m(z) \equiv 0(z^{r+m+1}), \tag{6.29}$$

where the symbol $0(z^n)$ denotes that powers z^l, $l < n$, are absent. Without affecting the theoretical development, it is usually assumed that

$$b_0 = 1. \tag{6.30}$$

Then, equating coefficients of $z^0, z^1, \ldots, z^{r+m}$ on both sides of (6.29), one gets the system of linear equations as in (3.77) and (3.78), where $h(k)$'s, q_k's, and p_k's are replaced, respectively, by h_k's, b_k's, and a_k's for the case under study here. Furthermore, as before,

$$h_k = 0, \quad k < 0. \tag{6.31}$$

The counterparts of (3.77) and (3.78) are (6.32) and (6.33):

$$\begin{bmatrix} h_{-m} & h_{-m+1} & \cdots & h_{-1} & h_0 \\ h_{-m+1} & h_{-m+2} & \cdots & h_0 & h_1 \\ \cdots & \cdots & & \cdots & \cdots \\ h_{r-m} & h_{r-m+1} & \cdots & h_{r-1} & h_r \end{bmatrix} \begin{bmatrix} b_m \\ b_{m-1} \\ \vdots \\ b_1 \\ 1 \end{bmatrix} = \begin{bmatrix} a_0 \\ a_1 \\ \vdots \\ a_{r-1} \\ a_r \end{bmatrix} \tag{6.32}$$

$$\begin{bmatrix} h_{r-m+1} & h_{r-m+2} & \cdots & h_r & h_{r+1} \\ h_{r-m+2} & h_{r-m+3} & \cdots & h_{r+1} & h_{r+2} \\ \cdots & \cdots & & \cdots & \cdots \\ h_r & h_{r+1} & \cdots & h_{r+m-1} & h_{r+m} \end{bmatrix} \begin{bmatrix} b_m \\ b_{m-1} \\ \vdots \\ b_1 \\ 1 \end{bmatrix} = \begin{bmatrix} 0 \\ 0 \\ \vdots \\ 0 \\ 0 \end{bmatrix} \tag{6.33}$$

The second system of linear equations is first solved for $b_m, b_{m-1}, \ldots, b_1$, and then the remaining unknowns, $a_0, a_1, \ldots, a_r$, can be obtained from the first system of linear equations by direct substitution. It may be noted that the set of linear equations in (6.33) may be rewritten as

$$
\begin{bmatrix}
h_{r-m+1} & h_{r-m+2} & \cdots & h_r \\
h_{r-m+2} & h_{r-m+3} & \cdots & h_{r+1} \\
\cdots & \cdots & & \cdots \\
h_r & h_{r+1} & \cdots & h_{r+m-1}
\end{bmatrix}
\begin{bmatrix}
b_m \\
b_{m-1} \\
\vdots \\
b_1
\end{bmatrix}
= -
\begin{bmatrix}
h_{r+1} \\
h_{r+2} \\
\vdots \\
h_{r+m}
\end{bmatrix}
$$

$$(6.34)$$

and attention will be directed toward the solution of this system of linear equations, noting that the coefficient matrix possesses a certain structure, called the Hankel matrix structure, defined next.

Definition 6.3. A matrix $Q = [q_{ij}]$, where q_{ij} is an element in the ith row and jth column, $1 \le i, j \le (m+1)$, is a Hankel matrix, provided it has only a maximum of $(2m+1)$ independent elements, defined by

$$q_{ij} = q_{i+j-1}.$$

Note that a Hankel matrix is symmetric. Also, since it has equal elements along lines perpendicular to the principal diagonal, it is also referred to as an orthosymmetric matrix.

Before discussing the recursive algorithm to solve the system of linear equations, it is necessary to state the following fact and introduce the definition for a normal power series.

Fact 6.2. *The coefficient matrix in (6.34) is singular, if and only if the $[r/m]$-order Padé approximant $(A_r(z)/B_m(z)$ in (6.28)–(6.30)) to the power series $h(z)$ in (6.27) has a nontrivial common factor in its numerator and denominator polynomials.*

Definition 6.4. When the power series $h(z)$ is such that for arbitrary $r \ge 0$, $m \ge 0$, the $[r/m]$-order Padé approximant does not have a nontrivial factor common to its numerator and denominator polynomials (i.e., the coefficient matrix in (6.34) is invertible), then $h(z)$ is called a normal power series.

Henceforth, only normal power series will be considered. Results on nonnormal power series are beyond the scope of this book. The algorithm summarized in Figure 6.1 (where Q is a symmetric Toeplitz matrix of order m) can be used to solve a system of linear equations,

$$H\mathbf{b} = \mathbf{f}, \tag{6.35}$$

where $\mathbf{b}$ is the unknown $(m \times 1)$ column vector, $\mathbf{f}$ is a specified $(m \times 1)$

vector, provided the Hankel matrix H is persymmetric (see Definition 6.5). From the assumption made, note that H is nonsingular.

Definition 6.5. A square matrix, $H = [h_{ij}]$, of order m, where h_{ij} is the element in the ith row and jth column of H, $1 \le i$, $j \le m$ is persymmetric, provided

$$h_{ij} = h_{(m+1-j),(m+1-i)}. \tag{6.36}$$

Note that, when H is symmetric, the preceding condition for a matrix to be persymmetric yields

$$h_{ij} = h_{ji} = h_{(m+1-i),(m+1-j)}. \tag{6.37}$$

Note that any persymmetric matrix is symmetric about its secondary diagonal and that any Toeplitz matrix is necessarily persymmetric. A $(m \times m)$ persymmetric matrix can have $m(m+1)/2$ independent elements.

It is next shown how the system of linear equations in (6.35), when H is a $(m \times m)$ persymmetric Hankel matrix, can be solved by solving a system of linear equations using the algorithm of Figure 6.1. Again, let h_{ij} be the element in the ith row and jth column of H, for $1 \le i$, $j \le m$. Since H is a persymmetric Hankel matrix of order m, it has at most m independent elements, $h_1, h_2, \ldots, h_m$. Let $[h_1 \quad h_2 \quad \cdots \quad h_m]$ be the first row of H. Then h_{ij}'s are related to h_i's by

$$h_{ij} = h_{(m+1-i),(m+1-j)} = h_{i+j-1}$$

for $1 \le i$, $j \le m$ and $2 \le (i+j) \le (m+1)$. Introduce the substitution

$$h_{i+j-1} \rightarrow q_{|m-(i+j-1)|}, \quad 1 \le i, \ j \le m. \tag{6.38}$$

Let $\mathbf{b} = [b_1 \quad b_2 \quad \cdots \quad b_m]^t$. Introduce another substitution,

$$b_i \rightarrow a_{m-i+1}, \quad 1 \le i \le m. \tag{6.39}$$

Then, if in (6.35),

$$\mathbf{f} = [f_1 \quad f_2 \quad \cdots \quad f_m]^t,$$

one gets, for $i = 1, 2, \ldots, m$,

$$f_i = \sum_{j=1}^{m} h_{(i+j-1)} b_j$$

$$= \sum_{j=1}^{m} h_{(i+j-1)} a_{m-j+1} \quad \text{(using (6.39))}$$

$$= \sum_{k=1}^{m} h_{m+i-k} a_k \quad \text{(substituting k for $(m-j+1)$)}$$

$$= \sum_{k=1}^{m} q_{|k-i|} a_k \quad \text{(using (6.38))}.$$

Therefore,

$$Q\mathbf{a} = \mathbf{f}, \tag{6.40}$$

where $\mathbf{a} = [a_1 \quad a_2 \quad \cdots \quad a_m]^t$ and Q is a symmetric nonsingular (this can be proved, using the fact that H is invertible) Toeplitz matrix of order m. The first row of Q is $[q_0 \quad q_1 \quad \cdots \quad q_{(m-1)}]$ and if q_{ij} is the element of Q in its ith row and jth column, $1 \le i$, $j \le m$, then $q_{ij} = q_{|j-i|}$. Equation (6.40) can be solved for the unknown vector $\mathbf{a}$ by applying Levinson's algorithm, requiring $0(m^2)$ operations, and the unknown vector $\mathbf{b}$ in (6.35) can be obtained from $\mathbf{a}$ via appropriate mapping of indices (see 6.39). Vector $\mathbf{b}$ in (6.35) can also be solved for directly from

$$Q\mathbf{b} = \mathbf{g}, \tag{6.41a}$$

where $\mathbf{g} = [g_1 \quad g_2 \quad \cdots \quad g_m]$ is obtained via the substitution,

$$f_i \to g_{m-i+1}, \quad i = 1, 2, \ldots, m. \tag{6.41b}$$

The preceding facts are illustrated by an example.

Example 6.3. Consider the system of linear equations (similar to (6.35)),

$$\begin{bmatrix} 1 & 2 & 3 & 4 \\ 2 & 3 & 4 & 3 \\ 3 & 4 & 3 & 2 \\ 4 & 3 & 2 & 1 \end{bmatrix} \begin{bmatrix} b_1 \\ b_2 \\ b_3 \\ b_4 \end{bmatrix} = \begin{bmatrix} 5 \\ 6 \\ 7 \\ 8 \end{bmatrix},$$

where the coefficient matrix of order 4 is a persymmetric Hankel matrix. This matrix is generated by the elements $[h_1 \quad h_2 \quad h_3 \quad h_4] = [1 \quad 2 \quad 3 \quad 4]$ in its first row. From (6.38),

$$h_1 = 1 \to q_3$$
$$h_2 = 2 \to q_2$$
$$h_3 = 3 \to q_1$$
$$h_4 = 4 \to q_0.$$

Similarly, from (6.39),

$$b_1 \to a_4, \quad b_2 \to a_3, \quad b_3 \to a_2, \quad b_4 \to a_1.$$

Therefore, the equivalent system of linear equations with a symmetric Toeplitz coefficient matrix is

$$\begin{bmatrix} 4 & 3 & 2 & 1 \\ 3 & 4 & 3 & 2 \\ 2 & 3 & 4 & 3 \\ 1 & 2 & 3 & 4 \end{bmatrix} \begin{bmatrix} a_4 \\ a_3 \\ a_2 \\ a_1 \end{bmatrix} = \begin{bmatrix} 5 \\ 6 \\ 7 \\ 8 \end{bmatrix}.$$

Another equivalent system of linear equations is obtained (similar to (6.41a) using the substitution in (6.41b)):

$$
\begin{bmatrix}
4 & 3 & 2 & 1 \\
3 & 4 & 3 & 2 \\
2 & 3 & 4 & 3 \\
1 & 2 & 3 & 4
\end{bmatrix}
\begin{bmatrix}
b_1 \\
b_2 \\
b_3 \\
b_4
\end{bmatrix}
=
\begin{bmatrix}
8 \\
7 \\
6 \\
5
\end{bmatrix}.
$$

When H in (6.35) or the coefficient matrix in (6.34) is an arbitrary invertible Hankel matrix (i.e., it is not necessarily persymmetric), there are various approaches to solve for the unknowns $b_1, b_2, \ldots, b_m$, which are undetermined coefficients of the polynomial

$$
B_m(z) = b_0 + b_1 z + \cdots + b_m z^m, \quad b_0 = 1. \tag{6.42}
$$

$B_m(z)$ may be computed explicitly via evaluation of determinants in the expression below, though this is by no means a computationally expedient manner to obtain the rational approximant in (6.28) to a specified power series in (6.27):

$$
B_m(z) = \frac{1}{\left(\det H_{m-1}^{(r)} \right)} \det
\begin{bmatrix}
h_{r-m+1} & h_{r-m+2} & \cdots & h_{r+1} \\
h_{r-m+2} & h_{r-m+3} & \cdots & h_{r+2} \\
\vdots & \vdots & & \vdots \\
h_r & h_{r+1} & \cdots & h_{r+m} \\
z^m & z^{m-1} & \cdots & 1
\end{bmatrix},
$$

$$\tag{6.43}$$

where $H_{m-1}^{(r)}$ is the coefficient matrix in (6.34) and is written below explicitly for the sake of clarity. $H_{m-1}^{(r)}$ is of order m.

$$
H_{m-1}^{(r)} =
\begin{bmatrix}
h_{r-m+1} & h_{r-m+2} & \cdots & h_r \\
h_{r-m+2} & h_{r-m+3} & \cdots & h_{r+1} \\
\vdots & \vdots & & \vdots \\
h_r & h_{r+1} & \cdots & h_{r+m-1}
\end{bmatrix}. \tag{6.44}
$$

Therefore,

$$
B_m(z) = \frac{1}{\det H_{m-1}^{(r)}} \det
\left[
\begin{array}{ccc:c}
 & & & h_{r+1} \\
 & H_{m-1}^{(r)} & & h_{r+2} \\
 & & & \vdots \\
 & & & h_{r+m} \\
\hdashline
z^m & z^{m-1} & \cdots & 1
\end{array}
\right]. \tag{6.45}
$$

The coefficient of the numerator polynomial $A_r(z)$ in (6.28) can be easily calculated from (6.32) after $B_m(z)$ has been found. The Hankel structure of

H can be exploited to obtain a recurrence relation between approximants of different orders, and this is discussed next.

6.3.1. Recursive Computation of Padé Approximants

A recurrence relation between the denominator polynomials of Padé approximants of different orders will be set up. For notational brevity, set

$$t_k = h_{r-m+k}, \quad k = 1, 2, \ldots, m+1, \ldots. \tag{6.46}$$

It will be assumed that the conditions for existence and uniqueness (in fact both of these are guaranteed, since only normal power series are considered) of approximants of required orders are satisfied. The system of equations involving the coefficients of $B_m(z)$ in (6.33) in the $[r/m]$-order approximation stage is given in (6.47):

$$\begin{bmatrix} t_1 & t_2 & \cdots & t_{m+1} \\ t_2 & t_3 & \cdots & t_{m+2} \\ \vdots & \vdots & & \vdots \\ t_m & t_{m+1} & \cdots & t_{2m} \end{bmatrix} \begin{bmatrix} b_m^{(m)} \\ b_{m-1}^{(m)} \\ \vdots \\ b_1^{(m)} \\ 1 \end{bmatrix} = \begin{bmatrix} 0 \\ 0 \\ \vdots \\ 0 \\ 0 \end{bmatrix}. \tag{6.47}$$

The additional superscript $(m+k)$ (in (6.47), $k=0$), on the b_i's, $i = 1, \ldots, m+k$, defining $B_{m+k}(z)$ (of course, $b_0^{(m+k)} = 1$), has been introduced to underscore the fact the $B_{m+k}(z)$ is the denominator polynomial associated with the $[r+k/m+k]$-order Padé approximant, $k = 0, \pm 1, \pm 2 \ldots$. Then consider the relevant polynomials $B_m(z)$ and $B_{m-1}(z)$, associated, respectively, with the $[r/m]$- and $[r-1/m-1]$-order approximants, and use (6.47) to arrive at (6.48).

$$\begin{bmatrix} t_1 & t_2 & \cdots & t_{m+2} \\ t_2 & t_3 & \cdots & t_{m+3} \\ \vdots & \vdots & & \vdots \\ t_{m+1} & t_{m+2} & \cdots & t_{2m+2} \\ t_{m+2} & t_{m+3} & \cdots & t_{2m+3} \end{bmatrix} \begin{bmatrix} b_{m-1}^{(m-1)} & b_m^{(m)} & 0 \\ \vdots & \vdots & b_m^{(m)} \\ \vdots & \vdots & \vdots \\ b_1^{(m-1)} & b_2^{(m)} & \\ 1 & b_1^{(m)} & b_2^{(m)} \\ 0 & 1 & b_1^{(m)} \\ 0 & 0 & 1 \end{bmatrix} = \begin{bmatrix} 0 & 0 & 0 \\ 0 & 0 & 0 \\ \vdots & \vdots & \vdots \\ d_{m-1} & 0 & d_m \\ e_{m-1} & d_m & e_m \\ \mathsf{x} & e_m & \mathsf{y} \end{bmatrix}. \tag{6.48}$$

In (6.47) it is straightforward to verify by direct computation that, for $k = m$, $m - 1$,

$$d_k = t_{k+1} b_k^{(k)} + t_{k+2} b_{k-1}^{(k)} + \cdots + t_{2k} b_1^{(k)} + t_{2k+1}, \tag{6.49a}$$

$$e_k = t_{k+2} b_k^{(k)} + t_{k+3} b_{k-1}^{(k)} + \cdots + t_{2k+1} b_1^{(k)} + t_{2k+2}, \tag{6.49b}$$

and scalars x, y, not of interest, may also be found by direct calculation. If the $[r+1/m+1]$-order approximant is to be devoid of a common factor in its numerator and denominator polynomials (see Fact 6.2), the matrix in (6.50) must be nonsingular. This matrix is obtained after replacing m by $(m+1)$ in the coefficient matrix in (6.47) and then disregarding the last column:

$$\begin{bmatrix} t_1 & t_2 & \cdots & t_{m+1} \\ t_2 & t_3 & \cdots & t_{m+2} \\ \vdots & \vdots & & \vdots \\ t_{m+1} & t_{m+2} & \cdots & t_{2m+1} \end{bmatrix}. \tag{6.50}$$

This implies that $d_m \neq 0$; otherwise there exists a scalar $z \neq 0$ such that

$$d_m z = \left[t_{m+1} b_m^{(m)} + t_{m+2} b_{m-1}^{(m)} + \cdots + t_{2m} b_1^{(m)} + t_{2m+1} \right] z. \tag{6.51a}$$

Combining (6.51a) and (6.47), the nonsingularity of the matrix in (6.50) would imply that $z = 0$, resulting in a contradiction. Therefore, $d_m \neq 0$ and d_m^{-1} exists. Similarly, it can be shown that d_{m-1}^{-1} exists. Postmultiply both sides of (6.48) by the matrix in (6.51):

$$\begin{bmatrix} -d_{m-1}^{-1} d_m \\ d_m^{-1} \left(e_{m-1} d_{m-1}^{-1} d_m - e_m \right) \\ 1 \end{bmatrix}. \tag{6.51b}$$

Define

$$v_m \triangleq \left(e_{m-1} d_{m-1}^{-1} d_m - e_m \right) \tag{6.52a}$$

$$B_1^{(m-1)} \triangleq \left(b_{m-1}^{(m-1)} \quad b_{m-2}^{(m-1)} \quad \cdots \quad b_1^{(m-1)} \quad 1 \quad 0 \quad 0 \right)^t \tag{6.52b}$$

$$B_2^{(m)} \triangleq \left(b_m^{(m)} \quad b_{m-1}^{(m)} \quad \cdots \quad b_2^{(m)} \quad b_1^{(m)} \quad 1 \quad 0 \right)^t \tag{6.52c}$$

$$B_3^{(m)} \triangleq \left(0 \quad b_m^{(m)} \quad \cdots \quad b_3^{(m)} \quad \cdots \quad b_1^{(m)} \quad 1 \right)^t. \tag{6.52d}$$

Then, following the postmultiplication of both sides of (6.48) by (6.51b), delete the last row on each side of the resulting equation to obtain

$$\begin{bmatrix} t_1 & t_2 & \cdots & t_{m+2} \\ t_2 & t_3 & \cdots & t_{m+3} \\ \vdots & \vdots & & \vdots \\ t_{m+1} & t_{m+3} & \cdots & t_{2m+2} \end{bmatrix} \left[-B_1^{(m-1)} d_{m-1}^{-1} d_m + B_2^{(m)} d_m^{-1} v_m + B_3^{(m)} \right]$$

$$= \begin{bmatrix} 0 \\ 0 \\ \vdots \\ 0 \\ 0 \end{bmatrix}. \tag{6.53}$$

Let

$$B_4^{(m+1)} \triangleq \left(b_{m+1}^{(m+1)} \quad b_m^{(m+1)} \quad \cdots \quad b_1^{(m+1)} \quad 1 \right),$$

where $b_k^{(m+1)}$, $k = 1, \ldots, m+1$, and $b_0^{(m+1)} = 1$ are the coefficients of the denominator polynomial of the $[r+1/m+1]$-order Padé approximant. Then, the counterpart of (6.47) for the $[r+1/m+1]$-order approximant is

$$\begin{bmatrix} t_1 & t_2 & \cdots & t_{m+2} \\ t_2 & t_3 & \cdots & t_{m+3} \\ \vdots & \vdots & & \vdots \\ t_{m+1} & t_{m+2} & \cdots & t_{2m+2} \end{bmatrix} B_4^{(m+1)} = \begin{bmatrix} 0 \\ 0 \\ \vdots \\ 0 \\ 0 \end{bmatrix}. \qquad (6.54)$$

Comparison of both sides of (6.53) and (6.54) yields

$$B_4^{(m+1)} = B_3^{(m)} + B_2^{(m)}d_m^{-1}v_m - B_1^{(m-1)}d_{m-1}^{-1}d_m. \qquad (6.55)$$

(6.55) provides the desired recurrence relation between the coefficients of the denominator polynomials of the $[r+1/m+1]$-, $[r/m]$-, and $[r-1/m-1]$-order Padé approximants. The coefficients of the denominator polynomial

$$B_{m+1}(z) = \sum_{k=0}^{m+1} b_k^{(m+1)}z^k, \qquad b_0^{(m+1)} = 1,$$

associated with the $[r+1/m+1]$-order Padé approximant have thus been obtained. From the sequence of polynomials $\{B_{m+k}(z)\}$, associated with approximants of order $[r+k/m+k]$, $k = 0, \pm 1, \pm 2 \ldots$, obtain another sequence of polynomials, $\{R_{m+k}(z)\}$, whose elements are defined by

$$R_m(z) = z^m B_m(z^{-1}). \qquad (6.56)$$

Define

$$k_m = e_m d_m^{-1} - e_{m-1}d_{(m-1)}^{-1} \qquad (6.57a)$$

$$\lambda_m = d_{m-1}^{-1}d_m, \qquad m = 1,2,\ldots \qquad (6.57b)$$

$$\mu_m = k_m, \qquad m = 1,2,\ldots, \text{ and } \mu_0 = d_0^{-1}h_1. \qquad (6.57c)$$

Henceforth the sequence of polynomials $\{B_m(z)\}$, associated with Padé approximants of order $[(m-1)/m]$, will be considered because certain interesting properties follow, especially when the power series $h(z)$ being approximated is restricted to be positive (see Definition 6.6 later in this section).

Theorem 6.1. *The elements of the sequence of polynomials $\{R_m(z)\}$ (see (6.56)), associated with Padé approximants of order $[m-1/m]$ to a normal power series $h(z)$, satisfy the following recurrence relation for $m = 0,1,2,3,\ldots$:*

$$R_{m+1}(z) = R_m(z)[z - \mu_m] - R_{m-1}(z)\lambda_m, \qquad (6.58a)$$

with

$$R_{-1}(z) = 0 \qquad \text{and} \qquad R_0(z) = 1, \qquad (6.58b)$$

where λ_m and μ_m are defined, respectively, in (6.57b) and (6.57c).

The proof of the above theorem follows quite easily from the recurrence relation arrived at in (6.55), and is omitted for the sake of brevity. Next, a positive power series is defined and the results given are with respect to notations introduced.

Definition 6.6. The power series $h(z)$ in (6.27) is positive, provided the Hankel matrix $H_{m-1}^{(m-1)}$ (in (6.44) with $r = m - 1$) is positive definite for all integers $m > 0$ (or $m - 1 \geq 0$).

Theorem 6.2. *If $\{R_m(z)\}$ is associated with a $\{[m - 1/m]\}$-order Padé approximant to a positive power series, then the zeros of $R_m(z)$ are real and simple.*

PROOF: When $r = m - 1$, from (6.46),

$$t_k = h_{k-1}, \quad k = 1, 2, \ldots, 2m, \ldots.$$

Also, from (6.49a), after replacing t_k by h_{k-1},

$$d_m = h_m b_m^{(m)} + h_{m+1} b_{m-1}^{(m)} + \cdots + h_{2m-1} b_1^{(m)} + h_{2m}.$$

Combining the preceding equation with (6.47), in which t_k is replaced by h_{k-1}, one gets

$$H_m^{(m)} M_m = [0 \quad 0 \quad \cdots \quad 0 \quad d_m]^t, \tag{6.59}$$

where

$$M_m \triangleq \left[b_m^{(m)} \quad b_{m-1}^{(m)} \quad \cdots \quad b_1^{(m)} \quad 1 \right]^t.$$

Therefore, premultiplying both sides of (6.59) by M_m^t,

$$M_m^t H_m^{(m)} M_m = d_m. \tag{6.60}$$

From Definition 6.6, $H_m^{(m)}$ is positive definite for all $m \geq 0$. Therefore, from (6.60),

$$d_m > 0, \quad m = 0, 1, 2, \ldots. \tag{6.61}$$

Let $z = \alpha_j$ be a zero of $R_m(z)$, that is, $R_m(\alpha_j) = 0$. Then from (6.58a) and (6.58b) one gets the following system of equations:

$$\begin{aligned}
&\left[R_0(\alpha_j) \quad R_1(\alpha_j) \quad \cdots \quad R_{m-1}(\alpha_j) \right] A_m \\
&= \left[R_0(\alpha_j) \quad R_1(\alpha_j) \quad \cdots \quad R_{m-1}(\alpha_j) \right] \alpha_j,
\end{aligned} \tag{6.62a}$$

where A_m is the tridiagonal matrix in (6.62b),

$$A_m = \begin{bmatrix}
\mu_0 & \lambda_1 & & & & \\
1 & \mu_1 & \lambda_2 & & & \\
& 1 & & \ddots & & 0 \\
0 & & \ddots & & \lambda_{m-1} & \\
& & & 1 & \mu_{m-1}
\end{bmatrix}. \tag{6.62b}$$

Using (6.57a), (6.57b), and (6.57c) and the fact that $d_m > 0$, $m = 0, 1, \ldots,$

[see (6.61)] one can decompose A_m into the following matrix product:

$$A_m = D_m^{-1} C_m, \tag{6.63}$$

where

$$D_m = \text{diag}\,[d_0 \quad d_1 \quad \cdots \quad d_{m-1}], \tag{6.64a}$$

$$C_m = \begin{bmatrix} h_1 & d_1 & & & & \\ d_1 & \mu_1 d_1 & d_2 & & \mathbf{0} & \\ & d_2 & \mu_2 d_2 & \cdot\,\cdot & & \\ & & \cdot\,\cdot & \cdot\,\cdot & d_{m-1} & \\ \mathbf{0} & & & d_{m-1} & \mu_{m-1} d_{m-1} \end{bmatrix}. \tag{6.64b}$$

Since D_m is positive definite and C_m is a symmetric matrix, it follows from a standard result in matrix theory that the eigenvalues of A_m in (6.63) are real. Note that, from (6.62a), α_j is clearly an eigenvalue of A_m. Therefore, the zeros of $R_m(z)$ are all real. It is not difficult to conclude that each zero is also of multiplicity 1.

6.4. Some Applications of the Discrete Hilbert Transform (DHT) and Circulants

6.4.1. Digital Filter Stabilization via DHT

In Chapter 3 it was seen that digital recursive filters are often specified in terms of the magnitude of the frequency response. Particular characteristics (like maximal flatness, equiripple behavior, etc.) define particular classes of filters (like Butterworth, Chebyshev, etc.). The phase response cannot be chosen arbitrarily if a stable and causal minimum-phase filter is desired. This is because, for classes of rational functions called minimum-phase functions, which are devoid of poles and zeros on and outside the unit circle, the phase associated with the transfer function is uniquely specified by its magnitude. In fact, the phase and magnitude of a minimum-phase transfer function form a Hilbert transform pair. Suppose, now, that a designed filter is found to be unstable. It may be desirable to stabilize this unstable filter without appreciable change in the magnitude of its frequency response. This may be achieved by constructing the frequency response magnitude function from the unstable filter transfer function, obtaining the phase function by taking the Hilbert transform of this magnitude function, and then characterizing the desired stabilized filter by the magnitude function and the calculated phase function. Theoretically, then, not only would the original filter be stabilized, but also no change in the magnitude of the frequency response would occur. This idea was exploited by Read and Treitel [3] to develop a stabilization procedure for unstable digital filters via the discrete Hilbert transform (DHT), which was introduced in Section 2.4. The subsequent development is based on the approach suggested by Read and Treitel.

Let $\{b(k)\}_{k=0}^{N-1}$ denote a finite sequence of length N; here we restrict the sequence to be real. In problems where the procedure is intended for application, the sequences encountered will be real; nevertheless, if necessary, complex sequences can be handled by making simple and obvious modifications. The case when N is even will be fully considered; again, the case when N is odd can be handled without trouble if the reader follows the treatment for the N even case. If $b_e(k)$ and $b_o(k)$ denote, respectively, the symmetric and antisymmetric components of $b(k)$, then

$$b(k) = b_e(k) + b_o(k), \tag{6.65}$$

where for $k = 1, 2, \ldots, N-1$,

$$b_e(k) = \tfrac{1}{2}[b(k) + b(N-k)] \tag{6.66a}$$
$$b_o(k) = \tfrac{1}{2}[b(k) - b(N-k)] \tag{6.66b}$$

and $b_e(0) = b(0)$, and $b_o(0) = 0$. Clearly,

$$b_e(k) = b_e(N-k) \tag{6.67a}$$
$$b_o(k) = -b_o(N-k) \tag{6.67b}$$

for $k = 1, \ldots, N-1$.

Definition 6.7. The finite sequence, $\{b(k)\}_{k=0}^{N-1}$, will be called discrete causal if $b(k) = 0$, $k \geq N/2$, for N even. (Note that any finite sequence may be made discrete causal by suitable addition of zeros). For a finite causal pulse, $\{b(k)\}_{k=0}^{N-1}$,

$$b_e(k) = \begin{cases} b(k), & k = 0 \\ \dfrac{1}{2}b(k), & 0 < k < \dfrac{N}{2} \\ 0, & k = \dfrac{N}{2} \\ \dfrac{1}{2}b(N-k), & \dfrac{N}{2} < k \leq N-1, \end{cases} \tag{6.68a}$$

$$b_o(k) = \begin{cases} 0, & k = 0 \\ \dfrac{1}{2}b(k), & 0 < k < \dfrac{N}{2} \\ 0, & k = \dfrac{N}{2} \\ -\dfrac{1}{2}b(N-k), & \dfrac{N}{2} < k \leq N-1. \end{cases} \tag{6.68b}$$

Define

$$\delta_N(k) \triangleq \begin{cases} 1, & k = 0 \\ 0, & \text{otherwise} \end{cases} \tag{6.69}$$

and

$$(\text{sgn})_N(k) \triangleq \begin{cases} 0, & k = 0, \ \dfrac{N}{2} \\ 1, & 0 < k < \dfrac{N}{2} \\ -1, & \dfrac{N}{2} < k \le N - 1. \end{cases} \tag{6.70}$$

Then, from (6.68a) and (6.68b), respectively, (6.71) and (6.72) follow.

$$b_e(k) = (\text{sgn})_N(k) b_o(k) + b(k) \delta_N(k), \tag{6.71}$$

$$b_o(k) = (\text{sgn})_N(k) b_e(k). \tag{6.72}$$

Let $\{b(r)\} \leftrightarrow \{B(k)\}$ denote an N-point DFT pair. Define

$$B(k) = B_r(k) + jB_i(k), \quad k = 0, 1, \dots, N - 1, \tag{6.73}$$

where $B_r(k)$ and $B_i(k)$ denote, respectively, the real and imaginary parts of $B(k)$. Since $\{b(r)\}$ is a real sequence,

$$\text{DFT}\{b_e(r)\} \triangleq \{B_r(k)\} \tag{6.74a}$$

$$\text{DFT}\{b_o(r)\} \triangleq \{jB_i(k)\}. \tag{6.74b}$$

Obviously, then

$$\{b_e(r)\} = \text{IDFT}\{B_r(k)\} \tag{6.75a}$$

$$\{b_o(r)\} = \text{IDFT}\{jB_i(k)\}. \tag{6.75b}$$

First, substituting (6.75a) and (6.75b) in (6.71) and (6.72), respectively, and then using (6.74a), (6.74b), one is able to relate the real and imaginary parts of the DFT of a finite causal pulse.

$$\{B_r(k)\} = \text{DFT}\{(\text{sgn})_N(r)\text{IDFT}[\{jB_i(k)\}] + b(r)\delta_N(r)\} \tag{6.76a}$$

$$\{B_i(k)\} = -j\text{DFT}\{(\text{sgn})_N(r)\text{IDFT}[\{B_r(k)\}]\}. \tag{6.76b}$$

On using the definitions for DFT and IDFT (see Definition 2.1), the preceding equations lead to (6.77a) and (6.77b):

$$B_i(k) = \frac{2}{N} \sum_{l=0,2,4,\dots}^{N-2} B_r(l) \cot \frac{\pi}{N}(k - l), \quad k \text{ odd}, \tag{6.77a}$$

$$B_i(k) = \frac{2}{N} \sum_{l=1,3,5,\dots}^{N-1} B_r(l) \cot \frac{\pi}{N}(k - l), \quad k \text{ even}. \tag{6.77b}$$

Equations (6.77a) and (6.77b) are in agreement with (2.40a) and (2.40b). Therefore, $\{B_i(k)\}$ may be viewed as the discrete Hilbert transform of $\{B_r(l)\}$. It is simple to verify that the relation between $\{B_i(k)\}$ and $\{B_r(l)\}$ is also expressible as

$$B_i(k) = \frac{1}{N} \sum_{l=0}^{N-1} B_r(l)\left[1-(-1)^{k-l}\right]\cot\frac{\pi}{N}(k-l) \qquad (6.78)$$

for $k = 0,1,\ldots,N-1$. Thus, the DHT of $\{B_r(l)\}$ is obtained as a circular convolution of $\{B_r(l)\}$ and the impulse response sequence of the discrete Hilbert transformer, given in (6.79).

$$\left\{\left[1-(-1)^l\right]\cot\frac{\pi}{N}l\right\}. \qquad (6.79)$$

It is well known [4, p. 248] that the phase $\theta(e^{j\omega})$ and the logarithm of the magnitude of the frequency response, $\log|H(e^{j\omega})|$, of a minimum-phase real rational function $H(z)$, are related through the Hilbert transform

$$\theta(e^{j\omega}) = -\frac{1}{2\pi}\int_0^{2\pi}\log_e|H(e^{j\phi})|\cot\frac{\omega-\phi}{2}\,d\phi. \qquad (6.80a)$$

In (6.80a) the integral is interpreted as Cauchy principal value in order to exercise care in evaluating the integral in the neighborhood of the singular points of the integrand. By taking a finite number, N, of discrete samples of $\theta(e^{j\omega})$ and $\log_e|H(e^{j\omega})|$, it is possible to relate those samples $\theta(k) \triangleq e^{j(2\pi kT/N)}$, $k = 0,1,\ldots,N-1$, and $\log_e|H(l)|$,[1] $l = 0,1,\ldots,N-1$, by a relation similar to (6.76b), so that $\{\theta(k)\}$ is the DHT of $\{\log|H(l)|\}$. This relation is given in (6.80b):

$$\{\theta(k)\} = -j\,\mathrm{DFT}\{(\mathrm{sgn})_N(r)\,\mathrm{IDFT}[\{\log|H(k)|\}]\}. \qquad (6.80b)$$

The relation in (6.80b) is key to an approach toward stabilization of unstable recursive filters without appreciable change in the frequency response magnitude via the DHT technique, as outlined in the flowchart of Figure 6.3. It is emphasized, however, that (6.80b) is only an approximation to the Hilbert transform relationship in (6.80a). So there could be some change in the frequency response magnitude, and though stabilization is usually attained, it cannot, theoretically, be guaranteed. In Figure 6.3, the inputs are the denominator coefficients $b(0), b(1), \ldots, b(n)$ of an unstable filter, characterized by the transfer function

$$H(z) = \frac{\displaystyle\sum_{k=0}^{m} a(k)z^{-k}}{\displaystyle\sum_{k=0}^{n} b(k)z^{-k}}.$$

---

[1] $\log_e|H(l)| \triangleq \log_e|H(e^{j(2\pi lT/N)})|$.

382

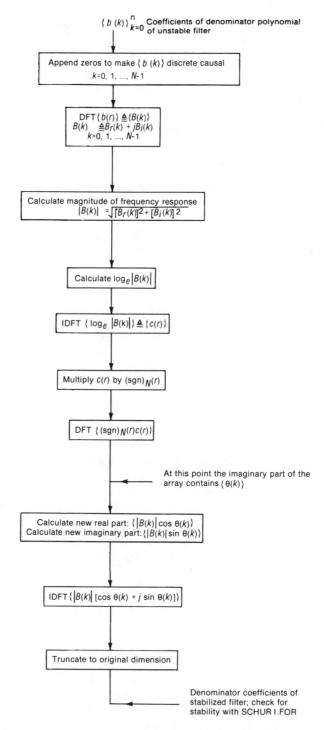

Figure 6.3. Flowchart of DHT Stabilization Procedure.

Zeros are added to the sequence $(b(0) \quad b(1) \quad \cdots \quad b(n))$ so that a causal sequence $\{b(k)\}_{k=0}^{N-1}$ results. After application of the various steps in the flowchart, the output gives the coefficients of the new filter denominator polynomial and this denominator polynomial is tested for stability. The next example illustrates the procedure.

Example 6.4. Let

$$H(z) = \frac{K}{z^{-3} + 5.5z^{-2} + 8.5z^{-1} + 3}$$

be the transfer function of an unstable filter, where K is a normalizing constant. Then

$$b(0) = 3, \quad b(1) = 8.5,$$
$$b(2) = 5.5, \quad b(3) = 1.$$

Choose $b(4) = b(5) = b(6) = b(7) = 0$, so that $\{b(k)\}_{k=0}^{7}$ is an 8-point discrete causal sequence. The denominator coefficients of the filter obtained after implementing the steps summarized in the flowchart of Figure 6.3 are

$$\hat{b}(0) = 5.9936, \quad \hat{b}(1) = 8.0076,$$
$$\hat{b}(2) = 3.4849, \quad \hat{b}(3) = 0.5447.$$

Before truncation, an 8-point sequence, $\{\hat{b}(k)\}_{k=0}^{7}$ was obtained, where

$$\hat{b}(4) = \quad 0.0889, \quad \hat{b}(5) = -0.0361,$$
$$\hat{b}(6) = -0.0674, \quad \hat{b}(7) = -0.0162.$$

The filter obtained was tested for stability via use of SCHUR1.FOR. It was found to be stable. The normalized plots of the magnitude of the frequency responses of the unstable and stabilized filters are shown in Figure 6.4A, B, respectively.

The simple Example 6.4 has been used for convenience in illustration. The reader can check that the offending pole, responsible for rendering $H(z)$ unstable, is at $z = -2$. Therefore, applying the scheme of Section 6.2.1, it is possible to obtain a stabilized filter with an identical magnitude of frequency response by replacing the pole at $z = -2$ with a pole at $z = -1/2$, while keeping the poles of $H(z)$ which are within the unit circle unchanged. The transfer function of the stable filter is, then,

$$\hat{H}(z) = \frac{1}{0.5z^{-3} + 3.5z^{-2} + 8z^{-1} + 6}.$$

The coefficients 5.9936, 8.0076, 3.4849, and 0.5447 in the preceding example, obtained by applying the numerical technique described in Figure 8.3, are quite close to the coefficients 6, 8, 3.5 and 0.5 of $\hat{H}(z)$. Some motivation for using the stabilization procedure considered in this section might lie in

384

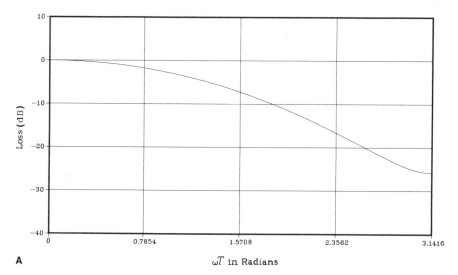

A

ωT in Radians

Figure 6.4A. Normalized plot of the magnitude of frequency response of unstable filter in Example 6.4.

Figure 6.4B. Normalized plot of the magnitude of frequency response of filter stabilized via DHT in Example 6.4.

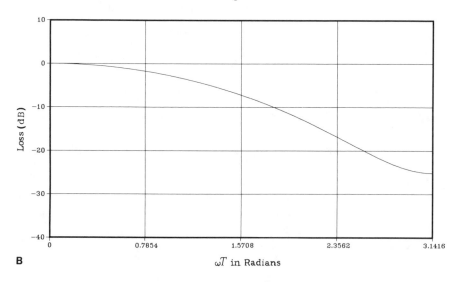

B

ωT in Radians

the availability of conveniently implementable software packages to attain the desired goal. Of course, if accurate root-finders are available, the engineer has the choice of using either the DHT stabilization scheme, the PLSI stabilization scheme or the conceptually simple procedure detailed in Section 6.2.1.

6.4.2. Digital Filtering Using the DHT

Sabri and Steenart [15] investigated the scope of using the DHT in digital filtering problems. Let $x(t) \leftrightarrow X(j\omega)$ be a Fourier transform pair representing the input to an ideal analog lowpass filter characterized by the frequency response,

$$W(j\omega) = \begin{cases} 1, & |\omega| \leq \omega_c \\ 0, & |\omega| > \omega_c, \end{cases}$$

whose Fourier inverse transform is (see Definition 1.1)

$$\begin{aligned} w(t) &= \frac{1}{2\pi} \int_{-\omega_c}^{\omega_c} 1 e^{j\omega t} \, d\omega \\ &= \frac{1}{\pi} \frac{\sin \omega_c t}{t}. \end{aligned} \tag{6.81}$$

If the Fourier transform of the lowpass filter output, $y(t)$, is denoted by $Y(j\omega)$, then

$$Y(j\omega) = X(j\omega)W(j\omega)$$

and

$$y(t) = \frac{1}{\pi} \int_{-\infty}^{\infty} x(\tau) \frac{\sin \omega_c(t - \tau)}{(t - \tau)} \, d\tau. \tag{6.82}$$

Using the trigonometric identity,

$$\sin(x - y) = \sin x \cos y - \cos x \sin y$$

in (6.82), it is possible to rewrite $y(t)$ as

$$y(t) = \left[\frac{1}{\pi} \int_{-\infty}^{\infty} \frac{x(\tau)\cos \omega_c \tau}{t - \tau} \, d\tau \right] \sin \omega_c t - \left[\frac{1}{\pi} \int_{-\infty}^{\infty} \frac{x(\tau)\sin \omega_c \tau}{t - \tau} \, d\tau \right] \cos \omega_c t. \tag{6.83}$$

Noting the notion of a Hilbert transform (H) operator, introduced in (2.31a) and (2.33), it is possible to express $y(t)$ in (6.83) as

$$y(t) = H\big[x(t)\cos \omega_c t\big]\sin \omega_c t - H\big[x(t)\sin \omega_c t\big]\cos \omega_c t. \tag{6.84}$$

To obtain the discrete equivalent of the analog filtering operation described in (6.84), t is replaced by kT, where k is a nonnegative integer and T is the sampling period, normalized for convenience to unity, the Hilbert transform is substituted by the DHT and $\omega_c t$ is replaced by $2\pi\beta k/N$, where N is the number of frequency samples, so that β the filter cutoff frequency expressed as the number of sample points in the frequency domain corresponds to an angle of

$$\alpha = \frac{2\pi\beta}{N} \text{ rad.} \tag{6.85}$$

Denote the sampled input and output, respectively, by the sequences

$$\{x(k)\} = (x(0) \quad x(1) \quad \cdots \quad x(N-1)) \triangleq \mathbf{x}^t \tag{6.86a}$$

$$\{y(k)\} = (y(0) \quad y(1) \quad \cdots \quad y(N-1)) = \mathbf{y}^t. \tag{6.86b}$$

Introduce the matrices

$$A = \text{diag}[1 \quad \cos\alpha \quad \cos 2\alpha \quad \cdots \quad \cos(N-1)\alpha] \tag{6.87a}$$

$$B = \text{diag}[0 \quad \sin\alpha \quad \sin 2\alpha \quad \cdots \quad \sin(N-1)\alpha]. \tag{6.87b}$$

Also, let $H = [h_{ks}]$ denote the DHT matrix, introduced in (2.46), over N points. Then, the discrete equivalent of (6.84) may be written as

$$\mathbf{y} = [BHA - AHB]\mathbf{u}, \tag{6.88}$$

where

$$LP \triangleq BHA - AHB \tag{6.89}$$

is the discrete lowpass filter matrix of order N. The definition for a circulant matrix is given next.

Definition 6.8. A $(m \times m)$ circulant matrix has the m elements in any one of its rows independent. The elements of the remaining rows are cyclic permutations of these independent elements. Specifically, if $(c_0 \quad c_1 \quad c_2 \quad \cdots \quad c_{m-1})$ is the first row of a circulant matrix C, then $(c_{m-1} \quad c_0 \quad c_1 \quad \cdots \quad c_{m-2})$, $(c_{m-2} \quad c_{m-1} \quad c_0 \quad \cdots \quad c_{m-3})$ $\cdots$ $(c_1 \quad c_2 \quad \cdots \quad c_{m-1} \quad c_0)$ form, respectively, the second row, the third row, and the mth row of C. (Note that a circulant becomes a special case of a Toeplitz matrix.)

In reference [5], it was shown that the LP matrix, defined in (6.89) and explicitly described in (6.90), is sparse. Remember that H is a circulant matrix. For even values of N,

$$LP = \begin{bmatrix} 0 & a_1 & 0 & a_3 & \cdots & a_{N-1} \\ a_1 & 0 & a_1 & 0 & \cdots & 0 \\ 0 & a_1 & 0 & a_1 & \cdots & a_{N-3} \\ \vdots & \vdots & \vdots & \vdots & & \\ a_{N-1} & 0 & a_{N-3} & 0 & \cdots & 0 \end{bmatrix}, \tag{6.90}$$

where

$$a_k = \begin{cases} \dfrac{2}{N} \cot \dfrac{\pi k}{N} \sin \dfrac{2\pi\beta k}{N}, & k \text{ odd} \\ 0, & k = 0 \text{ and } k \text{ even.} \end{cases} \qquad (6.91)$$

Furthermore,

$$a_k = a_{N-k}. \qquad (6.92)$$

The LP matrix can also be easily written for N odd and is seen to have properties similar to the DHT in (2.46). Clearly, LP is a sparse symmetric Toeplitz matrix and may be completely generated by approximately a quarter of the elements in the first row. Note that because of (6.92), LP is also a sparse circulant and has properties very similar to the DHT matrix H (see Section 2.4).

The LP matrix is more sparse when N is even than when N is odd and the interdependence of matrix element values as exhibited, for example, by (6.92), can be exploited to achieve reduced multiplicative complexity for implementing the filtering described in (6.88) and (6.89) in a manner analogous to that illustrated for computing the DHT in (2.48). This is because the LP matrix has properties very similar to the DHT matrix discussed in Section 2.4. Dutta Roy and Agarwal [5] showed that in addition to reduced multiplicative computational complexity and reduced storage, superior noise performance is also attained from the alternative formulation suggested in [7, Chapter 2] and (2.48) for the DHT. Furthermore, the minimal multiplicative complexity theory outlined in Section 2.5 is conveniently adaptable for the situation under consideration here.

Figure 6.5A. Normalized magnitude of frequency response of lowpass filter in Example 6.5.

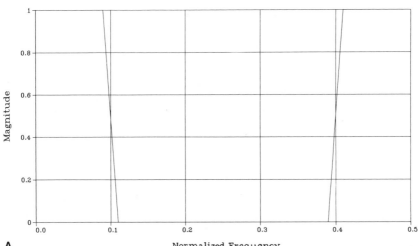

A

Let LP_1 and LP_2 be two discrete lowpass filter matrices associated with cutoff frequencies (expressed as the number of sample points) β_1 and β_2, respectively, with $\beta_1 > \beta_2$. Then, the discrete bandpass filter matrix is

$$BP = LP_1 - LP_2. \tag{6.93}$$

Sabri and Steenart [15] also suggested the use of

$$HP = I - LP \tag{6.94}$$

and

$$BS = I - BP \tag{6.95}$$

as the discrete highpass and discrete bandstop filter matrices, respectively, where I is an identity matrix of appropriate order.

Example 6.5. Let $N = 100$ and $\beta = 10$. It is required to plot the magnitude of the frequency characteristics of the lowpass filter. The input sequence, $\{x(k)\}$, may be taken to be the IDFT of frequency samples each of value 1 between $-0.5 \le f \le 0.5$, where f is the normalized frequency (with respect to the sampling frequency). Clearly,

$$\alpha = \frac{2\pi\beta}{100} = 0.2\pi \text{ rad}$$
$$= 0.1 \text{ Hz.}$$

The matrix LP can be formed. The magnitude of the normalized frequency response of the lowpass filter is plotted in Figure 6.5A and also in decibels in Figure 6.5B. Note that unlike conventional digital filters, the magnitude

Figure 6.5B. Normalized magnitude of frequency response in decibels of lowpass filter in Example 6.5.

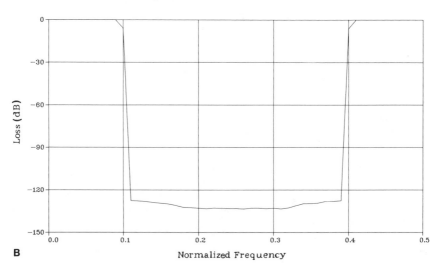

B   Normalized Frequency

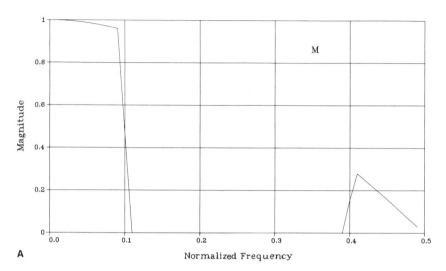

A

Figure 6.6A. Counterpart of Figure 6.5A when zero elements are dropped in the LP matrix.

Figure 6.6B. Counterpart of Figure 6.5B when zero elements are dropped in the LP matrix.

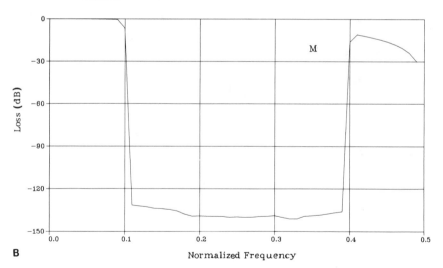

B

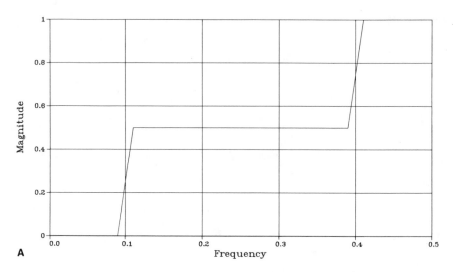

Figure 6.7A. Normalized magnitude of frequency response of highpass filter.

characteristics are symmetric around $f = 0.25$ (i.e., quarter of the sampling frequency) instead of $f = 0.50$. Sabri and Steenart suggested the dropping of the zero elements from the LP matrix for N even in order to remedy this problem. Figure 6.6A, B is the counterpart of Figure 6.5A, B, after the zero elements in the LP matrix were disregarded (this fact is identified in the plots by the additional letter M).

Figure 6.7B. Normalized magnitude of frequency response in decibels of highpass filter.

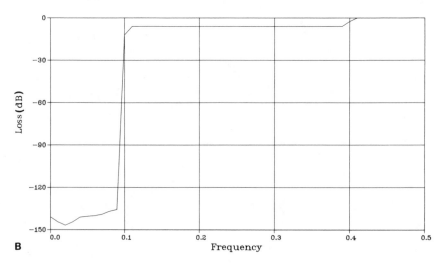

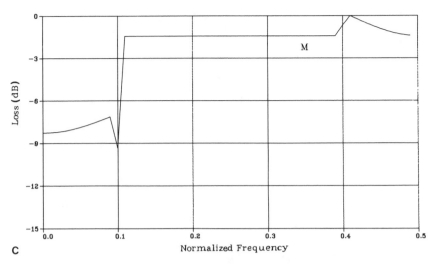

C

Figure 6.7C. Normalized magnitude of frequency response of highpass filter when zero elements are dropped in the LP matrix of Example 6.5.

The results of applying (6.94) and (6.93) with $\beta_1 = 20$ and $\beta_2 = 10$, and (6.95) with $\beta_1 = 30$ and $\beta_2 = 10$ in the preceding example, are reflected in the magnitude of the frequency plots of the highpass, bandpass, and bandstop filters in Figures 6.7A, B, C, 6.8A, B (to get Figure 6.8B the zero elements in the LP_1 and LP_2 matrices were dropped), and Figure 6.9A, B, respectively (to get Figure 6.9B, the zero elements in the LP_1 and LP_2

Figure 6.8A. Normalized magnitude of frequency response of band-pass filter.

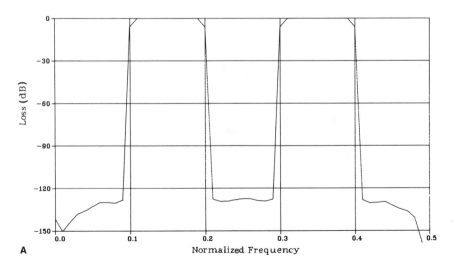

A

392

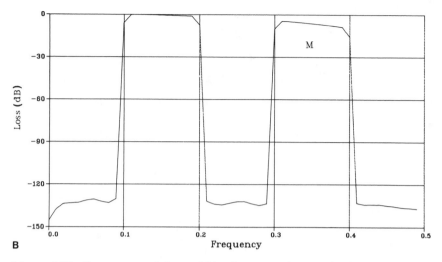

B

Figure 6.8B. Counterpart of figure 6.8A when zero elements in the LP$_1$ and LP$_2$ matrices are dropped.

Figure 6.9A. Normalized magnitude of frequency response of bandstop filter $\beta_1 = 30$, $\beta_2 = 10$.

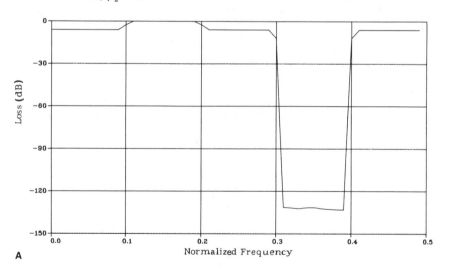

A

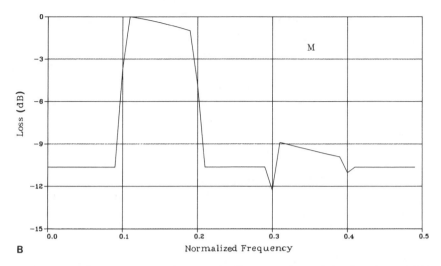

Figure 6.9B. Counterpart of Figure 6.9A when zero elements in the LP_1 and LP_2 matrices are dropped; $\beta_1 = 30$, $\beta_2 = 10$.

matrices were again disregarded). Note that, for the sake of simplicity only, all plots have been drawn as continuous lines instead of discrete points. It is acknowledged that there could be interpolation problems between discrete frequency points.

6.5. Conclusions and Suggested Readings

The exploitation of the structure of matrices characterizing a variety of digital filtering problems is crucial to the development of fast algorithms. The most commonly encountered matrices in engineering and scientific applications are probably the Toeplitz, Hankel, and circulant matrices. Over the years a rich body of results has been developed for these types of matrices. The monograph by Grenander and Szegö [6] is still the authoritative reference on Toeplitz matrices in an analytic setting. Algebraic properties of Toeplitz as well as Hankel matrices are documented in the recent monograph by Iohvidov [7]. The algebraic theory of circulant matrices is documented in the book by Davis [8]. For concise exposition of simple facts concerning Toeplitz and related matrices, the reader is referred to the paper by Roebuck and Barnett [9], which also contains a list of references pertaining to applications, especially in systems theory.

A detailed discussion of digital allpass filters is contained in reference [10], and this reference also includes a comprehensive account of planar least-squares inverse filtering, the Levinson recursion, and their applications in geophysical signal analysis. The recursion used to solve a Toeplitz system of linear equations generates polynomials which, when the Toeplitz matrix is positive definite, form an orthogonal sequence over the unit circle. These orthogonal polynomials, referred to as Szegö polynomials, have several interesting properties which are documented in detail in reference [11]. The counterparts of these polynomials, occurring in the recursion used to solve a Hankel system of linear equations, are referred to as Lanczos polynomials. The Lanczos polynomials are orthogonal over the real line. The derivation of the Lanczos recursion is given in detail. A generalization of this result to the matrix case is available in reference [12]. The techniques to prove certain properties of Lanczos polynomials are scalar versions of the matrix results in reference [13]. It is emphasized that, unlike the Szegö polynomials, Lanczos polynomials do not provide any keys to system stability. A flowchart implementing an algorithm for solving a Hankel system of linear equations can be developed similar to that for solving a Toeplitz system as in Figure 6.1. See reference [14].

The Levinson algorithm for solving a Toeplitz system of linear equations, n in number, requires time proportional to n^2 and memory proportional to n. This is useful because of the variety of applications which require repeatedly the solution of a Toeplitz system of simultaneous linear equations. For example, in a particular oil company dedicated to the interpretation of seismic exploration data, it is usually necessary to solve more than 100,000 sets of Toeplitz linear equations at very regular intervals. The design of prediction error filters required in many fields of applications concerned with prediction of a time series from its specified autocorrelation sequence depends on the solution of Toeplitz systems. In high-resolution spectral analyses, the important problem of power spectrum estimation from a sample of restricted temporal or spatial data, Toeplitz systems again play an important role. The interested reader is referred to the problems at the end of this chapter for more information on some of these applications. The advanced reader who wants to know more about the mathematical properties like orthogonality of the sequence of polynomials generated as a result of the Levinson recursion is referred to reference [17].

Problems

1. Hankel and Toeplitz matrices can be related through use of the square matrix J, which has 1's along the secondary diagonal (i.e., the diagonal from the top right-hand to the bottom left-hand corner) and zeros elsewhere. Note that

   $$J^2 = I \quad \text{and} \quad J^t = J,$$

   where I is an identity matrix of appropriate order. After studying the effects of premultiplying and postmultiplying, respectively, a specified matrix by J, confirm the fact that if H is a Hankel matrix, then JH and HJ are each Toeplitz. Furthermore, verify that

   $$(JH)^t = HJ.$$

2. Show that a persymmetric matrix H can be characterized by

$$JH'J = H,$$

where J is the symmetric matrix defined in the previous problem. If H is nonsingular, verify that H^{-1} is also persymmetric.

3. Is a Toeplitz matrix necessarily persymmetric? Is a persymmetric matrix necessarily Toeplitz? Justify your answers.

4. A square matrix $C = [c_{ij}]$ of order n, where c_{ij} is the element in its ith row and jth column is centrosymmetric, provided

$$c_{ij} = c_{n+1-i, n+1-j}, \quad i, j = 1, \ldots, n.$$

Centrosymmetric matrices occur in some digital filtering applications (e.g., see the paper by E. I. Jury, V. R. Kolavennu, and B. D. O. Anderson entitled, "Stabilization of certain two-dimensional recursive digital filters," published in Proc. IEEE, 65, June 1977, pp. 887–892). Confirm that a centrosymmetric matrix C of order n has $\frac{1}{2}n^2$ independent elements when n is even and $\frac{1}{2}n(n+1)$ independent elements when n is odd. Confirm also that C can be characterized by

$$JCJ = C,$$

where J has been defined in an earlier problem.

5. Show that a persymmetric Hankel matrix must be centrosymmetric.

6. The driving-point impedance or admittance of a one-port network containing positive inductors and capacitors is a reactance function. A reactance function, $F(s)$, has the property: $\operatorname{Re} F(s) \gtreqless 0$, for $\operatorname{Re} s \gtreqless 0$. Furthermore, if $F_1(s)$ and $F_2(s)$ are two reactance functions, then the composite function, $F_1[F_2(s)]$ is also a reactance function. These functions are suited for use in frequency transformations for design of various types of analog filters from a lowpass prototype. Can stable digital allpass functions be viewed as the appropriate counterparts in digital filter theory of what reactance functions are in analog filter theory? Justify your answer with respect to the properties stated for a reactance function. You might like to refer to Problem 1 at the end of Chapter 3 and appropriate sections in Chapter 3.

7. What is the simplest way to stabilize a filter, characterized by the transfer function

$$H(z^{-1}) = \frac{\left(1 + \frac{1}{2}z^{-1}\right)\left(2 + \frac{3}{4}z^{-1}\right)}{\left(1 + 2z^{-1}\right)\left(3 + 4z^{-1}\right)},$$

so that the magnitude of the frequency response is unchanged? Obtain the transfer function of the stabilized filter and state whether or not it is minimum phase.

8. Compare and contrast the results of stabilizing the transfer function in the previous problem via
   a. the planar least-squares inverse technique,
   b. and the discrete Hilbert transform technique,
   from the standpoints of computational complexity in implementation and the quality of approximation measured in terms of an error criterion based on the difference between the magnitude of the frequency responses of the stabilized filter and the original unstable filter.

9. Let $H(z)$ be a stable digital allpass rational function having real coefficients. Show that
   a. $|H(z)| > 1$ for $|z| < 1$
   b. and $|H(z)| < 1$ for $|z| > 1$
   Recall that you used these properties in Theorem 3.1. The above result is the discrete counterpart of the fact that if $H(s)$ is a reactance function, then
   a. $Re\,H(s) > 0$ for $Re\,s > 0$
   b. and $Re\,H(s) < 0$ for $Re\,s < 0$
   Of course,
   $$Re\,H(s) = 0 \quad \text{for} \quad Re\,s = 0.$$

10. Let $\dfrac{P_r(z)}{Q_m(z)}$ denote the $[r/m]$ order Padé approximant to the function e^z. Show that,
   $$\frac{P_0(z)}{Q_0(z)} = 1$$
   $$\frac{P_0(z)}{Q_1(z)} = \frac{1}{1-z}$$
   $$\frac{P_1(z)}{Q_1(z)} = \frac{2+z}{2-z}.$$
   Compute $\dfrac{P_2(z)}{Q_0(z)}, \dfrac{P_3(z)}{Q_3(z)}, \dfrac{P_2(z)}{Q_3(z)}$ and $\dfrac{P_5(z)}{Q_5(z)}$. Check that the approximants you computed satisfy the known closed form expression for the approximant
   $$\frac{P_r(z)}{Q_m(z)} = \frac{\displaystyle\sum_{k=0}^{r} \frac{(r+m-k)!\,r!\,z^k}{(r+m)!\,k!\,(r-k)!}}{\displaystyle\sum_{k=0}^{m} \frac{(r+m-k)!\,m!\,(-z)^k}{(r+m)!\,k!\,(m-k)!}}$$

11. Consider the problems of estimating the point $x(k)$ given the k preceding points, $x(0), x(1), \ldots, x(k-1)$. The minimum mean-square linear predictor is used in a variety of applications, including speech production, model fitting, and seismic signal analysis. For this predictor the estimate $\hat{x}(k)$ is given by
   $$\hat{x}(k) = \sum_{i=1}^{k} h_i^{(k)} x(k-i),$$
   where the predictor coefficients $h_i^{(k)}$ for $i = 1, 2, \ldots, k$ of the kth order predic-

tor have to be determined by minimizing

$$E\left[(x(k)-\hat{x}(k))^2\right],$$

where the symbol E denotes what is referred to as the mathematical expectation operator.

Define

$$r(l)=E[x(k+l)x(k)].$$

According to the orthogonality principle, the least-squares linear prediction error is orthogonal to each preceding point (or data point), that is, for each integer l in $1\le l\le k$,

$$E\left[\left\{x(k)-\sum_{i=1}^{k}h_i^{(k)}x(k-i)\right\}x(k-l)\right]=0.$$

a. Using the properties of the expectation operator,

$$E[x_1(k)+x_2(k)]=E[x_1(k)]+E[x_2(k)]$$

and

$$E[cx(k)]=cE[x(k)],\text{ for a constant }c,$$

show that the predictor coefficients $h_i^{(k)}$ may be calculated by solving a Toeplitz system of equations similar to that encountered in (6.21). You may use the fact that $r(l)=r(-l)$, $l=1,2,\ldots,k$ in order to apply LEVINS.FOR in any specific problem of this type.

b. Find an expression for the prediction error

$$E\left[\{x(k)-\hat{x}(k)\}^2\right]$$

in terms of $r(l)$, $l=0,1,\ldots,k$, and the predictor coefficients.

12. Consider the problem of solving (6.21), assuming that the conditions in (6.19) are valid. In many applications the Toeplitz coefficient matrix is positive definite; suppose this is, indeed, the case. Let $a_i^{(l)}$, $i=1,2,\ldots,l$ be the coefficients in the lth iteration for $l=1,2,\ldots,m$ (note that $-a_i^{(m)}=a_i$ in (6.21) for $i=1,2,\ldots,m$), which are calculated in the Levinson recursion. Denote the mean-square-error in the lth iteration by $E^{(l)}$. Justify that the following steps lead to the solution of the Toeplitz system of equations under consideration via the Levinson recursion. The parameter $\rho^{(l)}$ introduced below is called the reflection coefficient in the lth iteration.

$l=1$

$$\rho^{(1)}=\frac{q_1}{q_0},\quad a_1^{(1)}=\rho^{(1)},\quad E^{(1)}=q_0\left[1-\left(\rho^{(1)}\right)^2\right]$$

$2\le l\le m$

$$\rho^{(l)}=\frac{1}{E^{(l-1)}}\left[q_l-\sum_{i=1}^{(l-1)}a_i^{(l-1)}q_{l-i}\right]$$

$$a_l^{(l)}=\rho^{(l)}$$

$$a_i^{(l)}=\left[a_i^{(l-1)}-\rho^{(l)}a_{l-i}^{(l-1)}\right],\quad i=1,2,\ldots,l-1$$

$$E^{(l)}=E^{(l-1)}\left[1-\left(\rho^{(l)}\right)^2\right].$$

It can be proved that when the Toeplitz coefficient matrix is positive definite, then $|\rho^{(l)}| < 1$, $l = 1, 2, \ldots, m$.

13. In the previous problem, the coefficients $a_i^{(l)}$ are calculated by the Levinson recursion. Denote

$$A^{(l)}(z^{-1}) = 1 - \sum_{i=1}^{l} a_i^{(l)} z^{-i}.$$

Verify that

$$A^{(l)}(z^{-1}) = A^{(l-1)}(z^{-1}) - \rho^{(l)} z^{-l} A^{(l-1)}(z)$$

for $l = 1, 2, \ldots, m$, where $A^{(0)}(z^{-1}) \triangleq 1$.
Can you infer from the statements in the text and at the end of the preceding problem that when $|\rho^{(l)}| < 1$, $l = 1, 2, \ldots, m$ then $A^{(l)}(z^{-1}) \neq 0$ in $|z| \geq 1$, for $l = 1, 2, \ldots, m$?

14. A real-valued time series $\{x(k)\}$ is defined to be an autoregressive (AR) process of order m if it is generated by the recurrence

$$x(k) + \sum_{i=1}^{m} b(i) x(k - i) = a(0) \varepsilon(k),$$

where $\{\varepsilon(k)\}$ is a white noise process. The autocorrelation sequence $\{r(k)\}$ of the time series is specified by

$$r(l) = E\{x(k + l) x(k)\},$$

where the symbol E denotes mathematical expectation. The Fourier transform

$$S(e^{j\omega}) = \sum_{k=-\infty}^{\infty} r(k) e^{-jk\omega}$$

of the autocorrelation sequence is the power spectral density function. Also,

$$r(k) = \frac{1}{2\pi} \int_{-\pi}^{\pi} S(e^{j\omega}) e^{j\omega k} \, d\omega.$$

In the spectrum estimation problem, one seeks to estimate $S(e^{j\omega})$, the power spectral density, from only a finite set of autocorrelations, $r(0), r(\pm 1)$, $\ldots, r(\pm m)$ with $r(-k) = r(k)$. The AR model fitting of this specified set of autocorrelations is done by multiplying both sides of the first equation by $x(k - l)$ and then operating with E to get

$$r(l) + \sum_{i=1}^{m} b(i) r(l - i) = \begin{cases} |a(0)|^2, & l = 0 \\ 0, & l = 1, 2, \ldots, m. \end{cases}$$

Rewrite the above set of equations in the form shown in (6.21), and hence convince yourself of the applicability of Levinson's algorithm in this situation. The estimated power spectral density function is

$$S(e^{j\omega}) = \left| \frac{a(0)}{1 + b(1) e^{-j\omega} + \cdots + b(m) e^{-j\omega m}} \right|^2.$$

15. Consider the following spectral factorization problem. Consider a finite, causal, real sequence $(x(0), x(1), \ldots, x(N - 1))$. Its autocorrelation is a sequence

$(r(-N+1), r(-N+2), \ldots, r(-1), r(0), r(1), \ldots, r(N-2), r(N-1))$, where

$$r(k) = \sum_{i=0}^{N-1} x(i) x(k+i), \quad k = 0, \pm 1, \ldots, \pm(N-1).$$

Clearly $r(k) = r(-k)$. The spectrum $S(z)$ is the z-transform of the autocorrelation sequence

$$S(z) = \sum_{k=-(N-1)}^{N-1} r(k) z^{-k}$$

a. Show that $S(z)$ is a real-valued function on $|z| = 1$.
b. It is stated that $S(z)$ is positive on $|z| = 1$. (In general, recall from Chapter 1 that $S(z)$ is nonnegative on $|z| = 1$.) It is required to find the inverse $A(z^{-1}) = \dfrac{1}{B(z^{-1})}$ of the spectral factor $B(z^{-1})$ in

$$S(z) = B(z) B(z^{-1}),$$

where $B(z^{-1})$ is a polynomial in z^{-1} of degree $(N-1)$.

Show that a polynomial (of degree $N-1$ in z^{-1}) approximant to $A(z^{-1})$, might be obtained by solving a Toeplitz system of linear equations.

16. Consider the formal power series

$$H(z) = \sum_{k=0}^{\infty} h_k z^k.$$

Associate the Hankel matrices H_n and $\hat{H}_n$, shown below, with $H(z)$.

$$H_n = \begin{bmatrix} h_0 & h_1 & h_2 & \cdots & h_n \\ h_1 & h_2 & & & \vdots \\ h_2 & h_3 & & & \\ \vdots & \vdots & & & \\ h_n & \cdots & & & h_{2n} \end{bmatrix},$$

$$\hat{H}_n = \begin{bmatrix} h_1 & h_2 & h_3 & \cdots & h_n \\ h_2 & h_3 & & & \vdots \\ h_3 & h_4 & & & \\ \vdots & \vdots & & & \\ h_n & \cdots & & & h_{2n-1} \end{bmatrix}.$$

The series $H(z)$ is called a Stieltjes series if H_n is positive definite for $n = 0, 1, 2, \ldots$, and $\hat{H}_n$ is negative definite for $n = 1, 2, \ldots$.
a. Justify whether or not the power series expansion of the function

$$F(z) = \frac{1}{\sqrt{z}} \tanh\sqrt{z}$$

about the point $z = 0$ is a Stieltjes series.
b. Calculate the Padé approximants $[\frac{3}{4}]$, $[\frac{4}{4}]$, and $[\frac{5}{5}]$ to the series you obtained in (a).
c. Repeat parts (a) and (b) for the function

$$G(z) = \frac{1}{\sqrt{z}} \coth\sqrt{z}.$$

Note that $F(z)$ and $G(z)$ are driving-point impedances of uniformly distributed RC transmission lines, short-circuited and open-circuited, respectively, at one end.

References

1. Levinson, N. 1947. The Wiener RMS (root mean square) error in filter design. J. Math. Phys. 25:261–278.

2. Baker, G. A. 1975. Essentials of Padé Approximants. Academic Press, New York.

3. Read, R. R., and Treitel, S. 1973. The stabilization of two-dimensional recursive filters via the discrete Hilbert transform. IEEE Trans. Geoscience Electronics, 11:153–207.

4. Gold, B., and Rader, C. M. 1969. Digital Processing of Signals. McGraw-Hill Book Co., New York.

5. Dutta Roy, S. C., and Agrawal, A. 1978. Digital lowpass filtering using the discrete Hilbert transform. IEEE Trans. Acoustics, Speech, Signal Proc., 26:465–467.

6. Grenander, U., and Szegö, G. 1958. Toeplitz Forms and Their Applications. University of California Press, Berkeley, CA.

7. Iohvidov, I. S. 1982. Hankel and Toeplitz Matrices and Forms: Algebraic Theory, Birkhäuser, Boston.

8. Davis, P. K. 1979. Circulant Matrices. John Wiley Interscience, New York.

9. Roebuck, P. A., and Barnett, S. 1978. A survey of Toeplitz and related matrices. Int. J. Systems Sci, 9:921–934.

10. Robinson, E. A., and Treitel, S. 1980. Geophysical Signal Analysis. Prentice-Hall, Englewood Cliffs, NJ.

11. Szegö, G. 1967. Orthogonal Polynomials. 3rd Ed. American Mathematics Society, Providence, RI.

12. Bose, N. K., and Basu, S. 1980. Theory and recursive computation of 1-D matrix Padé approximants. IEEE Trans. Circuits Systems, 27:323–325.

13. Basu, S., and Bose, N. K. 1983. Matrix Stieltjes series and network models. SIAM. J. Math. Anal., 14(2):209–222.

14. Rissanen, J. 1974. Solution of linear equations with Hankel and Toeplitz matrices. Numer. Math., 22:361–366.

15. Sabri, M. S., and Steenart, W. 1977. Discrete Hilbert transform filtering. IEEE Trans. Acoustics, Speech, Signal Proc., 25:452–454.

16. Rabiner, L. A., and Schafer, R. W. 1978. Digital Processing of Speech Signals. Prentice-Hall, Englewood Cliffs, NJ.

17. Markel, J. D., and Gray A. H. Jr. 1976. Linear Prediction of Speech. Springer-Verlag, Berlin, Heidelberg.

18. Akhiezer, N. I., 1965. The Classical Moment Problem. Oliver and Boyd, London.

19. Durbin, J., 1960. The fitting of time series models. Rev. Int. Inst. Stat., 28:233–244.

20. Schroeder, M. R., 1984. Linear prediction, entropy, and speech analysis. IEEE Acoustics, Speech, Signal Proc. Magazine, 1:3–11.

Chapter 7
Special Topics

7.1. Introduction

The presentation of the topic of digital filters within the general framework of digital signal processing has been focused on systems that are linear with respect to the field of real numbers. However, a body of literature has proliferated on the analysis and design of systems that are linear over a finite field Z_q of q elements, $0, 1, 2, \ldots, q-1$, starting with the richly documented results in the theory and applications of shift register sequences [1]. Such sequences have been used extensively to construct error-correcting codes, in multiple address coding, encipherment, counting, scrambling, fault detection, synchronization, as test sequences for system identification [2], in range-radar applications [3], in detecting sonar signals embedded in intense background noise [4], in the design of multilevel sequences for frequency-hopping patterns, and in numerous other problems occurring in ranging systems, radar systems, and spread-spectrum communications [5]. The binary maximal-length linear feedback shift register sequences, commonly referred to as pseudorandom sequences, may be viewed as impulse response sequences of digital filters, whose inputs, outputs, and multiplier coefficients are all elements of the finite field $Z_2 = \{0, 1\}$. These digital filters, referred to as binary digital filters, are introduced in Section 7.2. The structure and properties of the impulse response sequence or such filters are identified with the remarkable properties of pseudorandom sequences, exploited in numerous applications. It is indicated how certain properties (like irreducibility or primitivity, in the context of Section 7.2) of the denominator polynomial of the filter transfer function of specified order affect the period of the impulse response sequence (this sequence is, of necessity, periodic).

Section 7.3 is devoted to a brief exposition of the elements of spatio-temporal filter analysis and synthesis; specifically, the two-dimensional filters are considered. Two-dimensional (2-D) filters, which are linear over the field of real numbers, are becoming increasingly popular for processing

2-D digital data in applications such as image deblurring, medical diagnosis, geophone arrays for seismic applications, scene analysis, weather forecasting, and processing of sonar and radar arrays [6]. A detailed presentation of the topic is beyond the scope of this book; the objective, instead, is to select some topics of interest that either can be viewed as generalizations of one-dimensional (1-D) results to which the reader has already been exposed, or which are based on the foundation provided by the 1-D theory via the use of some simple, but elegant, artifices.

7.2. Binary Digital Filters (Filtration over Z_q, when $q = 2$)

In this section, the theory of one-dimensional discrete-time systems which are linear time-invariant with respect to the finite field Z_q of q elements, $0, 1, 2, \ldots, q - 1$ with q prime, is developed. Of particular interest is the binary field Z_2. For brevity in exposition, and because practically all the results to be discussed have their natural generalization to the case of arbitrary q (with the only constraint that q be a prime number), the case when $q = 2$ will be emphasized. The digital signal is regarded as a sequence of ONE's and ZERO's, and the signal then is referred to as a binary signal. In digital communications, the ONE's could represent pulses and the ZERO's may represent no pulses and a sequence of ONE's and ZERO's is a kind of amplitude-modulated square wave. Binary sequences have been extensively used in digital communications [1, 3] and in identification of control systems [2]. Here, an exposition of the important properties of a type of binary sequence, referred to as a maximum length sequence or pseudorandom sequence, is undertaken in the context of binary digital filtering, where the filter is linear with respect to a finite field of two elements, 0 and 1.

Definition 7.1. A binary digital filter is a circuit that has for its input (and output) a sequence of 0's and 1's, which are ordered in time, and the filter is realizable entirely in terms of logical **and** scalar multiplication, modulo-2 addition, and delays. The input, output, and multiplier coefficients are all elements of Z_2 and the output, in general, is a linear combination of past and present values of the input and past values of the output.

An element $y(k)$ of the output sequence of a causal binary digital filter is then expressible as

$$y(k) = \sum_{r=0}^{m} a(r)x(k-r) + \sum_{r=1}^{n} b(r)y(k-r), \qquad (7.1)$$

where $\{x(k)\}$ and $\{y(k)\}$ are, respectively, the input and output sequences. For a causal sequence, $\{x(k)\}$, $x(k) = 0$, $k < 0$, and in our discussions here, only causal input sequences are considered. On denoting the (one-sided) z-transforms of $\{y(k)\}$ and $\{x(k)\}$ by $Y(z)$ and $X(z)$, respectively, as in Chapter 1, (7.2) follows from (7.1):

$$Y(z) = \left[\sum_{r=0}^{m} a(r)z^{-r} \right] X(z) + \left[\sum_{r=1}^{n} b(r)z^{-r} \right] Y(z)$$
$$+ \sum_{r=1}^{n} [b(r)z^{-r}[y(-r)z^r + \cdots + y(-1)z]]. \tag{7.2}$$

In (7.2), $y(-n)$, $y(-n+1), \ldots, y(-1)$ are the initial conditions and, on rearranging terms after noting that $-1 \equiv 1 \bmod 2$, (7.3) follows (see Problem 1 at the end of this chapter):

$$Y(z) = \frac{\displaystyle\sum_{r=0}^{m} a(r)z^{-r}}{1 + \displaystyle\sum_{r=1}^{n} b(r)z^{-r}} X(z) + \frac{\displaystyle\sum_{r=1}^{n} [b(r)z^{-r}[y(-r)z^r + \cdots + y(-1)z]]}{1 + \displaystyle\sum_{r=1}^{n} b(r)z^{-r}}$$

$$\tag{7.3}$$

The transfer function $H(z)$ is obtained from (7.3), after making all initial conditions zero:

$$H(z) \triangleq \left.\frac{Y(z)}{X(z)}\right|_{\substack{\text{zero initial} \\ \text{conditions}}} = \frac{\displaystyle\sum_{r=0}^{m} a(r)z^{-r}}{1 + \displaystyle\sum_{r=1}^{n} b(r)z^{-r}}. \tag{7.4}$$

$H(z)$ is, thus, a rational transfer function, whose coefficients are either 0 or 1. Strictly proper rational functions, which satisfy the condition in (7.5), will henceforth be of interest.

$$m < n. \tag{7.5}$$

The transfer function $H(z)$ in (7.4) is realizable as in Figure 7.1 (see Problem 2 at the end of this chapter). This realization is, of course, not minimal in the number of delay elements. Various possible realizations are discussed in reference [7].

7.2.1. Impulse Response of Binary Digital Filters

The impulse response sequence $\{h(k)\}$ of a causal binary digital filter is obtainable from the power series expansion of $H(z)$ about $z^{-1} = 0$:

$$H(z) = \sum_{k=0}^{\infty} h(k)z^{-k}. \tag{7.6}$$

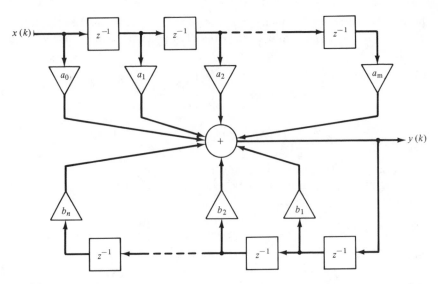

Figure 7.1. A realization of the transfer function in (7.4). Note that $-1 \equiv 1 \bmod 2$, and also for notational convenience a_k, b_k are used to denote $a(k)$, $b(k)$, respectively.

To obtain $\{h(k)\}$ from $H(z)$ in (7.4), first define in (7.4)

$$A(z^{-1}) \triangleq \sum_{r=0}^{n-1} a(r)z^{-r} \tag{7.7a}$$

$$B(z^{-1}) \triangleq \sum_{r=0}^{n} b(r)z^{-r}, \quad b(0)=1. \tag{7.7b}$$

In (7.7a), the constraint of (7.5) has been imposed so that $a(r)=0$, $r > n-1$. Define a causal sequence $\{s(k)\}$,

$$\frac{1}{B(z^{-1})} = \sum_{k=0}^{\infty} s(k)z^{-k}. \tag{7.8}$$

Then,

$$H(z) = \frac{A(z^{-1})}{B(z^{-1})} = A(z^{-1}) \sum_{k=0}^{\infty} s(k)z^{-k}. \tag{7.9}$$

Substituting (7.7a) in (7.9), and making simple algebraic manipulations, using the fact that (z^{-1} is treated as a unit delay operator)

$z^{-r}s(k) = s(k-r)$,

$$H(z) = \sum_{k=0}^{n-1} \left[\sum_{r=0}^{k} a(r)s(k-r) \right] z^{-k} + \sum_{k=n}^{\infty} \left[\sum_{r=0}^{n-1} a(r)s(k-r) \right] z^{-k}.$$

(7.10)

Comparing (7.10) with (7.6), one gets

$$h(k) = \sum_{r=0}^{k} a(r)s(k-r), \quad k = 0,1,\ldots,n-1 \qquad (7.11a)$$

$$h(k) = \sum_{r=0}^{n-1} a(r)s(k-r), \quad k = n,n+1,\ldots. \qquad (7.11b)$$

Substitute (7.7b) in (7.8) and cross-multiply to get

$$s(0) + \sum_{k=1}^{n-1} \left[s(k) + \sum_{r=1}^{k} b(r)s(k-r) \right] z^{-k}$$

$$+ \sum_{k=n}^{\infty} \left[s(k) + \sum_{r=1}^{n} b(r)s(k-r) \right] z^{-k} = 1.$$

On equating coefficients of z^{-k} on both sides of the preceding equation for $k = 0,1,2,\ldots$, and using the fact that $-1 \equiv 1 \bmod 2$,

$$s(0) = 1 \qquad (7.12a)$$

$$s(k) = \sum_{r=1}^{k} b(r)s(k-r), \quad k = 1,2,\ldots,n-1 \qquad (7.12b)$$

$$s(k) = \sum_{r=1}^{n} b(r)s(k-r), \quad k = n,n+1,\ldots. \qquad (7.12c)$$

Since $\{s(k)\}$ in (7.8) is a causal sequence, $s(k) = 0$, $k < 0$, and (7.12) can be written in a slightly more compact form:

$$s(0) = 1 \qquad (7.13a)$$

$$s(k) = \sum_{r=1}^{n} b(r)s(k-r), \quad k = 1,2,\ldots. \qquad (7.13b)$$

Since $\{s(k)\}$ satisfies a linear difference equation of order n having coefficients in Z_2, it can be realized by what is known as a linear feedback shift register (LFSR) as shown in Figure 7.2.

Definition 7.2. A shift register of degree n is a device consisting of n consecutive 2-state memory units. It shifts the content of each memory

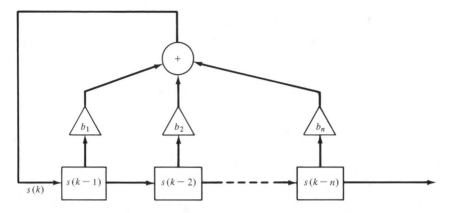

Figure 7.2. A linear feedback shift register realizing the linear difference equation in (7.13b). Note that b_k is used to denote $b(k)$ for notational convenience.

stage to the next memory stage at each clock pulse. A feedback shift register is a shift register including a feedback loop which computes as new term for the first stage a function $f(x_1, x_2,\ldots, x_n)$ of all the present terms in the binary storage elements $x_1, x_2,\ldots, x_n$ (see Figure 7.3). If the feedback function $f(x_1, x_2,\ldots, x_n)$ is expressible in the form

$$f(x_1, x_2,\ldots, x_n) = k_1 x_1 \oplus k_2 x_2 \oplus \cdots \oplus k_n x_n,$$

where each k_i is either 0 or 1 for $i = 1, 2,\ldots, n$ and $\oplus$ denotes modulo-2 addition, then the feedback shift register is a LFSR.

An authoritative account of shift register sequences has been given in reference [1]. Since the properties of the impulse response sequence of a

Figure 7.3. An n-stage nonlinear feedback shift register.

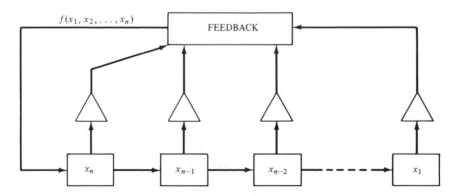

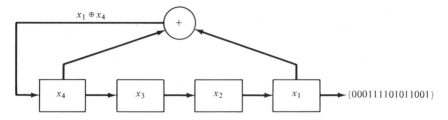

Figure 7.4. The 4-stage linear feedback shift register in Example 7.1.

binary digital filter are identical to the properties of the output sequence of a shift register, a brief discussion of these relevant properties will be given.

Fact 7.1. *The output sequence of the n-stage shift register in Figure 7.2 is periodic with a maximum period of* $2^n - 1$.

The above important property of periodicity is illustrated by an example.

Example 7.1. The initial state of the 4-stage LFSR (degree 4) shown in Figure 7.4 is (proceeding from the left to the right)

$$1 \quad 0 \quad 0 \quad 0.$$

The feedback function for the LFSR is

$$f(x_1, x_2, x_3, x_4) = x_1 \oplus x_4.$$

Starting from the given initial state, it is simple to verify that the succession of states is

1	0	0	0	← Initial state
1	1	0	0	
1	1	1	0	
1	1	1	1	
0	1	1	1	
1	0	1	1	
0	1	0	1	
1	0	1	0	
1	1	0	1	
0	1	1	0	
0	0	1	1	
1	0	0	1	
0	1	0	0	
0	0	1	0	
0	0	0	1	← End of one period
1	0	0	0	
1	1	0	0.	

The sequence of states for the last stage (given by the last column in the above matrix of numbers) in one period is

$$0 \quad 0 \quad 0 \quad 1 \quad 1 \quad 1 \quad 1 \quad 0 \quad 1 \quad 0 \quad 1 \quad 1 \quad 0 \quad 0 \quad 1,$$

which is a sequence of length 15—the maximum possible length over one period obtainable by a LFSR of degree 4. In fact, the sequence of states for each stage has the same period.

Having established the periodicity of $\{s(k)\}$, assume that the period of $\{s(k)\}$ is p, that is,

$$s(k) = s(k+p), \quad \text{for all } k \geq 0. \tag{7.14}$$

Note that since $\{s(k)\}$ is causal, its periodicity is interpreted to be over the domain $0 \leq k < \infty$ and not over $-\infty < k < \infty$. From (7.13) and (7.14) and use of simple arguments it readily follows that

$$s(p-n+1) = s(p-n+2) = \cdots = s(p-1) = 0. \tag{7.15}$$

Fact 7.2. *The period of $\{s(k)\}$ is the smallest integer p such that $B(z)$, obtained by replacing z^{-1} with z in (7.7b), divides $1 + z^p$. (Note that $B(z)$ is obtained from $B(z^{-1})$ after replacing z^{-1} by z and powers of z^{-1} by corresponding powers of z.)*

It is clear, then, that $p \geq n$, where n is the degree of polynomial $B(z)$. Therefore, from (7.11b),

$$h(p+k) = \sum_{r=0}^{n-1} a(r)s(p+k-r), \quad k = 0,1,\ldots. \tag{7.16}$$

From (7.16), (7.11b), and (7.14) it follows that

$$h(p+k) = \sum_{r=0}^{n-1} a(r)s(k-r) = h(k), \quad k > (n-1). \tag{7.17}$$

Similarly, from (7.16), (7.11a), (7.14), and (7.15) it follows that

$$h(p+k) = \sum_{r=0}^{k} a(r)s(p+k-r) + \sum_{r=k+1}^{n-1} a(k)s(p+k-r)$$

$$= \sum_{r=0}^{k} a(r)s(p+k-r) \tag{7.18}$$

$$= h(k), \quad k \leq n-1.$$

Combining (7.17) and (7.18),

$$h(p+k) = h(k), \quad k = 0,1,2,\ldots. \tag{7.19}$$

Thus, the periodicity of the sequence $\{h(k)\}$ has also been established. This important result is summarized next. Again, periodicity is interpreted to be over the domain of causality.

Theorem 7.1. *The impulse response sequence* $\{h(k)\}$ *associated with a binary digital filter characterized by a transfer function*

$$H(z) = \frac{A(z^{-1})}{B(z^{-1})},$$

where $A(z^{-1})$ *and* $B(z^{-1})$ *are defined in* (7.7), *is periodic.*

$B(z^{-1})$, a polynomial in variable z^{-1}, plays an important role in the determination of the period of $\{h(k)\}$ in the preceding theorem, and is often referred to as a characteristic polynomial. $A(z^{-1})$, again viewed as a polynomial in z^{-1}, might affect the period of $\{h(k)\}$, for example, when it has a factor polynomial in z^{-1} common with $B(z^{-1})$. For brevity in notation and exposition, the irreducibility and relative primeness conditions will be given with respect to polynomials $A(z)$ and $B(z)$ (mentioned in Fact 7.2) explicitly introduced in (7.20), instead of $A(z^{-1})$ and $B(z^{-1})$ in (7.7):

$$A(z) \triangleq \sum_{r=0}^{n-1} a(r)z^r \tag{7.20a}$$

$$B(z) \triangleq \sum_{r=0}^{n} b(r)z^r, \quad b(0) = 1. \tag{7.20b}$$

Then, using Fact 7.2 and method of proof based on the principle of contradiction, Fact 7.3, stated below, can be verified. (See Problem 3 at the end of this chapter.)

Fact 7.3. *The sequence* $\{h(k)\}$ *associated with* $H(z) = A(z^{-1})/B(z^{-1})$, *where* $A(z^{-1})$ *and* $B(z^{-1})$ *are defined in* (7.7), *has the same period as the sequence* $\{s(k)\}$ *associated with* $1/B(z^{-1})$, *if the polynomials* $A(z)$ *and* $B(z)$, *defined in* (7.20) *are relatively prime.*

Figure 7.5 lists the program GCF.FOR that can be used to determine the greatest common factor (or greatest common divisor) of two polynomials (and therefore can also be used to test two polynomials for relative primeness) whose coefficients are in Z_q, q prime—the finite field of q elements, $\{0,1,2,\ldots,q-1\}$. Figure 7.6A lists the program GFOD1.FOR that can be used to obtain the periodic sequence $\{s(k)\}$ that is associated with $1/[B(z^{-1})]$ in (7.8) via use of (7.13). For the sake of brevity, the period or exponent of $\{s(k)\}$ is often referred to as the period or exponent of $B(z^{-1})$. Figure 7.6B lists the program GFOD2.FOR that allows one to obtain the periodic sequence $\{h(k)\}$, associated with $H(z) = A(z^{-1})/B(z^{-1})$, via use of the recurrence relations in (7.11) and (7.13).

```
C
C
C       ***********************************************************************
C
C       THIS PROGRAM DETERMINES THE GREATEST COMMON FACTOR
C       BETWEEN TWO POLYNOMIALS A(Z) AND B(Z) WITH COEFF. IN GF(2)
C       N IS THE DEGREE OF B,        M IS THE DEGREE OF A
C       TAKE N GREATER THAN M
C       READ THE COEFFICIENTS IN INCREASING POWER OF Z
C
C       ***********************************************************************
C
        IMPLICIT INTEGER (A-Z)
        DIMENSION A(0:50),B(0:50),C(0:50)
        WRITE(6,10)
10      FORMAT(1X,'THE DEGREES OF B(Z) AND A(Z) ARE;',1X)
        READ(5,11) N,M
11      FORMAT(2I)
        WRITE(6,12)
12      FORMAT(1X,'THE COEFF. OF B(Z) ARE;',1X)
        READ(5,13) (B(I),I=0,N)
13      FORMAT(1X,50I1)
        WRITE(6,13) (B(I),I=0,N)
        WRITE(6,14)
14      FORMAT(1X,'THE COEFF. OF A(Z) ARE:',1X)
        READ(5,13) (A(I),I=0,M)
        WRITE(6,13) (A(I),I=0,M)
C
C       CALCULUS OF THE COMMON FACTOR
C
1       K=N-M
        DO 50 I=K,N
        IF((B(I)-A(I-K)).EQ.0) GO TO 51
        C(I)=1
        GO TO 50
51      C(I)=0
50      CONTINUE
        IF(K.LT.1) GO TO 65
        DO 60 I=0,K-1
60      C(I)=B(I)
65      DO 70 J=1,N
        IF(C(N-J).EQ.1) GO TO 71
70      CONTINUE
        GO TO 1100
71      NN=N-J
        IF(NN.EQ.0) GO TO 2000
        IF(NN.GE.M) GO TO 100
        N=M
        M=NN
        DO 80 I=0,N
```

Figure 7.5 *(Partial listing, continued.)*

```
80        B(I)=A(I)
          DO 90 I=0,M
90        A(I)=C(I)
          GO TO 1
100       N=NN
          DO 110 I=0,N
110       B(I)=C(I)
          GO TO 1
1100      WRITE(6,500)
500       FORMAT(1X,'THE GREATER COMMON FACTOR IS:',1X)
          WRITE(6,13) (A(I),I=0,M)
          GO TO 1500
2000      WRITE(6,600)
600       FORMAT(1X,'THE POLYNOMIALS ARE RELATIVELY PRIME IN GF(2)',1X)
1500      STOP
          END
```

Figure 7.5. Listing of program GCF.FOR to determine the greatest common factor of two polynomials having coefficients in Z_2.

In Table 7.1, one period of sequence $\{s(k)\}$ (denoted by $\{s(k)\}_p$) associated with a transfer function of the type $1/B(z^{-1})$ is given when the polynomial $B(z^{-1})$ is of degree n with respect to variable z^{-1}. For a fixed value of n, those polynomials which generate a periodic sequence of maximum period, $2^n - 1$, are important. A detailed study of these types of polynomials, their relationship to irreducible polynomials, and properties of maximum length sequences are undertaken in the next section. We terminate this section by including an example, which illustrates the result summarized in Fact 7.3, especially with respect to the role of $A(z^{-1})$ in influencing the period of the sequence $\{h(k)\}$ associated with $A(z^{-1})/B(z^{-1})$, which may or may not be in reduced form.

Example 7.2. 1. Let

$$B(z^{-1}) = 1 + z^{-5}$$
$$A(z^{-1}) = 1 + z^{-1} + z^{-2} + z^{-3} + z^{-4}.$$

By use of GCF.FOR $A(z)$ divides into $B(z)$. By use of GFOD2.FOR the sequence $\{h(k)\}$ is described by

$$\{h(k)\}_p = 1,$$

while from Table 7.1 the sequence $\{s(k)\}$ over one period is

$$\{s(k)\}_p = 1 \quad 0 \quad 0 \quad 0 \quad 0.$$

Note that element values over one period only are given for the periodic sequences. The subscript underscores this fact.

```
C
C       **************************************************************
C
C       THIS PROGRAM EVALUATES THE SEQUENCE GENERATED BY A
C       TRANSFER FUNCTION OF THE FORM 1/B(Z) WITH COEFFICIENTS
C       IN GF(Q).
C       N2 IS THE DEGREE OF B(Z) AND NQ IS THE MODULUS OF GF(Q)
C       THE SEQUENCE GENERATED BY 1/B(Z) IS STORED IN NS(I)
C
C       **************************************************************
C
        DIMENSION NB(0:50),NS(0:500)
        READ(5,100) NQ,N2
100     FORMAT(2I)
110     FORMAT(I)
        DO 4 I=0,N2
4       READ(5,110) NB(I)
        WRITE(6,150)
150     FORMAT(1X,'THE DENOMINATOR POLYNOMIAL IS:',1X)
        WRITE(6,140) (NB(I),I=0,N2)
C
C       EVALUATE THE SEQUENCE GENERATED BY 1/B(Z)
C
        NPERD=NQ**N2-1
        NS(0)=1
        DO 5 J=1,N2
        NS(J)=0
        DO 6 JJ=1,J
        NS(J)=NS(J)+NB(JJ)*NS(J-JJ)
6       CONTINUE
C
C       MOD Q ALGEBRA
C
        M=NS(J)/NQ
        NS(J)=NQ-NS(J)+M*NQ
        IF((NQ-NS(J)).EQ.0) GO TO 11
        GO TO 5
11      NS(J)=0
5       CONTINUE
        DO 7 J=N2+1,NPERD
        NS(J)=0
        DO 8 JJ=1,N2
        NS(J)=NS(J)+NB(JJ)*NS(J-JJ)
8       CONTINUE
C
C       MOD Q ALGEBRA
C
        M=NS(J)/NQ
        NS(J)=NQ-NS(J)+M*NQ
        IF((NQ-NS(J)).EQ.0) GO TO 12
        GO TO 7
```

Figure 7.6A. *(Partial listing, continued.)*

```
12        NS(J)=0
7         CONTINUE
          WRITE(6,130) NQ
          WRITE(6,120)
120       FORMAT(1X,'THE SEQUENCE GENERATED BY 1/B(Z) IN GF(Q) IS:',1X)
130       FORMAT(1X,'WE ARE WORKING IN GF MOD:',2X,I2)
          WRITE(6,140) (NS(I),I=0,NPERD-1)
140       FORMAT(1X,52OI3)
          STOP
          END
```

Figure 7.6A. Listing of program GFOD1.FOR to obtain the impulse response of casual filter characterized by $\dfrac{1}{B(z^{-1})}$.

2. Let

$$B(z^{-1}) = 1 + z^{-1} + z^{-4}$$
$$A(z^{-1}) = 1 + z^{-1} + z^{-2}.$$

Then it can be verified that

$$\{h(k)\}_p = 1\ \ 0\ \ 1\ \ 1\ \ 0\ \ 0\ \ 1\ \ 0\ \ 0\ \ 0\ \ 1\ \ 1\ \ 1\ \ 1\ \ 0,$$

which is a shifted version of the corresponding $\{s(k)\}_p$ in Table 7.1 because $A(z)$ and $B(z)$ are relatively prime.

3. Let

$$B(z^{-1}) = 1 + z^{-6}$$
$$A(z^{-1}) = 1 + z^{-1}.$$

Here,

$$\{h(k)\}_p = 1\ \ 1\ \ 0\ \ 0\ \ 0\ \ 0,$$

and $\{h(k)\}$ happens to have the same period as $\{s(k)\}$ associated with $1/B(z^{-1})$ even though $A(z)$ and $B(z)$ are not relatively prime.

7.2.2. Primitive Polynomials

In this section the relation between $B(z^{-1})$ and the period of the sequence $\{s(k)\}$ generated from

$$\frac{1}{B(z^{-1})} = \sum_{r=0}^{\infty} s(r)z^{-r}$$

is considered, especially when maximum length sequences are desired. Again for the sake of brevity in notation and exposition, the above equation

```
C      ************************************************************************
C
C      THIS PROGRAM EVALUATES THE SEQUENCE GENERATED BY A
C      TRANSFER FUNCTION OF THE FORM A(Z)/B(Z) WITH COEFFICIENTS
C      IN GF(Q).
C      N1 IS THE DEGREE OF THE NUMERATOR
C      N2 IS THE DEGREE OF THE DENOMINATOR
C      THE SEQUENCE GENERATED BY 1/B(Z) IS STORED IN NS(I)
C      THE SEQUENCE GENERATED BY A(Z)/B(Z) IS STORED IN NR(I)
C
C      ************************************************************************
C
       DIMENSION NA(0:50),NB(0:50),NS(0:500),NR(0:500)
       READ(5,100) NQ,N1,N2
100    FORMAT(3I)
       DO 1 I=0,N1
1      READ(5,110) NA(I)
110    FORMAT(I)
       WRITE(6,120)
120    FORMAT(1X,'THE NUMERATOR POLYNOMIAL IS:',1X)
       WRITE(6,230) (NA(I),I=0,N1)
       DO 2 I=0,N2
2      READ(5,110) NB(I)
       WRITE(6,150)
150    FORMAT(1X,'THE DENOMINATOR POLYNOMIAL IS:',1X)
       WRITE(6,230) (NB(I),I=0,N2)
C      EVALUATE THE SEQUENCE GENERATED BY 1/B(Z)
       NPERD=NQ**N2-1
       NS(0)=1
       DO 5 J=1,N2
       NS(J)=0
       DO 6 JJ=1,J
       NS(J)=NS(J)+NB(JJ)*NS(J-JJ)
6      CONTINUE
C      MOD Q ALGEBRA
       M=NS(J)/NQ
       NS(J)=NQ-NS(J)+M*NQ
       IF((NQ-NS(J)).EQ.0) GO TO 11
       GO TO 5
11     NS(J)=0
5      CONTINUE
       DO 7 J=N2+1,NPERD
       NS(J)=0
       DO 8 JJ=1,N2
       NS(J)=NS(J)+NB(JJ)*NS(J-JJ)
8      CONTINUE
C      MOD Q ALGEBRA
       M=NS(J)/NQ
       NS(J)=NQ-NS(J)+M*NQ
       IF((NQ-NS(J)).EQ.0) GO TO 12
       GO TO 7
```

Figure 7.6B. *(Partial listing, continued.)*

```
12        NS(J)=0
7         CONTINUE
          IF(N1.EQ.0) GO TO 27
C         EVALUATE THE SEQUENCE GENERATED BY THE TRANSFER FUNCTION
C         STORE THE RESULT IN NR(I)
          DO 21 J=0,N1
          NR(J)=0
          DO 22 JJ=0,J
          NR(J)=NR(J)+NS(J-JJ)*NA(JJ)
22        CONTINUE
C         MOD Q ALGEBRA
          M=NR(J)/NQ
          NR(J)=NR(J)-M*NQ
21        CONTINUE
          DO 23 J=N1+1,NPERD
          NR(J)=0
          DO 24 JJ=0,N1
          NR(J)=NR(J)+NA(JJ)*NS(J-JJ)
24        CONTINUE
C         MOD Q ALGEBRA
          M=NR(J)/NQ
          NR(J)=NR(J)-M*NQ
23        CONTINUE
27        WRITE(6,210) NQ
210       FORMAT(1X,'WE ARE WORKING IN GF MOD: ,2X,I2)
          WRITE(6,220)
220       FORMAT(1X,'THE SEQUENCE GENERATED BY 1/B(Z) IS .1X)
          WRITE(6,230) (NS(I),I=0,NPERD-1)
230       FORMAT(1X,51013)
          IF(N1.EQ.0) GO TO 28
          WRITE(6,240)
240       FORMAT(1X,'THE SEQUENCE GENERATED BY A(Z)/B(Z) IS:',1X)
          WRITE(6,230) (NR(I),I=0,NPERD-1)
28        STOP
          WRITE(6,330) NQ
          WRITE(6,320)
320       FORMAT(1X,'THE SEQUENCE GENERATED BY 1/B(Z) IN GF(Q) IS:',1X)
330       FORMAT(1X,'WE ARE WORKING IN GF MOD:',2X,I2)
          WRITE(6,140) (NS(I),I=0,NPERD-1)
140       FORMAT(1X,52012)
          STOP
          END
```

Figure 7.6B. Listing of program GFOD2.FOR to obtain the impulse response of casual filter characterized by $\dfrac{A(z^{-1})}{B(z^{-1})}$.

Table 7.1. One Period of Sequence $\{s(k)\}$ in (7.8)

Degree n	Coefficients of $B(z^{-1})$ in ascending powers of z^{-1}	One period of $\{s(k)\}$ computed via use of GFOD1.FOR
1	1 0	1
1	1 1	1
2	1 0 1	1 0
	1 1 1	1 1 0
3	1 1 1 1	1 1 0 0
	1 0 1 1	1 0 1 1 1 0 0
	1 1 0 1	1 1 1 0 1 0 0
	1 0 0 1	1 0 0
4	1 1 1 1 1	1 1 0 0 0
	1 0 0 0 1	1 0 0 1
	1 0 0 1 1	1 0 0 1 1 0 1 0 1 1 1 1 1 0 0 0
	1 0 1 0 1	1 0 1 0 0 0
	1 1 0 0 1	1 1 1 1 0 1 0 1 1 0 0 1 0 0 0
5	1 1 1 1 1 1	1 1 0 0 0 0
	1 0 0 0 0 1	1 0 0 0 0
	1 0 0 1 0 1	1 0 0 1 0 1 1 0 0 1 1 1 1 1 0 0 0 1 1 0 1 1 1 0 1 0 1 0 0 0 0
6	1 0 0 0 0 0 1	1 0 0 0 0 0
	1 0 0 1 0 0 1	1 0 0 1 0 0 0 0 0
	1 1 1 1 1 1 1	1 1 0 0 0 0 0

will be considered with z^{-1} replaced by z, that is,

$$\frac{1}{B(z)} = \sum_{r=0}^{\infty} s(r)z^r, \qquad (7.21)$$

and results will be discussed with respect to polynomial $B(z)$ in (7.20b). Also, the period of $\{s(k)\}$ will be referred to also as the period or exponent of $B(z)$.

Fact 7.4. *If $B(z)$ can be factored as*

$$B(z) = B_1(z)B_2(z),$$

where $B_1(z)$ and $B_2(z)$ are relatively prime polynomials, then the period of $B(z)$ is the least common multiple of the periods of $B_1(z)$ and $B_2(z)$.

Example 7.3. Let

$$B_1(z) = z + 1$$
$$B_2(z) = z^4 + z^3 + z^2 + z + 1.$$

From Table 7.1, the periods of $B_1(z)$ and $B_2(z)$ are, respectively, 1 and 5. The period of $B(z) = B_1(z)B_2(z) = 1 + z^5$ is 5.

Fact 7.5. *If $B_1(z)$ is irreducible and $B(z) = [B_1(z)]^n$ for some positive integer n, the period p of $B(z)$ is a multiple $e(n)$ of the period p_1 of $B_1(z)$, that is,*

$$p = e(n)p_1,$$

where

$$e(n) = 2^k$$

and

$$2^{k-1} < n \leq 2^k.$$

Example 7.4. Let

$$B(z) = (z+1)^4 = 1 + z^4.$$

The period of $B_1(z) = 1 + z$ is $p_1 = 1$. Here, $n = 4$, and $2^{2-1} < 4 \leq 2^2$, so that $k = 2$. Therefore, $e(4) = 2^2 = 4$ and the period of $B(z)$ by use of Fact 7.5 is

$$p = 4 \cdot 1 = 4,$$

which agrees with the result in Table 7.1.

To be able to generate a sequence of maximum possible period, $2^n - 1$, from a polynomial $B(z)$ of a specified degree n, it is necessary that $B(z)$ be an irreducible polynomial. However, the irreducibility of $B(z)$ is not sufficient for it to generate a maximum length sequence.

Example 7.5. It is well known that a polynomial $(1 + z^p)$ factors into a product of two factors with coefficients in Z_2. To wit,

$$(1 + z^5) = (1 + z)(1 + z + z^2 + z^3 + z^4),$$

where $1 + z + z^2 + z^3 + z^4$ is an irreducible polynomial of degree $n = 4$. From Table 7.1, however, the period associated with this polynomial is only 5 and not 15 (or $2^4 - 1$).

The necessity of irreducibility for a polynomial to generate a sequence of maximum length can be easily demonstrated. (See Problem 4 at the end of this chapter.) When $2^n - 1$ is a prime number, then an irreducible polynomial of degree n will generate a maximum length sequence; in general, when $2^n - 1$ is not a prime number the period of an irreducible polynomial of degree n is a factor of $2^n - 1$, and therefore is always an odd number.

Definition 7.3. A polynomial $B(z)$ of degree n whose period is $2^n - 1$ is called a primitive polynomial.

A result of fundamental importance in the construction of error-correcting codes is that there exists at least one irreducible polynomial of every degree. In fact, the number of irreducible polynomials, $\psi(n)$, of specified degree n is

$$\psi(n) = \frac{1}{n} \sum_{k|n} 2^k \mu\left(\frac{n}{k}\right), \qquad (7.22)$$

where $\mu(k)$ is the Moebius function defined by

$$\mu(k) = \begin{cases} 1 & \text{if } k = 1 \\ 0 & \text{if } k \text{ has any repeated prime factor} \\ (-1)^m & \text{if } k \text{ is a product of } m \text{ distinct primes} \end{cases} \qquad (7.23)$$

and the summation in (7.22) is over all positive divisors k of n (including 1 and n).

Example 7.6. It is required to calculate $\psi(6)$. The positive integers that divide 6 are 1, 2, 3, 6. Clearly, on using (7.23),

$$\mu(6) = 1, \quad \mu(3) = -1, \quad \mu(2) = -1, \quad \mu(1) = 1.$$

Applying (7.22),

$$\psi(6) = \frac{1}{6}\left[(2)(1) + (4)(-1) + (8)(-1) + (64)(1)\right] = 9.$$

There also exists at least one primitive polynomial of every degree. In fact, the number $\lambda(n)$ of primitive polynomials of degree n is

$$\lambda(n) = \frac{\phi(2^n - 1)}{n}, \qquad (7.24)$$

where $\phi(k)$ is the Euler totient function,

$$\phi(k) = \prod_r p_r^{(e_r - 1)}(p_r - 1), \qquad (7.25a)$$

with

$$k = \prod_r p_r^{e_r} \qquad (7.25b)$$

being the factorization of k as a product of irreducible factors.

Example 7.7. It is required to calculate $\lambda(10)$. Note that

$$2^{10} - 1 = 1023 = (3)(11)(31)$$
$$\phi(1023) = (2)(10)(30) = 600.$$

Applying (7.24),

$$\lambda(10) = \frac{600}{10} = 60.$$

It is mentioned that a primitive polynomial of degree n is an irreducible factor of the cyclotomic polynomial $C_p(z)$, $p = 2^n - 1$.

$$C_p(z) = \prod_{k|p}(z^k + 1)^{\mu(p/k)},$$

Table 7.2. Number of Primitive Polynomials
of Degree n, $2 \leq n \leq 25$

n	$2^n - 1$	$\lambda(n)$
2,	3,	1
3,	7,	2
4,	15,	2
5,	31,	6
6,	63,	6
7,	127,	18
8,	255,	16
9,	511,	48
10,	1023,	60
11,	2047,	176
12,	4095,	144
13,	8191,	630
14,	16383,	756
15,	32767,	1800
16,	65535,	2048
17,	131071,	7710
18,	262143,	7776
19,	524287,	27594
20,	1048575,	24000
21,	2097151,	84672
22,	4194303,	120032
23,	8388607,	356960
24,	16777215,	276480
25,	33554431,	1296000

where the product is extended over all positive divisors of p and the exponent, of course, involves the Moebius function. Table 7.2 gives $\lambda(n)$ for $n = 2, \ldots, 25$, obtained from computer program NUPREP.FOR, a listing for which is also provided in Figure 7.7.

7.2.3. Pseudorandom Sequences

In the previous section, it was pointed out that there exist periodic sequences of length $2^n - 1$ for a certain polynomial $B(z)$ of specified degree n. These maximum length (period) sequences, generated by primitive polynomials, satisfy the following "randomness properties." When generated by a shift register, the sequences are also deterministic and therefore not genuine random sequences. They are commonly referred to as pseudorandom sequences.

Fact 7.6. *One period of a maximum length sequence of length $2^n - 1$ contains exactly 2^{n-1} 1's and $(2^{n-1} - 1)$ 0's.*

```
C
C       A FORTRAN PROGRAM (NUPREP.FOR) TO COMPUTE THE NUMBER OF
C       PRIMITIVE BINARY POLYNOMIALS OF SPECIFIED DEGREE N.
C
        INTEGER IPR(10),IER(10)
C
        WRITE(5,110)
110     FORMAT(2X,'ENTER N, THE DEGREE OF  POLYNOMIAL')
        READ(5,120) N
120     FORMAT(I)
        K=2**N-1
        K1=K/2
        J=0
        L=1
        DO 150 I=2,K1
130     M=MOD(K,I)
        IF(M.NE.0) GO TO 140
        K=K/I
        J=J+1
        GO TO 130
140     IF(J.EQ.0) GO TO 150
        IER(L)=J
        IPR(L)=I
        L=L+1
        J=0
        IF(K.LE.2) GO TO 160
150     CONTINUE
160     IF(K.EQ.1)  GO TO 170
        IPR(L)=2**N-1
        IER(L)=1
        GO TO 180
170     L=L-1
180     WRITE(6,190)
190     FORMAT(//,10X,'2**N-1 IS FACTORED AS FOLLOW',//)
        WRITE(6,200) ((IPR(I),IER(I)),I=1,L)
200     FORMAT(15X,I8,'**',I2)
C
        IPROD=1
        DO 210 I=1,L
210     IPROD=IPROD*IPR(I)**(IER(I)-1)*(IPR(I)-1)
        LAMBDA=IPROD/N
        WRITE(6,220) N,LAMBDA
220     FORMAT(///,4X,'THE NUMBER, LAMBDA(',I2,
     1  '), OF PRIM. POLY. =',I16)
        STOP
        END
```

Figure 7.7. Listing of program NUPREP.FOR to compute the number of primitive binary polynomials of specified degree.

Fact 7.7. *In every period of a maximum length sequence, half the runs have length 1, one fourth have length 2, one eighth have length 3, one sixteenth have length 4,..., one 2^kth have length k..., as long as the number of runs (of successive 1's or successive 0's) so indicated is greater than 1. Thus, there are two runs of length k for each run of length $k + 1$.*

Fact 7.8. *If a window of width n is slid along a maximum period sequence, each of the $2^n - 1$ nonzero binary n-tuples is seen exactly once.*

Fact 7.9. *Let $S_0 \triangleq \{s(0), s(1),...\}$ be a maximum period sequence and let $S_k \triangleq \{s(k), s(k+1),...\}$ be a cyclic shift of S_0. Then, $S_0 + S_k = \{s(0) + s(k), s(1) + s(k+1),...\}$ is $S_j = \{s(j), s(j+1),...\}$ for some j, where S_j is also a maximum period sequence and is, furthermore, a cyclic shift of S_0.*

Fact 7.10. *If a period of a maximum length sequence is compared, term by term, with any cyclic shift of itself, the number of matchings differ from the number of nonmatchings by at most 1. In fact, if the 1's are replaced by -1's and the 0's by $+1$'s, the discrete autocorrelation function of a maximum length binary sequence $\{s(k)\}$ of period p is*

$$\rho(0) = 1$$
$$\rho(i) = -\frac{1}{p}, \quad 1 \le i \le p - 1.$$

See Problem 8 at the end of this chapter for the definition of discrete autocorrelation function.

Example 7.8. Consider a sequence in Table 7.1:

$$1 \ 0 \ 0 \ 1 \ 0 \ 1 \ 1 \ 0 \ 0 \ 1 \ 1 \ 1 \ 1$$
$$1 \ 0 \ 0 \ 0 \ 1 \ 1 \ 0 \ 1 \ 1 \ 1 \ 0 \ 1 \ 0$$
$$1 \ 0 \ 0 \ 0 \ 0$$

Since the above sequence is over one period of a primitive polynomial, it is a maximum-length sequence. In fact, since the polynomial is of degree $n = 5$ and the period is $31 = 2^5 - 1$, by Fact 7.1 the sequence must be a maximum length sequence. The following observations may be made:

The number of 1's	$= 16$.
The number of 0's	$= 15$.
The number of runs	$= 16$.
The number of length 1 runs $=$	8.
The number of length 2 runs $=$	4.
The number of length 3 runs $=$	2.

Also,

the number of length 4 runs $= 1$

and

the number of length 5 runs $= 1$.

Facts 7.6 and 7.7 are satisfied. It may be verified that Facts 7.8, 7.9, and 7.10 are also satisfied.

7.3. Two-Dimensional Linear Shift-Invariant Digital Filters

Processing of multidimensional data for extraction of useful or desired information is required in numerous applications. In the preceding chapters the temporal variable of time was the only independent variable, and a signal that is a function of time is a one-dimensional (1-D) signal. In many physical problems, however, signals may be functions of either two or more spatial variables or one or more spatial variables and the temporal variable. Such signals are n-D ($n \geq 2$) signals or multidimensional signals. The pictorial signal, for example, in a video transmission system is characterized by its brightness function, which is a function of two spatial variables and the time-variable. Seismic arrays are employed in event location, studies of geophysical structures, and in the process of differentiation between man-made events like low-yield underground nuclear explosion and natural events like earthquakes or tremors. In ordinary geophysical exploration and prospecting, man-made explosions are used to generate seismic signals which penetrate into land formations, having different characteristics for the sedimentary layers, with different velocities. The reflected signals, where the signal is a function of a spatial variable and the time variable, are recorded as traces by a sensor array. The received signal is usually corrupted by noise from a variety of causes. These include wind noise, noise due to reflections from unimportant soil layers, direct wave propagation from the source of explosion to the sensor array, and industrial noise. In order to filter out the noise, seismic data-processing requires the design of filters which discriminate between events based upon their velocities. These velocity filters are two-dimensional (when the velocity filtering is implemented from data collected from a one-dimensional or linear array of sensors). Other occurrences of multidimensional signals that have to be processed or filtered are included in the areas of digital imagery for medical x-rays, photographic data and nonphotographic remotely sensed data, where the signal corresponding to the pixel intensity is a function of two spatial variables, and array processing required in (besides seismology) sonar, radar, physiological and radioastronomy problems, where the multidimensional data from the sensor arrays have to be processed for the estimation of the speed and direction of the propagating wave. Radar and sonar problems call for location and classification of targets that radiate or reflect energy. Electromagnetic radiation from different regions of the universe are analyzed by radio astronomists while propagating wavefronts in the electroencephalogram are of interest to physiologists.

In the subsequent sections, attention is centered around the fundamentals of bidimensional (two-dimensional or 2-D) filtering of 2-D signals. The reader will then be able to record the similarities and differences between some of the 1-D results presented in earlier chapters and their corresponding 2-D counterparts. The more enthusiastic and motivated reader should be able to examine, question, and conclude from the 2-D results presented whether or not these results can be naturally generalized to $n - D$ $(n > 2)$ situations.

A 2-D discrete signal, $\{x(k_1, k_2)\}$, is usually obtained by sampling a 2-D continuous signal $x_c(t_1, t_2)$, where t_1 and t_2 may either be spatial variables or one of them may be a spatial variable and the other a temporal (or time) variable. Rectangular sampling of $x_c(t_1, t_2)$ is done by sampling $x_c(t_1, t_2)$ every T_1 seconds along the t_1 direction and every T_2 seconds along the t_2 direction. Then, for any set of integers (k_1, k_2),

$$x(k_1, k_2) \triangleq x_c(k_1 T_1, k_2 T_2).$$

For certain continuous signals, whose Fourier transform,

$$X_c(\omega_1, \omega_2) = \int_{-\infty}^{\infty} \int_{-\infty}^{\infty} x_c e^{-j(\omega_1 t_1 + \omega_2 t_2)} \, dt_1 \, dt_2$$

is band-limited, that is, $X_c(\omega_1, \omega_2) = 0$ outside a region of finite extent in the (ω_1, ω_2) plane, it is possible to select the sampling periods T_1, T_2 to be small enough so that $x_c(t_1, t_2)$ may be reconstructed from all the samples in the sequence $\{x(k_1, k_2)\}$. It can be shown that if

$$X_c(\omega_1, \omega_2) = 0, \quad |\omega_1| > \omega_{01}, \quad |\omega_2| > \omega_{02},$$

then by choosing

$$T_1 < \frac{\pi}{\omega_{01}}, \quad T_2 < \frac{\pi}{\omega_{02}},$$

the continuous signal $x_c(t_1, t_2)$ may be reconstructed from the samples $x(k_1, k_2)$ in the sequence $\{x(k_1, k_2)\}$ by using the reconstruction formula

$$x_c(t_1, t_2) = \sum_{k_1 = -\infty}^{\infty} \sum_{k_2 = -\infty}^{\infty} x(k_1, k_2) \left[\prod_{i=1}^{2} \frac{\sin(\pi/T_i)(t_i - k_i T_i)}{(\pi/T_i)(t_i - k_i T_i)} \right].$$

Note that the kernel enclosed in square brackets is a product separable 2-D function. The preceding result is a generalization of the 1-D sampling theorem (recall theorem 1.1) and the 1-D reconstruction formula, which were discussed in Section 1.2. The 2-D problem offers additional flexibilities. Indeed, the sampling raster may not be rectangular; it is possible to sample a continuous signal, whose Fourier transform is band-limited, over nonunique sampling geometries and still be able to reconstruct the continuous signal from the samples. A general treatment of the sampling strategies in the 2-D case is beyond both the scope and the intent of the subsequent exposition; the interested reader will find additional details in reference [26].

For our purpose here, it will be assumed that the 2-D digital signal available for processing was either self-existent (i.e. no analog-to-digital conversion was necessary) or was derived from a continuous signal via rectangular sampling. In the latter case it is not uncommon to denote the discrete sequence $\{x(k_1, k_2)\}$ by $\{x(k_1T_1, k_2T_2)\}$ in order to bring out explicitly the sampling periods in the underlying analog to digital conversion via rectangular sampling. Either notation is usable without any possibility of confusion.

7.3.1. Elementary Analysis

Let $\{x(n_1, n_2)\}$ be a two-dimensional (2-D) sequence x (more accurately, *bisequence*, but for brevity it will be referred to as sequence), where n_1, n_2 each belong to an index set of integers. This sequence may be obtainable, for example, by sampling a spatial process like a pictorial image.

Common examples of 2-D sequences are

Unit impulse:
$$\delta(n_1, n_2) = \begin{cases} 1 & n_1 = n_2 = 0 \\ 0 & \text{otherwise} \end{cases}$$

Unit step:
$$u(n_1, n_2) = \begin{cases} 1 & n_1 \geq 0, \ n_2 \geq 0 \\ 0 & \text{otherwise} \end{cases}$$

Sampled sinusoid: $s(n_1, n_2) = e^{j(\omega_1 n_1 + \omega_2 n_2)}$
$\qquad -\infty \leq n_i \leq \infty, \ i = 1, 2$ and ω_i is the angular frequency in radians per second.

Exponential:
$$e(n_1, n_2) = \begin{cases} a_1^{n_1} a_2^{n_2} & n_1 \geq 0, \ n_2 \geq 0 \\ 0 & \text{otherwise.} \end{cases}$$

The region in the (n_1, n_2) plane, where a sequence is nonzero, is the region of support of the sequence. The region of supports for typical sequences are shown in Figure 7.8. In this section attention will be centered around 2-D linear shift-invariant (LSI) filters defined next.

Definition 7.4. Let $\{x_i(n_1, n_2)\}$ and $\{y_i(n_1, n_2)\}$, be, respectively, the input and output sequences of a 2-D digital filter, for $i = 1, 2, \ldots$. This filter may be characterized, then, by an operator O, which maps the set of input sequences to the corresponding set of output sequences:

$$O\{x_i(n_1, n_2)\} = \{y_i(n_1, n_2)\}.$$

The 2-D digital filter is LSI if and only if

$$O\left\{\sum_i c_i x_i(n_1, n_2)\right\} = \sum_i c_i \{y_i(n_1, n_2)\}$$

for any admissible input sequence $\{x_i(n_1, n_2)\}$ and scalar constants c_i, and

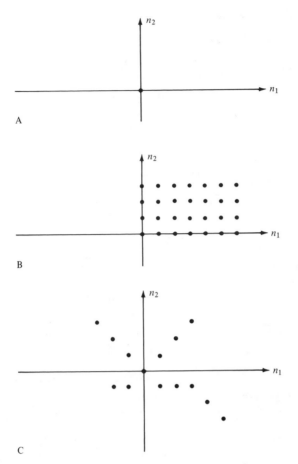

Figure 7.8. Regions of support of 2-D sequences; (A) the sequence is a unit impulse; (B) the sequence is a unit step; (C) a sequence with region of support in all four quadrants.

furthermore,

$$O\{x_i(n_1 - k_1, n_2 - k_2)\} = \{y_i(n_1 - k_1, n_2 - k_2)\}$$

for arbitrary integers k_1, k_2.

See Problem 9 at the end of this chapter for use of Definition 7.4.

Example 7.9. The following difference equation provides an input/output description of a system

$$y(n_1, n_2) + \tfrac{1}{2}y(n_1 - 1, n_2) + \tfrac{1}{4}y(n_1, n_2 - 1) = x(n_1, n_2).$$

It is required to determine whether or not the system is LSI.

Adopting the previous notations,

$$y_i(n_1, n_2) + \tfrac{1}{2}y_i(n_1 - 1, n_2) + \tfrac{1}{4}y_i(n_1, n_2 - 1) = x_i(n_1, n_2),$$

multiply the previous equation by c_i, $i = 1, 2$ and add to get

$$\sum_{i=1}^{2} \left[c_i y_i(n_1, n_2) + \tfrac{1}{2}c_i y_i(n_1 - 1, n_2) + \tfrac{1}{4}c_i y_i(n_1, n_2 - 1) \right] = \sum_{i=1}^{2} c_i x_i(n_1, n_2).$$

Therefore, the response to an input

$$x_3(n_1, n_2) \triangleq \sum_{i=1}^{2} c_i x_i(n_1, n_2)$$

is

$$y_3(n_1, n_2) \triangleq \sum_{i=1}^{2} c_i y_i(n_1, n_2).$$

Again, after replacing n_i by $(n_i - k_i)$, $i = 1, 2$ it can be seen from the characterizing difference equation that the response to an input $x_i(n_1 - k_1, n_2 - k_2)$ is $y_i(n_1 - k_1, n_2 - k_2)$. Therefore, the system is LSI.

Example 7.10. Consider the following input/output description of a system

$$y(n_1, n_2) + y(n_1, n_2 - 1) = a(n_1, n_2)x(n_1, n_2),$$

where $a(n_1, n_2)$ is a spatially varying coefficient. It is required to determine whether or not the system is shift-invariant.

After replacing n_i by $(n_i - k_i)$, $i = 1, 2$, the above difference equation becomes

$$y(n_1 - k_1, n_2 - k_2) + y(n_1 - k_1, n_2 - 1 - k_2)$$
$$= a(n_1 - k_1, n_2 - k_2)x(n_1 - k_1, n_2 - k_2).$$

However, shift invariance requires that the response to $x(n_1 - k_1, n_2 - k_2)$ be $y(n_1 - k_1, n_2 - k_2)$, satisfying the governing equation

$$y(n_1 - k_1, n_2 - k_2) + y(n_1 - k_1, n_2 - 1 - k_2)$$
$$= a(n_1, n_2)x(n_1 - k_1, n_2 - k_2).$$

Since the preceding two equations are not identical, the system is not shift-invariant, as expected. You can, however, establish that the system is linear. A LSI 2-D digital filter is characterizable by its unit impulse response sequence, $\{h(n_1, n_2)\}$, defined next.

Definition 7.5. The unit impulse response sequence $\{h(n_1, n_2)\}$ is the output sequence of a 2-D LSI digital filter when the input sequence is the unit impulse $\{\delta(n_1, n_2)\}$.

$$\{h(n_1, n_2)\} = O\{\delta(n_1, n_2)\}.$$

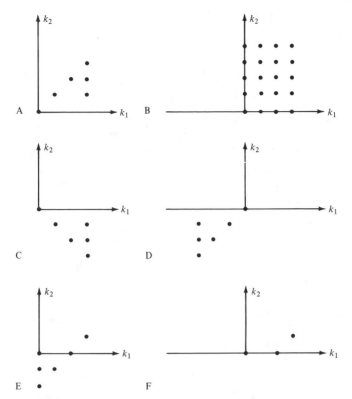

Figure 7.9. Illustrating the various steps in a 2-D convolution; (A) $\{h(k_1, k_2)\}$; (B) $\{x(k_1, k_2)\}$; (C) $\{h(k_1, -k_2)\}$; (D) $\{h(-k_1, -k_2)\}$; (E) $\{h(3 - k_1, 1 - k_2)\}$; (F) $y(3, 1)$ in (7.26b). Only regions of support are shown.

Let $\{x(n_1, n_2)\}$ be the input to a 2-D LSI digital filter whose unit impulse response sequence is $\{h(n_1, n_2)\}$. Then the output sequence $\{y(n_1, n_2)\}$ is obtainable by convolving $\{h(n_1, n_2)\}$ with $\{x(n_1, n_2)\}$.

$$\{y(n_1, n_2)\} = \{h(n_1, n_2)\} * \{x(n_1, n_2)\},$$

where

$$y(n_1, n_2) = \sum_{k_1} \sum_{k_2} h(k_1, k_2) x(n_1 - k_1, n_2 - k_2) \qquad (7.26a)$$

$$= \sum_{k_1} \sum_{k_2} x(k_1, k_2) h(n_1 - k_1, n_2 - k_2). \qquad (7.26b)$$

$\{h(n_1 - k_1, n_2 - k_2)\}$ is obtained from $\{h(k_1, k_2)\}$ by reflecting $\{h(k_1, k_2)\}$ about both axes and then translating each point n_1 units to the right and n_2 units up. (See Figure 7.9.) The output from a 2-D digital filter may also be computable in the transform domain as in the 1-D case using the 2-D z-transform defined next.

Definition 7.6. The z-transform, $X(z_1, z_2)$ of a sequence $\{x(n_1, n_2)\}$ is defined to be

$$X(z_1, z_2) = Z\{x(n_1, n_2)\} = \sum_{k_1 = -\infty}^{\infty} \sum_{k_2 = -\infty}^{\infty} x(k_1, k_2) z_1^{-k_1} z_2^{-k}.$$

$\{x(n_1, n_2)\} \leftrightarrow X(z_1, z_2)$ will be used to denote a z-transform pair. When the lower limits in the summation are $k_1 = k_2 = 0$, $\{x(n_1, n_2)\}$ will be called a first quadrant quarter-plane causal sequence. The values of z_1, z_2 for which the z-transform coverage absolutely form the region of convergence.

From 7.26a it follows that

$$\sum_{n_1} \sum_{n_2} y(n_1, n_2) z_1^{-n} z_2^{-n_2} = \sum_{n_1} \sum_{n_2} \sum_{k_1} \sum_{k_2} h(k_1, k_2) x(n_1 - k_1, n_2 - k_2) z_1^{-n_1} z_2^{-n_2}.$$

Therefore, from the definition of z-transform,

$$Y(z_1, z_2) = \sum_{n_1} \sum_{n_2} \sum_{k_1} \sum_{k_2} h(k_1, k_2) z_1^{-k_1} z_2^{-k_2} \times$$

$$x(n_1 - k_1, n_2 - k_2) z_1^{-(n_1 - k_1)} z_2^{-(n_2 - k_2)}.$$

After interchanging the order of summations on the right-hand side of the above equation, followed by simple algebraic manipulations, it can be shown that

$$Y(z_1, z_2) = H(z_1, z_2) X(z_1, z_2), \tag{7.27}$$

where

$$H(z_1, z_2) \leftrightarrow \{h(n_1, n_2)\}, \quad X(z_1, z_2) \leftrightarrow \{x(n_1, n_2)\}.$$

For a quarter-plane LSI 2-D filter, the transfer function $H(z_1, z_2)$ in (7.27), relating the output transform $Y(z_1, z_2)$ to the input transform $X(z_1, z_2)$, is a rational function of the form

$$H(z_1, z_2) = \frac{\displaystyle\sum_{r_1 = 0}^{m_1} \sum_{r_2 = 0}^{m_2} a(r_1, r_2) z_1^{-r_1} z_2^{-r_2}}{\displaystyle\sum_{r_1 = 0}^{n_1} \sum_{r_2 = 0}^{n_2} b(r_1, r_2) z_1^{-r_1} z_2^{-r_2}} = \frac{Y(z_1, z_2)}{X(z_1, z_2)}, \quad b(0,0) \neq 0$$

$$\tag{7.28}$$

($H(z_1, z_2)$ has been written as a rational function in z_1^{-1}, z_2^{-1} because it is sometimes convenient to work directly with the delay variables). In (7.28), $a(r_1, r_2)$'s and $b(r_1, r_2)$'s are *real constants* and without any loss of generality it is convenient to assign

$$b(0,0) = 1. \tag{7.29}$$

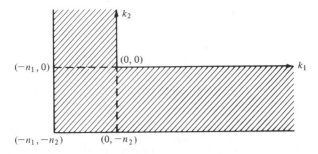

Figure 7.10. Boundary (or initial) conditions required for implementing the difference equation in (7.30) are contained in the hatched area.

Then $H(z_1, z_2)$ in (7.28) is obtained by applying the z-transform, under zero boundary conditions, to the 2-D difference equation relating the output sequence $\{y(k_1, k_2)\}$ to the input sequence $\{x(k_1, k_2)\}$ of the 2-D filter.

$$
\begin{aligned}
y(k_1, k_2) = \sum_{r_1=0}^{m_1} \sum_{r_2=0}^{m_2} a(r_1, r_2) x(k_1 - r_1, k_2 - r_2) \\
- \sum_{\substack{r_1=0 \\ r_1+r_2 \neq 0}}^{n_1} \sum_{r_2=0}^{n_2} b(r_1, r_2) y(k_1 - r_1, k_2 - r_2).
\end{aligned}
\tag{7.30}
$$

The preceding equation permits recursive computation along a certain direction as you will see. For recursive computations along other directions, (7.30) should be rewritten in other forms, which will be mentioned.

If the difference equation in (7.30) is to be used to compute $y(k_1, k_2)$ for all $k_1 \geq 0$, $k_2 \geq 0$, then a set of boundary conditions has to be specified. Specifically, the set of boundary conditions that are required is contained in the hatched region of Figure 7.10. When $y(k_1, k_2)$ is obtained via (7.30), using the boundary conditions in the hatched region of Figure 7.10, the filter is said to recurse in the $(+k_1, +k_2)$ direction. It is also possible to rewrite (7.30) so that $y(k_1 - n_1, k_2 - n_2)$ appears on the left-hand side of the equation and the rest appear on the right. When computation of the output sequence is based on this equation with the appropriate set of initial conditions, the filter is said to recurse in the $(-k_1, -k_2)$ direction. Also it is possible for the filter to recurse in the $(+k_1, -k_2)$ or $(-k_1, +k_2)$ directions. When the recursion is along the $(+k_1, +k_2)$ direction, the filter will be referred to as the first quadrant quarter-plane type of filter. Even for the first quadrant quarter-plane type of filter, the output sequences may be computed using different types of sequencing schemes. Four possibilities for sequencing are sketched in Figure 7.11. In the first two cases, the sequencing rules can be programmed simply, but the output points have to be com-

Figure 7.11. Different types of sequencing schemes for implementing a first quadrant quarter-plane filter.

puted serially, while in the last two cases the sequencing rules may be a bit more complex but carry along with them the advantage of the feasibility for parallel computation.

The frequency response of the 2-D LSI digital filter is obtained by evaluating $H(z_1, z_2)$ in (7.28) at various points on the distinguished boundary T^2 of the unit bidisc, where $|z_1| = |z_2| = 1$. The magnitude-squared function of the frequency response is

$$T\left(e^{j\omega_1 T_1}, e^{j\omega_2 T_2}\right) = H\left(e^{j\omega_1 T_1}, e^{j\omega_2 T_2}\right) H\left(e^{-j\omega_1 T_1}, e^{-j\omega_2 T_2}\right)$$
$$= |H\left(e^{j\omega_1 T_1}, e^{j\omega_2 T_2}\right)|^2. \qquad (7.31)$$

In (7.31), ω_1 and ω_2 are, respectively, the horizontal and vertical spatial frequencies in radians per seconds, while T_1 and T_2 are the sampling periods in the horizontal and vertical directions. The frequency response is doubly periodic with periods $2\pi/T_1$ and $2\pi/T_2$ along the horizontal and vertical directions, respectively.

The phase function associated with the frequency response of the filter is

$$\phi\left(e^{j\omega_1 T_1}, e^{j\omega_2 T_2}\right) = \frac{1}{2j} \log_e \frac{H\left(e^{j\omega_1 T_1}, e^{j\omega_2 T_2}\right)}{H\left(e^{-j\omega_1 T_1}, e^{-j\omega_2 T_2}\right)}. \qquad (7.32)$$

The group delay response functions, $\tau_1(e^{j\omega_1 T_1}, e^{j\omega_2 T_2})$ and $\tau_2(e^{j\omega_1 T_1}, e^{j\omega_2 T_2})$ are defined as

$$\tau_i\left(e^{j\omega_1 T_1}, e^{j\omega_2 T_2}\right) = -\frac{\partial\left[\phi\left(e^{j\omega_1 T_1}, e^{j\omega_2 T_2}\right)\right]}{\partial\omega_i}, \quad \text{for } i = 1, 2. \qquad (7.33)$$

As in the 1-D case, an IIR 2-D filter is realized via a recurrence relation by which the output samples of the filter are explicitly determined as a weighted sum of past output samples as well as past and/or present input samples. When the output samples of the filter are explicitly determined as a weighted sum of past and present input samples only, the 2-D discrete-space filter is nonrecursive. The transfer function of a 2-D LSI first-quadrant quarter-plane type of recursive filter, whose implementation is given by the difference equation of (7.30), was seen to be a rational function of the delay variables z_1^{-1}, z_2^{-1}. When this type of filter is implemented nonrecursively, the transfer function specializes to a polynomial in z_1^{-1}, z_2^{-1}. However, it should be borne in mind that unlike 1-D discrete-time filters, where the past, present, and future are uniquely defined, 2-D discrete-space filters provide flexibility in the choice of ordering, by which one can distinguish between past, present, and future samples. See Problems 13 and 14 at the end of this chapter for filters that are not of quarter-plane type. These filters, though recursively implementable, have a wider region of support for their impulse responses than quarter-plane filters.

The flexibility in ordering in the 2-D case can be exploited to realize 2-D recursible filters different from the first-quadrant quarter-plane type (for example, 2-D filters of the "asymmetric half-plane" type are known to have less restricted frequency responses than quarter-plane filters and therefore are more versatile in filter design; see Problem 13 at the end of this chapter for the region of support for the impulse response of an asymmetric half-plane filter).

Example 7.11. Let the difference equation relating the input and output sequences $\{y(k_1, k_2)\}$ and $\{x(k_1, k_2)\}$ of a first-quadrant quarter-plane 2-D recursive digital filter be (see Problem 15 at the end of this chapter for inference on stability)

$$y(k_1, k_2) - y(k_1 - 1, k_2) - y(k_1, k_2 - 1) = x(k_1, k_2), \ k_1 \geq 0, \ k_2 \geq 0.$$

Calculate the transfer function of the filter and also find the unit impulse response sequence.

Let $y(k_1, k_2) \leftrightarrow Y(z_1, z_2)$, $x(k_1, k_2) \leftrightarrow X(z_1, z_2)$. On z-transforming the given equation under zero boundary conditions,

$$Y(z_1, z_2)\left[1 - z_1^{-1} - z_2^{-1}\right] = X(z_1, z_2).$$

The transfer function is

$$H(z_1, z_2) = \frac{Y(z_1, z_2)}{X(z_1, z_2)} = \frac{1}{1 - \left(z_1^{-1} + z_2^{-1}\right)}.$$

The unit impulse response sequence $\{h(k_1, k_2)\}$ is obtained from

$$H(z_1, z_2) = \frac{1}{1 - \left(z_1^{-1} + z_2^{-1}\right)} = \sum_{k_1 = 0}^{\infty} \sum_{k_2 = 0}^{\infty} h(k_1, k_2) z_1^{-k_1} z_2^{-k_2}.$$

$$(7.34)$$

$\{h(k_1, k_2)\}$ is, in principle, computable via use of the inverse z-transform formula:

$$h(k_1, k_2) = \left(\frac{1}{2\pi j}\right)^2 \oint_{C_1} \oint_{C_2} H(z_1, z_2) z_1^{-1+k_1} z_2^{-1+k_2} \, dz_1 \, dz_2,$$

where the contour of integration for both integrals must lie completely within the region of convergence in the z_1, z_2 planes. Since the integral is, in general, difficult to evaluate (to appreciate the magnitude of difficulty, even in a relatively simple problem, consult Problem 15 at the end of this chapter), and since the partial-fraction expansion (see Problem 24 at the end of this chapter) method for inversion does not apply in the 2-D case, ad hoc schemes will be used to obtain $\{h(k_1, k_2)\}$. On taking a formal power series expansion for the left-hand side of (7.34) (i.e., without worrying about convergence properties of power series),

$$\sum_{k_1=0}^{\infty} \sum_{k_2=0}^{\infty} h(k_1, k_2) z_1^{-k_1} z_2^{-k_2} = \sum_{k=0}^{\infty} \left(z_1^{-1} + z_2^{-1} \right)^k$$

$$= \sum_{k=0}^{\infty} \sum_{r=0}^{\infty} \binom{k}{r} \left(z_1^{-1} \right)^k \left(z_2^{-1} \right)^{k-r},$$

$$\binom{k}{r} \triangleq \frac{k!}{r!(k-r)!}.$$

On substituting $r = k_1$, $k - r = k_2$, the right-hand side of the preceding equation becomes

$$\sum_{k_2=0}^{\infty} \sum_{k_1=0}^{\infty} \binom{k_1 + k_2}{k_1} z_1^{-k_1} z_2^{-k_2}.$$

Therefore,

$$h(k_1, k_2) = \frac{(k_1 + k_2)!}{k_1! k_2!}.$$

Next, the 2-D counterpart of a stability criterion encountered in the 1-D case (Theorem 1.2) is stated.

2-D BIBO Stability Criterion. *A 2-D LSI digital filter is BIBO stable if and only if its unit impulse response sequence, $\{h(k_1, k_2)\}$, is absolutely summable.*

For the 2-D first quadrant quarter-plane LSI filter, the counterpart of Theorem 1.4 required in the test for the BIBO stability criterion is given next.

Fact 7.11. *A 2-D first quadrant quarter-plane LSI filter, characterized by the transfer function in (7.28) is BIBO stable if (but not only if) the "characteristic polynomial" (the expression in the denominator of (7.28)) is nonzero in the domain described below.*

$$\sum_{r_1=0}^{n_1} \sum_{r_2=0}^{n_2} b(r_1, r_2) z_1^{-r_1} z_2^{-r_2} \neq 0, \quad |z_1| \geq 1 \quad \text{and} \quad |z_2| \geq 1$$

or, equivalently,

$$\sum_{r_1=0}^{n_1} \sum_{r_2=0}^{n_2} b(r_1, r_2) z_1^{r_1} z_2^{r_2} \neq 0, \quad |z_1| \leq 1 \quad \text{and} \quad |z_2| \leq 1.$$

The stability test in Fact 7.11, though more difficult to implement than in the 1-D case, can be checked for either algebraically or numerically. Various methods advanced for the purpose are documented in reference [8]. Since those methods are beyond the scope of this book, the interested reader is referred to the text [8] as well as the survey paper [9]. Unlike the 1-D case, the numerator of the transfer function might influence the BIBO stability property in the 2-D case due to the cancellation of common zeros on T^2 between two relatively prime polynomials. (The common zeros of the numerator and denominator polynomials of a rational function, after cancellation of common factors, are referred to as nonessential singularities of the second kind.) The zeros of the denominator polynomial of a bivariate rational function in reduced form are the nonessential singularities of the first kind.

Example 7.12. Let

$$H(z_1, z_2) = \frac{\left(1 - z_1^{-1}\right)^2 \left(1 - z_2^{-1}\right)^2}{2 - z_1^{-1} - z_2^{-1}} = \frac{(z_1 - 1)^2 (z_2 - 1)^2}{2z_1^2 z_2^2 - z_1^2 z_2 - z_2^2 z_1}.$$

The numerator and denominator polynomials in the rightmost expression above do not have any common factor; however, these polynomials have a common zero at $z_1 = 1$, $z_2 = 1$. Although the denominator, $2 - z_1^{-1} - z_2^{-1}$ has a zero on $|z_1| = |z_2| = 1$, it can be shown that the filter characterized by $H(z_1, z_2)$ is BIBO stable. Therefore, it is emphasized that the criterion in Fact 7.11 is sufficient but not necessary for BIBO stability.

In certain simple cases, the condition in Fact 7.11 may be expressible as a necessary and sufficient condition involving the coefficients of the polynomial under test. It is not very difficult to show, for example, that the polynomial with real coefficients

$$1 + az_1 + bz_2 + cz_1 z_2 \neq 0, \quad |z_1| \leq 1, |z_2| \leq 1$$

if and only if

$$|a| < 1 \tag{7.35a}$$

$$\left| \frac{1 + a}{b + c} \right| < 1 \tag{7.35b}$$

$$\left| \frac{1 - a}{b - c} \right| > 1. \tag{7.35c}$$

In the next section, you will be making use of the above inequalities.

The support of the impulse response sequence $\{h(k_1, k_2)\}$ of a first-quadrant quarter-plane 2-D digital filter is in the first quadrant. The following impulse response sequences generated from $\{h(k_1, k_2)\}$ are, respectively, associated with second-, third-, and fourth-quadrant quarter-plane filters.

$$h_2(k_1, k_2) = h(-k_1, k_2)$$
$$h_3(k_1, k_2) = h(-k_1, -k_2)$$
$$h_4(k_1, k_2) = h(k_1, -k_2).$$

The z-transforms $H_2(z_1, z_2)$, $H_3(z_1, z_2)$, $H_4(z_1, z_2)$ of $\{h_2(k_1, k_2)\}$, $\{h_3(k_1, k_2)\}$, $\{h_4(k_1, k_2)\}$ are, respectively, $H(z_1^{-1}, z_2)$, $H(z_1^{-1}, z_2^{-1})$, and $H(z_1, z_2^{-1})$, where $\{h(k_1, k_2)\} \leftrightarrow H(z_1, z_2)$ forms a z-transform pair. The counterpart of Fact 7.11 yielding the stability criterion for quarter-plane filters made to recurse along $(-k_1, +k_2)$, $(-k_1, -k_2)$, and $(+k_1, -k_2)$ directions can be easily obtained (see Problem 20 at the end of this chapter).

Quarter-plane filters made to recurse along different directions may be cascaded to produce zero phase filters. It is easy to verify that the following transfer functions characterize 2-D zero phase recursive filters:

$$H_5(z_1, z_2) = H(z_1, z_2) H(z_1^{-1}, z_2^{-1}) \tag{7.36a}$$

$$H_6(z_1, z_2) = H(z_1, z_2) H(z_1^{-1}, z_2) H(z_1^{-1}, z_2^{-1}) H(z_1, z_2^{-1}) \tag{7.36b}$$

$$H_7(z_1, z_2) = H(z_1^{-1}, z_2) H(z_1, z_2^{-1}), \tag{7.36c}$$

where $H(z_1, z_2)$ is the transfer function of a first-quadrant quarter-plane type of filter. The four-factor decomposition suggested by $H_6(z_1, z_2)$ has broad consequences. In fact, it was first shown by Pistor [16] that an unstable zero-phase 2-D LSI recursive filter with a finite number of coefficients can always be decomposed exactly into four stable 2-D LSI recursive filters, when made to recurse along the $(+k_1, +k_2)$, $(-k_1, +k_2)$, $(-k_1, -k_2)$, and $(+k_1, -k_2)$ directions, respectively. Ekstrom and Woods [17] extended the results to a broader class of functions. The main drawback of these results is that the factors are, in general, of infinite order and, before implementation, truncation and subsequent test for stability are necessary.

7.3.2. 2-D Recursive (IIR) Filter Design

Various schemes for designing 2-D recursive filters have been proposed. Some of these techniques that make good use of the various theoretical results presented earlier on 1-D filters will be discussed here.

7.3.2.1. Rotated Filters. It is possible to obtain a 2-D digital filter from a 1-D analog prototype after a rotation of the frequency axis followed by a continuous to discrete bilinear transformation. The details of the procedure may be summarized as follows [24]:

Algorithm

Step 1. Design a stable 1-D analog lowpass filter transfer function, $H_a(s)$, giving Butterworth, Chebyshev, or other desired types of frequency responses. Set s equal to either s_1 or s_2.

Step 2. From $H_a(s_2)$ [or $H_a(s_1)$] obtain $H_a(s_1, s_2)$.

$$H_a(s_1, s_2) \triangleq H_a(-s_1\sin\theta + s_2\cos\theta).$$

This operation results in a transfer function whose frequency response is obtained by rotating the frequency response of $\hat{H}_a(s_1, s_2) \triangleq H_a(s_2)$ [or $H_a(s_1)$] by the angle θ.

Step 3. Apply the double bilinear transformation

$$s_1 = k_1 \frac{z_1 - 1}{z_1 + 1}, \quad k_1 > 0$$

$$s_2 = k_2 \frac{z_2 - 1}{z_2 + 1}, \quad k_2 > 0$$

to give a 2-D digital filter having a transfer function

$$H(z_1, z_2) = H_a\left[-k_1 \frac{z_1 - 1}{z_1 + 1}\sin\theta + k_2 \frac{z_2 - 1}{z_2 + 1}\cos\theta\right].$$

Unless otherwise specified, set $k_1 = k_2 = k$. (Filters having circularly symmetric frequency responses are of interest here.) Then this step is also equivalent to frequency scaling the analog filter in Step 1 by replacing s with ks, and assigning $k_1 = k_2 = 1$ in Step 3.

Step 4. Cascade several rotated filters, as obtained above with values of θ equally spaced between 180° and 360°.

The constant k in Step 3 is positive and is used to readjust the cutoff frequency of the 2-D digital filter. Due to the double periodicity of the frequency response of 2-D digital filters, the desired frequency response of rotated filters is distorted to an extent that depends on the angle of rotation. The cutoff frequency and the shape of the magnitude response depends on the number of rotated filters being cascaded. The technique used to obtain a 2-D filter with a specified cutoff frequency in a given direction consists of an iteration that modifies the cutoff frequency of the original 1-D continuous filter until the desired cutoff frequency of the 2-D digital filter is attained. The primary disadvantage of the rotated filter discussed above is that it may be BIBO unstable, and slight perturbation of the coefficients of the filter transfer function is usually necessary to guarantee stability [10].

Example 7.13. Consider the normalized second-order Butterworth analog filter transfer function

$$H_a(s) = \frac{1}{s^2 + \sqrt{2}\,s + 1}.$$

$H_a(s)$ is formed from the poles of $1/(1+s^4)$ in the left half-plane (see Section 3.2). Form the 2-D digital filter transfer function

$$H(z_1, z_2) = H_a\left[-\frac{z_1 - 1}{z_1 + 1}\sin\theta + \frac{z_2 - 1}{z_2 + 1}\cos\theta \right].$$

The plots of $|H(e^{j\theta_1}, e^{j\theta_2})|$ vs θ_1, θ_2 in the range $-\pi \le \theta_i \le \pi$, $i = 1, 2$ are shown in Figure 7.12 for the cases when (a) $\theta = 315°$, (b) $\theta = 285°$, and (c) $\theta = 345°$.

Goodman [10] discussed the possibility of implementing the design technique described above by starting with a digital rather than an analog 1-D lowpass filter. Basically, instead of using the mapping

$$s \to -k\frac{z_1 - 1}{z_1 + 1}\sin\theta + k\frac{z_2 - 1}{z_2 + 1}\cos\theta,$$

as in Step 3 above with $k_1 = k_2 = k$, he considered the map

$$z^{-1} \to \frac{1 + k[(z_1 - 1)/(z_1 + 1)]\sin\theta - k[(z_2 - 1)/(z_2 + 1)]\cos\theta}{1 - k[(z_1 - 1)/(z_1 + 1)]\sin\theta + k[(z_2 - 1)/(z_2 + 1)]\cos\theta},$$

$$(7.37)$$

in order to arrive at a 2-D digital transfer function $H(z_1, z_2)$ from a designed 1-D digital filter transfer function $H(z)$. Either $H(z)$ is obtained as described in Section 3.2 or it is derived from an analog filter transfer function $H_a(s)$ after replacing s by $(z-1)/(z+1)$. In the latter case, since $z = (1+s)/(1-s)$, the mapping in (7.37) can be justified after the substitution of

$$s = -k\frac{z_1 - 1}{z_1 + 1}\sin\theta + k\frac{z_2 - 1}{z_2 + 1}\cos\theta$$

in $z^{-1} = (1-s)/(1+s)$. Let the expression to the right of the arrow in (7.37), after rearrangement and simplification, be denoted by

$$g(z_1, z_2) \triangleq \frac{cz_1z_2 + bz_2 + az_1 + 1}{z_1z_2 + az_2 + bz_1 + c} = \frac{c + bz_1^{-1} + az_2^{-1} + z_1^{-1}z_2^{-1}}{1 + az_1^{-1} + bz_2^{-1} + cz_1^{-1}z_2^{-1}},$$

where

$$b = \frac{1 - k\sin\theta - k\cos\theta}{1 - k\sin\theta + k\cos\theta}$$

$$a = \frac{1 + k\sin\theta + k\cos\theta}{1 - k\sin\theta + k\cos\theta} \qquad (7.38a)$$

$$\frac{1}{c} = \frac{1 - k\sin\theta + k\cos\theta}{1 + k\sin\theta - k\cos\theta}, \qquad c \ne 0.$$

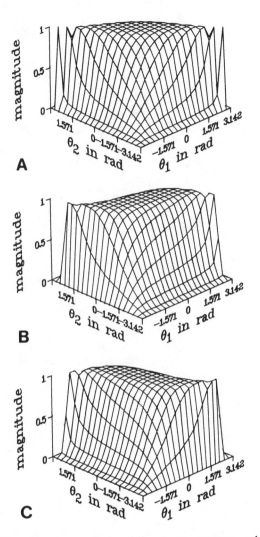

Figure 7.12A. The plot of magnitude of frequency response, when $\theta = 315°$ in Example 7.13. B. The plot of magnitude of frequency response, when $\theta = 285°$ in Example 7.13. C. The plot of magnitude of frequency response, when $\theta = 345°$ in Example 7.13.

Therefore,

$$H(z_1, z_2) = H\left[z^{-1} \rightarrow g(z_1, z_2)\right], \qquad (7.38b)$$

where $H(z)$ denotes the transfer function of a designed 1-D digital filter. The coefficients a, b, and c in (7.38a) are constrained by the relation $a + b - c = 1$, which implies that $g(z_1, z_2)$ has a nonessential singularity of the second kind at $z_1 = -1$, $z_2 = -1$. It can be shown via tedious arguments, not of interest here, that $H(z_1, z_2)$ in (7.38b) will be BIBO unstable. To resolve this problem, Goodman considered the frequency transformation function

$$g(z_1, z_2) = \frac{c + bz_1^{-1} + az_2^{-1} + z_1^{-1}z_2^{-1}}{1 + az_1^{-1} + bz_2^{-1} + cz_1^{-1}z_2^{-1}},$$

where, however, the coefficients a, b, c (assumed to be real) are *not* constrained by (7.38a). Note that $g(z_1, z_2)$ satisfies the relation

$$g(z_1, z_2) \cdot g(z_1^{-1}, z_2^{-1}) = 1,$$

implying that $|g(z_1, z_2)| = 1$ on the distinguished boundary $T^2 \triangleq (z_1, z_2 : |z_1| = 1 = |z_2|)$. A real rational function which satisfies the properties just mentioned is a 2-D allpass function. It can be shown that when $H(z)$ is BIBO stable, then the 2-D filter, characterized by $H[z^{-1} \to g(z_1, z_2)]$, is also BIBO stable, when it is recursively implemented along the $(+k_1, +k_2)$ direction, if

$$1 + az_1 + bz_2 + cz_1z_2 \neq 0, \quad |z_1| \leq 1, \ |z_2| \leq 1. \tag{7.39}$$

The preceding condition follows on applying Fact 7.11 to the 2-D transfer function $H[z^{-1} \to g(z_1, z_2)]$ generated from the 1-D transfer function $H(z)$ after replacing the delay variable z^{-1} by the frequency transformation function $g(z_1, z_2)$; this may be viewed as a 2-D generalization of the result in Problem 1 at the end of Chapter 3.

Goodman showed that the validity of the condition in (7.39) can be expressed directly in terms of the following linear inequalities involving the coefficients

$$a + b - c < 1 \tag{7.40a}$$
$$a - b + c < 1 \tag{7.40b}$$
$$-a + b + c < 1 \tag{7.40c}$$
$$-a - b - c < 1. \tag{7.40d}$$

The equivalence of the conditions in (7.35) and (7.40) is briefly substantiated next.

$(7.35) \to (7.40)$

$$|a| < 1 \leftrightarrow -1 < a < 1 \leftrightarrow (1 + a) > 0 \quad \text{and} \quad (1 - a) > 0$$

$$\left| \frac{1 + a}{b + c} \right| > 1 \quad \text{and} \quad (1 + a) > 0 \leftrightarrow -(1 + a) < (b + c) < (1 + a)$$

$$\left| \frac{1 - a}{b - c} \right| > 1 \quad \text{and} \quad (1 - a) > 0 \leftrightarrow -(1 - a) < (b - c) < (1 - a).$$

Clearly then, the conditions in (7.35) imply the conditions in (7.40).

$(7.40) \rightarrow (7.35)$

$$\left.\begin{array}{l} (7.40\text{a}) \text{ and } (7.40\text{b}) \rightarrow a < 1 \\ (7.40\text{c}) \text{ and } (7.40\text{d}) \rightarrow a > -1 \end{array}\right\} \leftrightarrow |a| < 1$$

$$\left.\begin{array}{l} (7.40\text{a}) \leftrightarrow (b - c) < (1 - a) \\ (7.40\text{b}) \leftrightarrow (b - c) > -(1 - a) \end{array}\right\} \leftrightarrow \left|\frac{1 - a}{b - c}\right| > 1 \quad (\text{Note that } (1 - a) > 0)$$

$$\left.\begin{array}{l} (7.40\text{c}) \leftrightarrow (b + c) < (1 + a) \\ (7.40\text{d}) \leftrightarrow (b + c) > -(1 + a) \end{array}\right\} \leftrightarrow \left|\frac{1 + a}{b + c}\right| > 1 \quad (\text{Note that } (1 + a) > 0).$$

Clearly, the conditions in (7.40) imply the conditions in (7.35).

Finally, combining the results proved above, it can be concluded that conditions in (7.35) and (7.40) are equivalent.

You are advised to verify the results in the example below.

Example 7.14. Using BUTTER.FOR, obtain the second-order Butterworth digital filter transfer function $H(z)$ such that $H(1) = 1$. Obtain $H(z_1, z_2)$ after replacing z^{-1} in $H(z)$ by $g(z_1, z_2)$ defined in the transformation below.

(a)
$$z^{-1} \rightarrow \frac{-0.2612 + 0.2612z_1^{-1} + 0.2612z_2^{-1} + z_1^{-1}z_2^{-1}}{1 + 0.2612z_1^{-1} + 0.2612z_2^{-1} - 0.2612z_1^{-1}z_2^{-1}}.$$

(b)
$$z^{-1} \rightarrow \frac{0.7154 - 0.7154z_1^{-1} - 0.7154z_2^{-1} + z_1^{-1}z_2^{-1}}{1 - 0.7154z_1^{-1} - 0.7154z_2^{-1} + 0.7154z_1^{-1}z_2^{-1}}.$$

Note that the coefficients of the transformations in (a) and (b) above satisfy (7.40).

The plots of $|H(e^{j\theta_1}, e^{j\theta_2})|$ vs θ_1, θ_2 in the range $-\pi \leq \theta_i \leq \pi$, $\theta_i \triangleq \omega_i T_i$, $i = 1, 2$ are shown in Figure 7.13A and B, respectively, for the above two cases. See also Figure 7.13C.

Chang and Agarwal [11] also considered the problem of design of rotated filters by using a 1-D digital filter as prototype. Specifically, they suggested the map

$$z \rightarrow z_1 z_2^{\beta/\alpha}, \quad \beta, \alpha \text{ integers} \tag{7.41}$$

in order to obtain $H(z_1, z_2)$ from $H(z)$. The above transformation in the frequency domain has the effect of rotation of the frequency response with contraction of the bandwidth, while in the spatial domain it is equivalent to the rotation of recursion direction with a new sampling interval. The stability property remains invariant, since the transformation does not change the filter coefficients but only the sample positions. The rotational effect of the frequency response, the property that categorizes this type of

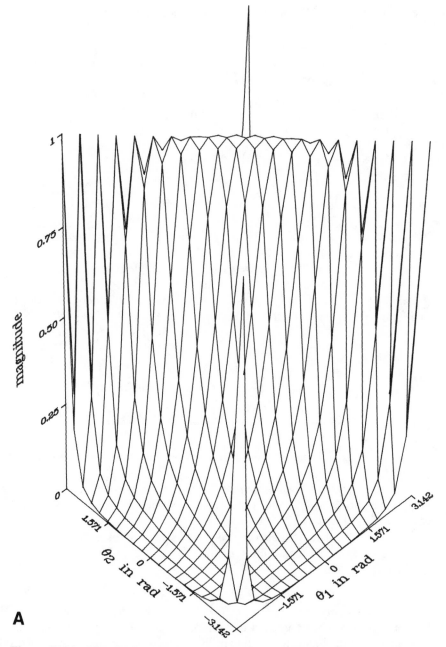

Figure 7.13A. Magnitude of frequency response for transformation in (a) of Example 7.14.

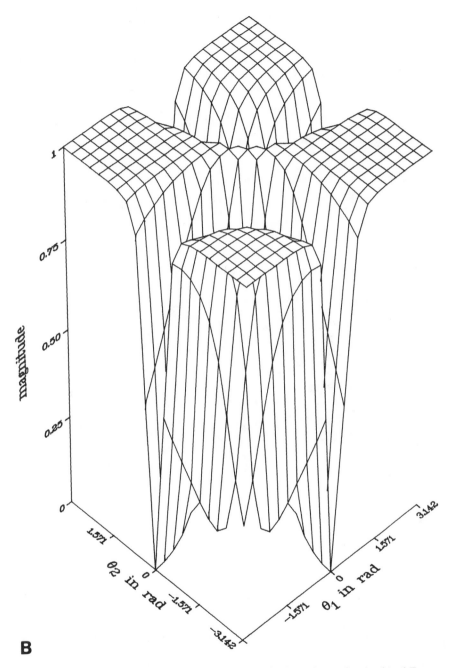

B

Figure 7.13B. Magnitude of frequency response for transformation in (b) of Example 7.14.

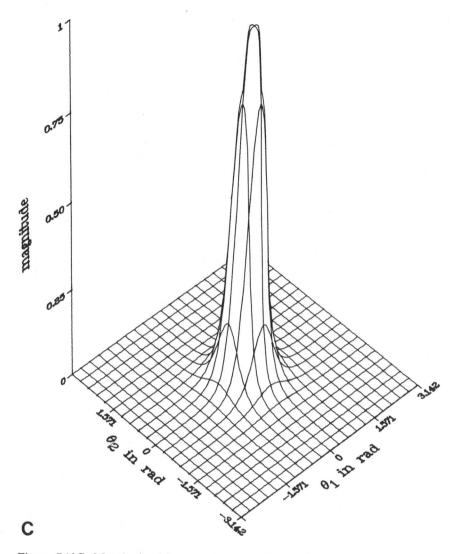

C

Figure 7.13C. Magnitude of frequency response for transformation in (a) of Example 7.14 in a contracted scale.

design among the class of rotated filter, becomes evident from the frequency transformation associated with the map in (7.41):

$$e^{j\omega T} \to e^{j\omega_1 T_1} e^{j(\beta/\alpha)\omega_2 T_2}$$

or

$$\omega T \to \omega_1 T_1 + (\beta/\alpha)\omega_2 T_2.$$

Rotated filters have been used to design velocity filters for seismic signals

[25]. For information on velocity filters and fan filters see Problem 23 at the end of this chapter.

7.3.2.2. Design Via Spectral Transformations. Spectral transformations will be understood to take the form of 2-D allpass functions and the generated functions using such transformations are stable if the original transfer function is stable. Thus, a stable filter design can be altered quickly to produce other stable designs via use of such transformations. This is especially useful in 2-D design, where stability tests are difficult, from the computational standpoint, to implement with infinite precision.

The problem is to find transformations

$$\hat{z}_1^{-1} \to g_1(z_1, z_2)$$
$$\hat{z}_2^{-1} \to g_2(z_1, z_2), \tag{7.42}$$

which will enable one to obtain $H(z_1, z_2)$ from a specified $\hat{H}(\hat{z}_1^{-1}, \hat{z}_2^{-1})$,

$$H(z_1, z_2) = \hat{H}(g_1(z_1, z_2), g_2(z_1, z_2)), \tag{7.43}$$

so that (1) $H(z_1, z_2)$ is a stable real rational function when $\hat{H}(\hat{z}_1^{-1}, \hat{z}_2^{-1})$ is so and (2) some important characteristics of the frequency response, like maximal flatness or equiripple properties in the pass or stopbands, remain invariant, while other characteristics like cutoff frequency or the number and shape of stop and passband regions, are altered in a desired manner. It is not difficult to prove that for these properties to be satisfied, the functions $g_i(z_1, z_2)$, $i = 1, 2$ in (7.42) must be stable 2-D real rational allpass functions, that is, $g_i(z_1, z_2)$ must be a stable real rational function with

$$g_i(z_1, z_2) g_i(z_1^{-1}, z_2^{-1}) = 1. \tag{7.44}$$

The general form of $g_i(z_1, z_2)$ in the spectral transformation is

$$g_i(z_1, z_2) = (\pm 1) \prod_{k=1}^{K_i} \frac{z_1^{n_k^{(i)}} z_2^{m_k^{(i)}} \left[1 + \displaystyle\sum_{\substack{n=0 \ m=0 \\ (n,m) \neq (0,0)}}^{n_k^{(i)} \; m_k^{(i)}} a_k^{(i)}(n, m) z_1^{-n} z_2^{-m} \right]}{\left[1 + \displaystyle\sum_{\substack{n=0 \ m=0 \\ (n,m) \neq (0,0)}}^{n_k^{(i)} \; m_k^{(i)}} a_k^{(i)}(n, m) z_1^{-n} z_2^{-m} \right]} \tag{7.45}$$

It is simple to verify that $g_i(z_1, z_2)$ in (7.45) satisfies (7.44). Also, $g_i(z_1, z_2)$ should be stable. When the spectral transformations in (7.42) specialize to

$$\hat{z}_1^{-1} \to g_1(z_1) \quad \text{or} \quad g_1(z_2)$$
$$\hat{z}_2^{-1} \to g_2(z_2) \quad \text{or} \quad g_2(z_1), \tag{7.46}$$

the relationships for the parameters can be obtained from the 1-D frequency transformations as discussed in Section 3.4, when the rational functions on the right side of (3.46) are of order 1 or 2. In this context, it is mentioned that

$$\sum_{k=1}^{K_i} \left(n_k^{(i)} + m_k^{(i)} \right), \quad i=1,2$$

determine the number of stopbands and passbands of the frequency response, after transformation, in the region $-\pi \le \omega_1 T_1 \le \pi, -\pi \le \omega_2 T_2 \le \pi$. The reader is referred to the paper by Pendergrass and associates [12] for further details concerning design via this method.

7.3.2.3. Another Frequency Transformation for 2-D Design [13]. Another procedure for frequency transformation for 2-D design dwells on the design of a 1-D lowpass digital filter prototype whose magnitude-squared function is known to be of the form

$$G(e^{j\omega T}) = |H(e^{j\omega T})|^2 = \frac{\sum_{k=0}^{m} p(k)\cos k\omega T}{\sum_{k=0}^{n} q(k)\cos k\omega T}. \tag{7.47}$$

The frequency transformation used is a specialization of the one introduced by McClellan for the design of 2-D nonrecursive filters (see Section 7.3.4). To obtain 2-D filters with approximate circular symmetry and zero phase, use the 1-D to 2-D transformation:

$$\cos \omega T \to \tfrac{1}{2}\cos \omega_1 T_1 + \tfrac{1}{2}\cos \omega_2 T_2 + \tfrac{1}{2}\cos \omega_1 T_1 \cos \omega_2 T_2 - \tfrac{1}{2}. \tag{7.48}$$

The transformation in (7.48) is seen to map $\omega = 0$ to $(\omega_1, \omega_2) = (0,0)$. On substituting the mapping in (7.48) to (7.47), one obtains a 2-D magnitude-squared function of the form

$$G(e^{j\omega_1 T_1}, e^{j\omega_2 T_2}) = \frac{\sum_{k=0}^{m} \sum_{r=0}^{m} s(k,r)\cos(k\omega_1 T_1)\cos(r\omega_2 T_2)}{\sum_{k=0}^{n} \sum_{r=0}^{n} t(k,r)\cos(k\omega_1 T_1)\cos(r\omega_2 T_2)}.$$

$$\tag{7.49}$$

In order to arrive at the form in (7.49) from (7.47) through use of (7.48), use either the trigonometric identity in (7.52) or Problem 19 at the end of this chapter. Next, in order to arrive at a real rational stable filter transfer function $H(z_1, z_2)$ whose magnitude-squared function equals $G(e^{j\omega_1 T_1}, e^{j\omega_2 T_2})$, it is necessary to implement the following 2-D spectral factorization:

$$G(z_1, z_2) = H(z_1, z_2) H(z_1^{-1}, z_2^{-1}), \qquad (7.50)$$

where $G(z_1, z_2)$ is derivable from (7.49) by replacing $e^{j\omega_i T_i}$ with z_i, or equivalently, by replacing $\cos k\omega_i T_i$ with $(z_i^k + z_i^{-k})/2$, $i = 1, 2$. Unlike the 1-D case, the 2-D spectral factorization is, in general, difficult, if at all possible, to solve. $G(z_1, z_2)$ could also be factored into a product of four quarter-plane filter transfer functions as in (7.36b) using cepstral techniques.

Example 7.15. The transfer function of a second-order digital Butterworth filter having a cutoff frequency of 2000 rad/sec and a sampling frequency of 9 kHz is (using BUTTER.FOR)

$$H(z) = \frac{K(z^{-1} + 1)^2}{1 + az^{-1} + bz^{-2}}, \qquad (7.51)$$

where $a = -1.6877905$, $b = 0.73033851$, and K is a real constant. This filter is to be treated as a prototype and the transformation in (7.48) is to be used to obtain a 2-D digital filter magnitude-squared frequency response. From (7.51),

$$H(z^{-1}) H(z) = K^2 \frac{4 + 4(z + z^{-1})(z + z^{-1})^2}{(1 + a^2 + b^2) + (a + ab)(z + z^{-1}) + b(z^2 + z^{-2})}.$$

The expression for $G(e^{j\omega T})$ of (7.47) specialized for this case is

$$G(e^{j\omega T}) = K \frac{6 + 8\cos \omega T + 2\cos 2\omega T}{(1 + a^2 + b^2) + 2(a + ab)\cos \omega T + 2b\cos 2\omega T}.$$

In order to be able to substitute (7.48) in (7.47), it is necessary, in general, to use the trigonometric identity

$$\cos n\theta = \sum_{k=0}^{m} \binom{n}{2k} (-1)^k \cos^{(n-2k)}\theta (1 - \cos^2 \theta)^k, \qquad (7.52)$$

where $m = n/2$ for even n and $m = (n-1)/2$ for odd n. In this problem, it is only required to express $\cos 2\theta$ in terms of $\cos \theta$, that is, $\cos 2\theta = 2\cos^2 \theta - 1$. Then one gets

$$G(e^{j\omega T}) = K \frac{4 + 8\cos \omega T + 4\cos^2 \omega T}{(1 + a^2 + b^2 - 2b) + 2(a + ab)\cos \omega T + 4b\cos^2 \omega T}.$$

Therefore,

$$G(e^{j\omega_1 T_1}, e^{j\omega_2 T_2}) = G(e^{j\omega T})\Big|_{\cos \omega T = \frac{1}{2}(\cos \omega_1 T_1 + \cos \omega_2 T_2 - 1 + \cos \omega_1 T_1 \cos \omega_2 T_2)},$$

and it is a simple matter to express this in the form (7.49).

Example 7.16. For a counterexample to illustrate the infeasibility of spectral factorization in the 2-D case, when the factors are constrained to be of finite order, consider the rational function

$$G(z_1, z_2) = \frac{(1+z_2)^2(1+z_1)^2}{4(z_1+z_2)^2}.$$

$G(z_1, z_2)$ is factorable as

$$G(z_1, z_2) = \frac{(1+z_2)(1+z_1)}{2(z_1+z_2)} \frac{(1+z_2)(1+z_1)}{2(z_1+z_2)}. \qquad (7.53)$$

It is possible to show that no other nontrivial finite-order factorization exists. Comparing the above equation with (7.50),

$$H(z_1, z_2) = \frac{(1+z_2)(1+z_1)}{2(z_1+z_2)}. \qquad (7.54)$$

Unfortunately, $H(z_1, z_2)$ is not a stable rational function and therefore is not a spectral factor. Also, it is easy to establish (by invoking the result of uniqueness, up to units, of factorization of bivariate polynomials as a product of irreducible factors) that the factorization in (7.53) is unique, when $H(z_1, z_2)$ of (7.50) or, in this case, of (7.54) is required to be a rational function with real coefficients, like $G(z_1, z_2)$.

7.3.3. Maximally Flat 2-D Recursive (IIR) Filters

The reader has, by now, encountered some of the fundamental difficulties encountered in 2-D filter design. In spite of these difficulties, there are some results in 1-D problems that generalize nicely to the 2-D case, sometimes accompanied by one or more interesting twists. In Chapter 3, the reader had come across the importance of maximally flat filters (Butterworth filters) in both analog and digital 1-D filter design. In this section, maximal flatness of 2-D filters and conditions to be satisfied in order to achieve this property will be briefly discussed. The development of the theory will parallel that recently put forward by Valenzuela and Bose [14].

Let the real rational function, $S_{21}(p_1, p_2)$, in the two independent complex variables p_1, p_2, characterize the transmission properties of a two-port lumped continuous-time filter. In case the two-port is distributed or discrete-time, this transfer function could be viewed as a real rational function in some transformed variables. The magnitude-squared function of the transmission characteristic is obtained by specializing the independent variables p_i to $p_i = j\omega_i$ (for all real values of ω_i, $i = 1, 2$) in the function

$$T(p_i, p_2) \triangleq S_{21}(p_1, p_2)S_{21}(-p_1, -p_2). \qquad (7.55)$$

Consider, then, the real rational function

$$T(j\omega_1, j\omega_2) = |S_{21}(j\omega_1, j\omega_2)|^2 \triangleq \frac{\displaystyle\sum_{i=0}^{m_1}\sum_{j=0}^{m_2} a_{i,j}\omega_1^i\omega_2^j}{\displaystyle\sum_{i=0}^{n_1}\sum_{j=0}^{n_2} b_{i,j}\omega_1^i\omega_2^j}, \quad (i+j) \text{ even.}$$

(7.56)

In (7.56), $(i + j)$ is even because $T(j\omega_1, j\omega_2) = T(-j\omega_1, -j\omega_2)$. Since $T(j\omega_1, j\omega_2)$ is nonnegative definite for all real values of ω_1, ω_2, it is necessary that n_1, n_2, m_1, m_2 be even.

Define

$$m \triangleq \max(i + j) \text{ such that } a_{i,j} \neq 0 \tag{7.57a}$$

$$n \triangleq \max(i + j) \text{ such that } b_{i,j} \neq 0. \tag{7.57b}$$

In the design of lowpass filters, the constraint $n > m$ has to be imposed and without loss of generality it may be assumed that

$$m_1 \geq m_2, \quad n_1 \geq n_2 \tag{7.58}$$

in (7.56). Then $T(j\omega_1, j\omega_2)$ in (7.56) can be rewritten as

$$T(j\omega_1, j\omega_2) = \frac{\displaystyle\sum_{k=0}^{m} P_k(\omega_1, \omega_2)}{\displaystyle\sum_{k=0}^{n} Q_k(\omega_1, \omega_2)}, \tag{7.59}$$

where the homogeneous polynomials $P_k(\omega_1, \omega_2)$, $Q_k(\omega_1, \omega_2)$, each of degree k are defined below. Homogeneous polynomials are also referred to as forms and each term in a form has the same total degree.

$$P_k(\omega_1, \omega_2) = \sum_{i=0}^{k} a_{k-i,i}\omega_1^{k-i}\omega_2^i, \quad k = 0, 2, \ldots, m_2. \tag{7.60a}$$

$$P_k(\omega_1, \omega_2) = \sum_{i=0}^{m_2} a_{k-i,i}\omega_1^{k-i}\omega_2^i, \quad k = m_2 + 2, \ldots, m_1. \tag{7.60b}$$

$$P_k(\omega_1, \omega_2) = \sum_{i=k-m_1}^{m_2} a_{k-i,i}\omega_1^{k-i}\omega_2^i, \quad k = m_1 + 2, \ldots, m. \tag{7.60c}$$

$$Q_k(\omega_1, \omega_2) = \sum_{i=0}^{k} b_{k-i,i}\omega_1^{k-i}\omega_2^i, \quad k = 0, 2, \ldots, n_2. \tag{7.60d}$$

$$Q_k(\omega_1, \omega_2) = \sum_{i=0}^{n_2} b_{k-i,i}\omega_1^{k-i}\omega_2^i, \quad k = n_2 + 2, \ldots, n_1. \tag{7.60e}$$

$$Q_k(\omega_1, \omega_2) = \sum_{i=k-n_1}^{n_2} b_{k-i,i}\omega_1^{k-i}\omega_2^i, \quad k = n_1 + 2, \ldots, n. \tag{7.60f}$$

To derive the conditions of maximal flatness for $T(j\omega_1, j\omega_2)$ in (7.56) at $(\omega_1, \omega_2) = (0,0)$, consider an arbitrary straight line through $(0,0)$; this is parametrizable by

$$\omega_1 = \alpha\omega, \qquad \omega_2 = \beta\omega \qquad (7.61)$$

for arbitrary but fixed real numbers α and β, not both zero. Then

$$T(j\alpha\omega, j\beta\omega) = \frac{\displaystyle\sum_{k=0}^{m} A_k(\alpha, \beta)\omega^k}{\displaystyle\sum_{k=0}^{n} B_k(\alpha, \beta)\omega^k}, \qquad (7.62)$$

where the homogeneous polynomials, $A_k(\alpha, \beta)$ and $B_k(\alpha, \beta)$ are defined in (7.63a) and (7.63b), respectively.

$$A_k(\alpha, \beta)\omega^k \equiv P_k(\alpha\omega, \beta\omega) \qquad (7.63a)$$

$$B_k(\alpha, \beta)\omega^k \equiv Q_k(\alpha\omega, \beta\omega). \qquad (7.63b)$$

Imposing the conditions for maximal flatness at $\omega = 0$ [and therefore $(\omega_1, \omega_2) = (0,0)$ from (7.61)] for a 1-D lowpass filter, one has (see reference [22] or refer to Section 3.2.1)

$$A_k(\alpha, \beta) \equiv B_k(\alpha, \beta), \quad k = 0, 2, \ldots, m \qquad (7.64a)$$

$$B_k(\alpha, \beta) \equiv 0, \quad k = m+2, \ldots, n-2 \qquad (7.64b)$$

$$B_n(\alpha, \beta) > 0 \quad \text{for any fixed } (\alpha, \beta) \text{ except } (0,0). \qquad (7.64c)$$

Equations (7.64a) and (7.64b) are consequences of maximal flatness property while (7.64c) guarantees attenuation along any radial direction through $(0,0)$ in the 2-D filter, characterized by (7.59). Therefore, the bivariate form, $B_n(\alpha, \beta)$ must be positive.

Note that the positivity test for an n-variate form is essentially equivalent [21] to the positivity test for an $(n-1)$-variate polynomial. Consider the grid in Figure 7.14. Let the point on the grid with coordinates (i, j) be associated with the monomial $\omega_1^i \omega_2^j$ and the coefficients $a_{i,j}, b_{i,j}$. Then, the grid-points on the antidiagonal, denoted by $k = r$ for $r = 0, 2, \ldots, n$, are associated with the coefficients of a homogeneous polynomial of degree r in indeterminates ω_1, ω_2. On equating coefficients of powers of monomials in α, β from the identities in (7.64a) and (7.64b) one has

$$a_{i,j} = b_{i,j} \qquad (i, j) \in I \triangleq \{(i, j): \quad i = 0, 1, \ldots, m_1$$

$$j = 0, 1, \ldots, m_2$$

$$\{i + j = 0, 2, \ldots, m\} \qquad (7.65a)$$

$$b_{i,j} = 0 \qquad (i, j) \notin I \quad \text{and} \quad i + j = m+2, \ldots, n-2. \qquad (7.65b)$$

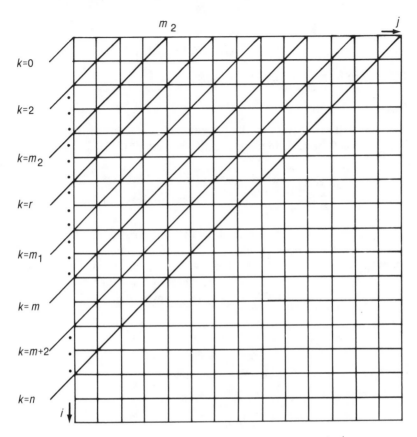

Figure 7.14. Grid associated with $a_{1,j}b_{1,j}$ and $\omega_1^i\omega_2^j$.

From (7.64c), since

$$B_n(\alpha,\beta) = \sum_{i=0}^{n} b_{n-i,i}\alpha^{n-i}\beta^i$$

is a positive definite form, it is clear that

$$b_{n,0} > 0 \quad \text{and} \quad b_{0,n} > 0. \qquad (7.66)$$

For the conditions in (7.66) to hold, the following equalities must be satisfied:

$$n = n_1 = n_2.$$

However, m, m_1, and m_2 may not be all equal.

Imposing, in addition, the condition for maximal flatness at $\omega = \infty$ on the 1-D lowpass filter described in (7.62) for an arbitrary but fixed real 2-tuple (α,β), different from $(0,0)$, one has [22, pp. 20–21]

$$A_k(\alpha,\beta) \equiv 0, \quad k = 2,4,\ldots,m \qquad (7.67)$$

or

$$a_{i,j} = 0, \quad (i,j) \in I - (0,0), \tag{7.68a}$$

and therefore from (7.65a),

$$b_{i,j} = 0, \quad (i,j) \in I - (0,0). \tag{7.68b}$$

Maximal flatness at $(0,0)$ of the magnitude-squared response function in (7.56) could have been defined by equating to zero at $(0,0)$ all derivatives of the form,

$$\frac{\partial^{i+j} T(j\omega_1, j\omega_2)}{\partial \omega_1^i \, \partial \omega_2^j}, \quad i+j = 1,2,\ldots,n-1. \tag{7.69}$$

It is reasonably straightforward to show that the conditions obtained from such a scheme are equivalent to the conditions in (7.64a) and (7.64b). To substantiate briefly this fact, first suppose that all the derivatives in (7.69) are equated to zero. For notational brevity, we henceforth denote $T(j\omega_1, j\omega_2)$ by $T(\omega_1, \omega_2)$. Then $T(\omega_1, \omega_2)$ can be expanded as a formal series,

$$T(\omega_1, \omega_2) = T(0,0) + \sum_{k=0,2,\ldots} T_{n+k}(\omega_1, \omega_2), \tag{7.70}$$

where $T_k(\omega_1, \omega_2)$ is a homogeneous polynomial in ω_1, ω_2 of degree k. In (7.70) $T_n(\omega_1, \omega_2) \neq 0$. Therefore,

$$T(\alpha\omega, \beta\omega) - T(0,0) = \sum_{k=0,2,\ldots} T_{n+k}(\alpha, \beta) \omega^{n+k}. \tag{7.71}$$

On equating (7.71) to (7.62), conditions in (7.64a) and (7.64b) follow. Conversely, suppose that the conditions in (7.64a) and (7.64b) hold. Then (7.62) can be expanded as in (7.71), with $T_n(\alpha, \beta) \neq 0$. Suppose that $T(\omega_1, \omega_2) - T(0,0)$ has the formal power series expansion

$$T(\omega_1, \omega_2) - T(0,0) = \sum_{k=2,4,\ldots} S_k(\omega_1, \omega_2), \tag{7.72}$$

where $S_k(\omega_1, \omega_2)$ is a homogeneous polynomial of degree k. By proof based on contradiction it can be established that $S_k(\omega_1, \omega_2) \equiv 0$, $k = 2,4,\ldots,n-2$, and $S_n(\omega_1, \omega_2) \equiv T_n(\omega_1, \omega_2) \neq 0$.

Consider, now, the magnitude-squared response function in (7.73) satisfying the conditions in (7.64) and (7.68)

$$T_a(\omega_1, \omega_2) = \frac{1}{1 + \omega_1^8 + \omega_2^8}. \tag{7.73}$$

Figure (7.15A) shows the plot of $T_a(\omega_1, \omega_2)$. Next consider the product separable function in (7.74) which has the same total degree for the

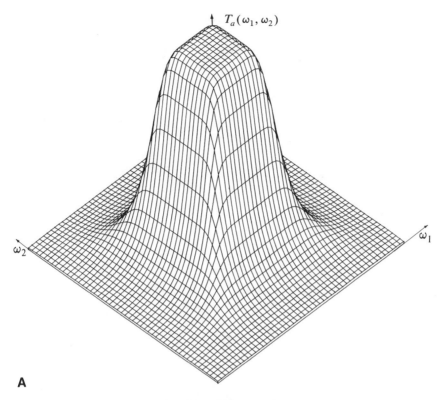

$$T_a(\omega_1, \omega_2)$$

A

Figure 7.15A. Plot of $T_a(\omega_1, \omega_2)$ versus ω_1, ω_2.

denominator polynomials as $T_a(\omega_1, \omega_2)$.

$$T_b(\omega_1, \omega_2) = \frac{1}{1 + \omega_1^4 + \omega_2^4 + \omega_1^4\omega_2^4} = \frac{1}{(1 + \omega_1^4)(1 + \omega_2^4)}. \quad (7.74)$$

Figure (7.15B) shows the plot of $T_b(\omega_1, \omega_2)$, which does not satisfy the conditions for maximal flatness in (7.64a) and (7.64b) as well as the condition in (7.64c). Some sacrifice in flatness is necessary to achieve the separability property which is useful in spectral factorization and possibly in final synthesis. Separability property is also required to achieve quadrantal symmetry of the magnitude-squared response in stable 2-D recursive filters [15]. Figure (7.15C) shows the plot of $T_c(\omega_1, \omega_2)$ in (7.75).

$$T_c(\omega_1, \omega_2) = \frac{1}{1 + \omega_1^8 + \omega_2^8 + \omega_1^8\omega_2^8} = \frac{1}{(1 + \omega_1^8)} \frac{1}{(1 + \omega_2^8)}. \quad (7.75)$$

$T_c(\omega_1, \omega_2)$ is product separable, but of higher order than $T_b(\omega_1, \omega_2)$ and has slightly better characteristics than $T_a(\omega_1, \omega_2)$ in (7.74). Finally, consider

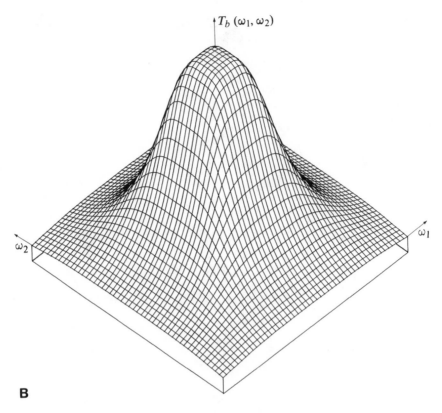

Figure 7.15B. Plot of $T_b(\omega_1, \omega_2)$ versus ω_1, ω_2.

$T_d(\omega_1, \omega_2)$ in (7.76)

$$T_d(\omega_1, \omega_2) = \frac{1}{1 + \omega_1^{2n}\omega_2^{2n}}.\tag{7.76}$$

Though $T_d(\omega_1, \omega_2)$ satisfies the conditions in (7.64a), (7.64b), and (7.68), it does not satisfy (7.64c) and, in fact, gives no attenuation along the $\omega_1 = 0$ and $\omega_2 = 0$ axes.

Example 7.17. First consider the magnitude-squared function $T_1(\omega_1, \omega_2)$ in (7.77) which corresponds to a product separable case that can be spectrally factored as in (7.78):

$$T_1(\omega_1, \omega_2) = \frac{1}{1 + \omega_1^6 + \omega_2^6 + \omega_1^6\omega_2^6}\tag{7.77}$$

$$T_1(\omega_1, \omega_2) = H_1(p_1, p_2)\cdot H_1(-p_1, -p_2),\tag{7.78}$$

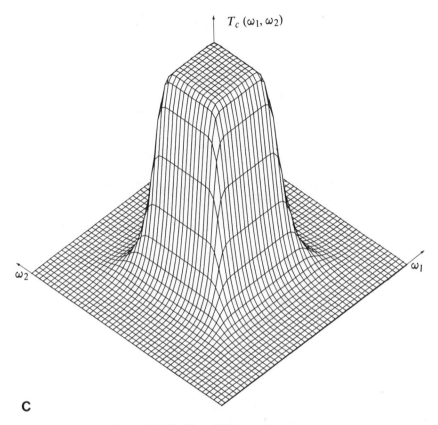

$T_c(\omega_1, \omega_2)$

C

Figure 7.15C. Plot of $T_c(\omega_1, \omega_2)$ versus ω_1, ω_2.

when $p_i = j\omega_i$, $i = 1,2$ yields

$$H_1(p_1, p_2) = \left(\frac{1}{1 + 2p_1 + 2p_1^2 + p_1^3}\right)\left(\frac{1}{1 + 2p_2 + 2p_2^2 + p_2^3}\right). \quad (7.79)$$

The transfer function $H_1(p_1, p_2)$ in (7.79) corresponds to a product of two maximally flat third-order (Butterworth) filters, one in each variable p_1 and p_2.

Figure 7.16A shows the magnitude response of $H_1(p_1, p_2)$ of (7.79). By substituting the bilinear transformation $p_i = (z_i - 1)/(z_i + 1)$, $i = 1,2$ in (7.79) the lowpass digital filter transfer function in (7.80) is obtained. You might like to use the results from Problem 21 at the end of Chapter 1 in order to implement the bilinear transformation using the Q_n matrix.

$$H_2(z_1, z_2) = .25\left(\frac{1 + 3z_1 + 3z_1^2 + z_1^3}{z_1 + 3z_1^3}\right)\left(\frac{1 + 3z_2 + 3z_2^2 + z_2^3}{z_2 + 3z_2^3}\right). \quad (7.80)$$

Figure 7.16B shows the magnitude response of $H_2(z_1, z_2)$ in (7.80).

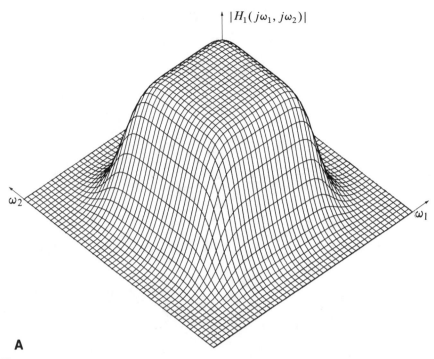

$$|H_1(j\omega_1, j\omega_2)|$$

A

Figure 7.16A. Plot of magnitude of frequency response $|H_1(j\omega_1, j\omega_2)|$ versus ω_1, ω_2.

Next, spectral transformations for digital filters are used in order to transform the digital lowpass prototype in (7.80) into a highpass, a bandpass, and a bandstop digital filter.

By substituting the lowpass to highpass transformation of (7.81) in (7.80) the highpass filter $H_3(z_1, z_2)$ in (7.82) is obtained.

$$z_i^{-1} \to -z_i^{-1}, \quad \text{or} \quad z_i \to -z_i, \quad i = 1,2 \tag{7.81}$$

$$H_3(z_1, z_2) = 0.25\left(\frac{1 - 3z_1 + 3z_1^2 - z_1^3}{-z_1 - 3z_1^3}\right)\left(\frac{1 - 3z_2 + 3z_2^2 - z_2^3}{-z_2 - 3z_2^3}\right). \tag{7.82}$$

Figure 7.16C shows the magnitude response of $H_3(z_1, z_2)$.

In order to transform the lowpass digital prototype in (7.80) into a bandpass digital filter the following transformation is used. (See Table 3.4.) For convenience, the sampling period T has been normalized to unity. Therefore, set $T = 1$ in the remainder of this example.

$$z_i \to -\frac{-[(1 - k_i)/(1 + k_i)] - [2\alpha_i k_i/(1 + k_i)]z_i + z_i^2}{1 - [2\alpha_i k_i/(1 + k_i)]z_i - [(1 - k_i)/(1 + k_i)]z_i^2}, \quad i = 1,2 \tag{7.83a}$$

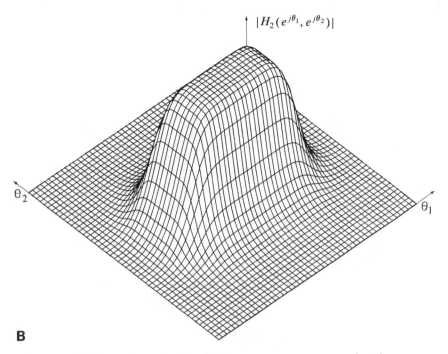

B

Figure 7.16B. Plot of magnitude of frequency response $|H_2(e^{j\theta_1}, e^{j\theta_2})|$ versus θ_1, θ_2; $\theta_i = \omega_i T_i$, $i = 1, 2$.

with

$$\alpha_i = \frac{\cos[(\omega_{i2} + \omega_{i1})/2]}{\cos[(\omega_{i2} - \omega_{i1})/2]}, \quad i = 1, 2 \tag{7.83b}$$

$$k_i = \cot\left(\frac{\omega_{i2} - \omega_{i1}}{2}\right)\tan\left(\frac{\omega_{ci}}{2}\right) \quad i = 1, 2, \tag{7.83c}$$

where ω_{i2} and ω_{i1}, $i = 1, 2$ in (7.83) are the upper and lower cutoff frequencies, in the passband of the desired bandpass filter, along the ω_i, $i = 1, 2$ directions; ω_{ci}, $i = 1, 2$ is the cutoff frequency of the lowpass digital prototype along the ω_i, $i = 1, 2$ direction.

Let the design of the bandpass filter be based on the following specification.

$$\omega_{c1} = \omega_{c2} = 0.5\pi$$

$$\omega_{12} = \omega_{22} = 0.5\pi + 0.2\pi$$

$$\omega_{11} = \omega_{21} = 0.5\pi - 0.2\pi.$$

Therefore, $\alpha_i = 0$, $k_i = 1.376$ for $i = 1, 2$. The transformation in (7.83a)

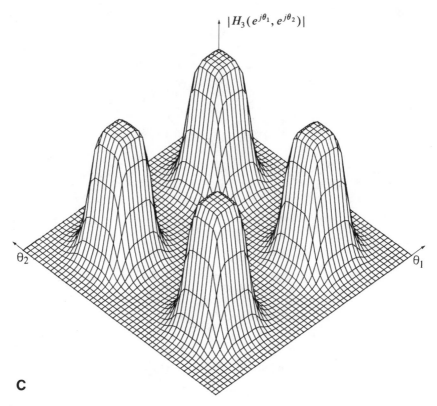

c

Figure 7.16C. Plot of the magnitude of frequency response $|H_3(e^{j\theta_1}, e^{j\theta_2})|$ versus θ_1, θ_2; $\theta_i = \omega_i T_i$, $i = 1, 2$.

becomes

$$z_i \rightarrow -\frac{0.15838 + z_i^2}{1 + 0.15838 z_i^2} \quad \text{or} \quad z_i^{-1} \rightarrow -\frac{0.15838 + z_i^{-2}}{1 + 0.15838 z_i^{-2}}. \tag{7.84}$$

Substituting (7.84) in (7.80), the bandpass filter $H_4(z_1, z_2)$ in (7.85) is obtained.

$$H_4(z_1, z_2) = \left(\frac{0.597 - 1.791 z_1^2 + 1.791 z_1^4 - 0.597 z_1^6}{6.050 z_1^6 + 3.484 z_1^4 + 2.549 z_1^2 + 3.40} \right)$$

$$\times \left(\frac{0.597 - 1.791 z_2^2 + 1.791 z_2^4 - .597 z_2^6}{6.050 z_2^6 + 3.484 z_2^4 + 2.549 z_2^2 + 3.40} \right). \tag{7.85}$$

Figure 7.16D shows the magnitude response of $H_4(z_1, z_2)$. In order to transform the digital lowpass prototype into a bandstop digital filter a similar transformation is used.

$$|H_4(e^{j\theta_1}, e^{j\theta_2})|$$

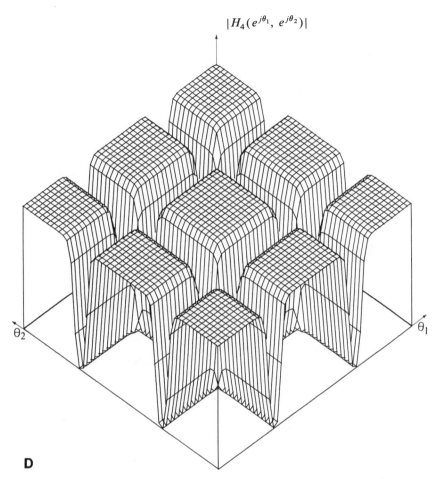

D

Figure 7.16D. Plot of the magnitude of frequency response $|H_4(e^{j\theta_1}, e^{j\theta_2})|$ versus θ_1, θ_2; $\theta_i = \omega_i T_i$, $i = 1,2$.

For the stopband given in (7.86),

$$0.5\pi - 0.2\pi \le \omega_i \le 0.5\pi + 0.2\pi, \quad i = 1,2, \tag{7.86}$$

the transformation reduces to

$$z_i \to \frac{0.15838 + z_i^2}{1 + 0.15838 z_i^2} \quad \text{or} \quad z_i^{-1} \to \frac{0.15838 + z_i^{-2}}{1 + 0.15838 z_i^{-2}}. \tag{7.87}$$

By substituting (7.87) in (7.80), the bandstop digital filter $H_5(z_1, z_2)$ ob-

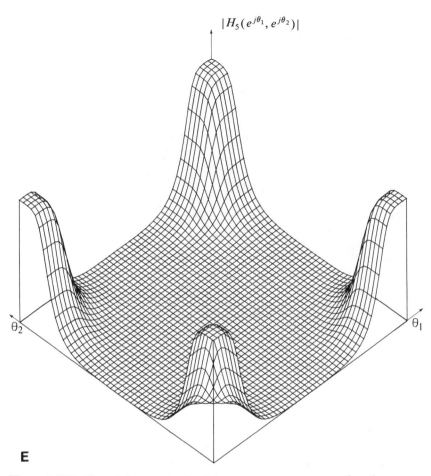

$|H_5(e^{j\theta_1}, e^{j\theta_2})|$

E

Figure 7.16E. Plot of the magnitude of frequency response $|H_5(e^{j\theta_1}, e^{j\theta_2})|$ versus $\theta_1, \theta_2; \ \theta_i = \omega_i T_i, \ i = 1,2.$

tained is

$$H_5(z_1, z_2) = \left(\frac{1.553 + 4.658z_1^2 + 4.658z_1^4 + 1.553z_1^6}{6.050z_1^6 + 3.484z_1^4 + 2.549z_1^2 + 3.40} \right)$$

$$\times \left(\frac{1.553 + 4.658z_2^2 + 4.658z_2^4 + 1.553z_2^6}{6.050z_2^6 + 3.484z_2^4 + 2.549z_2^2 + 3.40} \right). \qquad (7.88)$$

Figure 7.16E shows the magnitude response of $H_5(z_1, z_2)$.

The example illustrates that a low-order separable filter, even though not maximally flat, gives a reasonably good frequency response. Also, the shape of the magnitude response of the filter is maintained after the transformations. As mentioned before by increasing the order of the filter, better frequency response can be obtained. The separable transfer functions, $H_4(z_1, z_2)$ and $H_5(z_1, z_2)$ were checked for stability by using SCHUR1.FOR. Those functions were found to be stable because in (7.84) and also in (7.87), the delay variable z_i^{-1}, for $i = 1, 2$, was replaced by stable frequency transformation functions in order to generate the desired filters transfer functions from the lowpass prototype.

General Comments. First, it should be noted that in the r-dimensional case ($r > 2$), conditions for maximal flatness at $(\omega_1, \omega_2, \ldots, \omega_r) = (0, 0, \ldots, 0)$ can be obtained by equating to zero at $(0, 0, \ldots, 0)$ all derivatives which are the r-D counterpart of (7.69). Then, via parametrization of the type

$$\omega_i = \alpha_i \omega, \quad i = 1, 2, \ldots, r \tag{7.89}$$

for arbitrary but fixed real numbers α_i, $i = 1, 2, \ldots, r$, not all zero, the r-D counterparts of (7.64)–(7.66) are obtainable in a straightforward manner.

Second, it should be noted that the maximally flat property of the magnitude squared lowpass response function $|S_{21}(j\omega_1, j\omega_2)|^2$, at $(\omega_1, \omega_2) = (0, 0)$, is invariant at point or points in the (Ω_1, Ω_2) plane to which $(0, 0)$ is mapped under an analytic mapping of the type

$$\phi_i(\omega_1, \omega_2) = \Omega_i, \quad i = 1, 2. \tag{7.90}$$

The transformation in (7.90) has a natural extension to the r-D case ($r > 2$), thus making it feasible to apply the conditions for maximal flatness in the design of multidimensional filters having magnitude-squared response characteristics other than the lowpass type. Multidimensional filters having maximally flat response are sometimes required for the design of prefilters and interpolators for use with a time-varying image, which is a function of two independent spatial variables and the temporal variable [31, p. 519].

7.3.4. 2-D Nonrecursive (FIR) Filter Design

Various approaches have been proposed for the design of 2-D nonrecursive filters, where the computation of the present output is not based on the use of past outputs. Therefore, the design problem is essentially a polynomial approximation problem and involves the finding of a finite set of impulse response coefficients such that prescribed specifications are satisfied by the filter response characteristics. As in the 1-D case, no stability tests are required. Here, an efficient technique for the design of some classes of high order 2-D nonrecursive, linear phase filters will be discussed. This method was proposed first by McClellan [18] and involves the conversion of an already designed zero phase 1-D FIR filter to a zero phase 2-D FIR filter by a suitable frequency transformation. Specifically, consider the transfer function of a 2-D nonrecursive filter

$$H(z_1, z_2) = \sum_{k_1=0}^{N_1-1} \sum_{k_2=0}^{N_2-1} h(k_1, k_2) z_1^{-k_1} z_2^{-k_2}, \qquad (7.91)$$

where $\{h(k_1, k_2)\}$ is the impulse response sequence of finite support. On imposing the conditions that N_1 and N_2 are each odd integers and

$$h(N_1 - k_1 - 1, k_2) = h(k_1, k_2), \quad k_1 = 0, 1, \dots, \tfrac{1}{2}(N_1 - 1) \quad (7.92a)$$

$$h(k_1, N_2 - k_2 - 1) = h(k_1, k_2), \quad k_2 = 0, 1, \dots, \tfrac{1}{2}(N_2 - 1), \qquad (7.92b)$$

it is straightforward to verify that the frequency response $H(e^{j\omega_1 T_1}, e^{j\omega_2 T_2})$ of the filter can be written in the form

$$H(e^{j\omega_1 T_1}, e^{j\omega_2 T_2}) = U \sum_{k_1=0}^{(N_1-1)/2} \sum_{k_2=0}^{(N_2-1)/2} f(k_1, k_2) \cos k_1 \omega_1 T_1 \cos k_2 \omega_2 T_2, \qquad (7.93)$$

where

$$f(0,0) = h\left(\frac{N_1-1}{2}, \frac{N_2-1}{2}\right)$$

$$f(0, k_2) = 2h\left(\frac{N_1-1}{2}, \frac{N_2-1}{2} - k_2\right), \quad k_2 = 1, 2, \dots, \frac{N_2-1}{2}$$

$$f(k_1, 0) = 2h\left(\frac{N_1-1}{2} - k_1, \frac{N_2-1}{2}\right), \quad k_1 = 1, 2, \dots, \frac{N_1-1}{2}$$

$$f(k_1, k_2) = 4h\left(\frac{N_1-1}{2} - k_1, \frac{N_2-1}{2} - k_2\right), \quad k_1 = 1, 2, \dots, \frac{N_1-1}{2},$$

$$k_2 = 1, 2, \dots, \frac{N_2-1}{2} \qquad (7.94)$$

and

$$U \triangleq e^{[-j(((N_1-1)/2)\omega_1 T_1 + ((N_2-1)/2)\omega_2 T_2)]}. \qquad (7.95)$$

Again, a 1-D digital FIR filter has a transfer function of the form

$$G(z) = \sum_{k=0}^{N-1} g(k) z^{-k}. \qquad (7.96)$$

Imposing the condition that N in (7.96) is an odd integer and

$$g(k) = g(N - 1 - k), \quad k = 0, 1, \dots, \frac{N-1}{2}, \qquad (7.97)$$

it is easy to verify that

$$G(e^{j\omega T}) = V \left[\sum_{k=0}^{(N-1)/2} h(k)\cos k\omega T \right], \qquad (7.98)$$

where

$$h(0) = g\left(\frac{N-1}{2}\right)$$

$$h(k) = 2g\left(\frac{N-1}{2} - k\right), \quad k = 1, 2, \ldots, \frac{N-1}{2} \qquad (7.99)$$

and

$$V \triangleq e^{-j[(N-1)/2]\omega T}. \qquad (7.100)$$

If $x \triangleq \cos \omega T$, then

$$\cos k\omega T = \cos k(\cos^{-1} x) = T_k(x),$$

where $T_k(x)$ is the Chebyshev polynomial of first kind, order k (see Section 3.3.2). Therefore, the function in square brackets on the right-hand side of (7.98) can be written as a polynomial in $x = \cos \omega T$, that is,

$$\sum_{k=0}^{(N-1)/2} h(k)\cos k\omega T = \sum_{k=0}^{(N-1)/2} \hat{h}(k)[\cos \omega T]^k. \qquad (7.101)$$

Of course, the $\hat{h}(k)$'s in (7.101) can be easily determined from the $h(k)$'s. It is now logical to introduce the frequency transformation

$$\cos \omega T \to \sum_{k_1=0}^{r_1} \sum_{k_2=0}^{r_2} t(k_1, k_2)\cos k_1\omega_1 T_1 \cos k_2\omega_2 T_2, \qquad (7.102)$$

which is often referred to as the generalized McClellan transformation, since McClellan originally considered the case, $r_1 = r_2 = 1$. Then, the 2-D function obtained after introducing the above transformation is

$$\hat{H}(e^{j\omega_1 T_1}, e^{j\omega_2 T_2})$$

$$= \sum_{k=0}^{(N-1)/2} \hat{h}(k) \left[\sum_{k_1=0}^{r_1} \sum_{k_2=0}^{r_2} t(k_1, k_2)\cos k_1\omega_1 T_1 \cos k_2\omega_2 T_2 \right]^k.$$

The transformation parameters, $t(k_1, k_2)$'s in (7.102) have to be chosen so that the contours (ωT = constant) in the ($\omega_1 T_1, \omega_2 T_2$) or ($\theta_1, \theta_2$)-plane have some desired shape. The form of mapping involving the variables $\theta = \omega T$, $\theta_1 = \omega_1 T_1$, $\theta_2 = \omega_2 T_2$ in the case when $r_1 = r_2 = 1$ in (7.102) can be obtained from

$$\cos \theta_2 = \frac{\cos \theta - t(0,0) - t(1,0)\cos \theta_1}{t(0,1) + t(1,1)\cos \theta_1}.$$

As θ varies from 0 to π, $\cos \theta_2$ as a function of $\cos \theta_1$ is a family of curves

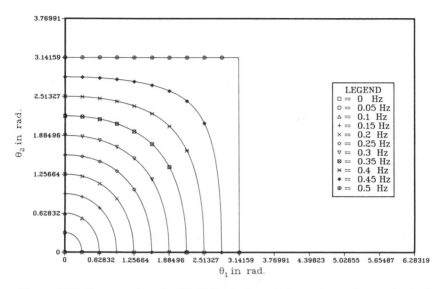

Figure 7.17. Contour plots for McClellan's frequency transformation employed in the design of FIR circularly symmetric filters; sampling period has been normalized to unity.

in the square $[0 \leq \theta_1 \leq \pi] \times [0 \leq \theta_2 \leq \pi]$. These curves are either monotone decreasing or monotone increasing, since the partial derivative

$$\frac{\partial[\cos \theta_2]}{\partial[\cos \theta_1]} = \frac{[-t(1,0)t(0,1) + t(1,1)t(0,0) - t(1,1)\cos \theta]}{[t(0,1) + t(1,1)\cos \theta_1]^2}$$

is sign-invariant as θ_1 varies from 0 to π for any fixed θ. If all the contours are monotonically decreasing, then the point $\theta = 0$ is mapped to the point $(\theta_1, \theta_2) = (0,0)$ and the point $\theta = \pi$ is mapped to the point $(\theta_1, \theta_2) = (\pi, \pi)$. These types of contours in the special case when

$$\cos \omega T \rightarrow \tfrac{1}{2}\cos \omega_1 T_1 + \tfrac{1}{2}\cos \omega_2 T_2 + \tfrac{1}{2}\cos \omega_1 T_1 \cos \omega_2 T_2 - \tfrac{1}{2}$$

are approximately circular, as shown in Figure 7.17. It is readily apparent that the approximation to the circular form is good, especially for small $r = (\omega_1^2 T_1^2 + \omega_2^2 T_2^2)^{1/2}$. If all the contours are increasing, then the point $\theta = 0$ is mapped to $(\theta_1, \theta_2) = (0, \pi)$ and the point $\theta = \pi$ is mapped to $(\theta_1, \theta_2) = (\pi, 0)$. McClellan used this type of monotonically increasing contours to design fan filters. A particular choice of coefficients that yield fan filter characteristics are

$$t(1,0) = -t(0,1) = 0.5$$
$$t(0,0) = t(1,1) = 0.$$

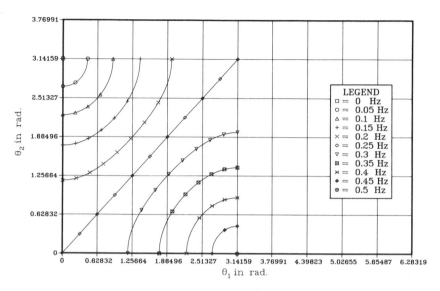

Figure 7.18. Contour plots for McCellan's frequency transformation employed in the design of FIR fan filters; sampling period has been normalized to unity.

The contour plots for the above set of parameter values are shown in Figure 7.18. The generalized transformation in (7.102) has been used in conjunction with fast algorithms to design and implement 2-D nonrecursive filters satisfying various specifications and characteristics in a computationally expedient manner.

7.3.5. 2-D DFT

The 2-D DFT is a straight forward generalization of the 1-D DFT. If $\{y(r_1, r_2)\}$ is a finite sequence over a rectangular grid, $0 \le r_1 \le N_1 - 1$, $0 \le r_2 \le N_2 - 1$, then the 2-D DFT $\{Y(k_1, k_2)\}$ is defined by

$$Y(k_1, k_2) = \sum_{r_1 = 0}^{N_1 - 1} \sum_{r_2 = 0}^{N_2 - 1} y(r_1, r_2) w_{N_1}^{r_1 k_1} w_{N_2}^{r_2 k_2}, \qquad (7.103)$$

where $w_{N_1} = e^{-j2\pi/N_1}$, $w_{N_2} = e^{-j2\pi/N_2}$. The 2-D IDFT is defined by

$$y(r_1, r_2) = \frac{1}{N_1 N_2} \sum_{k_1 = 0}^{N_1 - 1} \sum_{k_2 = 0}^{N_2 - 1} Y(k_1, k_2) w_{N_1}^{-r_1 k_1} w_{N_2}^{-r_2 k_2}. \qquad (7.104)$$

(7.103) and (7.104) are respectively the 2-D counterparts of (2.4) and (2.5). As in the 1-D case, the 2-D DFT is useful for implementing 2-D digital filters and for examining the frequency content of a 2-D signal. Note the product separability of the kernels $w_{N_1}^{r_1 k_1} w_{N_2}^{r_2 k_2}$ and $w_{N_1}^{-r_1 k_1} w_{N_2}^{-r_2 k_2}$ in (7.103) and (7.104), respectively. This property of the kernels is very useful for fast computation of 2-D DFT using the 1-D FFT technique, especially when N_1 and N_2 are highly composite numbers.

7.4. Linear Shift-Variant (LSV) Digital Filters

The property common to the digital filters discussed so far in this text is linear shift-invariance. In this final section, the restriction of shift-invariance is removed. The motivation for undertaking the study, admittedly brief, of linear shift-variant (LSV) digital filters stems from some filtering applications, where the underlying system model is required to be linear but shift-variant. Such models occur, for example, in the study of digital-to-digital sampling rate conversion, where the objective is to convert a sequence obtained from sampling an analog waveform at a certain sampling rate, to another sequence that is obtainable from sampling the same analog waveform at a different rate. When the sampling rate is required to be decreased from the original sampling rate, the process is called decimation and when the sampling rate is required to be increased the process is called interpolation. The need for sampling rate conversion occurs in a variety of applications including speech processing, digital communications and radar systems. In digital filtering, the pulse code modulated type of signal representation is usually preferred. On the other hand, delta modulation, which is basically a high rate, one-bit coding technique, is used in analog-to-digital conversion because it is simple and economic. Therefore, decimation is necessary for conversion from the delta modulated format to the pulse code modulated format and interpolation is required for the reverse process.

7.4.1. Characterization of LSV Systems in Time Domain

A discrete-time LSV system is characterizable by the superposition sum

$$y(k) = \sum_{r=-\infty}^{\infty} h(k,r)x(r), \qquad (7.105)$$

where $\{y(k)\}$ and $\{x(k)\}$ are, respectively, the output and input sequences; $h(k,r)$ is the output at time k due to a unit impulse input applied at time r and therefore it is the unit impulse response of the system. For a causal LSV system, $h(k,r) = 0$ for $k < r$ (i.e., the output cannot appear before the application of the input or the system is nonanticipatory) and the input/output relation becomes

$$y(k) = \sum_{r=-\infty}^{k} h(k,r)x(r). \qquad (7.106)$$

When the input $x(r) = 0$ for $r < 0$, the input/output relation of the causal LSV system further simplifies to

$$y(k) = \sum_{r=0}^{k} h(k,r)x(r). \qquad (7.107)$$

A subclass of discrete-time causal LSV systems is characterizable by the LSV difference equation

$$\sum_{i=0}^{n} b_i(k)y(k-i) = \sum_{i=0}^{m} a_i(k)x(k-i), \qquad (7.108)$$

where $\{b_i(k)\}$ and $\{a_i(k)\}$ are sequences of functions of k and $b_0(k) \neq 0$, for all k. It can be shown that a discrete-time causal LSV system can be characterized by a LSV difference equation of order p if and only if the unit impulse response sequence $\{h(k,r)\}$ of the system is a pth-order degenerate sequence, defined next.

Definition 7.7. A sequence $\{h(k,r)\}$ is a pth-order degenerate sequence if $h(k,r)$ can be expressed as a finite sum (containing p terms) of product separable functions,

$$h(k,r) = \sum_{i-1}^{p} w_i(k)v_i(r),$$

where $w_i(k)$ and $v_i(r)$, $i = 1,2,\ldots,p$ are linearly independent functions of integer variables k, r, respectively.

7.4.2. Characterization of LSV Systems in Frequency Domain

Let $\{h(k,r)\}$ denote the impulse response sequence of a discrete-time LSV system.

Definition 7.8. The generalized transfer function, $H(z,k)$ for discrete-time LSV systems is defined as

$$H(z,k) \triangleq \sum_{r=-\infty}^{\infty} h(k,r)z^{-(k-r)}. \qquad (7.109)$$

On letting $k - r = l$, (7.109) becomes

$$H(z,k) = \sum_{l=-\infty}^{\infty} h(k,k-l)z^{-l}. \qquad (7.110a)$$

Therefore,

$$h(k,k-l) = \frac{1}{2\pi j}\oint H(z,k)z^{l-1}\,dz, \qquad (7.110b)$$

where the closed contour of integration lies with the region of convergence for the series in (7.110a).

Note that when the system is linear shift-invariant, $h(k, r)$ is a function of $(k - r)$ only; therefore, in that case the generalized transfer function is independent of k. Next, let the LSV system, characterized by its unit impulse response $\{h(k, r)\}$, be subjected to an input excitation,

$$x(k) = e^{j\omega k}, \quad -\infty < k < \infty.$$

Then, the output from the superposition sum (7.105) is

$$y(k) = \sum_{r=-\infty}^{\infty} h(k, r) e^{j\omega r}$$

$$= e^{j\omega k} \sum_{r=-\infty}^{\infty} h(k, r) e^{-j(k-r)\omega}. \quad (7.111)$$

Using the definition for $H(z, k)$ in the preceding equation,

$$y(k) = e^{j\omega k} H(e^{j\omega}, k).$$

Therefore, since

$$H(e^{j\omega}, k) = \frac{\text{response of system to } e^{j\omega k}}{e^{j\omega k}},$$

$H(e^{j\omega}, k)$ is appropriately called the generalized frequency function, which is obtained by evaluating the generalized transfer function on $|z| = 1$. The generalized transfer function $H(z, k)$ relates the input and output transforms $X(z), Y(z)$, respectively, by the relation

$$Y(z) = H(z, k) X(z). \quad (7.112)$$

Equation (7.112) can be arrived at from the following simple arguments. Since

$$x(r) = \frac{1}{2\pi j} \oint X(z) z^{r-1} \, dz,$$

then, after substituting the above expression for $x(r)$ in (7.105) and interchanging the orders of integration and summation, one gets

$$y(k) = \frac{1}{2\pi j} \oint \sum_{r=-\infty}^{\infty} h(k, r) z^{-(k-r)} X(z) z^{k-1} \, dz$$

$$= \frac{1}{2\pi j} \oint H(z, k) X(z) z^{k-1} \, dz,$$

from which (7.112) follows, since $\{y(k)\} \leftrightarrow Y(z)$ is a z-transform pair.

The frequency-domain analysis of LSV systems suffers from a serious drawback which arises from the fact that a LSV system, characterized by a LSV difference equation, might have a generalized transfer function which is not rational in the complex variable z. Consider, for example,

$$H(z,k) = \frac{1}{2k+1} \sum_{r=0}^{\infty} e^{-(k-r)^2} z^{-r},$$

which is the generalized transfer function of a LSV system characterized by the difference equation [28]

$$y(k) - \frac{2k-1}{2k+1} y(k-1) = \frac{e^{-k^2}}{2k+1}.$$

Hector Valenzuela used the following arguments to prove that $H(z,k)$ is not rational in z. For this, it is sufficient to prove that $\hat{H}(z,k)$ in

$$H(z,k) = \frac{1}{2k+1}\left[e^{-k^2} + \hat{H}(z,k)\right]$$

is not rational in z. Associate with

$$\hat{H}(z,k) = \sum_{i=1}^{N} \hat{h}_i(k) z^{-i}, \qquad \hat{h}_i(k) = e^{-(k-i)^2}$$

the $N \times N$ Hankel matrix, $\hat{H}_N$.

$$\hat{H}_N \triangleq \begin{bmatrix} \hat{h}_1(k) & \hat{h}_2(k) & \cdots & \hat{h}_N(k) \\ \hat{h}_2(k) & \hat{h}_3(k) & \cdots & \hat{h}_{N+1}(k) \\ \vdots & & & \\ \hat{h}_N(k) & \hat{h}_{N+1}(k) & \cdots & \hat{h}_{2N-1}(k) \end{bmatrix}. \qquad (7.113)$$

Applying Fact 1.11, a function is rational if and only if the infinite Hankel matrix $\hat{H}_\infty$ associated with its power series expansion has finite rank. After substituting the $\hat{h}_i(k)$'s for the case under consideration in (7.113), it is routine to show that

$$\det \hat{H}_N = \left(\prod_{j=1}^{N} e^{-(k-j)^2}\right)\left(\prod_{j=1}^{N-1} e^{2j(k-1)-j^2}\right) \det V_N, \qquad (7.114)$$

with

$$V_N = \begin{bmatrix} 1 & 1 & \cdots & 1 \\ \lambda_1 & \lambda_2 & & \lambda_N \\ \lambda_1^2 & \lambda_2^2 & & \lambda_N^2 \\ \vdots & \vdots & & \vdots \\ \lambda_1^{N-1} & \lambda_2^{N-1} & & \lambda_N^{N-1} \end{bmatrix}, \qquad (7.115)$$

where

$$\lambda_i = e^{-2(i-1)} \quad i = 1, 2, \ldots, N. \qquad (7.116)$$

V_N in (7.115) is a Vandermonde matrix, and since $\lambda_i \neq \lambda_j$ for $i \neq j$ we know that for any finite N,

$$\det V_n \neq 0. \tag{7.117}$$

From (7.114) and (7.117), it can be concluded that $\det \hat{H}_N \neq 0$ for any finite N. Therefore, $\hat{H}_\infty$ does not have a finite rank which implies that $\hat{H}(z, n)$ is not rational in z.

The preceding result, which implies that not every LSV system characterized by a difference equation has a rational generalized transfer function associated with it, limits considerably the use of frequency-domain analysis for LSV systems.

7.4.3. Decimation and Interpolation

It has been mentioned that in several applications, the need to change the sampling rate $1/T$ (where T is the sampling period) of a discrete-time signal, $\{x(kT)\}$ (obtained after sampling a continuous-time signal, $x_a(t)$, every T seconds) arises. We first consider the process of sampling rate reduction called decimation. Suppose that the desired sampling period is $T_1 = MT$, where $M > 1$ is a positive integer. Let $X_a(\omega)$ denote the Fourier transform of the continuous-time signal $x_a(t)$, which is assumed to be band-limited. In that case, for a certain finite value ω_0 of ω,

$$X_a(\omega) = 0, \quad |\omega| > \omega_0.$$

The Fourier transform, $X(e^{j\omega T})$, of the discrete-time signal $\{x(kT)\}$, is

$$X(e^{j\omega T}) = \frac{1}{T} \sum_{k=-\infty}^{\infty} X_a\left(\omega + k\frac{2\pi}{T}\right), \tag{7.118}$$

and by the sampling theorem (Theorem 1.1), $x_a(t)$ may be reconstructed from $\{x(kT)\}$ provided $T < \pi/\omega_0$. The decimated sequence $\{x(kT_1)\}$ is obtained from $\{x(kT)\}$ by retaining only one out of each group of M consecutive samples in $\{x(kT)\}$. Again, by the sampling theorem, $x_a(t)$ may be reconstructed from the decimated sequence $\{x(kT_1)\}$ provided $T_1 < \pi/\omega_0$, or equivalently,

$$T < \frac{\pi}{M\omega_0}. \tag{7.119}$$

The process of decimation can be understood after the relationship between the Fourier transforms of the two discrete-time sequences $\{x(kT_1)\}$ and $\{x(kT)\}$ is obtained. Since, at this stage, we are concerned with discrete-time sequences, it is notationally convenient to remove the sampling periods when defining the two sequences. However, to distinguish between the decimated sequence and the original sequence, the former is denoted by $\{y(k)\}$ and the latter by $\{x(k)\}$. Then

$$y(k) = x(Mk).$$

The relationship between the Fourier transform, $Y(e^{j\omega T_1})$ of $\{y(k)\}$ and the Fourier transform $X(e^{j\omega T})$ of $\{x(k)\}$ (remember that $\{x(k)\}$ was obtained by sampling $x_a(t)$ every T seconds while $\{y(k)\}$ is the sequence that would be obtained if $x_a(t)$ were sampled every $T_1 = MT$ seconds) can be shown to be (see Problem 30 at the end of this chapter)

$$Y(e^{j\omega T_1}) = \frac{1}{M} \sum_{k=0}^{M-1} X(e^{j(\omega T - 2\pi k/M)}).$$

If the inequality in (7.119) holds, that is, if $T < \pi/M\omega_0$, then

$$Y(e^{j\omega T_1}) = \frac{1}{M} X(e^{j\omega T}), \quad |\omega| < \frac{\pi}{T_1}.$$

On using (7.118) and neglecting aliasing, the preceding equation becomes

$$Y(e^{j\omega T_1}) = \frac{1}{MT} X_a(\omega)$$

$$= \frac{1}{T_1} X_a(\omega), \quad |\omega| < \frac{\pi}{T_1}.$$

However, (7.119) may not hold. In that case, distortion due to aliasing can only be avoided by passing $\{x(k)\}$ through a digital lowpass filter whose frequency response $H(e^{j\omega T})$ should, ideally, be

$$H(e^{j\omega T}) = \begin{cases} 1, & |\omega T| \le \frac{\pi}{M} \\ 0, & \text{otherwise} \end{cases}. \tag{7.120}$$

The process of decimation is thus realized by, first, lowpass filtering $\{x(k)\}$ with a digital filter having, ideally, the frequency response of (7.120). If the unit impulse response sequence of this filter is denoted by $\{h(k)\}$, then the output sequence $\{v(k)\}$ from the filter is defined by

$$v(k) = \sum_{r=-\infty}^{\infty} x(r)h(k-r). \tag{7.121}$$

The decimated sequence $\{y(k)\}$ is then given by

$$y(k) = v(Mk). \tag{7.122}$$

The input/output relationship of the system, which takes $\{x(k)\}$ as input and produces $\{y(k)\}$ as output, is obtained by substituting (7.121) in (7.122):

$$y(k) = \sum_{r=-\infty}^{\infty} x(r)h(Mk-r). \tag{7.123}$$

The system characterized by (7.123) is clearly linear but shift-variant.

Interpolation may be viewed as the reverse process of decimation. Here the sampling rate of a continuous-time signal has to be increased by a factor of L where L is a positive integer. Therefore, if the original sampling period is T, the new sampling period is required to be $T_1 = T/L$, $L > 1$. Given the discrete-time sequence $\{x(k)\}$, obtained by sampling $x_a(t)$ every T seconds in order to obtain the interpolated sequence $\{y(k)\}$, it becomes necessary to "fill-in" $L - 1$ new sample values between each pair of successive sample values in $\{x(k)\}$. To do this, first consider the sequence,

$$v(k) = \begin{cases} x\left(\dfrac{k}{L}\right), & k = 0, \pm L, \pm 2L, \ldots \\ 0, & \text{otherwise} \end{cases}.$$

Note that $\{v(k)\}$ is obtained by inserting $(L-1)$ zero valued samples between two successive samples in $\{x(k)\}$. The z-transform, $V(z)$, of $\{v(k)\}$ is related to the z-transform, $X(z)$ of $\{x(k)\}$ by

$$V(z) = X(z^L).$$

Therefore, the relationship between the Fourier transforms for the two sequences $\{v(k)\}$ and $\{x(k)\}$ is

$$V(e^{j\omega T_1}) = X(e^{j\omega T}), \quad T = LT_1.$$

Next, consider the effect of lowpass filtering $\{v(k)\}$ by a digital filter, having a frequency response $H(e^{j\omega T_1})$ given by

$$H(e^{j\omega T_1}) = \begin{cases} L, & |\omega T| \leq \pi \\ 0, & \dfrac{\pi}{T} < |\omega| < \dfrac{\pi}{T_1} \end{cases}. \tag{7.124}$$

Then, the Fourier transform of the output from the lowpass filter is

$$Y(e^{j\omega T_1}) = LX(e^{j\omega T}).$$

Therefore, using (7.118),

$$Y(e^{j\omega T_1}) = \frac{L}{T} X_a(\omega)$$

$$= \frac{1}{T_1} X_a(\omega). \tag{7.125}$$

The sequence, $\{y(k)\}$, whose Fourier transform is $Y(e^{j\omega T_1})$ of (7.125), is the desired interpolated sequence. If $\{h(k)\}$ denotes the unit impulse response of the filter having the frequency response of (7.124), then

$$y(k) = \sum_{r=-\infty}^{\infty} v(r)h(k-r). \tag{7.126}$$

Substituting the value of $v(r)$ in terms of $x(r)$ in (7.126), one obtains the input/output relation for the interpolator:

$$y(k) = \sum_{r=-\infty}^{\infty} h(k-rL)x(r). \tag{7.127}$$

Define

$$r = \left\lfloor \frac{k}{L} \right\rfloor - m,$$

when $\lfloor k/L \rfloor$ denotes the largest integer less than or equal to k/L. Then (7.127) can be rewritten as

$$y(k) = \sum_{m=-\infty}^{\infty} h\left[k - \left\lfloor \frac{k}{L} \right\rfloor L + mL\right] x\left[\left\lfloor \frac{k}{L} \right\rfloor - m\right]. \qquad (7.128)$$

Clearly, the system whose input/output relationship is describable by (7.128) is linear but shift-variant; in fact, the impulse response sequence of this system is periodically shift-variant.

Sampling rate conversion by a rational number $\frac{M}{L}$, where M and L are positive integers can also be achieved; first increase the sampling rate by L and then decrease it by M in order for the cascaded process to have a sampling period that is $\frac{M}{L}$ times the original sampling period.

7.5. Conclusions and Suggested Readings

The impulse response sequence of a digital filter whose coefficients are drawn from a finite field as opposed to the field of real numbers is necessarily periodic. The structure and properties of the impulse response sequence of such filters are intimately related to characteristics of polynomials like irreducibility and primitivity. With proper choice of multiplier coefficients in the digital filter realization, impulse response sequences which are pseudorandom in nature may be constructed. For generalization of these results to the 2-D case, see reference [19], where potentials for applications to array design are also singled out.

Results on 2-D digital filters are widely documented. The collection of research papers compiled in reference [20] along with additional references and related comments should be helpful to the interested reader for the acquiring of additional insight into the important problem area. Special care has been taken to construct problems in this chapter, which in addition to the textual material, introduce the interested reader to some key concepts in 2-D filter theory. For ready reference, these key concepts, which underscore the possibilities and problems in generalizing 1-D results to the 2-D case, are summarized in Table 7.3. The difficulties notwithstanding, tremendous progress has been made towards the consolidation of 2-D digital filtering theory, motivated by diverse applications which benefit from such theory. Linear shift-invariant (LSI) filtering operations are vital in various problems of digital image processing. For example, the task of image enhancement is heavily dependent on spatial filtering concepts; 2-D LSI filters are often designed to enhance certain properties of interest in a given image, reduce

Table 7.3. Key Properties that Underscore the Difficulties in 2-D Digital Filter Theory in Comparison to the 1-D Case

No.	Properties	1-D Case	2-D Case
1	Polynomial factorization over the complex field of coefficients	Factorable as a product of linear factors	Factorable as a product of irreducible factors, each of which need not be linear
2	Continued-fraction expansion over an arbitrary but fixed field of coefficients	Always possible because Euclidean division algorithm holds	Not possible, in general
3	Partial-fraction expansion	Always possible	Not possible, in general
4	Polynomial zeros	Finite in number and isolated	Infinite in number and occur on continuous algebraic curves; zero-set is unbounded
5	Singularities of rational functions, with relatively prime numerator and denominator polynomials	Poles	Nonessential singularities of the first ("poles") and the second kinds
6	Causality	Unique ordering in time of past, present and future samples	Orderings in space of past, present and future samples are nonunique
7	Stability	Easy to test	Test complicated due to possible presence of nonessential singularities of the second kind on unit torus T^2
8	Contour integral evaluation of inverse z-transform	Easy to perform	Difficult to perform, except in very special cases
9	Chinese remainder theorem	Applies	Does not apply
10	DFT computation	Possible via FFT techniques	Possible via FFT techniques
11	Common zeros of two relatively prime polynomials	Absent	Present, but these are finite in number in the 2-D case
12	Frequency response Wavenumber response (Frequency-wavenumber response)	Obtained by evaluating over unit circle	Obtained by evaluating over unit torus (bidisc distinguished boundary)

noise, and compensate for distortions. Image formation processes are never perfect. A recorded image could be blurred due to optical aberrations or due to relative motion between the recorder and the object. The deblurring process of image restoration is often based on the design of a 2-D inverse filter. This filter restores the original image by "inverting" the effects of the degrading phenomenon. A very interesting application of 2-D digital filtering theory occurs in geophysical explorations, where the search for oil and gas takes place in the face of high uncertainties about the subsurface rock formations. Therefore, a geophysicist or a petroleum engineer needs to locate and determine accurately the boundaries of those rock formations which contain hydrocarbons. This is done by seismic reflection studies, where 2-D filters are designed to discriminate signals from noise due to their different velocities of propagation (see Problem 23 at the end of this chapter). 2-D filters are also frequently used in the interpretation of gravity and magnetic maps to distinguish between local and regional features. The input to this type of filter could be observational data collected over a grid on the earth's surface. Smoothing or lowpass filtering of the 2-D observational data can yield a set of improved measurements. The wavenumber (in cycles per unit distance, associated with each independent variable in the Fourier transform of the spatial data) response of a spatial filter may be designed for separating gravity and magnetic anomalies.

There are numerous design techniques for 2-D digital filters. No attempt has been made to be needlessly complete in the exposition of all these techniques, especially because that would lead us considerably astray and the bottomline is that the art of design is best undertaken by adhering to guidelines rather than stringent, inflexible rules. Instead, certain specific design techniques have been expounded not only because those are quite interesting and are based on the design of 1-D prototypes but also because those might stimulate the reader to explore the possibilities of discovering other techniques. For example, the reader might find the problem of investigating the scopes for generalizing all 1-D techniques to the 2-D case interesting. For instance, the window method is quite simple and popular in the design of 1-D FIR filters and it was seen in Chapter 4 that the design of good windows is as much an art as a science. The feasibility of obtaining good 2-D windows, in the design of circularly symmetric 2-D FIR filters, from good 1-D windows has been investigated [30]. A reader wanting more documentation on 2-D design techniques should see the collection of papers in reference [20] and the books [8], [26], and [27].

The restriction of shift-invariance has been eliminated in Section 7.4. In many important applications, the system is required to be shift-variant. 1-D LSV filters occur, for example, in the decimator-interpolator realization of narrow-band filters, realization of filters with finite precision coefficients, aspects of seismic data processing where, with passage of time, the dominant frequency of the reflected signal usually becomes lower and its bandwidth narrower suggesting the desirability for design of filters whose frequency characteristics vary with time, and also in short-time spectrum analysis for dynamic or time-varying filtering of speech signals. 2-D LSV filters are often required in problems associated with 2-D image processing. To confine the exposition within reasonable limits, only the theory of 1-D LSV systems is briefly presented. The characterizations based on the superposition sum and the generalized transfer function have been discussed. It has been pointed out that unlike in the 1-D LSI case, the input/output sequences of only a subclass of causal discrete LSV systems are representable by a finite-order difference equation with time-varying coefficients. The important area of application concerned with multirate sampling, involving the principles of decimation and interpolation, where knowledge of LSV systems proves to be useful, is briefly, but adequately, covered.

Problems

1. Derive equation (7.3) from equation (7.1).

2. Verify that Figure 7.1 is a realization of the transfer function in equation 7.4.

3. Prove the result stated in Fact 7.3.

4. Using Facts 7.4 and 7.5, show that if a sequence $\{s(k)\}$ has maximum length, then its characteristic polynomial is irreducible.

5. After replacing 2 by 3 in equation (7.24), write down a computer program that will enable you to evaluate the number of primitive ternary polynomials (i.e., those having coefficients in Z_3) of specified degree n, for $n \le 25$.

6. Using equation (7.22), write down a computer program that will enable you to evaluate the number of irreducible binary polynomials of specified degree n, for $26 \le n \le 50$.

7. Solve the preceding problem for the case of irreducible ternary polynomials of specified degree n, when $n \le 25$.

8. The discrete autocorrelation function of a binary sequence $\{s(k)\}$ of period p is defined by

$$\rho(k) = \frac{1}{p} \sum_{i=0}^{p-1} (-1)^{s(i)+s(i+k)}.$$

Using the property of pseudorandom sequences described in the first sentence of Fact 7.10, justify the values of the two distinct levels of the discrete autocorrelation function, given in Fact 7.10.

9. Let $\{x(n_1, n_2)\}$ be the input sequence to a 2-D digital system; the correspond-
    ing output sequence will be denoted by $\{y(n_1, n_2)\}$.
    a. If the input/output relation of the system is characterized by the equation

    $$y(n_1, n_2) = y(n_1 - 1, n_2) + x(n_1 - 1, n_2 - 1) + x(n_1, n_2),$$

    show that the system satisfies the definition of linear shift-invariance.
    b. If the input/output relation is

    $$y(n_1, n_2) = [a(n_1, n_2) x(n_1, n_2)]^2,$$

    show that the system is neither linear nor shift-invariant. ($a(n_1, n_2)$ is a
    spatially varying coefficient.)
    c. How will you categorize the system with respect to linearity and shift-invari-
    ance, if the equation

    $$y(n_1, n_2) + a(n_1, n_2) y(n_1 - 1, n_2) = x(n_1, n_2)$$

    provides the input/output description of the system? Again $a(n_1, n_2)$ is a
    spatially varying gain.

10. Let $\{u(n_1, n_2)\}$ denote the 2-D unit step, defined in Section 7.3.1. Show the
    region of support for each of the following convolutions.
    a. $\{u(n_1, n_2)\} * \{u(n_1, n_2)\}$
    b. $\{u(n_1, n_2)\} * \{u(-n_1, -n_2)\}$
    c. $\{u(n_1, n_2)\} * \{\delta(n_1, n_2)\}$, where $\delta(n_1, n_2)$ is the 2-D unit impulse

11. Calculate the z-transform of $\{x(n_1, n_2)\}$ in each of the following cases.
    $\delta(n_1, n_2)$, $u(n_1, n_2)$ are, respectively, the 2-D unit impulse and 2-D unit step.
    a. $x(n_1, n_2) = \frac{1}{2}\delta(n_1 - 2n_2) u(n_1, n_2)$
    b. $x(n_1, n_2) = \frac{1}{4}u(n_1) u(6 - n_2)$
    c. $x(n_1, n_2) = u(n_1 + n_2) u(n_1 - n_2)$
    Hint: Let $m_1 = n_1 + n_2$, $m_2 = n_1 - n_2$ and note that $x(n_1, n_2) \neq 0$ when
    $m_1 \geq 0$ and $m_2 \geq 0$.

12. The impulse response $h(n_1, n_2)$ of a 2-D digital system consists of unit
    impulses at coordinates $(-1,0), (0,0), (1,0), (-1,-1), (0,-1), (1,-1), (-1,1)$,
    $(0,1)$ and $(1,1)$ in the (n_1, n_2)-plane. It is stated that the frequency response of
    the system is the z-transform of $\{h(n_1, n_2)\}$ evaluated at $z_i = e^{-j\omega_i T_i}$, $i = 1,2$.
    Evaluate this frequency response.

13. It is stated that

    $$H(z_1, z_2) = 1 + \sum_{k=1}^{\infty} \frac{1}{100 k^3} \left(z_2^k + z_2^{-k} \right) z_1^{-1}$$

    is the z-transform of the impulse response sequence, $\{h(k_1, k_2)\}$, of an
    asymmetric half-plane recursive filter.
    a. Show the region of support of the impulse response sequence.
    b. It is known that any 2-D LSI filter is BIBO stable if and only if

    $$\sum_{k_1} \sum_{k_2} |h(k_1, k_2)| < \infty$$

where $\{h(k_1,k_2)\}$ is the impulse response sequence. Using this stability criterion, verify whether or not the filter in this problem is BIBO stable.

14. Consider the transfer function of a digital filter,

$$H(z_1,z_2) = \frac{Y(z_1,z_2)}{X(z_1,z_2)} = \frac{1}{1+0.5z_1z_2^{-1}+z_1^{-1}+z_2^{-1}+z_1^{-2}z_2}.$$

It is stated that the filter is not of the quarter-plane type; this fact will become apparent to you as you proceed to answer the questions posed.

a. Convince yourself of the fact that its impulse response sequence $\{h(k_1,k_2)\}$ may be found if it is possible to implement the following difference equation, recursively, with zero boundary conditions.

$$h(k_1,k_2) = \delta(k_1,k_2) - \tfrac{1}{2}h(k_1+1,k_2-1) - h(k_1-1,k_2)$$
$$- h(k_1,k_2-1) - h(k_1-2,k_2+1),$$

where $\delta(k_1,k_2)$ represents the 2-D unit impulse.

b. In order to implement the desired recursion mark as coordinates in the (k_1,k_2)-plane the integer 2-tuples (n_1,n_2) that can be associated with each monomial, $z_1^{-n_1}z_2^{-n_2}$ occurring in the denominator function of $H(z_1,z_2)$. You should then be able to get the following sector, where past, present and future points in the ordering required to implement the recursion are identified. The parallelogram shown is formed by lines PA and PB and the present point, (k_{1p},k_{2p}) where $h(k_{1p},k_{2p})$ is being computed.

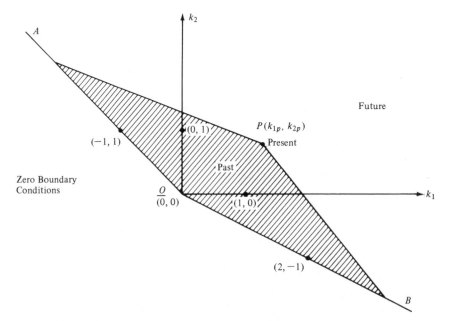

The sector, bounded by lines OA and OB, which forms an angle of less than 180° at vertex O, is the region of support of the filter impulse response.

Show that $\{h(k_1, k_2)\}$ may be computed recursively as follows: first obtain $h(k_1, k_2)$ on the boundary of the sector (or cone). You should be able to see that on this boundary

$$h(-k, k) = \left(-\tfrac{1}{2}\right)^k, \quad k \geq 0$$

and

$$h(2k, -k) = (-1)^k, \quad k > 0.$$

Continue the recursion to obtain the remaining elements in the desired sequence by forming a parallelogram shown in the above figure with the two lines on the boundary as sides and the point (k_{1p}, k_{2p}) (associated with $h(k_{1p}, k_{2p})$, the present state being computed) as a vertex.

15. Let

$$H(z_1, z_2) = \frac{2}{2 - z_1^{-1} - z_2^{-1}}$$

be the transfer function of a first quadrant quarter-plane IIR digital filter. Evaluate for $k_1 \geq 0$, $k_2 \geq 0$,

$$h(k_1, k_2) = \frac{1}{(2\pi j)^2} \oint_{C_1} \oint_{C_2} H(z_1, z_2) z_1^{-1+k_1} z_2^{-1+k_2} \, dz_1 \, dz_2.$$

Proceed as follows:

a. First show that $2 - z_1^{-1} - z_2^{-1} \neq 0$, $|z_1| > 1$, $|z_2| > 1$. Further, show the only zeros of $2 - z_1^{-1} - z_2^{-1}$ on $T^2 : |z_1| = |z_2| = 1$. C_1, C_2, may then be taken as $|z_1| = 1$, $|z_2| = 1 + \varepsilon$, $\varepsilon > 0$, respectively, in the region of convergence of $H(z_1, z_2)$.

b. Use the transformations

$$z_1^{-1} \to z_1, z_2^{-1} \to z_2$$

to show that

$$h(k_1, k_2) = \frac{1}{(2\pi j)^2} \oint_{|z_1|=1} \oint_{\substack{|z_2|=1-\delta \\ \delta > 0}} \frac{2}{2 - z_1 - z_2} z_1^{-(1+k_1)}$$

$$\times z_2^{-(1+k_2)} \, dz_1 \, dz_2.$$

c. Rewrite the integral in (b) as

$$h(k_1, k_2) = \frac{1}{2\pi j} \oint_{|z_2|=1-\delta} 2 z_2^{-(1+k_2)}$$

$$\times \left[\frac{1}{2\pi j} \oint_{|z_1|=1} \frac{1}{z_1^{1+k_1}(2 - z_1 - z_2)} \, dz_1 \right] dz_2$$

Noting that the only pole within the circle, $|z_1| = 1$, of the integral in square brackets (henceforth denoted by $I(z_2)$) occurs at $z_1 = 0$, show that this integral evaluates to

$$I(z_2) = \frac{1}{k_1!} \left[\frac{d^{k_1}}{dz_1^{k_1}} \left(\frac{1}{2 - z_1 - z_2} \right) \right]_{z_1 = 0}$$

d. Finally, show that

$$h(k_1, k_2) = \frac{1}{2\pi j} \oint_{|z_2| = 1 - \delta} 2I(z_2) z_2^{-(1+k_2)} \, dz_2$$

$$= \left(\tfrac{1}{2}\right)^{k_1 + k_2} \frac{(k_1 + k_2)!}{k_1! k_2!}.$$

e. Plot $h(k_1, k_2)$ vs k_1, k_2; can you infer whether or not the filter is BIBO stable?

16. Eising [23] defined a causality cone C_c as an intersection of two half-planes, $H_{p,r}$ and $H_{q,t}$, where

$$H_{p,r} = \left\{ (x_1, x_2) : (x_1, x_2) \in R \times R \triangleq R^2, px_1 + rx_2 \geq 0 \right\}$$

$$H_{q,t} = \left\{ (x_1, x_2) : (x_1, x_2) \in R \times R \triangleq R^2, qx_1 + tx_2 \geq 0 \right\},$$

where p, q, r, t are nonnegative integers satisfying $pt - qr = 1$. Furthermore, Eising defined a 2-D filter to be weakly causal if the support of its impulses response $\{h(k_1, k_2)\}$ is contained within a closed convex cone C in R^2 (i.e., supp$\{h(k_1, k_2)\} \subset C$), satisfying,
   (i) $C \cap (-C) = \{(0,0)\}$ (this condition merely implies that the cone makes an angle of less than π at $(0,0)$)
   (ii) $Q_1 \subset C$,
where Q_1 denotes the first quadrant. Eising also showed that there exists a C_c such that for any weakly causal filter, $C \subset C_c$. For the sake of brevity we denote C_c by $H_{p,q,r,t}$, where p, q, r, t are integers defined above. Show that
a. $H_{1,0,0,1}$ is the first quadrant, Q_1
b. Show that a causality cone, $H_{1,1,1,2}$ can be associated with the support of the impulse response for the filter of Problem 14 above.
c. Consider specified integer sets
   (i) $p = 3$, $r = 5$
   (ii) $p = 2$, $r = 7$
   (iii) $p = 7$, $r = 2$.
In each case, determine integers q, t such that

$$pt - qr = 1, \quad t < r \quad \text{and} \quad q < p.$$

Then in each case draw the causality cone $H_{p,q,r,t}$.
d. Referring to Problem 14 (above), convince yourself of the fact that a weakly causal filter can always be implemented recursively.
e. It is important to note that if Z denotes the set of integers and $Z^2 \triangleq Z \times Z$ (Cartesian product of Z with itself) then there exists a map

$$\phi : C_c \cap Z^2 \rightarrow Q_1 \cap Z^2,$$

given by

$$\phi(m, n) = (pm + rn, qm + tn),$$

which is bijective. Show that the points $(t, -q)$ and $(-r, p)$ in C_c are mapped by ϕ to points $(1,0)$ and $(0,1)$, respectively, in Q_1.

f. Suppose that the mapping

$$\phi: S_{p,q,r,t} \rightarrow S_{1,0,0,1},$$

from the causality cone C_c to the first quadrant Q_1, is given by

$$\phi H(\alpha,\beta) = \sum_{k_1=0}^{\infty} \sum_{k_2=0}^{\infty} h\big(\phi^{-1}(k_1,k_2)\big) \alpha^{k_1}\beta^{k_2},$$

where $\phi(m,n)$ is as in (e).
Show that the isomorphism ϕ can be described by the substitutions

$$z_1 = \alpha^p\beta^q, \quad z_2 = \alpha^r\beta^t$$

with inverse

$$\alpha = z_1^t z_2^{-q}, \quad \beta = z_1^{-r} z_2^p.$$

17. Let the transfer function of a first quadrant quarter-plane digital filter be

$$H(z_1,z_2) = \frac{1}{1 - az_1^{-1} - bz_2^{-1}}, \quad |a| + |b| < 1.$$

a. Show that the unit impulse response sequence for the filter is

$$h(k_1,k_2) = a^{k_1}b^{k_2}\frac{(k_1+k_2)!}{k_1!k_2!}.$$

b. The cepstrum of $\{h(k_1,k_2)\}$ is the sequence $\{\hat{h}(k_1,k_2)\}$, obtained by taking the inverse z-transform of $\log_e H(z_1,z_2)$. After setting up an appropriate contour integral as in Problem 15 above, evaluate it to obtain $\hat{h}(k_1,k_2)$.

18. Determine the locations of the nonessential singularities of the second kind in the finite z_1, z_2 planes for each of the rational functions given

a. $H(z_1,z_2) = \dfrac{z_1+z_2}{1+z_1z_2}$.

b. $H(z_1,z_2) = \dfrac{z_1+z_2}{1+z_1^2z_2^2}$.

c. $H(z_1,z_2) = \dfrac{1}{1+z_1^2+z_2^2}$.

19. Recall the definition for the Chebyshev polynomial, $T_k(x)$, of the first kind and kth order (see Section 3.2.2).

$$T_k(x) = \cos k(\cos^{-1}x).$$

a. Show that

$$\sum_{k=1}^{n} a_k \cos k\theta = \left(\sum_{k=1}^{n} a_k T_k(x) \right)\Bigg|_{x=\cos\theta}.$$

b. Express $\cos 9\theta$ and $\cos 10\theta$ as polynomials in $\cos\theta$.

20. Find the counterparts of the result stated in Fact 7.11 when considering quarter-plane filters which recurse, respectively, in the $(-k_1, +k_2)$, $(-k_1, -k_2)$, and $(+k_1, -k_2)$ directions.

21. The transfer function of a 2-D digital system is

$$H(z_1, z_2) = \frac{\left(1 - z_1^{-1}\right)^2 \left(1 - z_2^{-1}\right)^2}{2 - z_1^{-1} - z_2^{-1}}.$$

    a. Write down the difference equation relating the input sequence $\{x(k_1, k_2)\}$ to the output sequence $\{y(k_1, k_2)\}$.

    b. Does $H(z_1, z_2)$, when viewed as a rational function in z_1, z_2, have relatively prime numerator and denominator polynomials (i.e., polynomials devoid of a nonconstant common factor)?

    c. Find values of z_1, z_2 for which the numerator and denominator polynomials of $H(z_1, z_2)$, viewed as a rational function in z_1, z_2, are simultaneously zero.

22. Which of the following bivariate polynomials are irreducible over the coefficient field of complex numbers?

    a. $(z_1^2 + 1)(z_2 + 2)$.

    b. $z_1 + z_2 + 2$.

    c. $1 + z_1 z_2$.

    d. $1 + z_1 + z_2 + z_1 z_2$.

    e. $z_1^2 - z_2$.

23. Velocity filters provide a means to discriminate signals from noise due to their different apparent velocities. Consider a linear array of equispaced seismometers (sensors) as shown in the Figure in this problem.

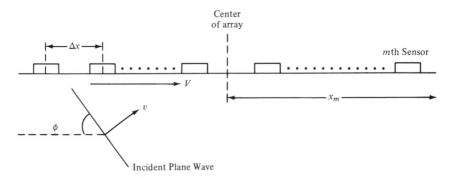

This array is used for recording a shock wave, $s(t, x)$, where t is time and x is the distance measured from the center of array. The shock wave propagates with an apparent (move-out) velocity, $V = v/\sin\phi$, on the earth's surface along the array line, where it is assumed that the plane wavefronts, traveling across the array sites at an angle ϕ with the horizontal approach, the array site with a velocity v, and successively excite individual array sensors. The seismic signal, therefore, may be describable by

$$s(t,x) = \begin{cases} s\left(t - \dfrac{x}{V}\right), & |x| \le x_m \\ 0, & |x| > x_m. \end{cases}$$

Clearly, $s(t - x/V)$ is a shifted version of $s(t) \triangleq s(t,0)$, the seismic record at the spatial origin; the shift equals x/V and represents the delay between the wave recorded at the center of the array and that recorded at a distance x from the center.

a. The Fourier transform

$$H(f,k) = \int_{-\infty}^{\infty} \int_{-\infty}^{\infty} s(t,x) e^{-j2\pi(ft + kx)} \, dt \, dx$$

(it is assumed that the shock wave is measured continuously in $[-x_m, x_m]$) is the 2-D spectrum of $s(t,x)$, and f is the ordinary "temporal" frequency in hertz, while the "spatial" frequency k is also referred to as the wavenumber. Let $S(f)$ denote the Fourier transform of $s(t)$. Show that the Fourier transform of $s(t,x)$, described above, is

$$2x_m S(f) \frac{\sin 2\pi[k + (f/V)] x_m}{2\pi[k + (f/V)] x_m}.$$

b. Consider a received wave that has a velocity V, corresponding to a straight line through the origin, whose slope is given by $k = -f/V$, on the (f,k) plane. Convince yourself of the fact that the main energy distribution of this wave is concentrated around this straight line.

c. Assume that it is desired to pass *only* waveforms with wavenumbers within the range, $-(|f|/V) < k < |f|/V$. The transfer function of the so-called fan filter (pie-slice filter) required for the purpose is

$$H(f,k) = \begin{cases} 1, & -\dfrac{|f|}{V} < k < \dfrac{|f|}{V} \\ 0, & \text{otherwise} \end{cases}.$$

Convince yourself of the fact that $H(f,k)$ defines a highpass velocity filter which passes signals with apparent velocities of magnitude greater than V and rejects signals with lower velocities.

d. Consider a velocity pass range from V_1 to V_2 with a center velocity V_c. The filter which meets these specifications is a bandpass velocity filter. Show that rotation in the (f,k) plane followed by highpass filtration may be used to replace bandpass filtering.

e. The result in (b) is based on the assumption of continuous spatial measurement. If that assumption is not made, show that

$$H(f,k) = 2S(f) \sum_{p=1}^{m} \cos\left[2\pi\left(k + \frac{f}{V}\right) p \Delta x\right]$$

How does the above expression change, if there is an additional sensor located at the array center?

24. Is it possible to implement the following partial fraction expansion?

$$\frac{1}{(1+z_1 z_2)(z_1 + z_2)} = \frac{a + b z_1 + c z_2 + d z_1 z_2}{(1 + z_1 z_2)} + \frac{e z_1 + f z_2 + g}{(z_1 + z_2)}.$$

In the preceding equation, a, b, c, d, e, f, g, are constants; some possibly zero.

25. Solve the 2-D difference equation

$$h(k_1, k_2) = h(k_1, k_2 - 1) + h(k_1 - 1, k_2)$$

for $k_1 \geq 0$, $k_2 \geq 0$, given that the boundary conditions are

$$h(0, -1) = \tfrac{1}{2}, \qquad h(-1, 0) = \tfrac{1}{2}$$

and

$$h(k_1, -1) = 0, \quad k_1 = 1, 2, \ldots, \infty$$
$$h(-1, k_2) = 0, \quad k_2 = 1, 2, \ldots, \infty.$$

It is desirable that you obtain an expression for $h(k_1, k_2)$ explicitly in terms of k_1, k_2. The z-transform method is recommended for obtaining the desired solution.

26. Can a 2-D rational transfer function of the form

$$H(z_1, z_2) = \frac{P(z_1, z_2)}{Q_1(z_1) Q_2(z_2)}$$

(where polynomial $P(z_1, z_2)$ has no factor common to polynomials $Q_1(z_1)$ or $Q_2(z_2)$) have any zero common to its numerator and denominator polynomials, $P(z_1, z_2)$ and $Q_1(z_1) Q_2(z_2)$? Very briefly justify your answer.

27. A state-space model of a 2-D single-input single-output system is given below. $\hat{x}(k_1, k_2)$ denotes the state vector, $x(k_1, k_2)$ is the input, and $y(k_1, k_2)$ is the output.

$$\hat{x}(k_1 + 1, k_2 + 1) = \begin{bmatrix} 0 & 0 \\ -0.1 & -0.5 \end{bmatrix} \hat{x}(k_1 + 1, k_2)$$

$$+ \begin{bmatrix} -0.1 & -0.5 \\ 0 & 0 \end{bmatrix} \hat{x}(k_1, k_2 + 1)$$

$$+ \begin{bmatrix} 0 \\ 1 \end{bmatrix} x(k_1 + 1, k_2) + \begin{bmatrix} 1 \\ 0 \end{bmatrix} x(k_1, k_2 + 1).$$

$$y(k_1, k_2) = [2 \quad 3] \hat{x}(k_1, k_2).$$

a. Applying z-transforms, determine the transfer function of the system.
b. Write down the difference equation relating the input and output sequences from the transfer function, calculated in (a).

28. A 2-D system is characterized by an impulse response sequence, $\{h(k_1, k_2)\}$, with

$$h(k_1, k_2) = k_1 - k_2 \lambda^{k_1}, \quad k_1 \geq 0, k_2 \geq 0,$$

and λ is a real parameter. Compute the system transfer function and determine whether or not it may be BIBO stable for a suitable value of λ.

29. Can the McClellan transformation be used to realize 2-D FIR filter magnitude characteristics of arbitrary shape? Justify your answer.

30. [29] In several applications the need to change the sampling rate $1/T$ (where T is the sampling period) of a discrete-time signal, $\{x(kT)\}$ arises. The process of sampling rate reduction is called decimation and the process of sampling rate increase is called interpolation. Suppose that the desired sampling period is $T_1 = MT$, where $M > 1$ is a positive integer. Let $X(e^{j\omega T})$ denote the Fourier transform of $\{x(kT)\}$. Show that the Fourier transform $Y(e^{j\omega T_1})$ of the decimated signal,

$$y(kT_1) = x(kMT),$$

obtained by retaining only one out of each group of M consecutive samples in $\{x(kT)\}$ (i.e., $y(kT_1) = x(kT_1)$ at the sampling instants of $y(kT_1)$, but is zero otherwise) is

$$Y(e^{j\omega T_1}) = \frac{1}{M} \sum_{r=0}^{M-1} X(e^{j(\omega T - 2\pi r)/M}).$$

Show that aliasing in the process of decimation can be avoided provided

$$T < \frac{\pi}{M\omega},$$

in which case

$$Y(e^{j\omega T_1}) = \frac{1}{M} X(e^{j\omega T}), \quad |\omega| < \frac{\pi}{T_1}.$$

References

1. Golomb, S. W. 1967. Shift Register Sequences. Holden-Day, San Francisco, CA.
2. Davies, W. D. T. 1970. System Identification for Self-Adaptive Control. Wiley-Interscience, New York.
3. Golomb, S. W. (ed.). 1964. Digital Communication with Space Applications. Prentice-Hall, Englewood Cliffs, NJ.
4. Stewart, J. L., and Westerfield, E. C. 1959. A theory of active sonar detectors. Proc. IRE, p. 872.
5. Sarwate, D. V., and Pursley, M. B. 1980. Crosscorrelation properties of pseudo-random and related sequences. Proc. IEEE, 68: 593–619.
6. Mersereau, R. M., and Dudgeon, D. E. 1975. Two-dimensional digital filtering. Proc. IEEE, 63:610–623.
7. Hoffman de Visme, G. 1971. Binary Sequences. The English Universities Press, London.
8. Bose, N. K. 1982. Applied Multidimensional Systems Theory. Van Nostrand Reinhold, New York.
9. Jury, E. I. 1978. Stability of multidimensional scalar and matrix polynomials. Proc. IEEE, 66:1018–1047.
10. Goodman, D. M. 1978. A design technique for circularly symmetric lowpass filters. IEEE Trans. ASSP, 26:290–304.

11. Chang, H., and Agarwal, J. K., 1977. Design of two-dimensional recursive filters by interpolation. IEEE Trans. Circuits Systems, 24:281–291.

12. Pendergrass, N., Mitra, S. K., and Jury, E. I. 1976. Spectral transformations for two-dimensional digital filters. IEEE Trans. CAS, 23:26–35.

13. Bernabo, F., Emiliani, P. L., and Cappellini, V. 1976. Design of two-dimensional recursive digital filters. Electronics Letters, 12:288–289.

14. Valenzuela, H. M., and Bose, N. K. 1983. Maximally flat rational approximants in multidimensional filter design. Circuits, Systems Signal Process 2(1):119–128.

15. Karivaratharajan, P., and Swamy, M. N. S. 1978. Quadrantal symmetry associated with two-dimensional digital transfer functions. IEEE Trans. CAS, 25:340–343.

16. Pistor, P. 1974. Stability criteria for recursive filters. Jan. 1974. IBM J. Res. Dev., pp. 59–71.

17. Ekstrom, M. P., and Woods, J. W. April 1976. Two-dimensional spectral factorization with applications in recursive digital filtering, IEEE Trans. ASSP, 24:115–128. Also see corrections in IEEE Trans. ASSP, Dec. 1976, 24:585.

18. McClellan, J. H. 1973. The design of two-dimensional digital filters by transformations. Proceedings 7th Annual Princeton Conference on Information Sciences and Systems, pp. 247–251.

19. Prabhu, K. A., and Bose, N. K. Feb. 1982. Impulse response arrays of discrete-space systems over a finite field. IEEE Trans. ASSP, 30:10–18.

20. Mitra, S. K., and Ekstrom, M. P. (ed.): 1978. Two-Dimensional Digital Signal Processing. Dowden, Hutchinson & Ross, Inc., Academic Press, Stroudsburgh, PA.

21. Bose, N. K., and Jury, E. I. 1975. Inner algorithm to test for positive definiteness of arbitrary binary forms. IEEE Trans. Auto Control, 20:700–702.

22. Rhodes, J. D. 1976. Theory of Electrical Filters. Wiley-Interscience, London.

23. Eising, R. 1980. State space realization and inversion of 2-D systems. IEEE Trans. Circuits Systems, 27:612–619.

24. Costa, J. M., and Venetsanopoulos, A. N. 1974. Design of circularly symmetric two-dimensional recursive filters. IEEE Trans. Acoustics, Speech, Signal Proc., 6:432–442.

25. Sako, B., and Hirano, K. 1984. Design of recursive digital filters for the selection or rejection of a particular velocity in seismic signals. Circuits, Systems Signal Proc., 3(2):177–191.

26. Dudgeon, D. E., and Mersereau, R. M. 1984. Multidimensional Digital Signal Processing. Prentice-Hall Englewood Cliffs, NJ.

27. Bose, N. K. (ed.): 1985. Multidimensional Systems Theory: Progress, Directions and Open Problems. D. Reidel Publishing Company, Dordrecht, Holland.

28. Huang, N. C., and Aggrawal, J. K. 1980. On linear shift-variant digital filters. IEEE Trans. Circuits Systems, 27:672–678.

29. Crochiere, R. E., and Rabiner, L. R. 1983. Multirate Digital Signal Processing, Prentice-Hall, Englewood Cliffs, NJ.

30. Huang, T. S. 1972. Two-dimensional windows. IEEE Trans. Audio Electroacoustics, 20:88–89.

31. Dubois, E., 1985. The sampling and reconstruction of time-varying imagery with application in video systems. Proc. of IEEE, 73:502–522.

Author Index

Subject Index